Environmental Security in Harbors and Coastal Areas

NATO Security through Science Series

This Series presents the results of scientific meetings supported under the NATO Programme for Security through Science (STS).

Meetings supported by the NATO STS Programme are in security-related priority areas of Defence Against Terrorism or Countering Other Threats to Security. The types of meeting supported are generally "Advanced Study Institutes" and "Advanced Research Workshops". The NATO STS Series collects together the results of these meetings. The meetings are co-organized by scientists from NATO countries and scientists from NATO's "Partner" or "Mediterranean Dialogue" countries. The observations and recommendations made at the meetings, as well as the contents of the volumes in the Series, reflect those of participants and contributors only; they should not necessarily be regarded as reflecting NATO views or policy.

Advanced Study Institutes (ASI) are high-level tutorial courses to convey the latest developments in a subject to an advanced-level audience

Advanced Research Workshops (ARW) are expert meetings where an intense but informal exchange of views at the frontiers of a subject aims at identifying directions for future action

Following a transformation of the programme in 2004 the Series has been re-named and re-organised. Recent volumes on topics not related to security, which result from meetings supported under the programme earlier, may be found in the NATO Science Series.

The Series is published by IOS Press, Amsterdam, and Springer, Dordrecht, in conjunction with the NATO Public Diplomacy Division.

Sub-Series

A. Chemistry and Biology	Springer
B. Physics and Biophysics	Springer
C. Environmental Security	Springer
D. Information and Communication Security	IOS Press
E. Human and Societal Dynamics	IOS Press

http://www.nato.int/science
http://www.springer.com
http://www.iospress.nl

Series C: Environmental Security

Environmental Security in Harbors and Coastal Areas

Management Using Comparative Risk Assessment and Multi-Criteria Decision Analysis

edited by

Igor Linkov
INTERTOX, Inc.
Boston, Massachusetts and Seattle,
Washington, U.S.A.

Gregory A. Kiker
University of Florida,
Gainesville, FL, U.S.A.

and

Richard J. Wenning
ENVIRON International Corporation
San Francisco, CA, U.S.A.

Springer

Published in cooperation with NATO Public Diplomacy Division

Based on papers presented at the NATO Advanced Research Workshop on
Management Tools for Port Security, Critical Infrastructure, and Sustainability
Thessaloniki, Greece,
March 2005

A C.I.P. Catalogue record for this book is available from the Library of Congress.

ISBN-10 1-4020-5801-2 (PB)
ISBN-13 978-1-4020-5801-1 (PB)
ISBN-10 1-4020-5800-4 (HB)
ISBN-13 978-1-4020-5800-4 (HB)
ISBN-10 1-4020-5802-0 (e-book)
ISBN-13 978-1-4020-5802-8 (e-book)

Published by Springer,
P.O. Box 17, 3300 AA Dordrecht, The Netherlands.

www.springer.com

Printed on acid-free paper

TABLE OF CONTENTS

Preface ix

Acknowledgements xiii

Part 1. Environmental Security: Regulatory Needs and Tools

Environmental Security, Critical Infrastructure and Risk Assessment: Definitions and Current Trends 3
D.A. Belluck, R.N. Hull, S.L. Benjamin, L. Alcorn, J. Linkov

Environmental Security: Approaches and Tools 19
R.J. Wenning, S.E. Apitz, D.A. Belluck, S. Chiesa, J. Figueira, Z. Filip, G. Ignatavičius , M.K. Kiker, R. Lahdelma, M. MacDonell, A. Mannini, B. Raissouni, D. Sinka, Y. Xenidis

Integrating Comparative Risk Assessment and Multi-Criteria Decision Analysis: Working through Wicked Problems and Their Impossible Solutions 37
G.A. Kiker, I. Linkov, T.S. Bridges

Environmental Security in Harbors and Coastal Areas: Management Using Comparative Risk Assessment and Multi-Criteria Decision Analysis Framework 53
B.G. Biggs, M.K. Kiker

Part 2. Coastal Areas: Challenges and Solutions

Environmental Security and Environmental Regulations in Israel 81
R. Bronfman

Managing Sediment Risk at the Basin Scale: European Frameworks in Support of Complex Decisions 85
S. Apitz, A. Oen, S. White

Multicriteria Decision Analysis and Strategic Uncertainties 97
C. Yoe

Ranking of Available Countermeasures Using MCDA Applied to Contaminated Environment 111
A. Grebenkov, A. Yakushau

Engaging the General Public and Other Stakeholder Groups 121
J. Skei

Risk and Decision Methods Applied to Aquatic Ecosystem Management: Considerations for Invasive and Endangered Species 127
B.S. Payne, A.C. Miller, B. Suedel

Stakeholder Preference Elicitation 149
R.M. Cooke

Multiobjective Risk/Cost Analysis of Artificial Marine Systems Using Decision Trees and Fuzzy Expert Estimations 161
E. Levner, J. Ganoulis, I. Linkov, Y. Benayahu

The *Bouchard-120* and Chalk Point Spill Responses: Objectives and Performance Metrics 175
S.P. Tuler, T.P. Seager, R. Kay, I. Linkov

Part 3. Applications of Multi-Criteria Decision Analysis

Comparative Assessment of Several Multi-Criteria Decision Analysis Tools for Management of Contaminated Sediments 195
I. Linkov, F.K. Satterstrom, B. Yatsalo, A. Tkachuk, G.A. Kiker, J. Kim, T.S. Bridges, T.P. Seager, K. Gardner

SMAA-TRI: A Parameter Stability Analysis Method for ELECTRE TRI 217
T. Tervonen, R. Lahdelma, J. Almeida Dias, J. Figueira, P. Salminen

Treating Uncertain Criteria and Stakeholder Preference Information in Environmental Multi-Criteria Decision Problems 233
R. Lahdelma, P. Salminen

Fuzzy vs. Probabilistic Methods for Risk Assessment of Coastal Areas 251
D. Angelides, Y. Xenidis

Decision Support Systems for the Management of Contaminated Sites: A Multi-Criteria Approach 267
S. Giove, P. Agostini, A. Critto, E. Semenzin, A. Marcomini

Visualization of Pareto Frontier in Environmental Decision Making 275
A.V. Lotov

Spatially-Explicit Population Models with Complex Decisions: Fish, Cattle, and Decision Analysis 293
R.A. Goodwin, V. Pandey, G.A. Kiker, J.B. Kim

Part 4. Applications of Risk Assessment

Risk Assessment in Detection and Prevention of Terrorist Attacks in Harbors and Coastal Areas 309
M.K. Zaidi

Complex Health Risk Assessment and Analysis from Exposure to Ionizing Radiation, Chemical Contaminants and Other Sources of Harm 317
V. Demin

Role of Risk Assessment in Environmental Security Planning and Decision-Making 329
R.J. Wenning, S. Della Sala, V. Magar

Humic Substances as a Natural Factor Lowering Ecological Risk in Estuaries 343
Z. Filip, K. Demnerova

Risk Assessment Applications in Croatia 355
D. Sinka, D. Skanata

Part 5. Case Studies in Environmental Security Analysis

Risk Assessment of the Effect of Wastewater Treatment and Desalination Processes on the Coastal Aquifer in Israel: Using Multi-Criteria Decision Analysis 369
N. Haruvy, S. Shalhevet

Evaluation of Drinking Groundwater for the Rural Areas Adjacent to the Nearby Desert of Giza Governorate of Greater Cairo, Egypt 379
M.M. Emara, I. El Sabagh, A. Kotb, A.S. Turkey, D. Husseen

Impact of Lithuanian Sea Harbors on Environmental Stability and Local Ecosystems 395
G. Ignatavičius, V. Vilutienė

The Caspian Sea Region: Environmental Security and Risk Assessment Applications in Oil and Gas Development Projects 409
E. Ostrovskaya, A. Kurapov, L. Anisimov

Investigation of Discharge of Fresh Water in the Canakkale Strait (Dardanelles -Turkey) 421
A. Baba, O. Deniz, M. Turkoglu, H. Ozcan

The South Mediterranean Coastal Sustainable Development Project: Environmental and Human Development Risk 429
B. Raissouni, M. El Bouyoussfi

Lake Sevan in Armenia: Socioeconomic Analysis for Secure Development Policies 439
V. Sargsyan

Multi-Criteria Decision Analysis and the Tar Creek Superfund Site: Initial Findings 451
C.J. Batchelor

Unmanned Sea Vehicles for Harbor Patrolling: Low-Cost Solutions to Enhance Coastal Security 465
S. Chiesa, N. Viola, A. Mannini, S. Olivero, F. Stesina

List of Participants 485

Author Index 491

ix–xi

Y20

Main Book Title:

PREFACE

Au: Igor Linkov, Greg Kiker, Richard Wenning

Throughout world history, great cities, nations, and empires have either collapsed or lost their dominance when local and regional natural resources were depleted to an extent that exceeded the ability of the environment to restore or replenish their supply. Diamond (2005)[1] identifies five factors that contribute to societies collapse: climate change, hostile neighbors, trade partners (that is, alternative sources of essential goods), environmental problems, and, finally, a society's response to its environmental problems. While different combinations of these factors have played a role in the demise of several societies, Diamond contends that the failure to recognize or respond to its environmental problems has consistently been a significant factor.

In this 21st century, our modern civilization may be facing the same challenges and crossroads that shaped earlier human history. Nearly 75% of the world's population lives within 100 km of the marine environment. Our demand for energy and appetite for raw materials has never been greater, placing enormous demands on land, air, water and marine resources and jeopardizing the flora and fauna that share our planet. An over reliance and, perhaps, misguided faith in technology to overcome the same environmental challenges that contributed to the collapse of earlier societies may only serve to delay the historically inevitable resource and political conflicts that occur when human demands exceed nature's ability to supply. We must be mindful of the 12 environmental problems Diamond (2005) claims have continuously challenged both ancient and modern societies: destruction of natural habitats (mainly through deforestation); reduction of wild foods; loss of biodiversity; erosion of soil; depletion of natural resources; pollution of freshwater; inhibition of natural photosynthetic resources; introduction of toxins and alien species to the environment; artificially induced climate change; and overpopulation.

Decision-making tools are urgently needed to support environmental management, particularly in the world's coastal areas where a delicate balance exists between human population demands for clean air, land, and water and the demands imposed by an increasingly global economy. Addressing environmental threats and identifying actions to mitigate those threats necessitates an understanding of the basic risk assessment paradigm and the tools of risk analysis to assess, interpret, and communicate risks. It also requires modification of the risk paradigm itself to incorporate a complex array of quantitative and qualitative information that shapes

---

1 Diamond, J. *Collapse: How Societies Choose to Succeed or Fail*, Viking Books, New York, 2005

the unique political and ecological challenges of different countries and regions around the world.

Over the past five years, the authors of this book and their colleagues have convened a series of meetings to explore the topic of environmental security and management. Each workshop has tried to demonstrate the power of risk assessment and decision analysis as tools that decision makers should use to address a broad range of difficult environmental problems. Risk analysis, these experts have argued, offers a relatively objective, unbiased, and rational approach to framing and solving complex problems. It imposes a certain degree of quantitative rigor, as opposed to the all-too-common tendency to make environmental recommendations based on assumptions and anecdotal evidence.

The first workshop in this series, entitled "Assessment and Management of Environmental Risks: Cost-efficient Methods and Applications"[2] (Lisbon, Portugal, October 2000), confirmed the role risk assessment could play as a platform for providing a scientific basis for environmentally sound and cost-efficient management policies, strategies, and solutions to various environmental problems. The second workshop, entitled "Comparative Risk Assessment and Environmental Management"[3] (Anzio, Italy, May 2002), explored the development and application of comparative risk assessment (CRA) and other risk-based decision-analysis tools in environmental management. The use of CRA was exceptional for facilitating decision making when various social, political, and economic activities compete for limited environmental resources. The third workshop, entitled "The Role of Risk Assessment in Environmental Security and Emergency Preparedness in Mediterranean Region"[4] (Eilat, Israel, April 2004), focused on environmental security challenges in the Middle East and how risk assessment could resolve some of the region pressing environmental needs.

This book is based on discussions and papers presented at a fourth workshop, entitled "Management Tools for Port Security, Critical Infrastructure, and Sustainability." The meeting was held in Thessaloniki, Greece in March 2005. Fifty-five international science, risk assessment, decision-making, environmental modeling and engineering experts from 11 countries explored environmental security issues in ports, harbors and coastal areas and the use of multi-criteria decision analysis (MCDA), in conjunction with risk assessment, to identify and rank environmental security threats, formulate responses to those threats, and evaluate the efficacy of different responses on threat reduction. This workshop, like those previous, was sponsored jointly by the Society for Risk Analysis and NATO.

---

2 Linkov, I., Palma Oliveira, J.M., eds. *Assessment and Management of Environmental Risks*, Kluwer, Amsterdam, 2001.

3 Linkov, I., Ramadan, A.B., eds. *Comparative Risk Assessment and Environmental Decision Making*, Kluwer, Amsterdam, 2004.

4 Morel, B., Linkov, I., eds. *Environmental Security: The Role of Risk Assessment*. Springer, Amsterdam, 2006.

This book is organized into five sections. The introductory section provides a fundamental understanding of environmental security and the environmental challenges at ports and harbors, and summarizes the different approaches to stakeholder involvement in decision-making. Part 2 reviews in more detail the challenges facing ports, harbors, and coastal areas. Environmental security is viewed increasingly by governments and international organizations as a critical issue; urban development and growth requirements and increasing environmental concerns are challenging current strategies regarding environmental protection and management. Part 3 reviews different MCDA approaches and tools that have emerged to evaluate natural and manmade environmental vulnerabilities at chemical manufacturing plants, energy plants, transportation networks, and other critical infrastructure located in heavily populated urban areas and coastal ports and harbors. Evaluation of the effectiveness of these approaches and tools in the current framework of environmental management is important to verify their appropriateness and identify possible future needs to address environmental security. Part 4 builds on the same issues, but focuses on the application of risk assessment tools to complement or inform the MCDA process. The book concludes with a section devoted to case studies.

The collection of papers presented in this book reflects the workshop conclusion that environmental resources in coastal areas will become severely challenged during the next few decades. The likelihood for social conflicts is high, assuming the current pace of world economic and population expansion continues or accelerates. Establishing, maintaining, or enhancing a sense of environmental security in different coastal regions and improving management of critical infrastructure in coastal areas will require (i) matching human demands with available environmental resources; (ii) recognition of environmental security threats and infrastructure vulnerabilities; and, (iii) identification of the range of available options for preventing and minimizing natural disasters, technological failures, and terror actions. These three considerations will require input from different stakeholder perspectives, and a broad range of quantitative and qualitative sociopolitical, environmental, and economic information.

This book emphasizes our beliefs that the convergence of seemingly disparate viewpoints from politicians, scientists, engineers, and the general public and often uncertain and limited information gathered from an equally broad range of disciplines is possible only by using one or more available risk assessment methodologies and decision-making tools. Much work remains to be done, and must be done, lest we experience the same fate as our ancestors.

Igor Linkov, Greg Kiker, and Richard Wenning
September 2006

ACKNOWLEDGEMENTS

The editors would like to thank Drs. Boris Yatsalo, Abou Bakr Ramadan, Jacques Ganoulis, David Belluck, Jongbum Kim, and Todd Bridges for their help in the organization of this workshop. We also wish to thank workshop participants for their contributions to the book and peer-review of manuscripts. We are deeply grateful for the commitment and excellent editorial and technical assistance provided by Deb Oestreicher, who oversaw the production of this book. We also wish to recognize Katie Quartano, Mindy Kiker, Bryce Biggs, Kyle Satterstrom, Eugene Linkov, and Elena Belinkaia for their efforts during the organization of the workshop. The workshop agenda was prepared in collaboration with the Society of Risk Analysis (SRA) and the U.S. Army Corps of Engineers. Financial support for the workshop was provided mainly by NATO. Additional support was provided by the U.S. Army Corps of Engineers, Cambridge Environmental Inc., and ENVIRON International Corporation.

Part 1

Environmental Security: Regulatory Needs and Tools

ENVIRONMENTAL SECURITY, CRITICAL INFRASTRUCTURE AND RISK ASSESSMENT

Definitions and Current Trends

D.A. BELLUCK
FHWA/USDOT
400 7th Street, S.W., Washington, D.C. 20590 USA

R.N. HULL
Cantox Environmental, Inc.
1900 Minnesota Court, Suite 130, Mississauga, Ontario L5N 3C9 CANADA

S.L. BENJAMIN
U.S. Department of Agriculture, Farm Service Agency
1400 Independence Avenue, S.W., Washington, D.C. 20501 USA

J. ALCORN
SAIC
11251 Roger Bacon Dr., Reston, VA 20190 USA

I. LINKOV
INTERTOX, Inc.
83 Winchester Street, Suite 1, Brookline, MA 02446 USA

Abstract

Population growth, needed economic growth, and social pressures for improved infrastructure coupled to the need for human health and ecological protection and environmental security make systematic and transparent environmental decision-making a complex and often difficult task. Evaluating complex technical data and developing feasible risk management options requires procedural flexibility that may not be part of existing evaluative structures. Experience has demonstrated that direct transposition of risk assessment and risk management frameworks (e.g. those developed in the United States and European Union) may not work in regions whose social, legal, historical, political and economic situations are not suitable or prepared for acceptance of these methodologies. Flexible decision-making, including the use and development of acceptable or unacceptable risk levels based on the critical nature of an infrastructure type, is one potential approach to assist risk managers in their decision-making. Unfortunately, the newness of the

I. Linkov et al. (eds.), Environmental Security in Harbors and Coastal Areas, 3–17.

discussions on the interrelatedness of environmental security and critical infrastructure has yet to produce a unified and comprehensive treatment of the fields. As a result, this paper will describe and define these terms in order to set the stage for discussions of human health and ecological risk assessment and risk management later in the paper. This paper reviews basic concepts defined in the field of risk assessment and extends its applicability to the areas of environmental security and critical infrastructure protection.

1. Environmental Security Defined

Environmental security has emerged as an increasingly important concern of governments and their defense establishments because of several trends that have the potential to threaten stability. These potential threat issues include: world population in 2015 will be 7.2 billion, up from 6.1 billion in year 2000; water scarcities and allocation will pose challenges to certain governments; groundwater depletion; contemporary environmental problems will persist and grow; globalization will be rocky, marked by chronic financial volatility and a widening economic divide; significant degradation of arable land; loss of tropical forests; greenhouse gas emissions will increase substantially; exacerbation of biological species loss; rapid urbanization; increasingly serious urban air and water quality problems; and global climate change induced glacial ice melt backs, sea level rise, and increasing storm frequency [22; 32].

"Environmental Security" is an ill-defined term [20] with many definitions whose two key elements are: repairing damage to the environment for human life support and for the moral value of the environment itself; and, preventing damage to the environment from attacks and other forms of human abuse [5]. Several definitions of environmental security exist and demonstrate that after more than two decades of discussion, the concept of environmental security still has no widely agreed upon formulation [20]. Examples include:

- "Environmental security (ecological security or a myriad of other terms) reflects the ability of a nation or a society to withstand environmental asset scarcity, environmental risks or adverse changes, or environment-related tensions or conflicts." [4]
- "Science-based case studies, which meld physical science with the discipline of political economy, are a suitable vehicle for forecasting future conflicts derived in some measure from environmental degradation." [20]
- "[T]hose actions and policies that provide safety from environmental dangers caused by natural or human processes due to ignorance, accident, mismanagement or intentional design, and originating within or across national borders." [5]
- "Environmental Security is a state of the target group, either individual, collective or national, being systematically protected from environmental risks caused by inappropriate ecological process due to ignorance, accident, mismanagement or design." [1]

- "Environmental security is protectedness of natural environment and vital interests of citizens, society, the state from internal and external impacts, adverse processes and trends in development that threaten human health, biodiversity and sustainable functioning of ecosystems, and survival of humankind." [1]
- "Environmental security is the state of protection of vital interests of the individual, society, natural environment from threats resulting from anthropogenic and natural impacts on the environment." [1]

The discipline of "environmental security" is neither a pure security issue nor an environmental issue [4]. However, environmental issues are often security concerns because, even without directly causing open conflict, they can result in environmental perturbations or triggers that can destabilize the status quo and result in a loss of regional, national, and local political, social, economic and personal security [32].

Environmental security concerns can be grouped into three general categories [1]: 1) security of the environment which is a good in itself; 2) security from environmental change that can create societal instability and conflict; and, 3) security from environmental change (e.g. water scarcity, air pollution, etc.) that would threaten the material well-being of individuals [1]. Common elements of environmental security definitions include: public safety from environmental dangers caused by natural or human processes due to ignorance, accident, mismanagement, or design; amelioration of natural resource scarcity; maintenance of a healthy environment; amelioration of environmental degradation; and, prevention of social disorder and conflict (promotion of social stability) [13].

Environmental security concerns include chemical/material releases to the environment. This is because, worldwide, an estimated one quarter to one third of disease burden is attributable to environmental factors [11]. Chemical or material releases to the environment or environmental alteration result in actual or perceived health risks that can result in societal conflicts between parties in support or opposition to the environmental perturbation.

2. Critical Infrastructure Defined

Critical infrastructures are complex societal systems [39] and also have many definitions. Examples include:

- "Civil and critical infrastructure systems such as transportation, communication, power, and financial systems have provided the foundation for modern society." [26]
- According to the U.S. Department of Defense, critical infrastructures are "Those systems and assets essential to plan, mobilize, deploy, and sustain military operations and transition to post-conflict military operations, and whose loss or degradation jeopardize the ability of the Department of Defense to execute the National Military Strategy." [30]

- "Critical infrastructure (or assets) in the highway transportation system include all of its components, including physical and cyber-based components, which are used in attaining transportation functions to serve national, regional and local objectives. Examples of these structures include the physical structures (roadways, bridges, tunnels), facilities (parking areas, toll complex), ITS (Intelligent Transportation Systems) components (signs and signals, network, control centers), and organizational components (personnel, procedures, communication)." [15]
- In Queensland, Australia "Critical infrastructure is defined as infrastructure which, if destroyed, degraded or rendered unavailable for an extended period, will significantly impact on social or economic well-being or affect national security or defence." [29]
- The United States Patriots Act states that critical infrastructures are: "Systems and assets, whether physical or virtual, so vital to the United States that the incapacity or destruction of such systems and assets would have a debilitating impact on security, national economic security, national public health or safety, or any combination of those matters." [21]

Regardless of the definition, it is clear that "Our society and modern way of life depend on a complex system of critical infrastructures" [35].

The convergence of critical infrastructure, environmental security, risk assessment and risk management is a function of the perception that these fields are inextricably interrelated and the need to make complex decisions based on multiple criteria as part of the risk management process. In order to fully understand how these two issues relate, one needs to understand how the concept of acceptable risk developed, its application to risk assessment and risk management, and its relationship to environmental security and critical infrastructure. The remainder of this paper will delve into these interrelationships.

3. Risk Assessment, Risk Management and Environmental Security

In order to survive and prosper, humans must alter their environment and use environmental resources. As modern societies increase in population, they must increase their use of renewable and nonrenewable resources to provide their citizens with essential goods, services and economic security. Expanding populations and economies build farms, homes, factories and transportation networks that use and release chemicals/materials to the environment purposefully, accidentally or incidentally. The field of risk assessment is developing to address anthropogenic risk.

3.1. ORIGINS OF RISK ASSESSMENT AND RISK MANAGEMENT

Releases of chemicals and materials, along with environmental alteration, is often monitored or regulated by government agencies. Government agencies use administrative tools to evaluate chemical/material releases and environmental alteration. A favored tool is quantitative risk assessment that has been described as

"An organized process used to describe and estimate the likelihood of adverse health outcomes from environmental exposures to chemicals" [28].

Quantitative risk assessment is the preferred tool in the United States and elsewhere to regulate or evaluate facilities, activities or processes that release chemicals to the environment. More than 25 years ago, the Inter-Agency Regulatory Liaison Group and the Office of Science & Technology Policy in the White House proposed an orderly set of activities under the headings of "hazard identification", "risk characterization", and "risk reduction". In 1983, the National Research Council published "Risk Assessment in the Federal Government: Managing the Process" [24] also known as the "Red Book" [7; 25]. The four steps of the current human health and ecological risk assessment process are hazard identification, dose-response assessment, exposure assessment, and risk characterization [28]. For both human health and ecological risk assessment, numerical risk assessment findings are compared to risk management and policy based acceptable risk levels (e.g. single point risk levels or risk ranges) to determine if there is a potential for significant or unacceptable risk.

Risk management is variously defined as "The process of analyzing, selecting, implementing, and evaluating actions to reduce risk" and "...the process of identifying, evaluating, selecting, and implementing actions to reduce risk to human health and to ecosystems" [28]. It has also been defined as a "...decision-making process involving considerations of political, social, economic, and technical factors with relevant risk assessment information relating to a hazard so as to develop, analyze, and compare regulatory and non-regulatory options, and to select and implement the optimal response for safety from that hazard [9]. The goal of risk management is scientifically sound, cost-effective, integrated actions that reduce or prevent risks while taking into account social, cultural, ethical, political, and legal considerations" [28].

Acceptable and unacceptable numerical risk levels (along with their synonymous terms) can vary by governmental unit or statute. The risk management paradigm allows risk managers at all levels of government to use the risk management decision-making process to allow chemical/material releases to the environment even though they have been found to exceed applicable acceptable risk levels. This type of risk management decision-making uses the numerical risk expression as a point of departure for decision-making rather than a bright line that cannot be exceeded. How this is done and why such a decision can be acceptable will be discussed in detail later in this paper. First, let us look at a very brief history of the evolution of the acceptable risk concept.

3.2. ACCEPTABLE/UNACCEPTABLE RISK CONCEPT

The assumption that public health could be protected by chemical risk management developed in the United States in the early 1900s for food additives. By 1958, an amendment to the U.S. Food, Drug, and Cosmetics Act put forth the concept that some chemicals might have no toxic threshold and prohibited the addition of any chemical that can cause cancer. It was also recognized that it was impossible to completely remove carcinogens from the food supply and, as a result, the U.S.

Food and Drug Administration put forth the proposal that if risks calculated under the no-threshold assumption were below some small value, the carcinogen was effectively absent in the food. A virtually safe dose (one in one hundred million or 10^{-8}) to limit cancer risk was proposed but was found to be an almost impossible burden on regulators. An alternative level for food additives was proposed at one in a million (10^{-6}), a level considered negligible by most people. This level became the criterion for acceptable risk in the United States when cancer risks from environmental exposures became recognized in the late 1960s and early 1970s. By the 1990s, it was recognized that the one-in-one-million risk level was very stringent and the idea of a lesser risk level of one-in-ten-thousand (10^{-4}) was introduced. In general, a risk above one-in-ten-thousand is considered excessive. The Clean Air Act Amendments of 1990 led to the development of a risk range of 10^{-4} to 10^{-6}. Thus, "If a cancer risk is judged to be *significant* or *unacceptable*, then it is generally expected that some action will be taken to reduce or eliminate the risk. In contrast, a *de minimis* or *essentially negligible* risk is one that is so small that no action needs to be taken. If a risk is judged to be *insignificant* or *acceptable*, however, this does not necessarily mean that it is *de minimis* or *negligible*" [16].

Defining acceptable and unacceptable risks is the foundation for determining regulatory compliance or the need for risk management interventions to reduce calculated risks. There are numerous definitions of acceptable or unacceptable risks that are discussed below. The concepts are illustrated in Table 1 and in the accompanying footnotes at the end of this paper. Chemical releases can result in human or ecological exposures and subsequent hazards or risks. Human and ecological risk assessment methods are used to calculate risks and hazards to potentially exposed receptors. Acceptable hazard or risk levels can vary by locality. For example, acceptable risk levels for human carcinogens can be set at a specific level (e.g. one-in-one-million) or range (e.g. one-in-one-million to one-in-ten-thousand). In Canada, the federal government does not, in general, recommend acceptable risk levels [16]; each province has adopted either 10^{-6} or 10^{-5} as the acceptable excess cancer risk level. Acceptable hazard levels for non-carcinogenic chemicals can be different for single chemicals (e.g. Hazard Quotient of 0.2 or 1.0) or multiple chemicals (e.g. Hazard Index of 1.0). Acceptable risks for ecological risk assessment can be a single value for individual organisms (e.g. Toxicity Index less than 1) and populations (e.g. 10% chance that no more than 20% of a population will be exposed above a benchmark value).

According to Vrijling et al. [40], "For complex societal systems as a whole, like a nation, one normally uses *individual risk* as a measure, which varies between 1 x 10^{-5} and 3.1 x 10^{-4} deaths per year for occupational, traffic and consumer risks respectively. The individual risk is then taken over the whole population at stake and a time period of one year. Although no general individual risk criteria are set for trivial risks either, one tends to measure those against the *de minimis* value of 10^{-6} or 10^{-5} deaths per year... indicating a potential low acceptable risk level for any individual [which] everybody can live with. In some cases of critical infrastructures, like high speed train links, individual risk criteria are set in the Netherlands ..." "The same is true for the zoning between hazardous chemical facilities and residential areas, at an individual risk contour of 10^{-6} deaths per year…" "For critical infrastructures sometimes also societal risks are defined. For social or group risks, the next step is to order the scenarios with increasing measure of potential consequences (mostly deaths). The cumulative probabilities (or frequencies) for exceeding a certain number of deaths are then derived from the probabilities of all scenarios contributing to that particular number of deaths." "The societal acceptable risk is judged at a national level by placing an upper-bound upon the expected number of fatalities per activity per year."

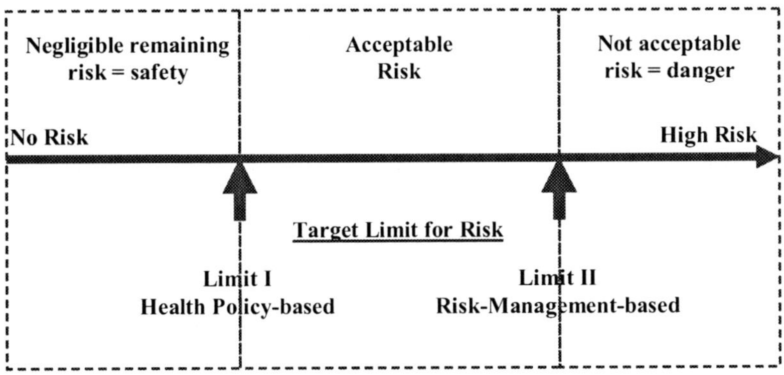

From: Bundesanstalt fur Arbeitsschutz und Abrbeitmedizin. [3], slightly modified.

Figure 1: Description of acceptable risk and risk ranges.

Ecological risk assessment, the study of non-human risks, is much more complex than human health risk assessment. The tremendous diversity of habitats and species makes the study of ecological risks challenging. According to a 2005 report prepared for the UK Department of Environment, Food and Rural Affairs (Defra) [39], there are no standards for establishing *ecological value* in ecological risk assessments. There needs to be much discussion by stakeholders as to what the "environmental value" is at each site that must be protected. The ecological value must reflect policy goals and societal values, which can range from protection of endangered or commercially or recreationally important species to preservation of ecosystem attributes for functional reasons (e.g. flood water retention by wetlands) or aesthetic reasons (e.g. visibility in the Grand canyon).

Deterministic methods are most often used to assess the potential for acceptable or unacceptable ecological risks. At its most basic level, actual toxicity data for the species of interest or a similar species are used to provide an ecologically protective numerical value for a given medium (e.g. surface water, sediment, soil) and ecological receptor. These calculated media-specific concentrations are associated with risks at the threshold of acceptable/unacceptable risk *to individual organisms*. When the ratio between a known or calculated contaminant level and the ecologically protective numerical value exceeds unity (or some other value established by a responsible governmental unit), then there is the potential for unacceptable risks, and either more detailed assessment is conducted or risk management is required. Conversely, not exceeding the acceptable ratio value usually equals acceptable ecological risks.

Other measures of acceptable ecological risk include those that do not allow more than a certain percentage of an exposed population to exceed a given concentration based on a selected toxicological value. Probabilistic methods and statistical models are also used to derive medium and receptor specific acceptable/unacceptable threshold concentrations. An important reason for moving away from concentration-effect data for single species is that ecological risk assessment generally is not concerned with protection of individual organisms, but rather protection of populations, communities and ecosystems [23].

Risk assessment numerical findings are not meant to be accurate or precise estimates of morbidity or mortality. They are merely numerical estimates using a systematic and transparent process based on elements of science, policy, law and professional judgment. Determining when these numerical risk or hazard estimates represent acceptable or unacceptable levels is not a scientific exercise. Rather, it is a risk management determination based on established acceptable risk levels (e.g. 10^{-6} or lower excess cancer risks are acceptable) or risk ranges (e.g. 10^{-4} to 10^{-6} risk ranges) that are in policy, guidelines, or statute. Risks greater than "a bright line" or "acceptable risk range" levels are not necessarily "unacceptable" because risk management practices allow management judgment in determining when a potentially "significant risk" becomes unacceptable. It is common practice to uniformly apply "bright lines" and risk ranges to all types of processes, activities, or facilities. In practice, risk assessors and managers often judge exceedence of a bright line or risk range as unacceptable risk unless risk management decisions, based on additional factors not related to the calculated risk value(s), are made to the contrary. It seems reasonable to assert, based on these considerations, that as risk assessment techniques evolve so should risk management techniques and along with them acceptable risk level expressions.

This advancement may be particularly important when estimating ecological risks. Ecological risk assessments are conducted using a tiered approach, where the finding of acceptable or unacceptable risks determines the move to higher levels of analysis or site remediation [27]. Unlike the assessment of human health risks, where it is considered more "acceptable" or "prudent" to err on the side of conservative (over-predicted) risk estimates, there is less support for overly conservative predictions of ecological risks and subsequent conservative risk management decisions. A benefit of ecological risk assessment is that it is possible to monitor, relatively easily, whether the ecological risk prediction was correct (or more importantly, incorrect), through the use of standard biomonitoring techniques. Acceptable/unacceptable ecological risk levels also may be defined less quantitatively than in human health risk assessment, and incorporate considerations such as probability of population (or community) persistence and restoration after some disturbance [18]. In addition, the area of disturbance relative to the remaining undisturbed habitat of suitable quality for the ecological receptors of interest can be factored in to the acceptable/unacceptable risk definition [34].

Findings of acceptable or unacceptable risk drive risk reduction efforts, regulatory compliance determinations, as well as complex and expensive litigation. It is for this reason that the subject of critical vs. non-critical infrastructure based acceptable risk levels is vital to understand and explicitly define.

Finally, acceptable and unacceptable risk levels can be determined by public perception and political decision-making processes. For example, Vogel [41] states: "The public's perception or tolerance of particular risks often differs from that of experts and in a democratic system the former's preferences – and values – often play an important role in the policy process. Thus governments can and frequently do chose to err on the side of caution, seeking to avoid or reduce particular risks that many citizens regard as unacceptable, even if the available scientific evidence does not or cannot prove evidence of harm. Vogel [41] cites work by T. Christoforou on the basic elements of the precautionary principle, noting that the public's perception or tolerance of particular risks often differs from that of experts and in a democratic system the former's preferences – and values – often play an important role in the policy process.

3.3. PRECAUTIONARY PRINCIPLE

The precautionary principle is often cited as an important doctrine to follow when making a risk management determination. However, this view is not universally accepted. The precautionary principle originated in Principle 15 of the Rio Declaration (1992); it states: "Where there are threats of serious or irreversible damage, a lack of full scientific certainty shall not be used as a reason for postponing cost-effective measures to prevent environmental degradation" [33]. The precautionary principle has been used to describe "[d]ecisions about the best ways to manage or reduce risks that reflect a preference for avoiding unnecessary health risks instead of unnecessary economic expenditures when information about potential risks is incomplete" [28]. While the precautionary principle has no formal legal effect in the United States, no country has so fully adopted the essence of the principle in domestic law [41]. In contrast, the European Union officially introduced the Precautionary Principle in Article 130 (the environmental section) of the 1993 Treaty of the European Union (Maastricht) and it has been referenced in 27 resolutions between 1994 and 1999 [41].

What happens when risk managers have insufficient scientific knowledge to make scientifically sound decisions? According to the Presidential/ Congressional Commission on Risk Assessment and Risk Management [28], "Decision-makers must balance the value of obtaining additional information against the need for a decision, however uncertain. Sometimes a decision must be made under the precautionary principle (decisions about the best ways to manage or reduce risks that reflect a preference for avoiding unnecessary health risks instead of unnecessary economic expenditures when information about potential risks is incomplete). Risk management determinations of acceptable risk often involve judging safety. Since safety is not absolute and is immeasurable, achieving acceptable risks means decision-makers must make tradeoffs between costs, both absolute and relative, with risks" [15].

4. Conclusions

It is clear from this discussion that environmental security is a very broad term used to encompass a wide variety of issues. For the purpose of this book, by *ensuring environmental security* we mean *guarding against environmental degradation in order to preserve or protect human, material, and natural resources at scales ranging from global to local*. The critical infrastructure concept is directly linked to environmental security. For environmental applications, critical infrastructure may be defined as *man-made structures constructed and maintained to assure human health, environmental protection, transportation networks, water supplies, clean air, food supplies and other critical elements necessary to maintain economic and national security.*

The question not yet fully addressed is how does one accommodate both the importance of critical infrastructure and of environmental security, given the current risk assessment/risk management paradigm? Many papers published in this volume address different aspects of this issue. For example, Belluck et al [2] defines environmental security in terms of chemical releases, risk assessment, and risk management, and proposes consideration of a flexible risk acceptability criteria to match the critical nature of a given type of infrastructure based on the use of a systematic and transparent risk management process that matches the rigor of the risk assessment on which it is based. Nevertheless, our review and current research shows that more work is necessary to address methodology and application of risk assessment and environmental security to emerging threats in general and specifically in the Middle East. As the fields of risk assessment, risk management, critical infrastructure, and environmental security merge, additional discussions will need to occur to define these interactions and their implications for environmental protection and regulatory activity.

5. Appendix: Definitions of Risk

- "Acceptable Risk Range – If the cumulative carcinogenic risk to an individual based on reasonable maximum exposure (RME) for both current and future land use is less than 1E-04 and the non-carcinogenic hazard index is less than 1, action generally is not warranted unless there are adverse environmental impacts." "*Note: The upper boundary of the risk range is not a discrete line at 1E-04, although EPA generally uses 1E-04 in making risk management decisions. A risk estimate that is greater than 1E-04 may be considered acceptable, if justified based on site-specific conditions. A risk manager may also decide that a baseline risk level less than 1E-04 is unacceptable due to site specific reasons and that remedial action is warranted.*" [38]

- "As in the *Benzene* case, the court did not define any particular method for EPA to use in determining what risks are acceptable. On remand, the agency, after taking comment on a number of possibilities, decided that it could not use any single metric as a measure of whether a risk is acceptable. Instead, it adopted a general presumption that a lifetime excess risk of cancer of approximately one in 10,000 (10-4) for the most exposed person would constitute acceptable risk and that the margin of safety should reduce the risk for the greatest possible number of persons to an individual lifetime excess risk no higher than one in 1 million (10-6)." [8])

- "...the published acceptable risk level does not necessarily represent the "safe level" but rather a target level with the expectation that the true risk to exposure is less than the published value." [12]
- "In general terms, a risk that is so small, whose consequences are so slight or whose associated benefits (perceived or real) are so great that persons or groups in society are willing to take or be subjected to that risk. In more technical terms, an arbitrary value denoting a very low probability of occurrence of a seriously adverse effect in persons exposed daily over a lifetime. The dose associated with this risk may be considered to have an insignificant impact on human health. *Synonyms: Tolerable Risk; Negligible Risk; Risk Level."* [17]
- "It is the Agency's responsibility to determine in the first instance what it considers to be a "significant" risk. Some risks are plainly acceptable and others are plainly unacceptable. If for example, the odds are one in a billion that a person will die from cancer by taking a drink of chlorinated water, the risk clearly could not be considered significant. On the other hand, if the odds are one in a thousand that regular inhalation of gasoline vapors that are 2 percent benzene will be fatal, a reasonable person might well consider the risk significant and take the appropriate steps to decrease or eliminate it. (I.U.D. v. A.P.I., 448 U.S. 607, 655). So a risk of (1/1000) (10(-3)) is clearly significant. It represents the uppermost end of the million-fold range suggested by the Court, somewhere below which the boundary of acceptable versus unacceptable risk must fall." "...free to use conservative assumptions in interpreting the data with respect to carcinogens, risking error on the side of overprotection rather than underprotection" (448 U.S. at 655, 656)." "Further guidance for the Agency in evaluating significant risk and narrowing the million-fold range described in the "Benzene Decision" is provided by an examination of occupational risk rates, legislative intent, and the academic literature on "acceptable risk" issues. For example, in the high-risk occupations of mining and quarrying, the average risk of death from an occupational injury or an acute occupationally-related illness over a lifetime of employment (45 years) is 15.1 per 1,000 workers. The typical occupational risk of deaths for all manufacturing industries is 1.98 per 1,000. Typical lifetime occupational risk of death in an occupation of relatively low risk, like retail trade, is 0.82 per 1,000. (These rates are averages derived from 1984-1986 Bureau of Labor Statistics data for employers with 11 or more employees, adjusted to 45 years of employment, for 50 weeks per year)." [37]
- "Tolerable risk. Risk level below which risks would be regarded as being widely acceptable, either because they are irreducible or because they compare with other risks routinely accepted (see comparative risk assessment). Some

sources distinguish tolerable risk from acceptable risk: the former being 'just acceptable' and kept under review." [36]

- Acceptable risk is the "…type of risk such that the benefits derived by an organism, a population, or an ecological system outweigh the adverse effects that might affect them as a result of being administered or exposed to a particular agent" [9].
- The determination of this "acceptable" or "tolerable" level of risk may have been prescribed before the risk assessment process begins - through societally determined acceptable levels of risk in the form of legislative environmental quality standards for instance, or industry derived "norms". In this case, risk management attempts to analyse which options for action based on the results of the risk assessment will produce these pre-determined risk levels. Where no acceptable risk standards exist, the risk management process will attempt to derive "acceptable" or tolerable risk on a case-by-case basis." "Decision making to determine "acceptable" or "tolerable" risk uses a number of approaches. The three major approaches to acceptable risk decisions are professional judgment where technical experts devise solutions, bootstrapping where historical precedent guides decision making and formal analyses where theory-based procedures for modeling problems and calculating the best decision are used. [10]
- "Any risk that is currently tolerated is considered to be acceptable" [19].
- "Defining an acceptable risk level gives meaning to the risk estimate generated from the risk assessment. There are few legislative, public policy, and judicial guidelines on how to define acceptable risk. Although "safe" has not been found to necessarily mean zero risk (State of Ohio v. EPA 997 F.2d 1520, 1533, D.C. Cir. 1993), the courts have not provided (1) a risk level above which risk management action must occur, (2) specific guidance as to what might be done to determine whether a risk is acceptable, or (3) workable definitions of acceptable, safe risk levels. The EPA currently "endorses" a risk range from 10^{-6} (one in a million) to 10^{-4} for one's lifetime risk from exposure to carcinogens and a hazard quotient of 1.0 for noncarcinogens. As our state survey shows, acceptable risk levels across the state regulatory agencies tend to mirror EPA guidance [6].
- "The level of Residual Risk that has been determined to be a reasonable level of potential loss/disruption" [21].

6. References

1. AC/UNU Millennium Project. No Date. Emerging International Definitions, Perceptions, and Policy Considerations. Environmental Security Study. http://www.acunu.org/millennium/es-2def.html.

2. Belluck DA, Hull RN, Benjamin SL, Alcorn J and Linkov I. 2005. Are Standard Risk Acceptability Criteria Applicable to Critical Infrastructure Based on Environmental Security Needs? In: Linkov, I., Morel, B., eds. Environmental Security and Risk Assessment. Kluwer Academic Publishers (this volume).

3. Bundesanstalt fur Arbeitsschutz und Abrbeitmedizin. 1998. Guidelines on Risk Management of the Hazardous Substances Committee Concerning Decisions with Far Reaching Consequences. November 15 1998. http://www.baua.de//prax/ags/ags_guidelines_rm.pdf.

4. Chalecki, E. L. No Date. Environmental Security: A Case Study of Climate Change. Pacific Institute for Studies in Development, Environment, and Security. http://www.pacinst.org/ environment_ and_security/env_security_and_climate_change.pdf.

5. Cheremisinoff, N.P. 2002. Environmental Security: The Need for International Policies. Pollution Engineering. 5/1/2002. http://www.pollutionengineering.com/CDA/ ArticleInformation/features/ BNP__Features__Item/0,6649,103962,00.html.

6. Commission on Geosciences, Environment and Resources. 1999. Environmental Cleanup at Navy Facilities: Risk-Based Methods http://books.nap.edu/books/0309063434/html/37.html#page_bottom.

7. Commission on Life Sciences. 1983. Risk Assessment in the Federal Government: Managing the Process. Committee on the Institutional Means for Assessment of Risks to Public Health, National Research Council. http://books.nap.edu/nap-cgi/chaphits.cgi?term=acceptable+risk&isbn=0309033497.

8. Commission on Life Sciences. 1994. Science and Judgment in Risk Assessment. Committee on Risk Assessment of Hazardous Air Pollutants, National Research Council.

9. Duffus, J.H. 2001. Risk Assessment Terminology. Chemistry International 23(2). March 2001. http://www.iupac.org/publications/ci/2001/march/risk_assessment.html.

10. European Environment Agency. No Date. Environmental Risk Assessment - Approaches, Experiences and Information Sources Environmental issue report No 4. http://reports. eea.eu.int/GH-07-97-595-EN-C2/en/riskindex.html.

11. European Environment Agency. 2003. Europe's Environment: The Third Assessment.

12. EXTOXNET. 1993. **Toxicology Information Briefs.** Cooperative Extension Offices of Cornell University, Oregon State University, the University of Idaho, and the University of California at Davis and the Institute for Environmental Toxicology, Michigan State University. September 1993. http://extoxnet.orst.edu/tibs/carcino.htm.

13. Glenn, J.C., Gordon, T.J., and R. Perelet. 1998. Defining Environmental Security: Implications for the U.S. Army. Editor: Molly Landholm. AEPI-IFP-1298

14. Army Environmental Policy Institute. December 1998. http://www.aepi.army.mil/Publications/-Defining%20Environmental%20Security%20-%20Implications%20for%20the%20US%20Army.pdf.

15. Haimes, Y.Y., Lambert, J.H., Horowitz, B.M., Kaplan,S., Pikus, I.M., Leung, M.F. and A.D. Mosenthal. 2004. Final Contract Report, Risk Assessment And Management Of Critical Highway Infrastructures: Executive Summary. February 2004. Center for Risk Management of Engineering Systems. University of Virginia. Virginia Transportation Research Council (A Cooperative Organization Sponsored Jointly by the Virginia Department of Transportation and the University of Virginia) Charlottesville, Virginia. VTRC 04-CR16. http://www.virginiadot.org/vtrc/main/online_reports/pdf/04-cr16.pdf.

16. Health Canada. 1998. Assessment And Management Of Cancer Risks From Radiological And Chemical Hazards. http://www.hc-sc.gc.ca/hecs-sesc/ccrpb/publication/98ehd216/ chapter4.htm.

17. Health Canada. 2000. Health Canada Decision-Making Framework for Identifying, Assessing, and Managing Health Risks. August 1, 2000. http://www.hc-sc.gc.ca/hpfb-dgpsa/ hcrisk_11_e.html.

18. Hobbs, R.J. and L.J. Kristjanson. 2003. Triage: How do we prioritize health care for landscapes? Ecological Management & Restoration 4(Suppl): S39-S45.

19. Hunter, P.R. and L. Fewtrell. 2001. Acceptable Risk. World Health Organization Water Quality Guidelines, Standards and Health. Chapter 10.

20. McNeil, F. 2000. Making Sense of Environmental Security. North-South Agenda. Paper. Thirty-Nine. http://www.miami.edu/nsc/publications/pub-ap-pdf/39AP.pdf

21. National Infrastructure Institute (NI^2) Center for Infrastructure Expertise and the University of New Hampshire. 2004. The Center for Infrastructure Expertise Library. July 22, 2004 (accessed). http://www.ni2ciel.org/Glossary/a_glossary.htm.

22. National Intelligence Council. 2000. Global Trends 2015: A Dialogue about the Future With Nongovernment Experts. December 2000. http://www.odci.gov/nic/PDF_GIF_global/globaltrend-2015.pdf.

23. Newman MC, Ownby DR, Me´Zin LCA, Powell DC, Christensen TRL, Lerberg SB, and B-A Anderson. 2000. Applying Species-Sensitivity Distributions in Ecological Risk Assessment: Assumptions of Distribution Type and Sufficient Numbers of Species. Environmental Toxicology and Chemistry, Vol. 19, No. 2, pp. 508–515.

24. NRC (National Research Council). 1983. Risk Assessment in the Federal Government: Managing the Process. Washington, D.C.: National Academy Press.

25. Omenn, G.S. 2003. Assessment of the Impact of the 1983 Red Book. W5.1. Society for Risk Analysis Annual Meeting 2003. W5 - Symposium: Impacts on Risk Policy and Practice, Red Book, Risk Commission, Understanding Risk. http://www.birenheide.com/sra/2003AM/program/singlese-ssion.php3?sessid=W5.

26. Oren, S. 2001. Market Based Risk Mitigation: Risk Management vs. Risk Avoidance. University of California at Berkeley, Presented at a White House OSTP/NSFWorkshop on Critical Infrastructure Interdependencies June 14-15, Washington DC. http://www.pserc.wisc.edu/ecow/get/ publicatio/-2001public/marketbasdriskmitigation-v2-oren.pdf.

27. Owens, R., Broderick, P, Charters, D., Sprenger, M, and T. Bernhard. 1999. Ecological Risk Assessments. Powerpoint. http://enviro.nfesc.navy.mil/erb/erb_a/support/rits/presentations/1999-5-era.pdf.

28. Presidential/Congressional Commission on Risk Assessment and Risk Management. 1997. *Framework for Environmental Health Risk Management.* Final Report, Washington, DC. Volume 1. http://www.spea.indiana.edu:8000/dhenshel/c311/EPAJAN.pdf.

29. Queensland Government. No Date. Securing Queensland's Critical Infrastructure: Guidelines for owners/operators. Department of the Premier and Cabinet. http://www.premiers.qld.gov. au/library/-pdf/infopackweb.pdf.

30. Secretary of the Navy. 2002. Department of Navy (DON) Critical Infrastructure Protection (CIP). Memorandum. June 16, 2002. http://neds.nebt.daps.mil/Directives/3501_1.pdf.

31. SERO (Southeast Region, Installation Management Agency). 2003. Sustainability 401: Part 1: Houston, we have a problem. Joint Services P2 & HWM Conference (14 August 2003). http://www.envquest.com/library-FORSCOM.asp.

32. Schwartz, P. and D. Randall. 2003. Abrupt Climate Change Scenario and Its Implications for United States National Security (October 2003). GBN: Global Business Network. http://www.gbn.com/ArticleDisplayServlet.srv?aid=26231.

33. Scottish Executive. No date. An Overview of Risk Guidance. http://www.scotland.gov.uk/library3/government/riskguidance.pdf.

34. Suter, G.W. II, B.W. Cornaby, C.T. Hadden, R.N. Hull, M. Stack and F.A. Zafran. 1995. An approach for balancing health and ecological risks at hazardous waste sites. Risk Analysis 15: 221-231.

35. The White House. 2003. National Strategy for the Physical Protection of Critical Infrastructures and Key Assets. February 2003. http://www.whitehouse.gov/pcipb/physical.html.

36. U.K. Department of the Environment, Transport and the Regions. 1999. Review of Technical Guidance on Environmental Appraisal A Report by EFTEC (Economics for the Environment Consultancy). April 30, 1999. http://www.defra.gov.uk/environment/economics/rtgea/22.htm.

37. United States Department of Labor. No Date. Occupation Exposure to 1,3-Buradiene. Regulations (Preambles to Final Rules. Section 7 – VII. Significance of Risk. http://www.osha.gov/-pls/oshaweb/owadisp.show_document?p_table=PREAMBLES&p_id=748.

38. United States Navy. 2001. U.S. Navy Human Health Risk Assessment Guidance. Chapter 12 – Risk Management. December 2001. http://www-nehc.med.navy.mil/HHRA/ guidancedocuments/-process/pdf/risk_mgmt.pdf.

39. United Kingdom Department of Environment, Food and Rural Affairs (UK Defra). 2005. The Economic, Social and Ecological Value of Ecosystem Services: A Literature Review. Report prepared by eftec, London, UK. January.

40. Vrijling, J. K., van Gelder, P.H.A.J.M., Goossens, L.H.J., Voortman, H.G., and M. D. Pandey. 2004. A Framework for Risk criteria for critical infrastructures: Fundamentals and Case Studies in the Netherlands. *Journal of Risk Research 7:569-679.*

41. Vogel, D. 2003. The Politics of Risk Regulation in Europe and the United States. Prepared for publication in The Yearbook of European Environmental Law, Volume 3, 2003.

19-36

Q51
Q57 D81
Q58 L92

ENVIRONMENTAL SECURITY

Approaches and Tools

R.J. WENNING
ENVIRON International Corporation
6001 Shellmound Street, Suite 700, Emeryville, CA 94608 USA

S.E. APITZ
SEA Environmental Decisions, Ltd.
1 South Cottages, The Ford, Little Hadham, Hertfordshire SG 11 2AT United Kingdom

D.A. BELLUCK
FHWA/USDOT
400 7th Street, S.W., Washington, D.C. 20590 USA

S. CHIESA
Politecnico di Torino
Dipartimento di Ingegneria Aeronautica e Spaziale, Corso Duca Degli, Abruzzi 24, 10120 Torino, Italy

J. FIGUEIRA
University of Coimbra and INESC-Coimbra
Faculty of Economics, Av., Dias da Silva, 165 Coimbra, 3004-512 Portugal

Z. FILIP
Institute of Chemical Technology Prague
Department of Biochemistry and Microbiology, Technicka 5, 16628 Prague 6, Dejvice, CZ 166-28 Czech Republic

G. IGNATAVIČIUS
University of Vilnius
Ozo 7-8 Vilnius, 05129 +370 Lithuania

M.K. KIKER
Zama Environmental
P.O. Box 142752, Gainesville, FL 32614-2752 USA

I. Linkov et al. (eds.), Environmental Security in Harbors and Coastal Areas, 19–36.

R. LAHDELMA
University of Turku
Department of Information Technology, Lemminkaisenkatu 14-A, Turku, 20520 Finland

M. MACDONELL
Argonne National Laboratory
9700 S. Cass Avenue, Argonne, IL 60439 USA

A. MANNINI
Dyria SistemiS.r.l.
via Veglia, 57/5/d Torino Italy

B. RAISSOUNI
Center for Environmental Issues & Regional Development
AUI CEIRD B.P. 1884, Ifrane 53000 Morocco

D. SINKA
Enconet International
Risk Assessment Department, Miramarska 20, 10.000 Zagreb, Croatia

Y. XENIDIS
Aristotle University of Thessaloniki
Department of Civil Engineering, Thessaloniki, 54124, Greece

Abstract

Within the past 5-10 years, several approaches and tools have emerged that may be useful for evaluating natural and human-made environmental vulnerabilities at chemical manufacturing plants, energy plants, transportation networks and other critical infrastructure located in heavily populated urban areas and coastal ports and harbors. The evaluation of the effectiveness of these approaches and tools in the current framework of environmental management is a crucial issue in order to verify the appropriateness of available techniques and methods and identify possible future needs to address environmental security. At the NATO Advanced Research Workshop, "Environmental Security at Ports, Harbors and Coastal Areas" held 17-21 April 2005, Thessalonica, Greece, a work group of international risk assessment, decision-making, environmental modeling and engineering experts from 11 countries explored the current state of different risk assessment and management tools and approaches for addressing environmental security in coastal ports and harbors. This chapter summarizes the findings of the work group and concludes with a summary of technical challenges and recommendations for future research.

1. Introduction

Environmental security is viewed increasingly by regulatory agencies and international organizations as a critical function of government [20]. Increases in human population and resource use has placed increasing pressure on natural resources, resulting in the destabilization of terrestrial and aquatic ecosystems in many parts of the world. Human activities have a direct impact on the environment. Excessive exploitation of natural resources due to demographic and economic growth has lead to the degradation of water, air and soil, as well as the decline of many species of wildlife, and the consequent loss of ecological integrity and the ability to support future population growth.

This degradation becomes obvious especially in highly populated coastal areas, which are challenged increasingly by problems connected to releases of municipal and/or industrial waste, population growth and industrial and tourism activities. Development generates pressures on the ecological, social and economic environment and often brings conflicts among different stakeholders with an impact to the environment. One of the most serious problems affecting coastal areas is the damage resulting from oil and chemical spills. Numerous and frequent spills demonstrate the extent of the damage inflicted on the environment when large volumes of chemicals or petroleum are released and, in particular, the susceptibility of coastal ecology to these spills. Effective strategies for management of coastal areas should, therefore, be considered as dynamic and multi-faceted processes capable of adapting to ever-changing ecological, social and economic pressures [68].

The decision-making process in environmental management, and particularly with regard to environmental security, is becoming increasingly complex because it must balance economic, environmental, human, political and other competing variables (Figure 1). In the absence of a systematic and transparent decision-making framework that offers clear solutions founded on sound scientific tools and approaches, the security of the population and the environment may be jeopardized. This chapter aims at presenting different approaches to preserve and/or enhance environmental security and understanding of the diversity of tools and applications available to address environmental challenges in ports, harbors and coastal areas.

2. Defining Environmental Security

Environmental security has been defined by many scientists, governments and organizations [17]; however there is not consensus on a single definition among them. Because environmental security has been described by many authors, this workgroup opted to reference the literature rather than generate yet another definition. Belluck et al. [4] offer a useful definition, which states that "environmental security involves actions that guard against environmental degradation in order to preserve or protect human, material, and natural resources at scales ranging from global to local."

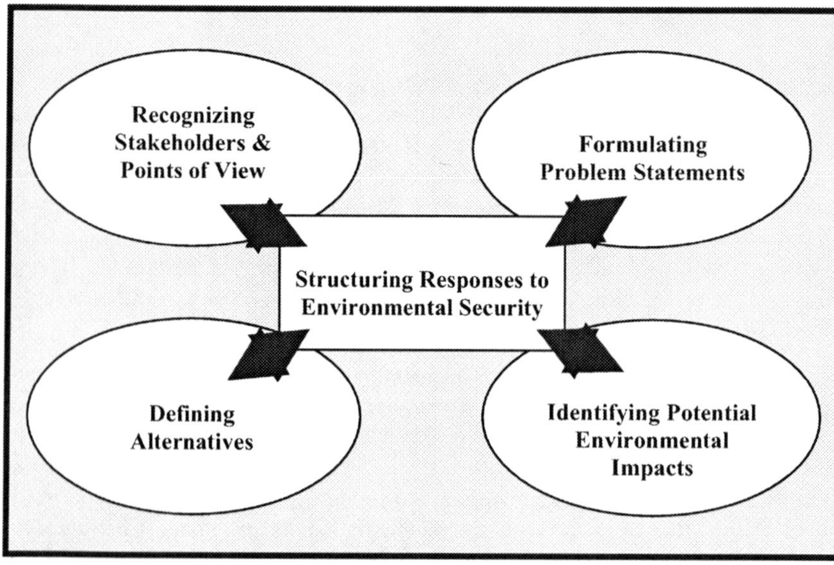

Figure 1: Balancing different perspectives and considerations in the decision-making process concerning environmental security.

This definition implies that the reasons for lack of environmental security should be explored between the social, economic, and environment parameters of security understanding by decision makers. The Belluck et al. [4] definition of environmental security (or insecurity) also implies connections between individuals, households, communities, regions, nation-states and globally, as well as connections across different tiers of ecological order.

This definition clearly distinguishes environmental security from environmental protection; the former addresses the balance between human socioeconomic needs and technology and the latter addresses the application of technical solutions to preserve a specific level of environmental quality. For example, sediment, soil and water quality criteria embodied in numerous environmental regulations are benchmarks intended to protect human health and wildlife from adverse health effects; the level of protection afforded by such benchmarks reflects society's goal to preserve or restore a certain level of ecological diversity, public safety and well-being.

In the same manner, the notion of environmental safety is another component of environmental security according to the Belluck et al. [4] definition. The focus of any safety concern (e.g., automobile safety, industrial safety, nuclear safety, public safety, etc.,) is on the function of machine equipment and tools, industrial facilities, power stations, agricultural areas and procedures for handling resources and waste materials in a manner that poses minimal risks to humans and does not negatively affect the environment through air, water and soil pollution.

3. Challenges to Environmental Security

Maintaining and enhancing environmental security requires the consideration of three elements: (1) understanding of basic human, ecological and environmental conditions; (2) predicting the various opportunities whereby security might be compromised; and (3) analyzing the range of options to enhance, prevent or minimize the contingency for such events to occur. Each of these elements must be examined in terms of the pressures imposed by three fundamental environmental security challenges: (i) demands on natural resources, (ii) environmental events, and (iii) human actions.

The first challenge for environmental security is the maintenance of resources (e.g., clean water, land, property, trees and crop viability) in the face of environmental changes such as global warming, sea level rise, build up of contaminants, etc., that erode both environmental and economic sustainability over time. These changes may be fundamentally human-made, but they are gradual and inexorable. The response to this challenge by the governments involves planning for both prevention and response and, perhaps, protection of limited resources from invasion/threat. Political decisions may involve efforts to prevent or remediate environmental problems, or they may involve efforts to protect resources in the face of change. For example, at the same time that international political efforts are underway to reduce global warming, efforts in several countries are focused on protection of coastal wetlands or mitigation of the effects of rising ocean levels.

A second environmental security challenge involves protection against the results of natural disasters. In such a case, one has to determine the possibility of a number of relatively predictable events such as major storms, earthquakes, etc., and protect against impacts such as dam breaks, chemical spills, explosions, etc. The forecasting of different possible scenarios is relatively straightforward; probabilities of different types of natural disasters are matched to impacts and possible outcomes, and then prevention technologies and their costs are considered. For the most unlikely scenarios, perhaps prevention is not a choice, but response contingencies should be considered (e.g., chemical plants might be moved away from storm tracks or faults, but still be built).

Finally, a third security challenge, which is increasingly viewed as having the potential for enormous environmental destruction and social upheaval, are human-made catastrophes, most notably terrorist attacks. These challenges are more difficult to predict when compared to storms, etc., because people can contrive to achieve improbable goals. Thus, the focus is more on identifying vulnerabilities and then figuring out how to prevent or respond to what are typically unforeseeable, or unpredictable, incidents. This third challenge is similar to the second type of security challenge, but it may involve an entirely different environmental management strategy.

With each challenge, risk managers identify vulnerabilities, examine the possible scenarios, with their associated ecological, social and economic threats, and then rank the threats using a severity (hazard) and probability (exposure) model. While this paradigm is familiar to practitioners of human health and ecological risk assessment and chemical exposure modeling, we need to expand our vocabulary to better embrace concepts of environmental security and the different challenges.

Each of the three challenges to environmental security may require entirely different sets of analysis and decision-making tools. For example, we may want to look for explosive devices in container ships. At the same time, we may be looking for environmental indicators that suggest climate changes or provide other early warnings of ecosystem changes. These and other environmental challenges require strategies that go beyond the promulgation or refinement of regulations, which deal with (a) environmental engineering solutions that seek to optimize input and output processes in the areas under stress; and (b) social programs that encourage education in environmental issues. While the former may help to cure local problems (and probably would require other resources and costs spending), the latter is perhaps more difficult to realize. Having recognized security challenges, there is a need to look at decisions about how to select prevention scenarios. This process requires responses to several issues such as: environmental managers advocating changes in economic practices; changes in resource use practices; building flood defenses; search for terrorists; implementation of technologies to reduce carbon emissions, etc. Another important issue is the orientation of environmental security in the event of a disaster, i.e., whether it should focus on moving populations inland, water desalination and/or environmental cleanup.

4. Structuring Responses to Environmental Security Challenges

4.1. INITIAL CONSIDERATIONS

Elliott [12] has proposed that achievement of successful marine environmental management requires fulfillment of six tenets: actions should be environmentally sustainable, economically viable, technologically feasible, legislatively permissible, achievable administratively and, lastly, socially desirable. That is to say, the management of environmental issues at ports such as municipal or industrial discharges or dredged material disposal to facilitate navigation should not be conducted in an environmentally deleterious manner but, rather, within the context of what is socially acceptable. However, it is increasingly apparent that the last tenet is not socially desirable but, instead, socially tolerable; that is, society increasingly copes with and tolerates changes achieved by management decisions, rather than actively wishing for that change. For example, a new sewage treatment plant that increases water charges to users and has few immediate perceptible benefits may be tolerated rather than demanded by society.

In theory, environmental decision-making to ensure environmental security should be performed through a systematic, transparent, logical and rational approach. In public decision-making processes, scientific or technical evaluations are often interspersed with political or risk management policy decisions that may confuse outside observers and result in a loss of trust and credibility. For rational individuals, decisions are based on the expected utility of a course of action. When rationality is extended to groups, three questions must be asked: (a) are group preferences consistent? (b) are the trade-offs constant?; and (c) are policy and technical issues clearly distinguished from one another and the final decision logical and supportable?

Environmental security problems involve shared resources, multiple perspectives, and group decision-making processes. The complexity of environmental systems and the uncertainty associated with estimates of risks, costs, benefits, and the goals, values, beliefs and objectives of different stakeholder groups make environmental decision-making particularly challenging. Problem solving without any methodology may distort the final results. Without the help of tools, the decisions tend to focus on a small subset of criteria, fixed opinions, insufficient information, miscalculated uncertainties and distorted motivations [62; 53].

4.2. RECOGNIZING STAKEHOLDERS AND DIFFERENT POINTS OF VIEW

Environmental impacts depend on stakeholders' points of view. The stakeholders consist of all the different people associated with the planning and decision process. In some cases, citizens should also be considered among the stakeholders since they bear the main consequences of environmental impacts from several activities. Especially in cases where there is opposition by the public opinion to specific decisions, which are assumed to have an environmental impact, the importance of including them among the stakeholders is obvious. At the outset, it is important to identify all stakeholders and explicitly determine who should participate in the planning process, in which phases, and to what extent. The stakeholders can be classified into standard stakeholders and interested groups. Standard stakeholders are those who have the legitimate responsibility to participate in the process. Standard stakeholders include the decision-makers, experts, and planners & analysts responsible for preparations and managing the process.

Special interested groups are often most active stakeholders. Interested groups typically include political parties, civil organizations or residents of a potentially impacted area. The interested groups have individual points of view for evaluating potential alternatives and often have different relational systems of preference [55; 2]. Depending on their interests, the groups will stand up for different alternatives and objectives, thus creating competition and conflicts based on misunderstanding, opposing interests or different social values [28; 3]. Interested groups add a sociopolitical dimension in the sense that those views and alternatives they find important must be taken into account when the actual decision is made.

Many tools exist for mapping stakeholder typologies [14; 60; 5; 46; 51]. For example, according to Susskind and Martin, stakeholders can be recognized as:

1. *Boosters* who see the issue as essential to their survival.
2. *Friends* to whom the issue is important, but not essential for survival.
3. *Guardians* who, in principle, are neutral and easily switch between points of view.

4. *Nonparticipants* who are totally uninterested or feel they have no power to influence decision-making.
5. *Hostiles* who have erroneous perceptions, and/or exhibit inconsistent behavior or fragile loyalty and often unknowingly act against their own interests.
6. *Preservationists* who will oppose all alternatives and favor, instead, the so-called zero alternative by rejecting problems and their solution.

Various techniques have been proposed for determining these points of view. Roy [57] assumes different points of view will emerge after thorough analysis of various consequences, taking into account the cultural background of the stakeholders involved. Hammond et al. [19] advocate an even-swap methodology for identifying stakeholders and their points of view and then making trade-offs among a range of objectives across a range of alternatives and possible impacts.

4.3. FORMULATING PROBLEM STATEMENTS

Environmental challenges can be classified into two categories, discrete and continuous. Continuous environmental challenges such as providing clean water, electricity or waste disposal require diligent and never-ending planning activities and decisions. Continuous challenges generally pose a numerous number of alternatives that change as population and resource demands change. In contrast, discrete environmental challenges are typically associated with unexpected or catastrophic events such as natural disasters and accidental chemical or oil spills. Discrete challenges generally pose a finite set of response alternatives, each having somewhat predictable outcomes.

In practice, one way of approaching continuous challenges is to discretize them before analysis. This can typically be implemented by multi-objective optimization [61; 45]. By doing so, goals and accomplishments can be recognized by stakeholders, despite the obvious need to periodically revise the same challenge in response to added pressures or changes. The application of the concept of adaptive management is suited well to addressing continuous environmental challenges [8, 9].

A common problem in decision-making is that decision makers are often unclear about their objectives. Multi-faceted issues such as environmental security exacerbate the decision-making process even more. Discrete multi-criteria decision problems are formulated in terms of a finite set of alternatives that are evaluated in terms of multiple criteria. The criteria provide numerical measures for all relevant impacts of different alternatives. The relevance of different impacts will depend on the stakeholders' points of view; thus, it is necessary to define precisely how each criterion and impact is measured.

Roy [57] presents a reasonable stepwise approach to resolving discrete multi-criteria decision problems, involving problem formulation or definition, choosing one or a small set of most-favorable alternatives, ranking the alternatives according to a defined set of preferences or requirements, and sorting of the preferred alternatives into pre-defined categories based on one or more likely outcomes such as costs, environmental impact and short- or long-term consequences. The process begins with the definition of objectives and the formulation of problem statements. What are we protecting? What challenges are we protecting against, and at what spatial and temporal scales? Which actions and events are controllable; which are not? Are we developing preventions, tracking changes, or selecting responses?

4.4. DEFINING ALTERNATIVES

An alternative is the object of a decision, and it may include proposing a-plan of action in the problem. For example, an alternative may consist of a particular way to construct a harbor, maintain a navigation channel or certain actions to improve safety. An alternative is not necessarily a physical activity; it can also be a particular management activity or a strategic plan.

The number of alternatives is highly situation dependent. In principle the number of potential alternatives is infinite, but the decision-making process requires the formation of a finite number of distinct alternatives. This set of alternatives may grow or shrink as a result of better understanding of the problem. It is often also possible to form new alternatives by combining the best parts from existing alternatives.

4.5. IDENTIFYING POTENTIAL ENVIRONMENTAL IMPACTS

Common environmental impacts associated with environmental security challenges include impacts on the soil, ground water, surface water, regional air quality, atmosphere, local flora and fauna, biodiversity and the landscape. Besides these environmental impacts, there are impacts related to the economy, employment, attainability and valuation of different areas, use of energy, services, safety and human health. Direct impacts on the physical and biological environment often indirectly generate significant social impacts, which can be classified into demographic changes, institutional conditions, community/area infrastructure, impacts on life-style, impacts on attitudes and conflicts between different social groups.

Impacts can be classified according to their temporal, spatial and regulatory properties. Temporal impacts can be classified as unique, recurrent or continuous. Continuous and recurrent impacts can be either short or long term. Spatially, impacts can be classified as local, regional, national, international or global. Impacts may be formally regulated or not regulated at all. For unregulated impacts, it should be decided how the temporal and spatial differences are taken into consideration [18; 6]. Finally, impacts may be classified as marginal or significant.

Four main approaches have been suggested to assist in the identification of environmental impacts: map overlays, impact checklists, impact matrices and cause-effect networks [26]. The impact matrices and cause-effect networks identify environmental impacts by establishing the important causal links between the source and the target of each possible impact associated with a particular problem or concern.

Environmental impacts can also be identified by remote sensing techniques. Such techniques are more and more implemented for many environmental problems, ranging from solid waste disposal to fire protection of forests. Remote sensing has the potential to prevail among the various techniques for identifying environmental problems on the ground. There are numerous applications at all levels (local, regional, national, and international) based on satellite images and aerial photos that present impacts of several pressures to the environment, either generated by man or nature (e.g. the Institute for Environmental Security, http://www.envirosecurity.net/index.php).

5. Approaches for Evaluating Environmental Security Challenges

Assuming that planning, prevention and response are the three critical aspects of environmental security in ports, harbors and coastal areas, according to Wenning et al. [66], the first step towards ensuring, maintaining or enhancing environmental security addresses the selection of appropriate assessment tools. Assessment tools should be capable of evaluating baseline, or current, environmental conditions and should be based on empirical data developed from information on current resource (biological, land, air and water) demands, population or economic requirements, industrial operations and environmental monitoring. The same, or different, set of assessment tools such as environmental models (e.g., air and hydrologic), economic models (e.g., cost-benefit, economy or facility growth and expansion) and ecological models should be capable of predicting future changes and challenges. Other considerations include assumptions and availability of information needed to develop different environmental models, as well as the costs required to undertake appropriately detailed and realistic risk analysis and any investments necessary to reduce uncertainties.

5.1. QUANTITATIVE RISK ASSESSMENT

Quantitative environmental risk assessment has evolved over the past 50 years from analysis of baseline ecological conditions and threats to wildlife and human health posed by chemical releases or oil spills to include comparative risk analysis of different remedy options for environmental cleanup at contaminated sites [52; 11; 66]. Risk assessment is used increasingly in the U.S. and other countries to evaluate the life cycle of new chemical substances and to weigh the net risks, costs and benefits to the environment associated with different environmental management strategies [40; 25; 29]. This evolution coincides with rapidly expanding knowledge of chemical environmental fate, ecology, geochemistry, human and wildlife toxicology and exposure modeling [48].

More recently, the application of environmental risk assessment as a predictive tool has further expanded in the U.S. and elsewhere to include analysis of environmental disaster scenarios at chemical manufacturing plants and in heavily populated urban areas and coastal ports and harbors where commercial shipping activities, power plants and chemical or petroleum transfer and storage facilities may be vulnerable to accidents or terrorism [20; 21; 69]. The assessment begins with a vulnerability assessment, comparable to the problem formulation step in U.S. ecological risk assessment, and serves as the foundation for all environmental security initiatives [64; 63; 1]. Risk assessment is increasingly viewed as an important tool for determining or ranking what should be protected, the likelihood of different types of threats, the probable health and environmental consequences of different threat outcomes, and how best to protect sensitive areas and to minimize damages [67].

Environmental risk assessment applied to environmental security and disaster decision-making and prevention planning has three components [67]. The first component involves the development of a baseline understanding of current environmental (air, soil, surface water, sediment and ecology) conditions. The information developed from environmental and ecology surveys can be used to evaluate and prioritize the seriousness of different environmental security challenges. The second component involves quantitative prediction with a high degree of confidence of the range of possible environmental impacts and the potential human health and ecological threats posed by different scenarios. The nature of this work itself poses a security challenge because of the potential sensitivity of the information that must be compiled and evaluated.

Another, and perhaps equally important, component that must be addressed is the use of the information and risk predictions generated by environmental risk assessment, cost-benefit analysis, multi-criteria decision analysis and other tools to describe the outcome of different possible environmental scenarios. A detailed chemical environmental fate and exposure risk analysis, along with careful uncertainty analysis of the factors used to predict spatial and temporal scales of the consequences, is needed to evaluate appropriate response actions and probabilities for successful prevention or mitigation. The results of a comprehensive environmental risk analysis must inform decision-makers on the prioritization of different disaster prevention/response action plans and the associated capital investments needed to minimize or prevent scenarios that are most or least likely to adversely impact the social, economic and environment functions.

5.2. MULTI-CRITERIA DECISION ANALYSIS (MCDA)

Multi-criteria decision analysis (often referred to as MCDA) tools can be classified into three main types of decision models. These are:

1. Value or utility function based methods such as Multi Attribute Utility Theory (MAUT); SMART; and, Analytic Hierarchy Process (AHP) [58]; MACBETH [13].
2. Stochastic Multi-criteria Acceptability Analysis methods such as SMAA [30]; SMAA-2 [32]; and, SMAA-O [45].
3. Outranking methods such as ELECTRE II [56]; ELECTRE III; ELECTRE IV [54]; PROMETHEE I and II methods; SMAA-3 [22]; and, SMAA-TRI.

Cost-benefit analysis (CBA) could also be classified as a multi-criteria method, but the philosophy is different; CBA is based on monetizing all causes and impacts, while MCDA methods acknowledge that decision criteria are incommensurate.

SMAA methods have been developed for discrete multi-criteria problems, where criteria data is uncertain, inaccurate or difficult to obtain accurate or weighted information [59]. In SMAA, uncertain or inaccurate criteria measurements are represented (as in MAUT) through probability distributions. In addition, partial or missing preference (weight) information can be modeled through probability distributions. This makes it easy to model, for example, mixed ordinal and cardinal criteria and preference information.

SMAA methods are based on exploring the weight space in order to describe the preferences that would make each alternative the most preferred one, or that would give a certain rank for a specific alternative. In the original SMAA method [30] the weight space analysis is performed based on an additive utility or value function and stochastic criteria data. The SMAA-2 method [28] generalizes the analysis to a utility or value function to include various kinds of preference information and to consider holistically all ranks. The SMAA-3 method [29] is based on so-called pseudocriteria similar to that used in the ELECTRE III decision-aid [65; 44]. The SMAA-O method [36] extends SMAA-2 for treating mixed ordinal and cardinal criteria in a comparable manner. The multivariate Gaussian distribution was first applied in conjunction to SMAA in [39]. Applications of different SMAA methods are described in detail elsewhere [59; 22; 23; 24; 31; 33; 27; 43].

6. Implementing Environmental Security Measures

In addition to assessment and decision analysis tools, another set of tools are needed to implement the concepts and decisions gained from security vulnerability assessments, environmental monitoring and understanding of stakeholder perspectives, regulatory requirements and industrial and economic goals. Different kinds of security technologies are needed according to different monitoring or control purposes and their application phase (i.e. steady state monitoring or emergency case control).

In order to meet the different requirements of the specific applications there are mobile, ground, marine or aerial tools such as sensors, vehicles, single or network (both manned and unmanned). At present, the state-of-technology has made unmanned survelliance equipment and vehicles remotely radio-controlled or predefined path capable much more affordable and cost-effective than in the past. A summary of tools is presented in Table 1.

A combination of different kinds of sensors and/or vehicles can be recommended according to the specific mission requirements. The hardware configuration of tools for monitoring and control process has a significant impact on the cost effectiveness of various options. The selection of the most appropriate solution (e.g., fixed or mobile sensors installed on board of vehicles, as well as type of vehicle) will come from the analysis of various security requirements and surveys of stakeholder/authority needs.

Table 1: Available tools and their features to support environmental security measures.

Tools	Features
Artificial Satellites	Extended monitored area; predefined survey paths; no environmental impacts.
Manned airplanes	High mission flexibility; large monitored area both on ground and sea; capability for active action (not just monitoring); hovering capability; highest cost; capable of collecting samples; low/medium visual impact; high acoustic impact
Ultralight manned aircraft	High mission flexibility; large monitored area both on ground and sea; low cost; no sampling capability; low/medium visual impact; medium/high acoustic impact
Micro UAVs	Large monitored area (with autonomous flight capability) both on ground and sea; lowest cost; medium mission flexibility; no sampling capability; low visual impact; low/ medium acoustic impact
Micro UAVs – rotary wing	Large monitored area (with autonomous flight capability) both on ground and sea; hovering capability; medium/high cost; medium mission flexibility; medium visual impact; medium/high acoustic impact
Manned boats	High mission flexibility; medium monitored area on water; capability for active action (not just monitoring) and sampling; high cost; visual impact
Unmanned Ship Vehicles	Medium monitored area (with autonomous navigation capability) on sea; capable of sampling; lowest cost; medium mission flexibility; low visual impact
Unmanned Airships	Large monitored area (with autonomous flight capability) both on ground and sea; lowest cost; hovering capability; medium mission flexibility; no sampling capability; visual impact
Fixed balloons	Medium/small monitored area both on ground and sea; lowest cost; low mission flexibility; no sampling capability; visual impact

7. Key Challenges and Future Research Needs

Societies that engage their citizens in technical discussions concerning environmental security require a realistic, robust and systematic way to hold such discussions. Risk assessment, MCDA and similar assessment and decision-making tools provide a suitable framework for conducting these discussions in a logical and fair manner. MCDA and similar decision tools allow for the simultaneous discussion of technical and values-based issues. While the human intellect can perform this function as part of an individual decision-making process, addressing environmental security challenges is not well suited to multi-stakeholder decision-making. Consequently, MCDA and similar decision tools are needed to provide the framework and structure to organize the broad range of environmental, economic and social information and to conduct societal discussions of complex issues in a transparent and organized manner.

Environmental security is a complex subject because of the varied interactions between human activities and the environment, the dynamic nature of environmental problems in coastal areas, the seemingly inconsistent nature of environmental data and societal demands and concerns and the economic implications of preventing or responding to security threats. These and other aspects of environmental security are sources of risk that define the appropriateness of different methods and tools available to manage the health and well-being of human populations and the integrity of the environment.

In this context, the consideration of different approaches initially suggests that quantitative risk assessment and MCDA models are appropriate to evaluating and understanding environmental security challenges. Future research is needed to introduce other methods to this field. A critical concern of this research should be the application and integration of both numerical and non-numerical data into a sound and acceptable background for decision makers who, often, base their approaches on non-quantitative and non-scientifically methods of decision-making.

8. References

1. Apostolakis, G.E., Lemon, D.M. (2005) A screening methodology for the identification and ranking of infrastructure vulnerabilities due to terrorism. *Risk Analysis* 25(2):361-376.

2. Bana e Costa, C. (1986) A multicriteria decision-aid methodology to deal with conflicting situations on the weights. *European Journal of Operational Research* 26, 22-34.

3. Banville C., Landry M., Martel J.M., Boulaire C. (1998) A stakeholder approach to MCDA, Systems Research and Behavioral Science, Vol. 15, pp. 15-32.

4. Belluck, D.A., Hull, R., Benjamin, S.L., Alcorn, J., Linkov, I. (2005) Environmental security, critical infrastructure and risk assessment: Definitions and current trends. In: Kiker, G. and Linkov, I. (eds). "Environmental Security and Environmental Decision-Making". Kluwer Academic Publishers (this volume).

5. Blair, J.D., Slaton, C.R., Savage, G.T. (1990) Hospital-physician joint ventures: a strategic approach for both dimensions of success. *Hospital and Health Services Administration*, 35, pp 3-16.

6. Bouyssou, D. (1990) Building criteria: a prerequisite for MCDA, in C.A. Bana e Costa (Ed.), Readings in Multiple Criteria Decision Aid (pp. 58-80), Springer-Verlag, Heidelberg.

7. Butler, J., Jia, J., Dyer, J. (1997) Simulation techniques for the sensitivity analysis of multi-criteria decision models. *European Journal of Operational Research* 103, 531-546.

8. Butler, K.F., Koontz, I.M. (2005) Theory into practice: implementing ecosystem management objectives in the USDA Forest Service. *Journal of Environmental Management* 35(2):138-50

9. Carlman, I. (2005) The rule of sustainability and planning adaptivity. *Ambio* 34(2):163-8.

10. Charnetski, J.R., Soland, R.M. (1978) Multiple-attribute decision making with partial information: The comparative hypervolume criterion. *Naval Research Logistics Quarterly* 25, 279-288.

11. Efroymson, R.A., Nicolette, J.P., Suter, G.W. (2004) A framework for net environmental benefit analysis for remediation or restoration of contaminated sites. *Journal of Environmental Management* 34(3):315-331.

12. Elliott, M., Cutts, N.D. (2004) Marine habitats: loss and gain, mitigation and compensation. *Marine Pollution Bulletin* 49:671-674.

13. Figueira, J., Greco, S., Ehrgott, M. (2005) "Multiple Criteria Decision Analysis: State of the Art Surveys". Springer-Verlag, New York.

14. Fottler, M.D, Blair, J.D., Whitehead, C.J., Laus, M.D., Savage, G.T. (1989) Assessing key stakeholders: who matters to hospitals and why? *Hospital and Health Services Administration*, 34, pp 525-546.

15. Furness, R.W., Greenwood, J.J.D, Jarvis, P.J. (1993) Can birds be used to monitor the environment? In: Furness, R.W and Greenwood, J.J.D. (eds). "Birds as Monitors of Environmental Change". pp. 1 - 41, Chapman & Hall, London.

16. Gerrits, L., Edelenbos, J. (2004) Management of sediments through stakeholder involvement: the risks and value of engaging stakeholders when looking for solutions for sediment-related problems. *Journal of Soils and Sediments* 4(4): 239-246.

17. Glen, J.C., Gordon, T.J. (2001) "State of the Future". AC/UNU Millennium Project Publications, August. ISBN:0-9657362-8-8

18. Grassin, N. (1986) "Constructing criteria 'population' for the comparison of different options of high voltage line routes", EJOR, Vol. 26, 42-47.Keeney, R.L. (1988), "Structuring objectives for problems of public interest". Operations Research, Vol. 36, 396-405.

19. Hammond, J.S., Keeney, R.L., Raiffa, H. (1998) Even swaps: a rational method for making trade-offs. *Harvard Business Review* 76(2):137-8, 143-8, 150.

20. Health and Safety Executive (HSE). (1992) Regulation 4 of the Offshore Installations (Safety Case) Regulations 1992 (SI 1992/2885) (SCR) United Kingdom. http://www.og.dti.gov.uk/regulation/guidance/reg_offshore/app10.htm. Accessed 25 July 2005.

21. Hendrickson, R.G. (2005) Terrorist chemical releases: assessment of medical risk and implications for emergency preparadness. Journal of Human Ecological Risk Assessment 11:487-499.

22. Hokkanen, J., Lahdelma, R., Miettinen, K., Salminen, P. (1998) Determining the implementation order of a general plan by using a multicriteria method. Journal of Multi-Criteria Decision Analysis 7 (5), 273–284.

23. Hokkanen, J., Lahdelma, R., Salminen, P. (1999) A multiple criteria decision model for analyzing and choosing among different development patterns for the Helsinki cargo harbor. Socio-Economic Planning Sciences 33, 1–23.

24. Hokkanen, J., Lahdelma, R., Salminen, P. (2000) Multicriteria decision support in a technology competition for cleaning polluted soil in Helsinki. Journal of Environmental Management 60, 339–348.

25. Johnson, B.B., Chess, C. (2003) How reassuring are risk comparisons to pollution standards and emission limits? Risk Analysis. 23(5):999-1007.

26. Julien, B., Fenves, S.J. & Small, M.J. (1992) An environmental impact identification system. Journal of Environmental Management (36): 167-184.

27. Kangas, J., Hokkanen, J., Kangas, A., Lahdelma, R., Salminen, P. (2003) Applying stochastic multicriteria acceptability analysis to forest ecosystem management with both cardinal and ordinal criteria. Forest Science 49 (6), 928–937.

28. Keeney, R.L. (1992) Value-Focused Thinking: A Path to Creative Decision Making. Cambridge, MA: Harvard University Press.

29. Knapp, K.C., Weinberg, M., Howitt, R., Posnikoff, J.F. (2003) Water transfers, agriculture, and groundwater management: a dynamic economic analysis. *Journal of Environmental Management* 67(4):291-301.

30. Lahdelma, R., Hokkanen, J., Salminen, P. (1998) SMAA - Stochastic multiobjective acceptability analysis. *European Journal of Operational Research* 106(1), 137-143.

31. Lahdelma, R., Salminen P., Hokkanen, J. 2000. Using multicriteria methods in environmental planning and management. *Journal of Environmental Management* 26(6), 595-605.

32. Lahdelma, R., Salminen, P. 2001. SMAA-2: stochastic multicriteria acceptability analysis for group decision making. *Operations Research* 49 (3), 444-454.

33. Lahdelma, R., Salminen, P., Simonen, A., Hokkanen, J. (2001) Choosing a reparation method for a landfill using the SMAA-O multicriteria method. In: "Multiple Criteria Decision Making in the New Millenium, Lecture Notes in Economics and Mathematical Systems". Volume 507. Springer-Verlag: Berlin, pp. 380–389.

34. Lahdelma, R., Salminen, P. (2002) Pseudo-criteria versus linear utility function in Stochastic Multicriteria Acceptability Analysis. *European Journal of Operational Research* 141(2), 454-469.

35. Lahdelma, R., Salminen, P., Hokkanen, J. (2002) Locating a waste treatment facility by using Stochastic Multicriteria Acceptability Analysis with ordinal criteria. *European Journal of Operational Research* 142(2), 345-356.

36. Lahdelma, R., Miettinen, K., Salminen, P. (2003)Ordinal criteria in stochastic multicriteria acceptability analysis (SMAA). *European Journal of Operational Research* 147(1), 117-127.

37. Lahdelma, R., Makkonen, S., Salminen, P. (2004) Treating dependent uncertainties in multi-criteria decision problems. In: Rao, M.R. and Puri, M.C. (eds). "Proceedings of. APORS'2003: Operational Research and Its Applications: Recent Trends". Volume II, New Delhi.

38. Lahdelma, R., Miettinen, K., Salminen, P., Tervonen, T. (2004) Computational Methods for Stochastic Multicriteria Acceptability Analysis. European Congress on Computational Methods in Applied Sciences and Engineering (ECCOMAS). P. Neittaanmaki, T. Rossi, S. Korotov, E. Oñate, J. Périaux, and D. Knorzer (eds.) pp.1-17.

39. Lahdelma, R., Makkonen, S., Salminen, P. (2006) Multivariate gaussian criteria in SMAA. European Journal of Operational Research, (in press).

40. Lin, B.L., Tokai, A., Nakanishi, J. (2005) Approaches for establishing predicted-no-effect concentrations for population-level ecological risk assessment in the context of chemical substances management. Environmental Science and Technology 39(13):4833-4840.

41. Macharis C., Brans, J.P., Mareschal, B. (1998) The GDSS PROMETHEE Procedure, Journal of Decision Systems, 7, 283-307.

42. Macharis, C., Springael, J., De Brucker, K., Verbeke, A. (2004) PROMETHEE and AHP: The design of operational synergies in multicriteria analysis: Strengthening PROMETHEE with ideas of AHP, European Journal of Operational Research, 153, 307-317.

43. Makkonen, S., Lahdelma, R., Asell, A.-M., Jokinen, A. (2003) Multicriteria decision support in the liberated energy market. Journal of Multi-Criteria Decision Analysis 12(1), 2003, 27-42.

44. Maystre, L.Y., Picted, J., Simos, J. (1994) "Méthodes Multicritéres". Presses Polytechiques et Universitaires Romandes, Lausanne, Switzerland.

45. Miettinen, K. (1999) Nonlinear multiobjective optimization. Kluwer, Dordrecht.

46. Mitchell, R., Agle, B., Wood, D. (1997) Towards a theory of stakeholder identification and salience: defining the principle of who and what really counts. Academy Management Review, No 4, pp 853-886.

47. Moffett, A., Sarkar S. (2006) Incorporating multiple criteria into the design of conservation area networks: A mini review with recommendations. Diversity and Distributions, (Diversity Distrib.) 12, 125–137

48. Monosson, E. (2005) Chemical mixtures: considering the evolution of toxicology and chemical assessment. Environmental Health Perspectives 113(4):383-390.

49. Ohenoja, E. (1992) Keminmaan Kallinkankaan suursienistö. Lutukka 8, 125-129.

50. Ozawa, C. and L. Susskind (1985) Mediating Science-Intensive Policy Disputes. Journal of Policy Analysis and Management 5(1):23-39.

51. Parker, R. (2004) "Standards Stakeholder Engagement, Update: Building Momentum". CIHI Partnership Meeting, Calgary, Canada.

52. Paustenbach, D.J. (1989) Important recent advances in the practice of health risk assessment: implications for the 1990s. Regulatory Toxicology and Pharmacology 10(3):204-243.

53. Payne, J.W., Bettman, J.R., Coupey, E., and Johnson, E.J. (1992) A Constructive Process View of Decision Making: Multiple Strategies in Judgment and Choice. Acta Psychologica, 80, 107-141.

54. Roy, B., Hugonnard, J.C. (1982) "Ranking of suburban line extension project of the Paris metro system by a multicriteria method," Transportation Research, Vol. 16A, no 4, 301-312.

55. Roy, B. and Ph. Vincke. (1984) Relational systems of preference with one or more pseudo-criteria: some new concepts and results. Management Science, 30(11):1323-1335.

56. Roy, B. (1991) "The outranking approach and the foundations of ELECTRE methods," Theory and Decision 31, 49-73.

57. Roy, B. (1996) Multicriteria methodology for Decision Aiding, Kluwer, Dordrecht.

58. Saaty T.L. (1980) The Analytic Hierarchy Process, NY, McGraw Hill.

59. Salminen, P., Hokkanen, J., Lahdelma, R. (1996) Multicriteria decision analysis project on environmental problems. University of Jyväskylä, Department of Mathematics, Laboratory of Scientific Computing, Report.

60. Savage, G.T., Nix, T.W., Whitehead, C.J., Blair, J.D. (1991) Strategies for assessing and managing organizational stakeholders. Academy of Management Executive, 5, pp 61-75.

61. Steuer, R. (1986) Multiple criteria optimization: theory, computation and application. John Wiley and Sons, Inc., New York.

62. Tversky, A., Kahneman D. (1986) Choice and the Framing of Decisions. Journal Of Business. 59:S251-S278.

63. United States Department of Homeland Security (USDHS) (2003) Vulnerability Assessment Methodologies Report. Phase I Final Report. Office for Domestic Preparedness, Washington, D.C. July.

64. United States Environmental Protection Agency (USEPA) (1996) Accidental Release Prevention Requirements: Risk Management Programs Under the Clean Air Act, Section 112(r)(7); List of Regulated Substances and Thresholds for Accidental Release Prevention, Stay of Effectiveness; and Accidental Release Prevention Requirements: Risk Management Programs Under Section 112(r)(7) of the Clean Air Act as Amended, Guidelines; Final Rules and Notice. Federal Register 61(120):31667-31720. 20 June.

65. Vincke, P.H. (1992) "Multicriteria Decision-Aid". John Wiley & Sons, Chichester, UK.

66. Wenning, R.J., Sorensen, M., Magar, V.S. (2005) Importance of implementation and residual risk analyses in sediment remediation. Integrated Environmental Assessment and Management 2(1):in press.

67. Wenning, R.J., Della Sala, S., Magar, V.S. (2005) Role of risk assessment in environmental security planning and decision-making. In: Kiker, G.A. and Linkov, I. (eds). "Environmental Security and Environmental Decision-Making". Kluwer Academic Publishers (this volume).

68. Xenidis, Y., Angelides, D.C., Tziavos, I., Koutitas, C., Savvaidis, P. (2005) An integrated framework for sustainable coastal zones management in Greece. IASME Transactions, Vol. 2, No 5, July, p.p. 813-822.

69. Zaidi, M. (2005) Risk assessment in detecting and preventing of terrorist attacks in harbors and coastal areas. In: Kiker, G.A. and Linkov, I. (eds). "Environmental Security and Environmental Decision-Making". Kluwer Academic Publishers (this volume).

INTEGRATING COMPARATIVE RISK ASSESSMENT AND MULTI-CRITERIA DECISION ANALYSIS

Working through Wicked Problems and Their Impossible Solutions

G.A. KIKER
University of Florida
Agricultural and Biological Engineering Department, PO Box 10570, Gainesville, FL 32611-0570 USA

I. LINKOV
INTERTOX, Inc.
83 Winchester Street, Suite 1, Brookline, MA 02446 USA

T.S. BRIDGES
U.S. Army Engineer Research and Development Center
Environmental Laboratory, 3909 Halls Ferry Rd, Vicksburg, MS 39180 USA

Abstract

"Wicked" problems emerge from the cross-disciplinary and multi-objective reality of current environmental challenges. Significant ecological risks and their uncertainty combine with conflicting stakeholder objectives and values to create a need for systematic risk and decision integration methods. Comparative Risk Assessment and Multi-Criteria Decision Analysis provide useful methods for integrating these diverse, decision-relevant factors. A typical wicked problem is realized in the combination of risk and decision factors within contaminated sediment challenges, such as those found in New York/New Jersey Harbor. In a larger context, we identify three essential decision ingredients, *People*, *Process* and *Tools* that should be carefully considered before prematurely embarking on a decision path.

1. Introduction

In the past, environmental managers worked primarily within a technical context, which simplified most complex challenges into simple, tamable objectives. Single objective problems of the past, such as flood control, have become complicated by multiple objectives and conflicting societal values [19]. The selection of alternatives that include significant ecological risks and uncertainties coupled with divergent stakeholder goals dominates resource managers' time and energy.

I. Linkov et al. (eds.), Environmental Security in Harbors and Coastal Areas, 37–51.

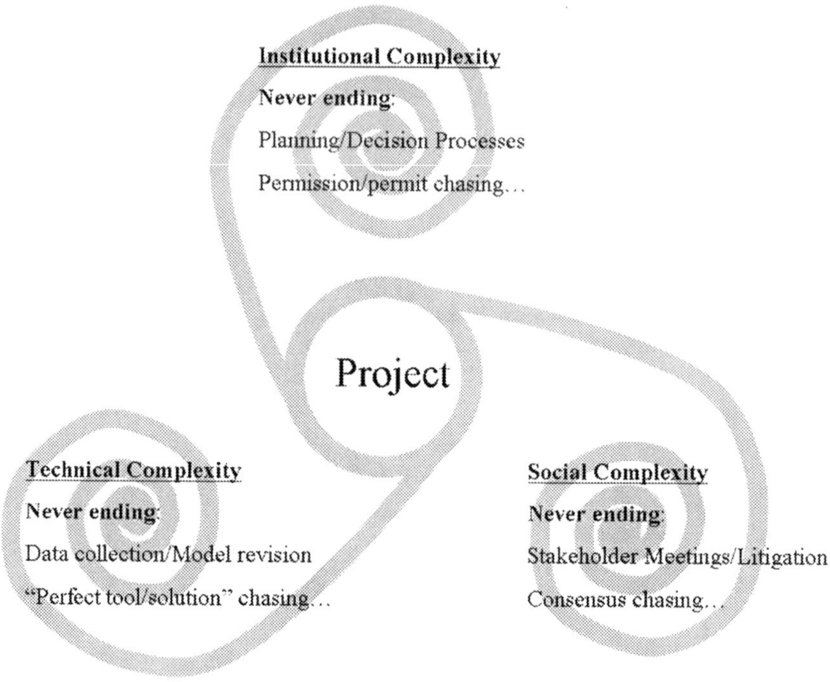

Figure 1: Project complexity in its three forms (Institutional, Social and Technical) can cause wasted energy, time and money in addressing wicked problems (adapted from Conklin [6]).

"Wicked" problems [21] stem from the cross-disciplinary and multiple objective nature of problems. This complex interaction of social, technical and environmental forces at different time and spatial scales was enigmatic enough to inspire the creation of a new word, *Panarchy* [9; 10] in an attempt to understand its effect on environmental management. Figure 1 stylizes the complexities that can draw a project into downward spirals of wasted energy, time and money (adapted from Conklin [6]). Seemingly endless institutional, technical or social iterations can occur when participants keep deciding that more planning, data collection or stakeholder meetings will somehow provide the project with enough inertia to break the impasse. In reality, projects devolve into chaos as the social, technical and institutional complexities become gridlocked: stakeholders who can't see how their values were taken into account litigate in the courts; scientists who are dissatisfied that their science wasn't used to aid in the decision complain about the management; and institutional managers who cannot understand each others perspectives withhold permissions or resources. A simplified solution to the problem is shown in Figure 2. All the social, technical and institutional perspectives are still present but are seen as important ingredients within the overall decision methodology. Environmental decision-makers are realizing that solutions to modern problems cannot simultaneously minimize risk and uncertainty while maximizing acceptability to all stakeholders.

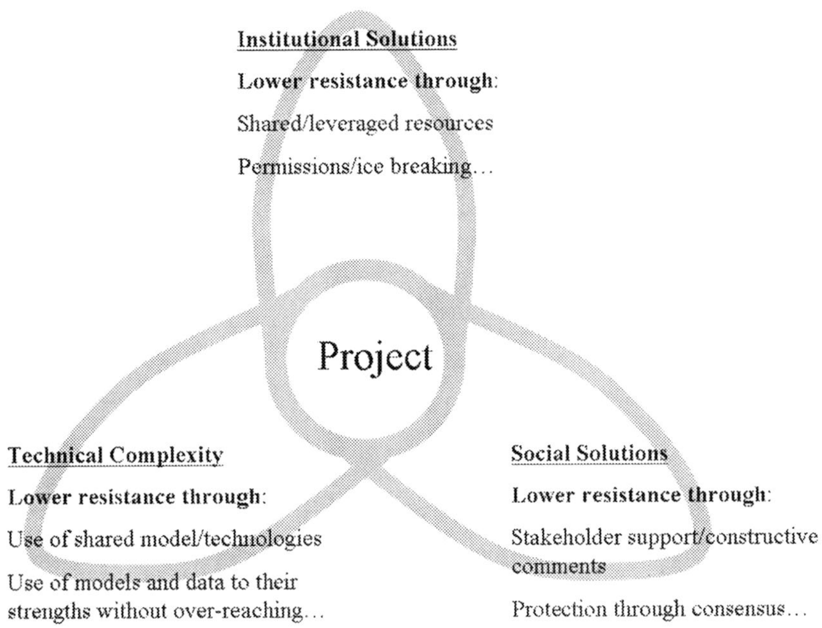

Figure 2 : Successful complex project navigation means using the three decision factors productively to address challenges.

One of the more confounding aspects of wicked problems is the inherent risk and uncertainty within our understanding of the system and proposed management alternatives. Risk is usually defined as the probability of an undesired event [23]. Within this definition, ecological risk describes potential effects to an ecosystem as a whole or a on a population within an ecosystem. Exposure and effects are the primary factors for describing what could happen to an organism or population under risk. Exposure describes the probability of encountering stress while the effect describes the impact resulting from that encounter. Uncertainty permeates throughout the concept of risk, reflecting both *ignorance* of ecosystem components and the *variability* of temporal, spatial and population heterogeneity [25].

In order to successfully manage the wide range of ecosystem and risk information, multi-criteria decision analysis (MCDA) provides a useful approach to create structured and defendable decisions [12; 11; 4; 8]. This paper reviews the role of risk analysis and decision analysis in addressing the complex environmental problems. In addition, an example case study from decision and risk analysis research in contaminated sediment management in the New York/New Jersey Harbor is included to highlight recent research applications.

2. Environmental Risk, Comparing Risks and Uncertainty

An important first step in risk analysis developing a hypothesis about a how a system works through the construction of a conceptual site model (CSM), as illustrated in Figure 3.

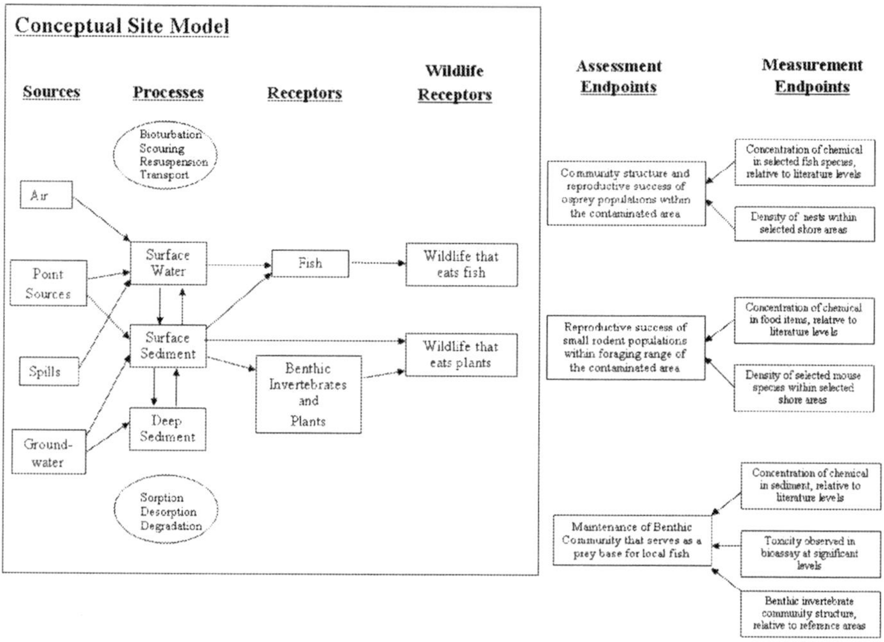

Figure 3: An example Conceptual Site Model along with assessment and measurement endpoints for Comparative Risk Analysis.

The CSM shows how the various physical, chemical and biological factors interact to produce risk to specific receptors. The CSM diagrams should point to various “endpoints” of interest that should be protected or restored. In addition, it should point to what “assessment endpoints” will be used to measure and present “evidence” to risk assessors for use in determining whether significant risks exist.

In translating the CSM and its various endpoints into a systematic risk assessment requires an approach that uses “lines of evidence” as a generic approach for addressing uncertainty [17]. The purpose of Comparative Risk Assessment (CRA) is to gather these multiple lines of evidence into a systematic analysis of risk, costs and benefits. Defining CRA as the “younger sub-topic of a still maturing discipline of environmental risk analysis”, Cura *et al.* [5] provides a useful review of varying definitions of CRA along with a summary describing the debate on how or whether non-expert valuation should be included. The regulatory nature of risk and its use in risk assessments creates a formalized and somewhat rigid structure for defining and calculating risk. Challenges abound in the risk assessment field in that controlled laboratory tests or limited field investigations must somehow describe complex ecosystem processes and their potential threat to humans and ecosystems [2].

Risk assessments are conducted under various levels of uncertainty and ecosystem variability, both of which must be addressed and managed to an acceptable level. Vorhees *et al.* [25] provide a systematic methodology for identifying and addressing uncertainty applied to dredged material assessment. Uncertainty is described in two ways: lack of knowledge and/or variability (including temporal or spatial heterogeneity). The USEPA [24] describes three sources of uncertainty: the inability to specify or characterize a problem (scenario uncertainty); the simplification imposed by representing real world relationships in terms of mathematical relationships (model uncertainty); and the lack of knowledge about the actual value of system factors (parameter uncertainty). Once identified, areas of uncertainty can be assessed to estimate both their magnitude and the relative ease their mitigation. The primary result is the systematic description and identification of the most critical areas and their potential impact on the results of the risk assessment.

3. Multi-Criteria Decision Analysis: Tools for Wicked Problem Analysis

Contaminated sediment management, along with its human and ecological risk challenges, presents an example wicked problem that combines risk assessment and uncertainty with a need for systematic decision analysis. A useful example is presented by contaminated sediment management issues within the greater New York/New Jersey (NY/NJ) Harbor area [26]. Several million cubic meters of sediments must be dredged each year to maintain navigation channels for harbor access. Due to long-term human use of the harbor area, significant contaminant concentrations have been recorded in certain areas. Additional challenges in sediment management have been created by the limitation of ocean disposal to only clean sediments and plans for deepening of existing channels to allow increased access of larger, transport vessels. Additional sediment management options, along with their associated risk and decision analysis, are required for contaminated sediments within the NY/NJ harbor area and need to be systematically explored for cost-efficient risk reduction. A screening level, comparative risk assessment [13] was developed for generalized areas within NY/NJ harbor. Eight sediment management alternatives (including no-action) were assessed according their potential performance in minimizing human health and ecological risks, as represented by six risk-focused criteria. The decision criteria were selected to represent the most important pathways derived from conceptual site models using each sediment management alternative. Kane-Driscoll et al. [13] provided quantitative estimates for these criteria with the resulting data are used to parameterize the table depicted in Figure 4. Figure 4 shows the eight dredged material remediation choices along with the four generalized criteria categories (cost, public acceptability, human health, and ecological health) and the seven more detailed sub-criteria that are established to aid decision-makers in judging the relative strengths of the alternatives.

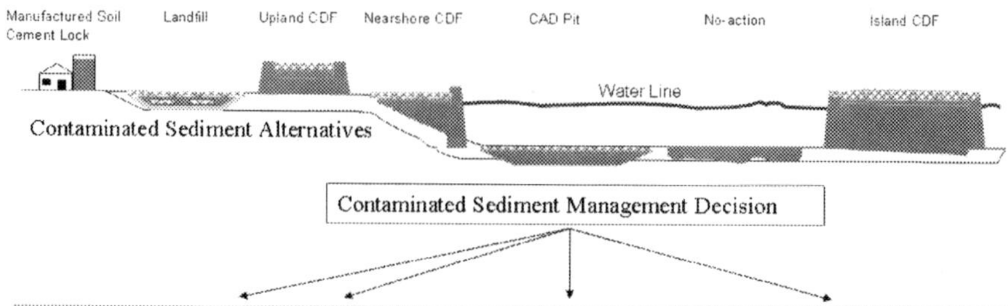

DM Alternatives	Cost[1] ($/CY)	Public Acceptability: Impacted Area/Capacity (acres / MCY)	Ecological Risk: Ecological Exposure Pathways	Ecological Risk: Magnitude of Ecological HQ	Human Health Risk: Human Exposure Pathways	Human Health Risk: Magnitude of Maximum Cancer Risk	Human Health Risk: Estimated Fish COC / Risk Level
CAD	5-29	4400	23	680	18	2.8 E-5	28
Island CDF	25-35	980	38	2100	24	9.2 E-5	92
Near-shore CDF	15-25	6500	38	900	24	3.8 E-5	38
Upland CDF	20-25	6500	38	900	24	3.8 E-5	38
Landfill	29-70	0*	0	0*	21	3.2 E-4	0*
No Action	0-5	0*	41	5200	12	2.2 E-4	220
Cement-Lock	54-75	0	14	0.00002	25	2.0 E-5	0*
Manufactured Soil	54-60	750	18	8.7	22	1.0 E-3	0*

• Listed in Driscoll et al. (2002) as "not applicable or not estimated" for some alternatives For subsequent MCDA computations, the criteria was assumed to have a minimal value.

[1] Cost data was estimated from the following sources (USACE, 1998; NRC, 1997) and expert interviews

Figure 4 . NY/NJ Harbor: an example decision structure with criteria and decision table.

To evaluate ecological risk, two criteria are selected. The number of complete ecological exposure pathways describes the number of uninterrupted source-to-ecological-endpoint paths that occur when using that alternative (as derived from conceptual site models). The maximum calculated hazard quotient describes largest ratio of chemical intake to a reference level from the alternative. To evaluate human health risk, two similar criteria are selected. The number of complete human exposure pathways describes the amount of uninterrupted source-to-human-endpoint paths. The maximum cancer risk calculated is the highest estimated cancer risk when using that alternative. The cost in dollars per cubic yard of sediment is used as a cost criterion. The footprint of the remediation alternative is the acres of land required to manage the sediment.

A table that contains each alternative's score in relation to each criterion is usually the final product of a Comparative Risk Analysis. The projects are then evaluated either by qualitatively comparing the alternatives scores on different criteria or by quantitatively aggregating criterion scores for each project and comparing the aggregate scores. Table 1 shows a summary of steps used to decide among options using an ad hoc approach to decision-making, Comparative Risk Analysis and Multi-Criteria Decision Analysis (MCDA) methods. Within both ad hoc and CRA methods, it can be unclear how to combine criteria to arrive at a ranking of project alternatives.

Table 1: A comparison of decision processes for *ad hoc* decision-making, comparative risk assessment, and multi-criteria decision analysis.

Elements of Decision Process	*Ad Hoc* Decision-Making	Comparative Risk Assessment	Multi-Criteria Decision Analysis
Define problems	Stakeholder input limited or non-existent. Therefore, stakeholder concerns may not be addressed by alternatives.	Stakeholder input collected after the problem is defined by decision-makers and experts. Problem definition is possibly refined based on stakeholder input.	Stakeholder input incorporated at beginning of problem formulation stage. Often provides higher stakeholder agreement on problem definition. Thus, proposed solutions have a better chance at satisfying all stakeholders.
Generate alternatives	Alternatives are chosen by decision-maker usually from pre-existing choices with some expert input.	Alternatives are generated through formal involvement of experts in more site-specific manner.	Alternatives are generated through involvement of all stakeholders including experts. Involvement of all stakeholders increases likelihood of novel alternative generation.
Formulate criteria by which to judge alternatives	Criteria by which to judge alternatives are often not explicitly considered and defined.	Criteria and subcriteria are often defined.	Criteria and subcriteria hierarchies are developed based on expert and stakeholder judgment.
Gather value judgments on relative importance of criteria	Non-quantitative criteria valuation weighted by decision maker	Quantitative criteria weights are sometimes formulated by the decision maker, but in a poorly justified manner.	Quantitative criteria weights are obtained from decision makers and stakeholders.
Rank/select final alternatives	Alternative often chosen based on implicit weights in an opaque manner.	Alternative chosen by aggregation of criteria scores through weight of evidence discussions or qualitative considerations.	Alternative chosen by systematic, well-defined algorithms using criteria scores and weights.

MCDA methods and tools can provide a systematic approach for integrating risk levels, uncertainty and valuation for differing criteria. A detailed analysis of the theoretical foundations of these decision methods and their comparative strengths and weaknesses is presented in Belton and Steward [1] while reviews of MCDA applications in various environmental areas is presented by Kiker *et al.* [15]. The common purpose of MCDA methods is to evaluate and choose among alternatives based on multiple criteria using systematic analysis that overcomes the limitations of unstructured individual or group decision-making. While the basic organization of criteria and alternatives is similar in most MCDA approaches, the methods differ in their synthesis of the information and strategy in ranking the alternatives by different means.

Table 2 summarizes three MCDA methods [20; 16] and describes their similarities and differences. Multi-attribute utility theory (MAUT) and the analytical hierarchy process (AHP) use optimization algorithms, whereas Outranking eschews optimization in favor of a dominance approach. The optimization approach employs numerical scores to communicate the merit of each option on a single scale. Scores are developed from the performance of alternatives with respect to individual criteria and then aggregated into an overall score. Individual scores may be simply summed or averaged, or a weighting mechanism can be used to favor some criteria more heavily than others. The goal of MAUT is to find a simple expression for the net benefits of a decision. Through the use of utility or value functions, the MAUT method transforms diverse criteria, such as those shown in Figure 5, into one common scale of utility or value. MAUT relies on the assumptions that the decision-maker is rational (preferring more utility to less utility, for example), that the decision-maker has perfect knowledge, and that the decision-maker is consistent in his judgments. The goal of decision-makers in this process is to maximize utility or value. Because poor scores on criteria can be compensated by higher scores on other criteria, MAUT is part of a group of MCDA techniques known as “compensatory” methods.

Similar to MAUT, AHP aggregates various facets of the decision problem using a single optimization function known as the objective function. The goal of AHP is to select the alternative that results in the greatest value of the objective function. Like MAUT, AHP is a compensatory optimization approach. However, AHP uses a quantitative comparison method that is based on pair-wise comparisons of decision criteria, rather than utility and weighting functions. All individual criteria must be paired against all others and the results compiled in matrix form. For example, in examining the choices in the remediation of contaminated sediments, the AHP method would require the decision-maker to answer questions such as, “With respect to the selection of a sediment alternative, which is more important, public acceptability or cost?” The user uses a numerical scale to compare the choices and the AHP method moves systematically through all pair-wise comparisons of criteria and alternatives. The AHP technique thus relies on the supposition that humans are more capable of making relative judgments than absolute judgments. Consequently, the rationality assumption in AHP is less rigid than in MAUT.

Table 2: Comparison of critical elements, strengths and weaknesses of several MCDA methods: MAUT, AHP, and outranking [20; 16].

Method	Important Elements	Strengths	Weaknesses
Multi-attribute utility theory	Expression of overall performance of an alternative in a single, non-monetary number representing the utility of that alternative Criteria weights often obtained by directly surveying stakeholders	Easier to compare alternatives whose overall scores are expressed as single numbers Choice of an alternative can be transparent if highest scoring alternative is chosen Theoretically sound — based on utilitarian philosophy Many people prefer to express net utility in non-monetary terms	Maximization of utility may not be important to decision makers Criteria weights obtained through less rigorous stakeholder surveys may not accurately reflect stakeholders' true preferences Rigorous stakeholder preference elicitations are expensive
Analytical hierarchy process	Criteria weights and scores are based on pairwise comparisons of criteria and alternatives, respectively	Surveying pairwise comparisons is easy to implement	The weights obtained from pairwise comparison are strongly criticized for not reflecting people's true preferences Mathematical procedures can yield illogical results. For example, rankings developed through AHP are sometimes not transitive
Outranking	One option outranks another if : "it outperforms the other on enough criteria of sufficient importance (as reflected by the sum of criteria weights)" and it "is not outperformed by the other in the sense of recording a significantly inferior performance on any one criterion" [20] Allows options to be classified as "incomparable"	Does not require the reduction of all criteria to a single unit Explicit consideration of the possibility that very poor performance on a single criterion may eliminate an alternative from consideration, even if that criterion's performance is compensated for by very good performance on other criteria	Does not always take into account whether over-performance on one criterion can make up for under-performance on another The algorithms used in outranking are often relatively complex and not well understood by decision makers

Unlike MAUT and AHP, outranking is based on the principle that one alternative may have a degree of dominance over another [14]. Dominance occurs when one option performs better than another on at least one criterion and no worse than the other on all criteria [20]. However, outranking techniques do not presuppose that a single best alternative can be identified. Outranking models compare the performance of two (or more) alternatives at a time, initially in terms of each criterion, to identify the extent to which a preference for one over the other can be asserted. Outranking techniques then aggregate the preference information across all relevant criteria and seek to establish the strength of evidence favoring selection of one alternative over another. For example, an outranking technique may entail favoring the alternative that performs the best on the greatest number of criteria. Thus, outranking techniques allow inferior performance on some criteria to be compensated by superior performance on others. They do not necessarily, however, take into account the magnitude of relative underperformance in a criterion versus the magnitude of over-performance in another criterion. Therefore, outranking models are known as "partially compensatory." Outranking techniques are most appropriate when criteria metrics are not easily aggregated, measurement scales vary over wide ranges, and units are incommensurate or incomparable [22].

3.1. ESSENTIAL DECISION ANALYSIS INGREDIENTS - PEOPLE, PROCESS AND TOOLS

While Comparative Risk Analysis and Multi-Criteria Decision Analysis are useful tools to contextualize risks and then place them into a functional decision structure, they still must fit into an even larger picture of decisions as seen in Figure 5 [15]. In the author's experience, successful environmental decision-making in complex settings depends on the integration of three key ingredients within the process: *People, Process and Tools*. All are essential for structured and defendable decisions.

Having an appropriate combination of *People* is the first essential element to the overall decision process. Different people have different roles and interests within a decision. The activity/involvement levels of three example groups of people (decision-makers, scientists/engineers and stakeholders) are symbolized by dark lines for direct involvement and dotted lines for less-direct involvement. While the actual membership and the function of these three base groups may intersect or vary, the roles of each are essential in gathering useful human input to the decision process. Each group has its own way of viewing the world, its own method of envisioning solutions and its own societal responsibility. Policy/Decision-Makers may spend most of their effort in defining the problem context and the overall constraints to the decision. In addition, they may have responsibility for the selection of the final decision and its implementation. Stakeholders may provide significant input to defining the problem but have a high degree of interaction in helping to formulate success criteria and contributing value judgments for weighting the various success criteria. Depending on the problem and regulatory context, stakeholders may have some responsibility in ranking and selecting the final option. Scientists and engineers often have a focused role in that they provide the measurements or estimations of the desired criteria that determine the success of various alternatives. While they may take a secondary role as stakeholders or decision-makers, their primary role is to provide the technical details as requested by the decision process.

Figure 5 places *Process* in the center of the overall decision process.

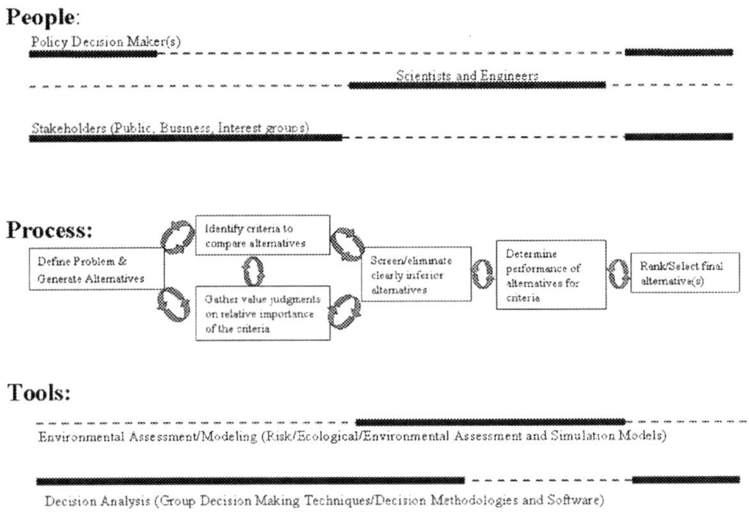

Figure 5: Essential decision ingredients: *People*, *Process* and *Tools*.

While it is reasonable to expect that the decision-making process may vary in specific details among regulatory programs or project types, emphasis should be given to designing an adaptable structure so that participants can modify aspects of the project to suit local concerns, while still producing a structure that provides the required outputs. The process depicted in Figure 5 follows two basic themes, (1) generating management alternatives, success criteria and value judgments and (2) ranking the alternatives by applying the criteria levels and value-weights.

The first part of the process generates and defines choices, success levels and preferences. The second part methodically prunes non-feasible alternatives by initially applying screening mechanisms (for example, overall cost, technical feasibility, general societal acceptance) followed by more detailed ranking of the remaining options by decision analytical techniques (AHP, MAUT, Outranking) that utilize the various criteria levels generated by environmental tools, monitoring or stakeholder surveys. The process is iterative and can be repeated as warranted by the needs of the decision makers or stakeholders.

The *Tools* used within group decision-making and scientific research are essential elements of the overall decision process. As with *People*, the applicability of the tools is symbolized by solid lines (direct/high utility) and dotted lines (indirect utility). Decision analysis tools help to generate and map preferences of stakeholder groups as well as individual value judgments into organized structures that can be generated through use of methodologies such as Soft Systems Methodology [3] and Collaborative Learning [7] linked with the other technical tools from risk analysis, modeling/monitoring, and cost estimations. The decision analysis software also provide useful graphical techniques and visualization methods to express the gathered information in understandable formats. When changes occur in the requirements or decision process, decision analysis tools can be used to respond efficiently to reprocess and iterate with the new inputs. The framework depicted in Figure 5 provides a focused role for the detailed scientific and engineering efforts invested in experimentation, environmental monitoring and modeling that provide the rigorous and defendable details for evaluating criteria performance under various alternatives. This symbiotic relationship between decision and scientific/engineering tools allows each to have a unique and valued role in the decision instead of attempting to apply either software tool past its intended target use.

4. Discussion

Current complex environmental solutions must be carefully designed for a balance of institutional, technical and societal factors without being drawn into the many pitfalls within wicked problems. Effective environmental decision-making requires a basic structure for jointly considering the environmental, ecological, technological, economic, and socio-political factors relevant to evaluating and selecting a management alternative. Integrating this heterogeneous information with respect to human aspirations and technical applications demands a systematic and understandable framework to organize people, processes, and tools for making a structured and defensible decision.

Stakeholder involvement is increasingly recognized as being an essential element of successful environmental decision making. The challenge of capturing and organizing that involvement as structured inputs to decision-making alongside the results of scientific and engineering studies can be met through the application of the tools reviewed in this paper. The current environmental decision-making context limits stakeholder participation within the "decide and defend" paradigm that positions stakeholders as constraints to be tested, rather than the source of core values that should drive the decision-making process. Consequently, potentially controversial alternatives are eliminated early. Instead, the final decision may be something to which no one objects too strenuously.

A fundamental element of Figure 5 is that the entire decision-making process can be cycled through many times in the course of arriving at a decision. The same basic *process* is used initially with rough estimates to sketch out the basic elements and challenges in the decision process with a few initial stakeholders and screening-level analysis or models. A first-pass effort may efficiently point out challenges that may occur, key stakeholders to be included or modeling/analysis studies that should be initiated. As these challenges become more apparent, one iterates again through the framework to explore and adapt the process to address the more subtle aspects of the decision. Each iteration reveals additional information that will benefit the overall decision. Thus, these initial iterations can provide fundamental guidance to the design of subsequent risk assessments and their success in providing useful information to the decision.

It must be remembered that these methods will not single out the "correct" decision, but will help improve understanding in a way that facilitates a decision-making process involving risk, multiple criteria, and conflicting interests. MCDA can visualize tradeoffs among multiple, conflicting criteria and can quantify the uncertainties necessary for comparison of available remedial and abatement alternatives. MCDA helps technical project personnel as well as decision makers and stakeholders systematically consider and apply value judgments to derive a favorable management alternative. MCDA also provides methods for participatory decision-making where stakeholder values are elicited and explicitly incorporated into the decision process.

Different MCDA methods have their associated strengths and limitations. No matter which analytical decision tool is selected, implementation requires complex tradeoffs. This complexity is probably one of the main reasons why MCDA is still not widely used. However, explicit, structured approaches will often result in a more efficient and effective decision process compared with the often intuition- and bias-driven decision processes that are currently used. Combining the strengths of risk assessment and multi-criteria decision analysis offers the most promise for taming wicked problems.

5. References

1. Belton, V. and Steward, T. 2002. *Multiple Criteria Decision Analysis An Integrated Approach.* Kluwer Academic Publishers: Boston USA.

2. Chapman, P.M., McDonald, B.G., and Lawrence, G.S. 2002. Weight-of-Evidence Issues for Sediment Quality (and Other) Assessments. Human and Ecological Risk Assessment 8(7):1489-1515.

3. Checkland, P. 1999. Systems Thinking, Systems Practice. John Wiley & Sons Inc. New York USA.

4. Clemen, R.T. and Riley, T. 2000. Making Hard Decisions. Duxbury Press. California, USA.

5. Cura, J.J., Bridges, T.S., McAdrle, M.E. 2004. Comparative risk assessment methods and their applicability to dredged material management decision-making. Human and Ecological Risk Assessment, 10:485-503.

6. Conklin, J. 2004. Wicked Problems and Social Complexity. http://cognexus.org/wpf/wickedproblems.pdf (last accessed 3/1/2005)

7. Daniels, S.E. and Walker, G.B. 2001. Working Through Environmental Conflict: the Collaborative Learning Approach. Praeger Publishers. Westport USA.

8. Figueira, J., Greco, S., and Ehrgott, M. 2005. Multiple Criteria Decision Analysis: State of the Art Surveys. Springer New York.

9. Gunderson, L.H. and Holling, C.S. 2002. Panarchy: understanding transformations in human and natural systems. Island Press. Washington. USA.

10. Gunderson, L.H., Holling, C.S. and Light, S.L. 1995. Barriers and Bridges to the Renewal of Ecosystems and Institutions. Columbia University Press. New York. USA.

11. Hammond, J.S., Keeney, R.L. and Raiffa, H. 1999. Smart Choices: a practical guide to making better life decisions. Broadway Books. New York. USA.

12. Keeney, R.L and Raiffa, H. 1976. Decisions with Multiple Objectives: Preferences and Value Tradeoff. John Wiley & Sons. New York. USA.

13. Kane Driscoll, S.B., W.T. Wickwire, J.J. Cura, D.J. Vorhees, C.L. Butler, D.W. Moore, T.S. Bridges. 2002. A comparative screening-level ecological and human health risk assessment for dredged material management alternatives in New York/New Jersey Harbor. Human and Ecological Risk Assessment 8: 603-626.

14. Kangas, J., Kangas, A., Leskinen, P., Pykalainen, J., 2001, MCDM Methods in strategic planning of forestry on state-owned lands in Finland: Applications and Experiences. Journal of Multi-Criteria Decision Analysis 10:257-271.

15. Kiker, G.A., Bridges T.S., Linkov, I., Varghese, A., and Seager, T. (2005). Application of Multi-Criteria Decision Analysis in Environmental Decision-Making. Integrated Environmental Assessment and Management 1(2):1-14.

16. Larichev, O.I., and Olson, D.L. (2001). *Multiple Criteria Analysis in Strategic Siting Problems*. Kluwer Academic Publishers. Boston. USA.

17. Menzie, C., Hope Henning, M., Cura, J., Finkelstein, K., Gentile, J., Maughan, J., Mitchell, D., Petron, S., Potocki, B., Svirsky, S., and Tyler, P. 1996. Special Report of the Massachusetts Weight-of-Evidence Workgroup: A Weight-of-Evidence Approach for Evaluating Ecological Risks. Human and Ecological Risk Assessment 2(2):277-304.

18. National Research Council, 1996. Understanding Risk: Informing Decisions in a Democratic Society. National Academy of Science, Washington, DC).

19. National Academy of Sciences (NAS), 2004. Adaptive Management for Water Resources Planning. National Academy of Science, Washington, DC).

20. Office of the Deputy Prime Minister (ODPM) (2004). *DLTR Multi-Criteria Decision Analysis Manual*. Available at http://www.odpm.gov.uk/stellent/groups/odpm_about/documents/page/-odpm_about_608524-02.hcsp. Downloaded March 3, 2005.

21. Rittel, H. and M. Webber. 1973. Dilemmas in a General Theory of Planning. Policy Sciences. Vol.4: 155-169.

22. Seager, T.P. 2004. Understanding industrial ecology and the multiple dimensions of sustainability. In *Strategic Environmental Management* by O'Brien and Gere Engineers. John Wiley & Sons: New York USA.

23. Suter, G. 1993. Ecological Risk Assessment. CRC Press. Boca Raton. USA.

24. USEPA (U.S. Environmental Protection Agency) 1997. Guiding Principles for Monte Carlo Analysis. EPA-630-R-91-001. Risk Assessment Forum. Washington DC, USA.

25. Vorhees, D.J., Kane Driscoll, S.B., von Stackelberg, K., Cura, J.J. and Bridges, T.S. 2002. An Evaluation of Sources of Uncertainty in a Dredged Material Assessment. Human and Ecological Risk Assessment 8(2): 369-389.

26. Wakeman, T.H. and Themelis, N.J. 2001. A basin-wide approach to dredged material in New York/New Jersey Harbor. Journal of Hazardous Materials 85:1-13.

ENVIRONMENTAL SECURITY IN HARBORS AND COASTAL AREAS

Management Using Comparative Risk Assessment and Multi-Criteria Decision Analysis Framework

B.G. BIGGS
Professional Training & Facilitation Services
PO Box 238 Richmond, KZN, 3780 South Africa

M.K. KIKER
Zama Environmental
224 SW 40th Terrace, Gainesville, Florida 32607 USA

Abstract

Practitioners of risk assessment (RA) and multi-criteria decision analysis (MCDA) typically apply their craft in contested settings. This requires a blend of high-level technical skills, combined with a clear understanding of the larger context of social, economic, and political concerns that influence problem situations. In this White Paper, we provide a review of the challenges faced by RA and MCDA practitioners involved in the management of complex environmental problems, specifically in relation to stakeholder engagement. Based on this review, six possible elements of best practice for stakeholder involvement are presented. We also provide a Directory of Tools and Methodologies which can be used by facilitators, with an indication of how each tool or methodology would be utilized to support stakeholder involvement within the context of RA and MCDA. A brief description of selected tools and methodologies that assist with the successful engagement of stakeholders in decision- making processes is detailed. The paper concludes with a discussion of key issues and future challenges.

1. Introduction

The requirements are many for decision makers involved in complex problem situations, especially when dealing with the threat of potential environment or health risks. Project managers must balance the demands of a wide variety of stakeholders with disparate needs, interests, and agendas. These stakeholders in turn must deal with information at varying levels of complexity from social, political, and technical sources. Charnley [9] presents RA as a way to include science in regulatory decision-making, especially when seeking to manage threats to health and the environment.

I. Linkov et al. (eds.), Environmental Security in Harbors and Coastal Areas, 53–77.

Risk assessment is appealing because it can provide a factual and defensible basis for the decision- making process. However, even the apparently factual nature of risk assessment is subject to stakeholder bias: "Risk is a social construct, with most people making decisions about risk based on a complex set of perceptions that include familiarity, harm, benefit, values, dread, voluntariness, and other factors." ([44] quoted in [9]: chapter 1, p 3). Risk assessment attempts to manage the uncertain nature of science, the subjectivity of stakeholder values, and the complexity of ambiguous problem situations. The strength of RA is dealing with the scientific or technical aspects of the decision, but the risk paradigm is not designed to easily handle value- related factors: "The challenge is to maintain a role for risk assessment and to preserve the integrity of science when decision-making is influenced by many non-technical factors.... Doing so is particularly challenging when risk management decisions are conducted as collaborative efforts among stakeholders with differing technical knowledge levels, interests, goals, and world views" ([9]: chapter 3, p 1).

As decision makers have been required to work more inclusively with a wide range of stakeholders, the dominance of technical factors over stakeholder values and broader social and political concerns represent an inherent limitation of RA. As a result, many decision makers have begun looking for other tools to handle this broader range of factors. Some suggest that MCDA pairs well with RA, as MCDA captures and integrates a wide variety of both quantitative and qualitative factors, including stakeholder values, into the decision making process. However, the quality of MCDA results is subject to the decision structure design and the information and values inputs, as described by Bouyssou:: "It is well known in statistics that the implementation of sophisticated data analysis methods cannot compensate for the weaknesses of the phase consisting in gathering and preparing the data. The same is true for MCDA: applying sophisticated aggregation procedures is of little use if the criteria have been built in an unconvincing way" ([6]: p 59).

Even though MCDA brings stakeholder values into the risk assessment context, it does not provide a methodology for working with stakeholders to generate criteria, establish weightings, and work through the results [3]. It was developed as an integration tool, not a "working-with-people" tool. Practitioners are beginning to suggest that MCDA needs to be paired with stakeholder analysis and collaborative planning methodologies in order to ensure that the process of deliberating over and selecting MCDA building blocks is transparent.

Practitioners recognize the need to combine a technical view of environmental decision analysis with a stakeholder view which takes into account social values. The National Research Council [36] has published a useful guide which addresses dealing with risk in a democracy, where decisions are informed by both scientific and technical information and also public concerns. The NRA sees risk characterization as a process that includes both analysis and deliberation, which work together to create knowledge about the problem situation and to make management decisions.

Risk characterization is the outcome of an analytic-deliberative process. Its success depends critically on systematic analysis that is appropriate to the problem, responds to the needs of the interested and affected parties, and treats uncertainties of importance to the decision problem in a comprehensible way. Success also depends on deliberations that formulate the decision problem, guide analysis to improve decision participants' understanding, seek the meaning of analytic findings and uncertainties, and improve the ability of interested and affected parties to participate effectively in the risk decision process. The process must have an appropriately diverse participation of representation of the spectrum of interested and affected parties, of decision makers, and of specialists in risk analysis, at each step [36].

Analysis is usually associated with scientific and technical expertise, while deliberation is associated with the decision making process which seeks to integrate expert and public interest. While much has been written to address the analytic aspect of risk characterization, less has been written to address deliberation. It has been suggested that decision analysis should integrate expert judgment with stakeholder values [42]. Stahl points out that both elements are needed for a robust decision making process, which can enable researchers to learn about social concerns, and at the same time allow stakeholders to understand more about the implications of expert data. The possible principles of best practice with associated tools and methodologies presented in this paper serve as a starting point to learn more about working with a wide variety of stakeholders, balancing the demands of analysis and deliberation in the decision making process.

McDaniels and Gregory [31] also emphasize the need to create a learning process within structured decision analysis through participation of stakeholders and scientists. They propose that learning is seen as an explicit objective of the decision making process. When learning becomes an objective, the participants are given an opportunity to make decisions knowing that the decisions will be revisited as action plans are implemented: "The prospect of adopting a policy for a period of time and then revisiting that decision when more is learned and the tradeoffs are better understood is attractive to many participants because it changes one-time decisions into iterative, sequential decisions" [31].

Stakeholders learn from one another as they embark upon a structured learning and decision making process, sharing their understanding of both the technical analysis and the social values that create their conceptualization of the problem situation.

Risk assessment and MCDA practitioners know that stakeholders are a vital element in any planning process, but how to effectively engage stakeholders remains an elusive goal. MCDA assists decision makers to integrate and to process diverse results of research from models, monitoring data, risk analysis, and stakeholder preferences. However, without a methodology to guide practitioners through this complexity, working with people can be frustrating: "Current practice treats stakeholder participation as a constraint – i.e. potentially controversial alternatives are eliminated early.... Ultimately, this process does little to serve the needs or interests of the people who must live with the consequences of an environmental decision" ([25]: p 27).

The strengths of both RA and MCDA are many, but without a methodology for stakeholder engagement, the structured decision making process will be unlikely to provide stable and substantively agreed outcomes that are required to take action for sustained environmental change. This White Paper presents three case studies which detail numerous challenges faced when working with stakeholders to understand and to improve complex environmental problem situations. Review of the case studies and the wider literature reveals six possible elements of best practice in a stakeholder engagement process designed to support the effective application of RA and MCDA.

2. Stakeholder Engagement Challenges: Three Case Studies

Stakeholder engagement has become a major element of both public and private sector strategic planning and policy formulation in recent years. In the environmental arena, stakeholder engagement has become an inevitable part of any significant environmental project. Increased access by communities to global environmental issues through the medium of television, democratization and the recognition of the rights of indigenous peoples, as well as the increasing professionalization of environmental groups, has fueled the process. Stakeholders are no longer satisfied with the "Three-'I' Model': 'inform, invite, and ignore'" ([11]: p 9). It is acknowledged that the results of a decision- making process have a better chance of being viewed as fair and credible when stakeholders have participated in the deliberation process [45]. Here though, issues of legitimacy and trust, inclusion and exclusion, as well as equality of access to the process, remain significant issues.

Stakeholder involvement has been widely documented across the globe. Three case studies are summarized below which challenge the notion that participation in and of itself is the panacea for problem solving in complex problem situations:

> Pursued simply for its own sake...the notion of community participation is potentially meaningless and its application likely to mask what continue to be decisions made in the interests of elite groups. In the absence of an explicit strategy for democratisation and capacity-building it is all too easy for state agencies and large industries to engage with other stakeholders in ways that ultimately have little connection to, or influence over, decision-making. While such engagement may lend an air of legitimacy to decisions in the short-term, it does little to address the underlying conflicts of interest that characterise natural resource use and the difficulties these create for the development of strategic and adaptive approaches to natural resource planning and management. [21]

In their study of coastal zone decision-making in the Lower Fitzroy and Port Curtis catchments of Central Queensland, Australia, Jennings and Lockie identify a range of issues that provide a challenge to the view that stakeholder engagement is a process that leads automatically to considered and democratic outcomes. The stakeholders involved in the decision making process were plagued by fundamental differences in their understanding of the problem situation, despite their participation in a cooperative planning process:

> ... Underlying an apparent consensus among the majority of stakeholders were deep divisions over what such values and aspirations meant in practice.
>
> Few stakeholders were able to articulate detailed aspirations for the long-term future of the coastal zone and their own use of it.
>
> Insufficient resourcing of stakeholder participation results often in the "burn-out" of volunteers faced with the double burden of assimilating vast amounts of information in areas where they have limited prior expertise, while earning no income as they do so.

In South Africa the coming of democracy has seen similar challenges. Ashley and Ntshona [1] present a study of the Wild Coast, South Africa, where stakeholders were brought together to plan the creation of a new national park that would simultaneously protect the environment and provide ready access for people:

> Within the institutional structure, there are champions of environmental concerns.... But there is no one whose mandate is to protect and enhance community rights.
>
> The approach to community involvement reveals deep ambiguities. On the one hand, this is supposed to be a park like no other: the first national park run in partnership with residents. On the other hand, the process and timetable simply did not allow for creation of such a community role.

In the U.S., the Fort Ord Army Base in California provides an example of chronic problems with a stakeholder involvement process [4]. A Restoration Advisory Board (RAB) was established by the Department of Defense (DOD) in order to make cooperative decisions about environmental clean-up of the Fort Ord surrounds; however, these representative stakeholder boards generated more conflict than consensus:

1. Participation in the RAB amplified the conflict of stake holders rather than bringing resolution. The DOD brought in outside facilitators to resolve conflicts that paralyzed the planning process, with no positive results.

2. RAB membership was plagued by a lack of understanding between technical and non-technical members, and an uncertainty about the its purpose.

3. The wider public was invited to attend RAB meetings; however, their participation was limited to questions presented at the beginnings of a meeting.

4. In the end, the Fort Ord RAB achieved few of the goals it was designed to achieve, despite the laudable intention to involve stakeholders.

Given the issues identified in the above cross-section of reports from different countries the question needs to be asked "Where to now with the stakeholder enga gement process?" For practitioners of RA and MCDA two questions present themselves: "When are the outcomes of the stakeholder engagement process stable enough to support the application of these techniques?" A second and related question is, "If stakeholder engagement processes do not result in stable and substantively agreed outcomes can this be factored into RA and/or MCDA?"

3. Emergent Best Practice for Stakeholder Engagement

Recent high profile disputes in the stakeholder engagement process may suggest that methodologies are inadequate. The restoration work carried out in the Florida Everglades provides an example rich in controversy and complexity, described in detail by Gunderson *et al* in *Barriers to Bridges* [15]. However, a review of the literature suggests that key elements of a best practice process are beginning to emerge [4; 3; 51; 38; 36; 42].

There are six possible elements of an emergent best practice for stakeholder engagement. The first element is the need to have a set of guiding principles for the process. This would include principles such as transparency, balance, continuity, and flexibility. Additionally, one should view the decision making process as a learning process. These may be seen as the core values that guide the process.

The second element addresses the rules of engagement – where are the boundaries, what are our terms of reference, etc? More specifically these can include the roles of participants and their mandates; a code of conduct; participant accountability; milestones or timelines; a definition of what comprises "agreement." A further useful addition here would be to identify the basis for decisions that stakeholders may be expected to make – such as rights-based decisions; values-based decisions; interest-based decisions; and power-based decisions.

The third key element is a set of broadly measurable outcomes or deliverables ("social" goals) that should be aime d for. These include educating and informing the public, incorporating public values into decision- making, increasing trust in institutions. These may be seen as by-products of the need to deliver a sustainable and concrete outcome to any process. However, they provide the "cement" that ensures that the outcome is embedded in a context that mutually reinforces the quality of the decision.

A fourth element that emerges is the need for a proactive and systematic stakeholder research program that precedes the formal engagement process. Who are the stakeholders in the process likely to be? Are we able to identify particular characteristics (archetypes or typologies)? What is their history of engagement in other processes, if any? Are there any "usual suspects?" This latter category of stakeholders may include global, national, or regional players with a record of "disruptive behavior." Is it possible to discern a particular strategy adopted by them? Can we identify appropriate interventions to balance their influence so as to prevent them dominating the process to its detriment and the detriment of other stakeholder groups?

A fifth element of best practice is the use of independent facilitators who enjoy the confidence and trust of most/all parties to the process. Given current low levels of trust in government institutions and agencies in many parts of the world, this credibility on the part of independent facilitators is crucial. Care should be taken not to use a facilitator who may be seen to be always an "expert witness for the prosecution."

A sixth element in the engagement process is often mentioned in the literature but does not appear to be explicitly addressed. This is the issue of ensuring the participation of less-advantaged participants through a "pay for play" formula. The volunteers in the engagement process – often representatives of local community or of aboriginal structures – are a key element but often lack the resources to participate fully over an extended period. Some form of travel and subsistence allowance, attendance fees, or other direct support should be considered. While this may be a controversial proposal, it will help to give weight to groups that could assist to bring balance to the process. A formula for such support would see the "paid" participants continuing to fund their own participation while "unpaid" participants received some minimal level of support to assist in keeping them involved.

These six possible elements of best practice begin to answer the questions proposed above: "When are the outcomes of the stakeholder engagement process stable enough to support the application of these techniques?" and "If stakeholder engagement processes do not result in stable and substantively agreed outcomes, can this be factored into RA and/or MCDA?" A review of selected tools and methodologies that can be used to support a stakeholder engagement process in line with the six elements of best practice for RA and MCDA is provided below.

4. Directory of Tools and Methodologies

Practitioners of RA and MCDA recognize the need to effectively engage with stakeholders. A wide and evolving variety of stakeholder engagement tools and methodologies are available. Each has its strengths and weaknesses, protagonists and detractors. Some of these approaches are relatively simple and are categorized as tools, while others are more complex and are categorized as methodologies – being more encompassing of the entire stakeholder engagement process. The choice of tools and methodologies is not straightforward. The facilitator is influenced by the specific context of the engagement and needs to take a wide variety of factors into account. A cross-section of these factors include:

- The relative sophistication of the stakeholder participants
- The information that is available on stakeholders and their opening stances
- The degree of learning implicit in the facilitative approach
- How familiar the participants are with the tools or methodologies
- The facilitator's own repertoire of tools and methodologies
- The degree of risk in using the tool or methodology – what if it fails?
- What the key issues are underlying the engagement?

- What it is hoped to accomplish through the engagement process?
- How much time is available to carry out the process?
- Is the facilitator acting as an independent party or does he/she have an agenda?

From this it can be seen that a contingency approach is appropriate. This avoids the risk of a one-size-fits- all engagement process. The facilitator needs to discuss the options with the conveners and, in the interests of transparency, the participants. Experience with facilitating such processes shows that the degree of trust the participants have in the facilitator(s) and the stage in the facilitation process have a significant bearing on the success of the engagement. Riskier tools and methodologies – those that go beyond the cognitive level of the process into the affective (emotional) and behavioral arena may become appropriate if and as trust and a sense of common purpose develop.

The Directory of Tools and Methodologies provided below is only a partial window into a wide range of tools and methodologies (see Table 1). The directory includes six "Stakeholder Engagement Essentials," which are highlighted as the core issues that facilitators will face when they work with stakeholders in the context of RA and MCDA, namely: (1) stakeholder mapping, (2) dealing with uncertainty, (3) dealing with conflict, (4) understanding participants' mental models, (5) negotiating trade-offs, and (6) transparency. The Stakeholder Engagement Essentials have been used to rank each tool and methodology, thus giving an indication of the degree to which each approach addresses these six elements of the facilitative process. This ranking exercise is subjective in nature, and is not presented as a final judgment of any technique; instead, it is intended that the ranking will provide a starting point for practitioners to research further any tool or methodology that interests them. While key words and categorization within the Directory may prove helpful in identifying appropriate tools and methodologies, they cannot substitute for sound judgment and experience on the part of the facilitator(s) and those convening the overall process. The Directory catalogues a range of tools and methodologies, which can be further explored through the references and citations provided.

5. Review of Specific Stakeholder Engagement Tools and Methodologies

In the two sections that follow, we provide a brief description of selected tools and methodologies to give the reader a sense of the range of possibilities available. Tools are distinguished from methodologies by the scope of implementation. Tools are smaller interventions that may be used to address a smaller element of a stakeholder process: such as, a conflict resolution technique, or a participative mapping exercise. Methodologies are larger interventions that can be used to structure or contain the whole of the stakeholder engagement process.

6. Tools

Three tools are described in brief: Stakeholder Mapping, Strategic Assumption Surfacing and Testing, and The Questions and Decisions Model.

A variety of two- and three-dimensional mapping tools are available to categorize stakeholders. These can be useful in strategizing around the degree of participation one should be seeking to secure and what level of difficulty particular stakeholders may pose. One typology of engagement [38] suggests that, in order of increasing levels of involvement, one should categorize stakeholders as follows: enlist for communication and dissemination; inform for understanding; consider for response or input; involve for commitment and expertise; and, consult for authority. This is one of many approaches to categorizing stakeholders in terms of their anticipated demands/needs for differing levels of involvement in the engagement process. Other ways of categorization include those used by Mitchel, Agle and Wood [33]. Their typology recognizes eight categories of stakeholder: Dormant, Discretionary, Demanding, Dominant, Dangerous, Dependent, Definitive, and Nonstakeholders—resulting from overlapping spheres representing Power, Legitimacy, and Urgency. Using a variety of such categorization tools can provide a rich picture of overlapping stake holder motivation and needs.

Strategic Assumption Surfacing and Testing (SAST), developed by Mason and Mitroff [29], is a useful tool for structuring stakeholder debate about the assumptions that underlie a policy or plan. Participants are assisted to create a map for exploring their assumptions through a five-step process, namely: (1) group formation; (2) assumption surfacing and rating; (3) debate within groups; (4) debate between groups; and (5) final synthesis. SAST is a useful tool to help channel strident debate into a productive structure. Groups can be formed by dividing people according to who advocates a particular strategy or who has a vested interest, thus assisting people who hold conflicting views to carefully define their position and debate their assumptions with other stakeholders in a constructive manner.

The final tool to be discussed briefly is the Questions and Decisions ™ (QnD™) screening model system, which is a tool used to integrate ecosystem, management, economic and socio-political factors into a user- friendly model/game framework [23]. QnD is written in object-oriented Java and can be deployed as a stand-alone program or as a web-based (browser-accessed) applet. The model can be constructed with any combination of detailed technical data or estimated interactions of the ecological/ management/social/economic forces influencing an ecosystem. The model development is iterative and can be initiated quickly through conversations with users or stakeholders in a scenario-style gaming process. Model alterations and/or more detailed processes can be added throughout the model development process. One particular application of QnD shows how the model can be used to support discussions with a multi-disciplinary scienti fic team. (Kiker and Linkov, 2005) In this case study, QnD created common ground to integrate specialized studies into the larger perspective of the team effort. As each specialist added his/her part into the whole, they were able to see their own area and other disciplines represented within the QnD design pictures. The QnD development process maintained each participant's attention on the larger problem situation and the objectives of the whole team, rather than the isolation and problem fragmentation that can be created by a specialist view.

6.1. USING THE DIRECTORY OF TOOLS AND METHODOLOGIES

6.1.1. Stakeholder Engagement Essentials

There are six essential tools and methodologies in a Stakeholder Engagement Toolbox. Each tool and methodology has been rated according to Stakeholder Engagement Essentials. These engagement essentials are seen to be the core issues that facilitators will face as they work with stakeholders in the context of RA and MCDA:

1. Stakeholder Mapping
2. Dealing with Uncertainty
3. Dealing with Conflict
4. Understanding Participants' Mental Models
5. Negotiating Trade-offs
6. Transparency

The number rating applied to each tool and methodology is subjective and open to debate. The effectiveness of a tool or methodology resides in at least two elements of any decision-making process: the purpose and design of the approach, and the skills and abilities of the facilitator using that approach.

6.1.2. Meaning of Symbols:

The Stakeholder Engagement Essentials boxes 1 -6 contain three symbols:

- -¦ = the technique strongly evidences the related stakeholder essential
- ? = the technique evidences the stakeholder essential only in part or at a minimal level
- a blank box, which means that the technique does not include that particular engagement essential

Two columns labeled "Ease"[2] and "Cost"[3] contain number rankings based on a scale of 1-10:

- For ease of use: 1 = the tool is easy to use; 10 = the tool is complex to use, requiring more start-up time
- For cost of use: 1 = least expensive; 10 = most expensive

Table 1: Directory of tools and methodologies: a selection of tools and methodologies for stakeholder engagement to support the use of RA and MCDA.

TOOLS	Stakeholder Essentials			Engagement			Ease	Cost
	1	2	3	4	5	6		
OPEN SPACE TECHNOLOGY [37]		?	¦	¦	¦	¦	4	3
Participants identify agenda; voting								
Participants determine agendas; powerful process								
ASSUMPTION SURFACING [29]	¦	¦	¦	?	¦	?	2	2
Make underlying visible								
Identify choice; assumptions; alternative assumptions								
BOUNDARY RELAXATION [28]		?	?	?	¦	?	2	2
Two stage: identify; explore relaxation								
Not-ing problem; exploring; boundary brainstorm								
BULLET PROOFING [22]		¦	?	?		?	42	2
Identify vulnerabilities and difficulties								
Informal Kepner Tregoe; brainstorm "what if"								

TOOLS	Stakeholder Essentials			Engagement			Ease	Cost
	1	2	3	4	5	6		
CATWOE [10]	¦	?		¦		¦	2	2
Issue-defining tool; system definition								
Customer; actors; transformation; worldview; owners; environment								
CONSENSUS MAPPING [17]		?	¦	¦	¦	?	5	6
How best to organize network of activities								
Generate ideas; form groups; cluster ideas; 'strawman' map								
CONSTRAINED BRAINWRITING [2]				?		?	2	2
How best to organize network of activities								
Primed sheets; individuals add ideas; sheets exchanged								
FACTORS IN 'SELLING' IDEAS [46]			?				2	2
Factors to bear in mind when selling an idea								
Context & content; timing, audience, idea champion, language, need								

TOOLS	Stakeholder Essentials Engagement 1	2	3	4	5	6	Ease	Cost
FOCUS GROUPS [43]	¦		?	¦	?		4	4
Free-wheeling discussion but focus & agenda								
Facilitated & recorded discussion; high facilitation demand								

2 The "Ease" of use column is used to indicate the relative simplicity or complexity of the tool or methodology presented. In general, tools are easier to use than methodologies because their scope is narrower and they deal with step-by-step processes.

3 The "Cost" column is used to indicate the relative expense involved in using a particular tool or methodology. The actual cost of implementation is dependent on numerous factors which would be project-specific, such as size of project or cost of consultants.

TOOLS	Stakeholder Essentials Engagement 1	2	3	4	5	6	Ease	Cost
FORCE-FIELD ANALYSIS [27]		¦		?	?	¦	2	2
Identify driving & restraining forces								
Identify forces; map; ways to remove restraining; increase driving								
FRESH EYES & NETWORKING [46]		?		?			3	2
Test clear problem statement on third parties								

TOOLS	Stakeholder Essentials			Engagement			Ease	Cost
	1	2	3	4	5	6		
Use of informal settings; network of contacts; possible internet groups								
GOAL ORIENTATION [39]		¦	?		?	?	2	2
Checklist approach identifying difficulties								
Problem description; needs; difficulties; constraints; clear problem statement								
HELP, HINDER [18]	¦	¦	?	?		?	3	3
Identify people/things that help/hinder								
Identify help/hindrances; how to build or circumvent; plan of action								
IDEA ADVOCATE [46]				?			2	2
Identify champion for a particular idea/approach								
Allocate; research; present; discuss & decide; avoid power/status imbalances								
IMPLEMENTATION CHECKLISTS [20]		¦	¦			?	2	2
Identify key elements of implementation								
Resources; motivation; resistance; procedures; structures; risk, etc.								
NOMINAL GROUP (IMPROVED) [13]		?	?				3	3

TOOLS	Stakeholder Essentials			Engagement			Ease	Cost
	1	2	3	4	5	6		
Generate ideas anonymously								
Pre-meet; initial ideas, add new ideas; transcribe ideas; discuss; vote								
INTERPRETIVE STRUCTURAL MODELLING [47]		?	?	?		?	2	3
Collection of items of data needing ranking								
Computer-aided paired comparison by each group member; priority; severity, etc.								
NEGATIVE BRAINSTORMING [49]	¦		¦	?	?		2	2
Brainstorming what could go wrong?								
Tear-down method; useful to identify how to deal with hostile criticism								
NOMINAL-INTERACTING TECHNIQUE [2]				?	?	?	3	3
Useful for ill-structured and obscure problems								
Private ideas; pooling of ideas; discussion of ideas; prioritization-initial then final								
QUESTIONS AND DECISIONS (QnD) [23]		¦	?		¦	¦	4	3
Web-based modeling platform; stakeholders discuss environmental implications								
Discuss problem situation; build quick model; iterate scenarios with stakeholders								

TOOLS	Stakeholder Essentials			Engagement			Ease	Cost
	1	2	3	4	5	6		
RISK COMMUNICATION [34]				¦		?	7	8
A mental models approach to developing risk communication								
Influence diagram; mental models (interviews & transcriptions); test communication								
TYPOLOGY OF STAKEHOLDERS [13; 40; 5]	¦		?	?			4	2
Stakeholders seen as threat and/or collaborators; suggestions to manage each one								
Maps and matrices, influence diagramming, and questionnaires								
CONFLICT ASSESSMENT FRAMEWORKS [8; 48; 50]	¦		¦	?			4	2
Conflict analysis chart and mapping, dynamics continuum								
Conflict assessment guide to include attempted management solutions to conflict								

METHODOLOGIES	Stakeholder Essentials 1	2	3	Engagement 4	5	6	Ease	Cost
SOFT SYSTEMS METHODOLOGY [10]	¦	¦	?	¦	?	¦	6	8
Structuring complex problem situation with attention to social constructs								
Finding out; purposeful activity models; debate; take action to bring improvement								
COLLABORATIVE LEARNING [11]	?	¦	¦	¦	?	¦	6	8
Policy development in context of environmental conflict								
Assessing conflict including stakeholders; mapping; training; design; facilitation								
ALTERNATIVE SCENARIOS [32; 46]		¦	?		?	?	5	6
Qualitatively different descriptions								
Identify specific decision & environmental forces								
CHARETTE [35]	?	?	¦	¦	¦	¦	8	8
Intensive two-week process								
Community plans – social, economic, physical; consensus – no votes								

METHODOLOGIES	Stakeholder Essentials			Engagement			Ease	Cost
	1	2	3	4	5	6		
COLLECTIVE NOTEBOOK (CNB) [16]	?			¦		?	6	5
Idea incubation and generation								
Participant idea for day; "priming" information; extended process								
DECISION SEMINAR [24]	¦	¦	¦	?	?	¦	7	6
A routinized social science research facility								
Clarify goals; describe trends; analyze conditions; project; select								
DIALECTICAL APPROACHES [29]	¦	?	¦	¦	?	¦	5	4
Use creative conflict to identify/challenge								
Proposal and counter-proposal; plan & counter-plan; present/probe								
METAPLAN INFORMATION MARKET [41]		?	¦			¦	6	7
Communication tools to develop common understanding								
Standard cads; presentation conventions; simple procedural rules								

7. Methodologies

Three methodologies are presented, which have been chosen because they may assist practitioners create the foundation needed for successful implementation of RA and MCDA: The Strategic Choice Approach, The Collaborative Learning Approach, and Soft Systems Methodology.

Friend and Hickling (1987) developed a methodology for dealing with complex choice situations which they called "the Strategic Choice Approach." The methodology is set out in their book *Planning Under Pressure: The Strategic Choice Approach*. The origins of the methodology came out of work they conducted at the Institute for Operational Research, a unit of the Tavistock Institute of Human Relations in London. The methodology has been extensively tested by the authors, particularly in contested local government settings. Early work in the Coventry Municipal Council in England led the authors to the view that:

- People held continually shifting views about issues and boundaries;
- Persistent pressures favored incremental or piecemeal approaches rather than comprehensive ones;
- There was a continuing dilemma of balancing urgency against uncertainty; and
- There were persistent difficulties in distinguishing technical from political aspects of the decision process.

The methodology involves seeing the choice process as involving working into problems initially as distinct from working towards decisions later. Four major skills are seen as being necessary in the choice process – shaping, designing, comparing, and choosing. People in choice situations are seen as favoring one of three broad approaches to clarifying the initial uncertainty in the choice process – "We need more information," "We need more coordination," "We need clearer objectives."

The methodology identifies a ten-stage process leading from "Foundations" through to "Horizons." An analytic method "Analysis of Interconnected Decision Areas" (AIDA) forms a key part of the process. AIDA is a formalized way of recording key elements of a decision situation and labeling them, which lends itself to being computerized (something the authors have subsequently done). The need to understand, communicate, and manipulate concepts such as "Decision Area," "Decision Link," Decision Graph," "Problem Focus," etc is an essential skill on the part of the facilitator. Further information on AIDA can be found in Nature (1965), the operational research literature [26] and in Cook and Morgan's *Participatory Democracy* (1971).

The methodology offers an apparently effective way of approaching and working through highly contested choice situations. However, an in-depth knowledge of the approach and superior facilitating skills appear to be key components in the use of the methodology.

Daniels and Walker [11] have developed a multi- methodology approach, presented in their book, *Working Through Environmental Conflict: The Collaborative Learning Approach*. The authors present four foundations of the collaborative learning approach, specifically -- conflict management, collaboration, the application of learning theory, and systems thinking. This approach has developed within the context of natural resource management and policy development, and several case studies are provided in the text. Theory and practice are both clearly presented with specific techniques for designing and implementing collaborative learning projects.

The Collaborative Learning Approach has numerous successful applications to its credit and presents a robust methodology. However, there are particular challenges faced by all participative methodologies. These flow from the belief that participants will give up their own agenda in favor of finding a common ground for all stakeholders. The facilitator proceeds from the standpoint that participants will cooperate with the process. Daniels and Walker express this sentiment as follows: "If the stakeholders will strive to give the process a chance, the facilitators will strive to conduct a process that is efficient, promotes civility, respects the knowledge and time that participants contribute, and fosters learning." ([11]: p16) Such methodologies, though they are participative, do not necessarily deal well with stakeholders who are single- minded in their goal to lobby for their agenda above all else, and who may be willing to sabotage the process if their agenda is not being achieved. For this reason, it is important to engage in a stakeholder research program to determine who the stakeholders are and what challenges they may present in planning sessions. Additionally, it is useful to ensure that the facilitators are well versed in conflict resolution.

The third methodology to be discussed in brief detail is Checkland's [10] Soft Systems Methodology (SSM). SSM has been widely used and thoroughly developed in a variety of contexts over the past thirty years. The methodology is around four main activities: (1) finding out about the problem; (2) formulating relevant purposeful activity models; (3) Debating the situation using the models; and (4) taking action in the situation to bring about improvement. Checkland provides numerous tools that support these four stages of analysis and planning. Amongst these tools is rich picture building which is useful as a starting point for exploratory conversations with stakeholders. SSM also proposes the use of the mnemonic CATWOE, useful for stimulating a critical look at the problem situation and deciding which elements should be included in models. The letters stand for Customers, Actors, Transformation, *Weltanschauung* (worldview), Owners, and Environment. By describing the problem situation in terms of CATWOE, participants are encouraged to think through the different perspectives and many assumptions that permeate the problem situation. Additionally, by clarifying the transformation that is desired, the scope of the problem situation at different levels of complexity is clarified and constructive debate is fostered.

SSM should be differentiated from hard systems modeling in which the model is seen to be a picture of reality. SSM is not designed to create a picture of the actual system. Instead, SSM helps to generate conversation and learning. The learning and insight that is generated is used to explore the problem situation, to find new insight, which in turn generates the ability to take action to improve the situation.

8. Discussion: Key Issues and Future Challenges

In order to make progress within the management of complex environmental challenges, we need to move forward from the impasses that seem to characterize current high-profile stakeholder engagement processes. This move forward will require a process of integration and evolution. At the same time a greater focus needs to be placed on learning and on proactive stakeholder approaches.

An initial setting out of four key priorities to be addressed on the path to more effective engagement and decision-making in contested public settings is proposed as follows:

1. Models and modeling/decision approaches such as RA and MCDA need to become more nuanced and transparent. The literature suggests that such a process has begun. Such approaches must incorporate the need to generate knowledge and understanding among participants. This is distinct from a technical approach that uses models to find the single best solution from a predetermined set of options. Where lay participants are concerned, the "what if" process needs to become an easily comprehended part of the interface between RA/MCDA practitioners and stakeholders.
2. The stakeholder-engagement facilitation process needs to be "professionalized." Certain of the case studies in this paper, as well as the wider literature, indicate that the failure of the stakeholder process can often be attributed to a poorly considered, poorly resourced and indifferently implemented facilitation process. The facilitation process needs to be seen as an integral and key component of any public process seeking to find common-ground solutions. As an adjunct to this – all stakeholder engagements processes should be followed by a rigorous debrief process to identify key learning that can inform future processes. Such learning should consider the techniques used, the outcomes secured, the stakeholder interaction process, the things that worked and the things that did not. In this way the capacity of institutions to run effective stakeholder engagement processes will be enhanced through a process of institutional learning.
3. Where public stakeholder engagement processes are concerned, careful thought needs to be given as to whom the ongoing custodian(s) of the process should be. In private organizations the management of the stakeholder interface is a key part of the institutionalized strategic management process. Specific managers are given the task of managing this interface on a routine basis. The seemingly episodic and project-based nature of public agency stakeholder engagement means that the engagement process is often the responsibility of project managers or project technical experts. Such a management process means that (institutional) learning from one process often dies with the project closeout. Public institutions need to see the engagement

process as a continuous process – periods of calm interspersed with areas of high activity. Suitable structures are needed to ensure that the institutional memory is regularly updated and maintained from one engagement process to the next.

4. Stock needs to be taken of the current status of the stakeholder engagement process. This paper suggests that an early form of best practice is emerging. Such best practice is not much concerned with the use of specific techniques but rather with what makes for a competent overall process. Identified elements of possible best practice should be subjected to a rigorous review process. If the elements survive this review, they should be incorporated into standard institutional operating procedures.

In conclusion it is important to note that the stakeholder engagement process is an evolving, not a static, field. Technical issues can be expected to become more complex and difficult to explain. Stakeholder groups can be expected to seek new ways to further their agendas. Under these circumstances the stakeholder engagement arena can be expected to remain a volatile and challenging one.

9. Acknowledgements

The authors are grateful to Dr Greg Kiker and Dr Igor Linkov for their advice and assistance in drafting this paper.

10. References

1. Ashley, C. and Ntshona, Z. (2003) Transforming Roles but not Reality? Private Sector and Community Involvement in Tourism and Forestry Development on the Wild Coast, South Africa. IDS. Research Paper 6. Sustainable Livelihoods in Southern Africa.

2. Backoff, R.W. and Nutt, P.C. (1988) A process for strategic management with specific application for the non- profit organisation, in Bryson, J.M. and Einsweiler, R.C. (eds.), Strategic Planning; Threats and Opportunities for Planners, Planners Press, Chicago.

3. Banville, C., Landry, M., Martel, J.M. Boulaire, C (1998) A Stakeholder Approach to MCDA. Systems Research and Behavioral Science, Vol 15, pp 15-32. Wiley and Sons.

4. Beierle, T.C. (1998) Public Participation in Environmental Decisions: An Evaluation Framework using Social Goals. Resources for the Future. Discussion Paper 99-06.

5. Blair, J.D., Slaton, C.R. and Savage, G.T. (1990) Hospital-physician joint ventures: a strategic approach for both dimensions of success. Hospital and Health Services Administration, 35, pp 3-16.

6. Bouyssou, D. (1990) Building criteria: a prerequisite for MCDA. In Bana e Costa, CA (ed.) Readings in Multiple Criteria Decision Aid, Springer- Verlag, Berlin, pp 58-80.

7. Bryson, J.M. (2003) What To Do When Stakeholders Matter: A Guide to Stakeholder Identification and Analysis Techniques. A paper presented at the London School of Economics and Political Science, 10 February 2003.

8. Carpenter, S. and Kennedy, W.J.D. (1988) Managing Public Disputes: A Practical Guide to Handling Conflict and Reaching Agreements. Jossey-Bass, San Francisco.

9. Charnley, G. (2000) Democratic Science: Enhancing the Role of Science in Stakeholder-Based Risk Management Decision-Making. Risk World.

10. Checkland, P. (1999) Systems Thinking, Systems Practice. Wiley Press, New York.

11. Cook, T.E. and Morgan, P.M. (1971) Participatory Democracy. Harper & Row Publishers, Inc., New York. Daniels, S.E. and Walker, G.B. (2001) Working Through Environmental Conflict: The Collaborative Learning Approach. Praeger Publishers, Connecticut.

12. EPA (2002) Community Culture and the Environment: A Guide to Understanding a Sense of Place. EPA 842-B-01-003). Office of Water.

13. Fottler, M.D., Blair, J.D., Whitehead, C.J., and Laus, M.D., and Savage, G.T. (1989) Assessing key stakeholders: who matters to hospitals and why? Hospital and Health Services Administration, 34, pp 525-546. Fox, W.M. (1987) Effective Group Problem Solving. Jossey-Bass, San Francisco.

14. Gable, C. and Shireman, B. (2005) Stakeholder Engagement: A Three-Phase Methodology Learning to Manage your Relationships with Stakeholders. Environmental Quality Management, Spring 2005.

15. Gunderson, L.H., Holling, C.S., Light, S.S. (1995) Barriers and Bridges to the Renewal of Ecosystems and Institutions. Columbia University Press, New York.

16. Haefele, J.W. (1962) Creativity and Innovation. van Nostrand Reinhold, New York.

17. Hart, S., Boroush, M., Enk, G. amd Hornick, W. (1985) Managing complexity through consensus mapping: technology for the structuring of group decisions. Academy of ManagementRreview, 10(3), pp 587-600.

18. Henry, J. and Martin, J.N.T. (1997) Creative Problem-solving Guide, B882 Residential School Booklet, The Open University

19. Herbst, T.D. and Yosie, T.F. (1998) Processes in Environmental Decisionmaking: An Evaluation of Lessons Learned, Key Issues, and Lessons Learned. Ruder Finn, Washington.

20. Isaksen, S.G., Dorval, K.B. and Treffinger, D.J. (1994) Creative Approaches to Problem Solving. Kendall-Hunt Publishing, Dubuque, IA.

21. Jennings, S.F. and Lockie, S (2002) Democratisation and Capacity Building in Coastal Zone Decision-Making in Australia: the Application of Stakeholder Analysis and Social Mapping. invited paper and paper publication, Coastal Zone Asia Pacific Conference, Bangkok, Thailand 12-16 March 2002.

22. Kepner, C.H. and Tregoe, B.B. (1981) The New Rational Manager. Kepner-Tregoe, Inc, Princeton, NJ. Kiker, G.A., Rivers-Moore, N.A., Kiker, M.K. and Linkov, I. (2005). QnD: A modeling game system for integrating environmental processes and practical management decisions. (in Review) (Chapter in Morel, B. Linkov, I., (Eds) "The Role of Risk Assessment in Environmental Security and Emergency Preparedness in the Mediterranean." Kluewer, Amsterdam 2005.

23. Kiker, G.A. and Linkov, I. (2005). The QnD Model/Game System: Integrating Questions and Decisions for Multiple Stressors. (in Review) (Chapter in Goncharova, N., Arapis, G., (Eds) "Ecotoxicology, Ecological Risk Assessment and Multiple Stressors " Kluewer, Amsterdam 2005.

24. Laswell, H.D. (1960) The Technique of Decision Seminars. Midwest Journal of Political Science, 4, pp 213-236.

25. Linkov, I., and Ramadan, eds. (2004) Multi-Criteria Decision Analysis: A Framework for Structuring Remedial Decisions at Contaminated Sites. Comparative Risk Assessment and Environmental Decision Making, Kulwer, pp 15-54.

26. Luckman, J. (1967) An Approach to the Management of Design, Operational Research Quarterly, 18 (4), pp 345-358.

27. Majaro, S. (1988) The Creative Gap. Longman, London.

28. Martin, J. (2000) Technique Library B822: Creativity, Innovation and Change. The Open University Business School. The Open University.

29. Mason, R.O., and Mitroff, I.I., (1981) Challenging Strategic Planning Assumptions: Theory, Cases, and Techniques. Wiley Press, New York

30. Mays, C. (2004) Stakeholder Involvement Techniques: Short Guide and Annotated Bibliography. Nuclear Energy Agency. ISBN92-64-02087-X.

31. McDaniels, T.L., and Gregory, R. (2004) Learning as an Objective within a Structured Risk Management Decision Process. Environmental Science and Technology, Vol 38, No 7.

32. Miller, W.C. (1987) The Creative Edge: Fostering Innovation Where You Work. Addison- Wesley, Reading, MA.

33. Mitchell, R. Agle, B. and Wood, D. (1997) Towards a Theory of Stakeholder Identification and Salience: Defining the Principle of Who and What Really Counts. Academy Management Review, No 4, pp 853-886.

34. Morgan, M.G., Fischoff, B., Bostrom, A. and Atman, C.J. (2002) Risk Communication: A Mental Models Approach. Cambridge University Press, New York.

35. Mossman, A. (1972) 'Charette: an approach to user participation on design decision making' dissertation for Liverpool School of Architecture, pp. 17-34.

36. National Research Council (1996) Understanding Risk: Informing Decisions in a Democratic Society. Edited by Stern, PC and Fineberg, HV. National Academy Press, Washington DC. Nutt, P.C. (1992) Managing Planned Change, Macmi llan

37. Owen, H. (1997) Open-Space Technology: A Users Guide. Berrett-Koehler, San Francisco.

38. Parker, R. (2004) Standards Stakeholder Engagement – Update – Building Momentum. CIHI Partnership Meeting, Calgary.

39. Rickards, T. (1974) Problem Solving Through Creative Analysis, Gower Publishing Company Limited, Epping.

40. Savage, G.T.; Nix, T.W., Whitehead, C.J. and Blair, J.D. (1991) Strategies for assessing and managing organizational stakeholders. Academy of Management Executive, 5, pp 61-75.

41. Schnelle, E. (1979) The Metaplan Method: Communication Tools for Planning and Learning Groups, Metaplan Series No. 7, from Metaplan GmbH, 2085 Quickborn, Goethestrasse 16, Germany.

42. Stahl, C., Cimorelli, A. and Chow, A. (2002) A New Approach to Environmental Decision Analysis: Multi- Criteria Integrated Resource Assessment (MIRA). Bulletin of Science, Technology & Society, Vol 22, No 6, December 2002, pp 443-459.

43. Stewart, D.W. and Shamdasani, P.N. (1990) Focus Groups: Theory and Practice. Applied Social Science Methods Series, Vol. 20, Sage.

44. Slovic, P. (1999) Trust, Emotion, Sex, Politics, and Science: Surveying the risk-assessment Battlefield. Risk Analysis, 19: pp 689-701.

45. Thibault, J.W. and Walker, W.L. (1975) Procedural Justice. A Psychological Analysis. Lawrence Erlbaum Associates, Inc. Hillsdale, NJ.

46. Van der Heijden, K. (1996) Scenarios: The Art of Strategic Conversation. Wiley, West Sussex, England. Van Gundy, A.B. (1988) Techniques of Structured Problem Solving (2nd ed). van Nostrand Reinhold, New York.

47. Warfield, J.N. (1982) Interpretive Structural Modelling (ISM). in Olsen S.A., Group Planning and Problem- solving Methods in Engineering Management, New York, Wiley.

48. Wehr, P. (1979) Conflict Resolution. Westview, Boulder, CO.

49. Whiting, C.S. (1958) Creative Thinking, Van Nostrand Reinhold, New York.

50. Wilmot, W.W. and Hocker, J.L. (1998) Interpersonal Conflict (5th ed). McGraw-Hill, Boston.

51. Yosie, T.F., and T.D. Herbst. (1988) Using Stakeholder Processes in Environmental Decision making: An Evaluation of Lessons Learned, Key Issues, and Future Challenges. Project supported by the American Industrial Health Council, the American Petroleum Institute, and the Chemical Manufacturers Association (September 1998).

Part 2

Coastal Areas: Challenges and Solutions

ENVIRONMENTAL SECURITY AND ENVIRONMENTAL REGULATIONS IN ISRAEL

R. BRONFMAN
The Democratic Choice Faction
The Knesset, Israel

Abstract

In order to reach trans-boundary cooperation on environmental regional issues resolution of conflict and definition of borders must come first – in order to unite we must first divide.

1. Water

Water, being the most important resource for life is in growing demand in the region. A steady and stark rise in the population, coupled with years of pollution of water by all players and countries in the region are making this precious natural resource even scarcer. In the year 2000, an Israeli citizen had 350 cubic meters of drinking water at his "disposal", compared to 3,300 for a French citizen. Had the year 2001 been a dry year, an expected shortage of 100 million cubic meters of drinking water could have been expected. These numbers are substantially lower for Palestinians and Jordanians. Solving the water shortage is a growing concern and I can mention three possible solutions:

1. The first solution is importation of water from Turkey using converted tankers. This is a local, Israeli only, solution and not a cheap one at that.
2. The second solution that has been discussed is desalination, either on a local Israeli level or in cooperation and coordination with Jordan. This is an energy intensive solution both in the desalination process and in the pumping process. For this solution, I can mention initial legislative initiatives that have began to gather momentum in the Israeli Parliament.
3. The final, and in my view, the most interesting solution is that of the Peace canal. This plan, drawn up by Mr. Boaz Wachtel, speaks of laying a 700 KM canal from the Sihan and Jihan rivers in Turkey to the region and thus supplying the water needs of Syria, Israel, Jordan, Palestine as well as solving the ever growing problem of the dropping level of the dead sea. Knowing that you will be holding a conference in Cairo, Egypt, on the topic of water, I can

I. Linkov et al. (eds.), Environmental Security in Harbors and Coastal Areas, 81–84.

highly recommend that you invite Mr. Wachtel to present his plan, which is both economical as well as environmentally sound. It does, however, require reaching a peace agreement between Israel and Syria.

2. Energy

The issue of Energy is a world problem these days. With the price of a barrel of Brent crude oil scraping 60$ (as of the writing of these lines) it is of utmost national interest of all countries to cut their energy consumption by conserving energy where possible and to encourage renewable energy resources in any way possible. Israel can be considered an energy power house. With a population of around 6 million, generation capacity reached, in 2002, over 43 billion kilowatt hours compared with 7.1 billion kilowatt hours in Jordan (with a similar population) and 75.2 billion kilowatt hours in Egypt with a population of nearly tenfold.

During the past two years I have been active in promoting energy conservation and renewable energy resources in Israel. To this end, I am in the process of legislating two central bills. The first is: "Conservation of Energy in Pubic Utilities" which will remedy a loophole in Israeli law whereby public utilities and government offices are not required to conserve energy. This bill has the potential of saving up to 20% of the government's energy consumption and nearly US$1billion per annum. The second bill, which was drafted in cooperation with the Public Utility Authority – Electricity, a regulatory body charged with fixing the price of electricity, granting of licenses for production of electricity and the such, is: "Combining Renewable Electricity in the Electricity Generation System". This bill gives tools to the Public Utility Authority, to give incentives by way of premium tariffs, net-metering and certificates to manufacturers of renewable energy. This will, to the best of my knowledge, be the first such system of incentives in the eastern basin of the Mediterranean.

3. Civil Society

The next issue is that of the roll of civil society environmental monitoring. With the ever-growing budget cuts put forward by a neo-liberal Israeli government which believes in the Washington Consensus in its extreme, the roll of civil society and NGOs is ever growing. In the past 10 years, Israel has seen an upsurge in the number of environmental NGOs. These organizations act, among other things, as observers who can sound the alarm in case of an environmental disaster or preferably, an impending disaster. NGOs use members of parliament such as myself to push forward issues and raise alerts as well as directly contact the ministry for the Environment.

As examples of cooperation between NGOs and myself I can mention a campaign that I took up having been contacted by such an NGO to battle against a plan of the Israeli Electric Company to lay high-voltage electricity cables across the Ramon Machtesh – one of five unique giant craters in the south of Israel planned to be declared world heritage biospheres by UNESCO. This campaign has not been concluded yet. Another example I can mention is whereby I was contacted by a group of citizens living in a lower class neighborhood who objected to paving a road that would involve ruining a small park they use on a regular basis. A final example is where I succeeded to impose on a gas company that supplies gas to government vehicles to install spillage monitors they were refusing to install till then. This was done by asking the Ministry of Finance to halt its agreement with the Paz Company until Paz agreed to install the monitor in all its stations.

4. Public Participation

Another important issue, which goes hand in hand with that of civil society monitoring, is enhancement of public participation. In this area, I have legislated to completion a bill that has made environmental committees in local councils and municipalities a statutory requirement. These committees, which include members of the public and members of Environmental organizations, will serve to raise the awareness of environmental issues in the local realm. They will act as environmental watchdogs on the local government level and will try to ensure that local government decisions are in line with environmental consideration and sustainable development.

An additional project I am working on is the amendment of the Israeli Planning and Building code so as to increase public participation in the planning aspect of the code. This, hopefully, will be done by way of a sub-committee of the Internal Affairs and Environmental Committee of the Knesset (the Israeli Parliament). Naturally, this is an area which will raise a strong opposition from interest groups. In fact, even the government has made inroads on even the very basic public participation in national planning. This was done by setting-up a fast-track committee of National Infrastructure which streamlines large scale projects in a more speedy and "efficient" way by making the process much shorter and limiting public criticism and scrutiny.

It is often said, in the environmental circles, that the Environment has no boundaries or borders. This is, undoubtedly, true. However, with the reality of the Israeli-Arab conflict, the lack of internationally recognized borders is a sever impediment to dealing with the environmental challenges we are facing in the region. In fact, this situation will probably serve as an aggravating factor to any environmental disaster. Just imagine a toxic spill in the north of Israel that threatens to affect Lebanon which is in a declared state of war with Israel, or a chemical accident at the south-western outskirts of Damascus that with the right wind could reach Israel in a less than an hour.

On the bright side, at least with Jordan and Egypt, Israel has diplomatic relations, and if put to the test, it could be expected that a good cooperation could be reached within a relatively short period of time. This said, it should be noted that last April, when a series of terrorist attacks rocked the Egyptian Sinai coast and the nearest modern rescue facilities were in Israel's southern city of Eilat, it took several hours before the Egyptian border police allowed Israeli rescue teams to rush into Egypt to do their job thus wasting precious time of rescue efforts being carried out.

Another example which can be given is an attack of Locust which came from Africa, through Egypt, into Israel and Jordan and was then halted by Israeli scientists who prevented even greater damage to all three countries. In this case, a belated warning by the Egyptian authorities allowed swarms of Locust to land in Israel and Jordan without sufficient preparation. Only urgent steps by the Israeli ministry of agriculture halted the threat and minimized the damage inflicted.

Between the Palestinians and the Israelis, the issue, is, as always, quite complex. On the one hand, I can mention a cooperation between the Palestinian city of Tulkarm and the Israeli regional council of Bat-Hefer. With support of the European commission, a joint sewage purification plant was established and the sewage on both sides of the green line are treated in a joint plant. On the other hand, during this past month, plans for an Israeli garbage dump near the Palestinian city of Nablus, were put in place. According to this plan, Israeli garbage would be "exported" to the Palestinian authority. Due to my involvement and that of other public figures, this plan has been halted.

5. Conclusions

It is evident, from the points that I made, added to the decades of human suffering on both sides of the conflict, that a lasting peace is needed. Such a peace should, in my view, follow the Arab peace initiative introduced by the Arab League in its summit in Beirut, Lebanon in 2002 and reiterated in Algiers just last month. Sadly, my government has, for various reasons which I will not go into here and which I strongly disagree with, not reacted to this initiative. However, with Prime Minister Sharon's disengagement plan, scheduled to be executed this coming June and with the new pragmatic Palestinian leadership, I can envisage the favorable falling of the die of destiny that would set in motion, of not already in motion, a dynamic that could substantially enhance chances for a real, honorable and fair peace between Israel and the Palestinians.

Such a resolution for the conflict would be a glaring opportunity for progressive programs of environmental cooperation and comprehensive Comparative Risk Assessment to manifest themselves in the region. Based on the estimated international funds that are expected to flow to the region, a positive and even ground-breaking opportunity could be envisioned. Such changes could resonate through the world and perhaps even instigate a paradigm shift in favor of sustainable development and betterment of the environment and the societies living in it. For people who are not afraid of vision, this is the time of reaching for the skies. At least from where I come from, it would seem that from the lowest and darkest reality we were in just a few months ago we have no choice but to look up and work for a better future.

MANAGING SEDIMENT RISK AT THE BASIN SCALE

European Frameworks in Support of Complex Decisions

S. APITZ
SEA Environmental Decisions
1 South Cottages, The Ford, Little Hadham, Hertfordshire SG11 2AT
United Kingdom

A. OEN
Norwegian Geotechnical Institute
Department of Environmental Engineering, P.O. Box 3930 Ullevaal Stadion, N-0806 Oslo, Norway

S. WHITE
Cranfield University, Silsoe
Institute of Water and Environment, Bedfordshire, MK45 4DT
United Kingdom

Abstract

The dynamic nature of sediments, from rivers to estuaries and the sea, calls for a holistic approach to sediment management that ensures that transport, quantity and quality are explicitly addressed throughout the framework. If sediments are hydrodynamically connected, it makes sense to prioritize those that pose risk downstream, and to do this in a manner that considers the entire sediment, contaminant and risk budget, from source to sink. Such an approach will provide insight into the highest impact potential changes in agricultural, industrial and development practices that may reduce sediment and contaminant inputs, and hence the cost of maintaining waterways and protecting the environment. Conceptual frameworks for basin-scale sediment management that provide an approach for addressing the complexities inherent in managing sediments at both a basin-wide and site-specific scale and their role in holistic European basin-scale sediment management decision making, are discussed.

1. Introduction

Sediment quality can be defined as the ability of sediment to support a healthy benthic population (the organisms that live in intimate contact with sediments at the bottom of water bodies). Quality can be affected by a number of physical, chemical and biological factors, but the focus of the discussion in this chapter is on

I. Linkov et al. (eds.), Environmental Security in Harbors and Coastal Areas, 85–96.

sediment quality as a function of the presence (and associated bioavailability) or absence (or non-availability) of toxic chemicals. Quality can be assessed by a number of methods, including combinations of chemical measurements (often compared to standards or benchmarks), toxicity tests and/or benthic community analyses. Because many contaminants have a tendency to associate with sediment particles [1, 2], they can accumulate in, and be transported by, sediments throughout a river basin.

Sediments are an essential part of the aquatic ecosystem, providing habitat and substrate for a variety of organisms, as well as playing a vital role in the hydrological cycle. Sediment quantity can, however, represent a risk to the well-being of a system, through imbalances, or through incompatible physical characteristics. Examples of risk include excesses or lack of sediments in rivers, estuaries, reservoirs, lakes and impoundments which can reduce storage and flow capacity, increase flood potential, damage hydro-power installations, degrade habitats, erode river channels downstream of sediment "blockages", and undermine the stability of channels and infrastructure (e.g. erosion of bridge piers). Examples of benefits include sediment supply to the nearshore environment (with implications for longshore drift/coastal stability), the provision or sustenance of wetland and aquatic habitats, sediment extraction for use in building/road industries, and beneficial use/capping of contaminants.

There is a need to broaden thinking on sediment management and risk assessment to a basin-wide approach that addresses both *quantity* and *quality* of sediment, as well as a need to recognize the fact that actions or changes in conditions in one part of the basin, whether the result of anthropogenic or natural processes, may well have impacts elsewhere in the basin. Sediment quantity and quality issues must be considered together, not separately as entirely different issues, as the interdependence of quality and quantity in dynamic river basins demands a holistic approach, precluding a decoupling of these facets.

Contaminant inputs and effects on sediment quality come from various sources and types, and via various pathways. As noted above, "properly functioning" river systems, in both ecosystem and socioeconomic terms, are dependent upon a balance of the aspects of sediment quality and sediment quantity. Both an excess and a lack of sediments, either due to past, present or future natural or anthropogenic processes, can put various functions of a river at risk. Thus, in a river basin, both sediment and contaminant sources are manifold and their respective locations, potential source strength (a function of sediment and contaminant quantity stored) and amenability to erosion, under current and projected conditions, must be determined. A description and model, whether conceptual or quantitative, of the mass flow of water, contaminants and particles (and thus risk) within a river basin can be termed a Conceptual Basin Model, or CBM [3]

Contaminants can partition, transfer and move through a dynamic river ecosystem through various media, including air, sediment, soil, water and biota. Management of risk in such an ecosystem, or within a given river basin, suggests that sediment management should be integrated into water and soil management. Therefore, an integrated approach to soil, sediment, water and biota should be developed. Thus, a conceptual appraisal of any proposed sediment management framework in light of water- and biota- focused perspectives is required. A decision-making hierarchy, which encompasses priority setting at a basin scale down through site-specific risk assessment at a local level, is a necessary approach for managing water, soil, sediment and biota, as well as point and diffuse contaminant sources.

2. The Need for Basin-Scale Management

Although most guidance documents have been generated for specific aspects of sediment management (such as dredged material disposal or environmental management [4]) a basin-scale approach must integrate various sediment goals and provide a universal framework. Different nations, organizations and stakeholders have different objectives when they address sediments, and frameworks must be devised that allow goals and priorities to be balanced in a transparent way. The goal of sustainable sediment management demands that sediments are managed, not one unit of sediment at a time, but instead with the interactions between that unit and all current or potential sources or sinks within a river basin, in mind [5].

One of the main drivers for European river basin management and for the SedNet initiative is the EC WFD. The WFD (Annex VII) requires member nations to develop River Basin Management Plans (RBMPs) which are required to deliver the WFD objective of good, or improving, ecological status in all water bodies. These RBMPs will need to consider many aspects of basin-scale management within the socio-economic environment of the region, country and continent. The concept of basin-scale *sediment* management, which is the focus of this chapter, is just one facet of such an RBMP, with other facets including water resources, flooding, nutrient management, priority substances and biota. As with sediment management, many of these other facets will also require understanding of the hydrodynamic continuum, and thus any plan derived for sediment management must be compatible with other requirements of the RBMP.

To achieve sustainable sediment management in a dynamic river basin, the various practitioners of sediment management must come to the table before any sediment management decisions are made and develop the sediment-specific aspects of RBMPs that will balance the environmental, economic, social and regulatory needs throughout the basin. To achieve such a balance, a common language must be developed such that priorities can be established and understood by all parties involved, information needs can be defined and filled, and sediment can be managed in a sustainable way.

The complex, multi-scale and multivariate nature of holistic sediment management requires the involvement of many "layers" of political, technical, scientific, economic and environmental analysis, which can be difficult to integrate and unify. Whilst many of these processes involve very different drivers, organizations and approaches (whether on site-specific or river-basin scale), holistic sediment management requires that the relationships between these processes, including their points of interaction, intersection and information exchange, be clearly defined. To achieve these goals, it is necessary to develop strategic, conceptual and process frameworks that identify these interactions, define common issues and terms, and help define and thus expedite information exchange in support of effective sediment management. This paper looks at the relationship between various types of decisions, appraisals and assessments in basin-scale sediment management. A definition of what sort of decisions and actions are informed and affected by what information should then lead to discussions of how such complex data will be combined, processed and handled, the focus of much of this book.

3. Frameworks for Basin-Scale Management

Figure 1 portrays a conceptual framework defining the relationship between basin-scale and site-specific considerations in river basin management. Note that this diagram need not be sediment-specific or risk-specific, but rather can be used to address management options for various media. As described above, management of river basins will require an evaluation of all relevant processes (both natural and anthropogenic) at the basin-scale using all available data. Whilst the focus of this chapter is on sediment risk management, this can only be achieved within the context of its broader milieu. However, it is important to reiterate that the selection, prioritization, implementation and evaluation of any management action is dependent upon a broad range of factors (including, but not limited to, economic appraisal, technical assessment, risk assessment and environmental assessment), all of which must be evaluated at every level of the decision-making and management process.

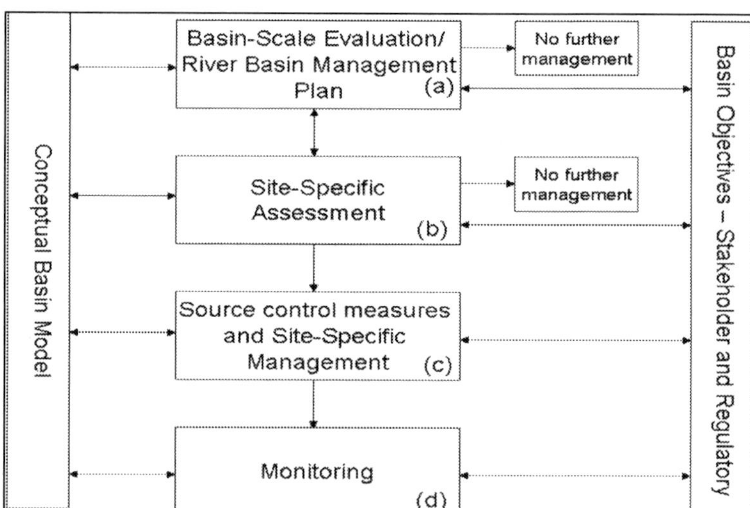

Figure 1: Conceptual diagram of the relationship between basin-scale and site-specific assessment and management in a river basin.

Figure 2, adapted from a UK approach to flood defense [6], shows the various types of appraisals necessary at each step in the decision process, and how information from each feeds into other steps. Whilst this paper will focus on risk appraisals, it should not be forgotten that these are only one part of the appraisal and decision-making process. How these other appraisals are combined in support of decisions is the focus of tools such as decision support systems (DSS) and Multi-Criteria Decision Analysis (MCDA). For these to be useful, however, it is important that both the developers and users of decision tools clearly understand how and why decisions are and should be made.

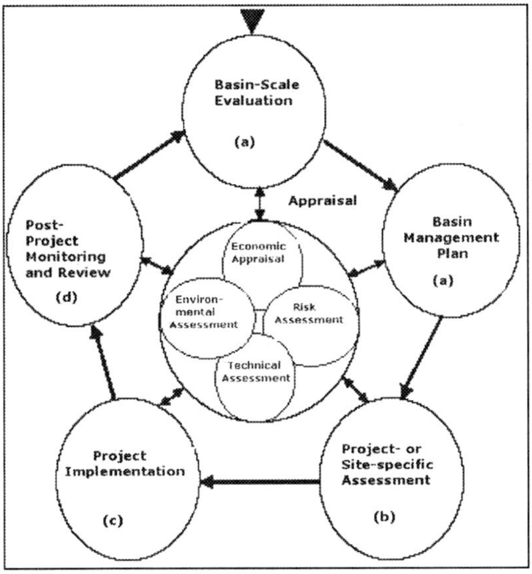

Figure 2: The iterative appraisal process at various phases of a basin scale/site-specific evaluation process (adapted from MAFF 2001). Note a, b, c and d in circles correspond to process steps in Figure 1a.

Evaluation at the basin scale may result in the elimination of some proposed management actions. However, for those specific actions, projects or control measures identified as a result of a RBMP, site-specific assessment must be carried out, followed as appropriate by management and monitoring.

4. Sediment Risk Management at the Basin and Site-Specific Scale

Clearly, risk should be assessed in at least two different spatial scales: the basin scale and the local (site-specific) scale, as well as at various stages in the decision-making process. Though some of the fundamental factors being evaluated may be the same for both scales (such as risk to human health, the environment and to river basin objectives), the methods, degree of detail and information available will differ. For example, prioritizing sites or management actions at the basin scale involves setting priorities for sediment units throughout the basin that account for several kinds of risk and management objectives, and prioritizes these sediment units for both socioeconomic and ecological management actions.

The term "Risk Prioritization" can be used to describe basin-scale risk-relevant factors used to prioritize sites in a basin in terms of relative risk. This, then, separates the risk factors from other factors, while still recognizing that many other technical, socioeconomic and regulatory factors, all outside the scope of this paper, will be brought to bear before sites are ultimately prioritized for the sediment-specific aspects of an RBMP. The term sediment Basin Management Plan (sBMP) can be used to refer to these sediment-specific aspects that will inform the RBMP or other basin-scale management activities.

Because of the extensive analyses required to provide sufficient information to prioritize site risks at a basin-scale, much of the data used in the analysis will necessarily be based upon screening-level information, as well as literature or generic criteria. Thus, sites that are given a high-priority status based upon screening-level risk evaluation will require further, site-specific and detailed analysis of risk before being subjected to potentially costly management actions. These sites are thus subject to site-specific risk assessment which can be defined as the evaluation of individual sediment parcels to determine and rank their risk relative to benchmarks, site- or basin-specific criteria. A further risk evaluation at the local scale (although with a significant basin-scale component) is needed to characterize risk to the environment for any given management option (dredging, for example), or to compare several options. This risk evaluation can be termed Project Risk Appraisal. When several options are being evaluated, this is often termed a Comparative Risk Assessment.

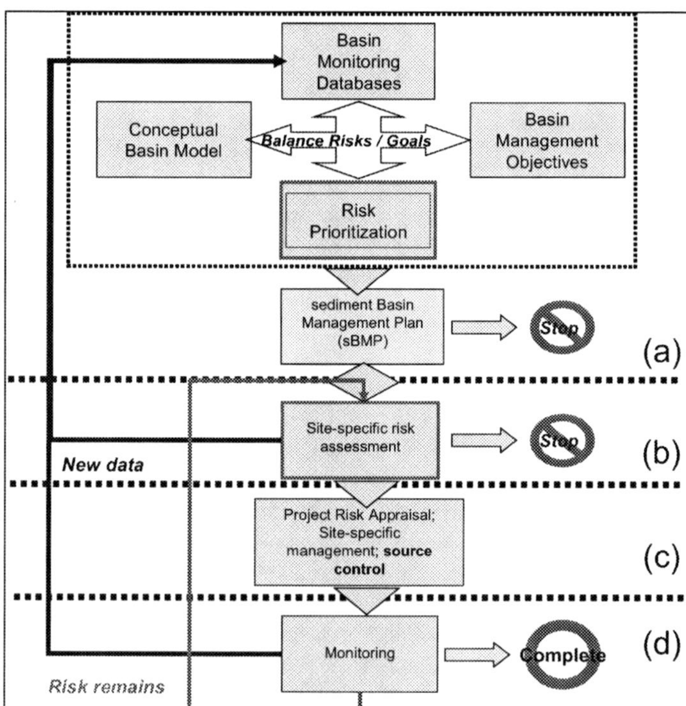

Figure 3: Process diagram for basin-scale and site-specific sediment risk management. A manager may "enter" the process at the basin scale (in level "a"), or at the site-specific scale (in level "b").

Figure 3 presents a more detailed process diagram for basin-scale and site-specific risk management, which details the risk appraisal aspects of the conceptual framework in Figure 1a. The letters denoting various hierarchical levels of the diagram (Figure 3) correspond to the same processes in Figure 1. This diagram is comprised of the three principal levels of risk evaluation described above.

Risk appraisal is only one aspect in the development of an sBMP but it is the risk-related aspects of such a process that are emphasized in this process diagram, and which represent a risk-specific "cross-section" through Figures 1 and 2. In risk-related terms, basin-scale assessment begins with the processing and synthesis of information derived from basin monitoring and from databases to develop a Conceptual Basin Model (CBM), and an assessment of the various objectives that are desired for the river basin as a whole and for sites within the river basin in particular (Basin Management Objectives). The CBM should consider the mass flows of water, particles and contaminants, and a screening level assessment of sediment quality and archived data.

This information is then used to evaluate, if possible, the relative risk associated with various sites, with a consideration of the potential risk implications of those management actions being considered to achieve socioeconomic goals (such as dredging, construction, etc).

A number of other appraisals (see Figure 2) should then be carried out to generate a sBMP, which should present a prioritized list of proposed projects and strategies, some to achieve ecological goals, and some, such as dredging, to support socioeconomic goals. It should be pointed out that an important part of any sustainable sBMP is a contaminant and sediment source control plan. The success of such a control plan is largely dependent upon a CBM that adequately accounts for diffuse and point sources throughout the basin. Low-priority sites or projects (which can be eliminated based on appraisals) are then set aside for no further action, while relatively higher-priority sites or projects are subject to more detailed assessment. Sites given high priority based upon screening risk will undergo a detailed risk assessment, referred to herein as site-specific risk assessment.

The results of such risk assessment may determine that, based upon more detailed information, management of the site based on risk is unnecessary, or the assessment may instead demonstrate the need for risk management. Site-specific risk assessment is characterized by tiered assessment and the determination of site-specific risk. Site-specific management options should be evaluated using a project risk appraisal, and comparative risk assessments. This approach will be used to evaluate and compare the potential risks and benefits of various remedial, management and/or disposal options based upon site-specific impacts on BMOs, site-specific risk, technical feasibility and regulations, as well as potential impacts upon other sites within the system (as informed by the CBM). Proposed projects in support of socioeconomic (rather than remedial) goals will still be subject to site-specific risk assessment and a project risk appraisal, though often at a much lower level, if there is no evidence of site-specific risk. They may then, however, go straight to management for socioeconomic goals without a remedial step. All projects should be subject to post-project monitoring, the results of which can both be used to evaluate the efficacy of any project as well as to help revise the CBM.

Not all projects have broad (e.g. basin-wide) implications, and current as well as past regulatory frameworks and decision makers do not and have not acted at the basin scale. Thus, in many cases, the process diagram shown as Figure 3 will be entered into at point (b), with a site-specific risk assessment and a Project Risk Appraisal. A properly conducted Project Risk Appraisal will consider potential impacts of a given management action upon adjacent and downstream sites, and will ideally consider available upstream solutions, thereby effectively addressing some basin-scale considerations. In all cases, many aspects of these evaluations are iterative, with new information informing various levels of the process.

To date, agencies, decision makers and governments have not yet developed all the policy or infrastructure for making basin-scale evaluations and decisions (although in Europe this should be changing with the advent of the WFD). Furthermore, the methodologies and procedures necessary to carry out such broad-reaching evaluations and decisions – regardless of who is involved - are still in very early stages of development. However, implementation of a basin-scale approach will be necessary if river basins are to be managed holistically and in line with emerging policy, therefore, iterations of such an approach must first be laid out in clear fashion, then refined and debated.

5. Setting Basin Management Objectives

It is not only components of the CBM that will drive a prioritization of sites and management actions. Societies also have a number of socio-economic goals for sediments and river basins which include regulatory, economic, aesthetic, recreational and ecological factors. In order to sustain economies, evaluate, prioritize and improve ecological status of sediments, stakeholders must decide on management objectives, and how they are to be measured and balanced. The EC Water Framework Directive mandates "Good ecological status of water bodies" (WFD, 2000/60/EG). Those who work with sediments have no trouble understanding and communicating the link between sediments and the ecological status of water bodies, but the specific definition and measure of "good ecological status" in sediments are still being refined throughout Europe.

The EC Habitats Directive mandates that there is a "Duty to demonstrate no harm" (Council Directive 92/43/EEC). Whether this "duty" applies to future, ongoing or planned anthropogenic activities, the consequences of such activities, and/or to various environmental media, must still be clarified. Most international bodies, including SedNet, speak of the goal of "sustainability". SedNet defines this term as *"The multipurpose management of sediment with full attention to adverse effects, so as to enhance the utility of river basins in the future"* [7].

As an additional complicating factor, EU programs driven by economic goals such as the Common Agricultural Policy (CAP) currently subsidize practices that may conflict with environmental sustainability. For example, wheat and maize are now being grown in the UK on highly erosive and sloping soils. Such practices are causing greater flooding, the filling of streets and gardens with sediments [8, 9], and severely silted watercourses. Consequently, regional, national, international and/or basin-wide objectives must be defined. A series of questions may be posed in this regard: Is the management goal to reduce the mass balance of contaminants in a river basin, to limit the exposure of these contaminants to the food chain, to protect benthic organisms, fisheries, shipping or farmers, and/or to achieve the "right" amount of sediment of the right type for ecological requirements? How can the degree of goal achievement be measured? How do these goals fit into current (or future planned) policy or frameworks?

There must therefore be a designation of objectives for the management of a given river basin, which can be defined as Basin Management Objectives. Objectives to be met during management generally fall under the following categories: 1) meeting regulatory criteria (e.g., WFD, Habitats Directive, North Sea Treaties, National and local legislation), 2) maintaining economic viability (e.g., navigation, fisheries, flood control, recreation), 3) ensuring environmental quality and nature development and 4) securing quality of human life. To develop Basin Management Objectives, all stakeholders must come together to identify and define site-specific and regional goals for a given basin. Such stakeholders may come from many fields (regulators, dredgers, fishermen, shippers, environmentalists and the general public, among others) and all should have input. The CBM, with as much information as is currently available should provide structure and insight into how various site-specific actions taken in one basin location might affect other basin sites and their potential uses.

One way in which the interaction between these objectives and various management actions can be communicated is with the European Environment Agency DPSIR approach (DPSIR stands for Drivers, Pressures, States, Impacts and Responses). Particularly useful for policy-makers, DPSIR builds on the existing Organization for Economic Co-operation and Development (OECD) model and offers a basis for analyzing the inter-related factors that impact the environment. The aim of such an approach is threefold: 1) to be able to provide information on all of the different elements in the DPSIR chain; 2) to demonstrate interconnectedness of the elements, and 3) to estimate the effectiveness of Responses [10, 11].

Once defined, Basin Management Objectives will control how we prioritize sites, how we assess and manage them, as well as how extensively we include an evaluation of land-based practices in a sediment management strategy. Inevitably, goals will differ from place to place and even from time to time, and not all objectives may be achieved. The question of who decides on the necessary compromises is also an important one. As the EC WFD mandates the involvement of all stakeholders throughout the decision process, a clearly communicated framework for defining and balancing BMOs is essential to successful management.

The manner in which potentially competing goals related to sediment management are balanced is not a purely technical decision. Rather, decision-making additionally requires consideration of socio-economic and political appraisals. Acceptance and implementation of the approach ultimately chosen and followed will require significant work, both technical and political. Different nations, organizations and stakeholders have different goals for sediment management. While SedNet is an open organization, and seeks input from all European nations and from academia, business, government and NGOs, representation is not uniform. Whether a single framework can successfully provide the tools for these disparate agendas and stakeholders to be reconciled remains to be seen. However, the definition of BMOs, and the subsequent development of a Risk Prioritization and sBMP are critical parts of sustainable sediment management. This chapter does not address how these choices will be made, since significant work must still be done on defining river basin-wide and/or Europe-wide methods, goals and priorities, but rather seeks to suggest a framework within which to clarify the dialogue.

6. Important Issues for Sediment Management Option Appraisal and Selection

Site-specific risk assessment should be carried out on sites prioritized for management. Approaches for site-specific risk assessment are addressed in detail in many papers, guidance documents and books [12-16, 17-19], and will not be discussed here. For site-specific assessment to be meaningful within a basin-scale framework, the CBM, BMOs and the issue of source control must be considered at each step of the assessment process.

Proposed management options must be evaluated, both in terms of site-specific and basin-scale risks and benefits, in a Project Risk Appraisal; the methodologies of which are outside the scope of this chapter. However, it should be pointed out that management options for sites being managed to reduce risk as well as for low-risk sites that are prioritized to be managed to meet non-remedial BMOs must be evaluated in terms of the CBM, overall BMOs and source-control issues, and in terms of the other appraisal factors described in Figure 2. For contaminated sites, the most cost-effective solution for managing a site may be upstream source control. On the other hand, some options that address an objective at one site may have negative impacts for other sites in the basin. These issues are particularly important in dynamic river basins. To address this, questions that should be asked and carefully considered during selection of the most appropriate options for managing contaminated sediments include:

- Is the site erosive or depositional?
- Will implementing certain management options change erosive vs. depositional status, and how will that impact other sites?
- Can sediments entering a parcel be considered as part of a solution that reduces risk, e.g. via burial, mixing or attenuation?
- Are there land-use changes planned for upstream locations that would reduce sediment loading to downstream locations?

- Does sediment entering an area bring with it new or more contaminants?
- Do remedial options increase risks downstream?
- Is the sediment a resource needed in the basin, e.g., in order to provide habitat or prevent channel erosion?
- Are the characteristics (e.g., grain size, organic matter content) of sediment arriving at a site appropriate to the objectives for the site?
- Is source control technologically and politically feasible?

After management actions have been selected and applied, monitoring must continue until risks are deemed to have reached acceptable levels. CBMs should be either continuously updated or periodically reviewed, and re-balanced in terms of changing BMOs and/or monitoring results. This process will support the WFD requirement of periodic reviews of RBMPs. Basin-scale sediment management should be an iterative process, and, if done properly, resources can be allocated for maximum benefit.

7. Conclusions

The proposed conceptual approach to basin-scale sediment management provides possible frameworks for addressing the complexities inherent in concurrently managing sediments at both a basin-wide and a site-specific scale. Acceptance and implementation of a basin-wide management approach will require significant work, both technical and political. Successful development of a basin-scale decision framework should provide a basis for parties with very different goals to come together in support of sustainable sediment management. Because all stakeholders are ultimately stewards of the same ecological and economic resources, breaking down technical, political, and socioeconomic barriers, whether real or perceived, should ultimately help to balance the various groups' often disparate objectives for sediment management in a mutually agreeable, beneficial and sustainable way.

8. References

1. Förstner, U., Sediment-associated contaminants - an overview of scientific bases for developing remedial options. Hydrobiologia, 1987. **149**: p. 221-246.

2. Kersten, M. and U. Forstner, Chemical fractionation of heavy metals in anoxic estuarine and coastal sediments. submitted to Wales Science and Technology, 1985.

3. Apitz, S.E. and S. White, A Conceptual Framework for River-Basin-Scale Sediment Management. Journal of Soils and Sediments, 2003. **3**(3): p. 132-138.

4. Apitz, S.E. and B. Power, From Risk Assessment to Sediment Management: An International Perspective. Journal of Soils and Sediments, 2002. **2**(2): p. 61-68.

5. Förstner, U., Sediments and the European Water Framework Directive. Journal of Soils and Sediments, 2002. **2**(2): p. 54.

6. MAFF, Flood and Coastal Defence Project Appraisal Guidance. Overview (including general guidance). 2001, Ministry of Agriculture, Fisheries and Food.

7. Brils, J. and S. participants, The SedNet Strategy Paper, in *http://www.sednet.org/materiale/SedNet_SP.pdf.* 2003.

8. Boardman, J., R. Evans, and J. Ford, Muddy floods on the South Downs, southern England: problem and responses. Environmental Science and Policy, 2003. **6**(1): p. 69-83.

9. Evans, R. and J. Boardman, Curtailment of muddy floods in the Sompting catchment, South Downs, West Sussex, southern England. Soil Use and Management, 2003. **19**: p. 223-231.

10. Elliott, M., The Role of the DPSIR aproach and conceptual models in marine environmental management: and example for offshore wind power. Marine Pollution Bulletin, 2002. **44**: p. iii-vii.

11. Joziasse, J., et al., Sediment management objectives and risk indicators, in Sediment Risk Management and Communication, S. Heise, Editor. 2005, in press, Elsevier.

12. Apitz, S.E., et al., Assessing and managing contaminated sediments: Part II, Evaluating risk and monitoring sediment remedy effectiveness. Integrated Environmental Assessment and Management (online-only), 2005. **1**(1): p. e1-e14.

13. Apitz, S.E., et al., Assessing and Managing Contaminated Sediments: Part I, Developing an Effective Investigation and Risk Evaluation Strategy. Integrated Environmental Assessment and Management, 2005. **1**: p. 2-8.

14. Chapman, P.M., et al., General guidelines for using the sediment quality triad. Marine Pollution Bulletin, 1997. **34**(6): p. 368-372.

15. Ingersoll, C.G. and D.D. MacDonald, Guidance manual to support the assessment of contaminated sediments in freshwater, estuarine, and marine ecosystems in British Columbia: Volume III Interpretation of the results of sediment quality investigations. 2003, British Columbia Ministry of Environment, Lands, and Parks: Victoria, British Columbia.

16. Wenning, R.J. and C.G. Ingersoll. Summary of the SETAC Pellston Workshop on Use of Sediment Quality Guidelines and Telated Tools for the Assessment of Contaminated Sediments. in 17-22 August 2002. 2002. Fairmont, Montana, USA: Society of Toxicology and Chemistry (SETAC). Pensacola, FL, USA.

17. Council, N.R., Contaminated Marine Sediments-Assessment and Remediation. 1989, Washington, D.C.: National Academy Press. 493 p.

18. Committee on Contaminated Sediments, M.B., Commission of Engineering and Technical Systems, National Research Council, Contaminated Sediments in Ports and Waterways: Cleanup Strategies and Technologies. 1997, Washington, D.C.: National Academy Press. 295.

19. Council, N.N.R., A risk-management strategy for PCB-contaminated sediments. 2001, Washington, D.C.: National Academy Press.

MULTICRITERIA DECISION ANALYSIS AND STRATEGIC UNCERTAINTIES

C. YOE
College of Notre Dame of Maryland
4701 North Charles Street, Baltimore, MD 21210-4010 USA

Abstract

Coastal environments present a great variety of complex environmental problems that lack objectively "correct" answers due, in part, to significant uncertainties. Traditional planning and analytical tools are not well-suited to address these uncertainties. Scenario planning is suggested as a useful tool for addressing an uncertain future that can be bounded by identifying key drivers. This amounts to using multiple without condition scenarios instead of one. Multi-criteria decision analysis (MCDA) techniques can then be applied to evaluate all plans in each scenario.

1. Introduction

Multicriteria decision analysis (MCDA) enjoys growing support as a decision support tool. Many advanced techniques have been developed to enable practitioners to address uncertainty in decision problems in a wide variety of ways [3]. These techniques are not transparent to non-practitioners of MCDA. More importantly, few of these techniques are appropriate for handling drivers of the future environment that are subject to a wide range of significantly different outcomes, such as climate change or no climate change, that decision makers or stakeholders can understand and use appropriately. This paper suggests that using MCDA in a scenario planning approach to solving coastal restoration problems presents one transparent and useful approach to such situations.

2. Coastal Problems

Coastal problems are problems without boundaries. They are of global importance and affect billions of people. Coastal environments present a great variety of complex problems that lack objectively "correct" answers. Complexity alone insures that there are no right or wrong answers but only better solutions or worse solutions. These sorts of "wicked" problems are not well served by the

I. Linkov et al. (eds.), Environmental Security in Harbors and Coastal Areas, 97–109.

deterministic optimization techniques of traditional planning and analytical methods. While these techniques are well suited to single objective planning problems or multiobjective problems that can be reduced to a single metric, few coastal problems fall in these categories.

2.1. COASTAL LOUISIANA

Nearly 40 percent of the coastal wetlands of the continental US are located in the Louisiana coastal area. Louisiana's millions of acres of wetlands are among the world's most diverse and productive ecosystems. This productive and fragile environment is disappearing (Figure 1). For about the last 50 years Louisiana has lost 34 square miles of wetlands per year. From 1932 to 2000 Louisiana lost approximately 1900 square miles of coastal land, an area the size of the state of Delaware. From 2000 to 2050 Louisiana is projected to lose another 700 square miles, an area the size of the greater Washington, DC and Baltimore metropolitan areas. If the problem continues unabated much of the seventh largest deltaic system existing in the world will be lost and the shoreline will advance inland as much as 33 miles in some areas.

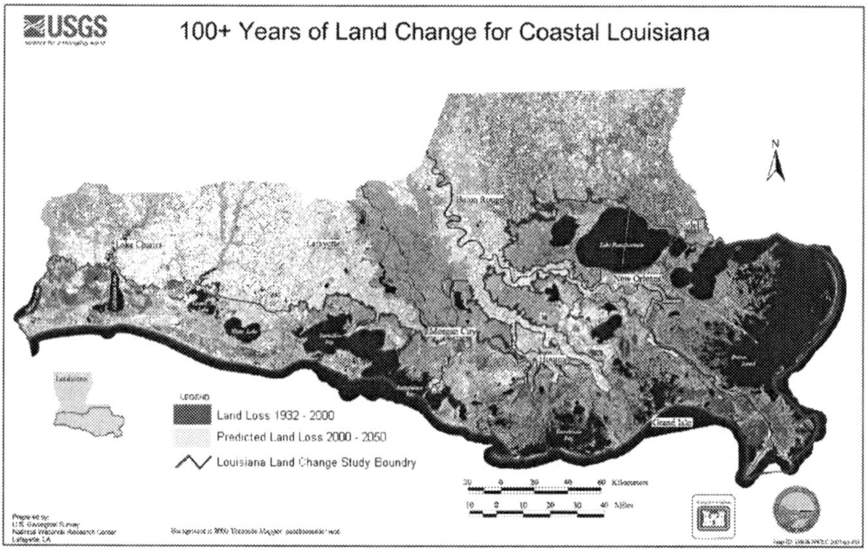

Figure 1: Land changes in coastal Louisiana.

The causes of coastal land loss in Louisiana are many and are conveniently grouped into natural and human causes. Natural causes includes subsidence, sea-level rise, erosion, salt water intrusion, barrier island degradation, and storms. Human causes include drainage for agriculture, mineral extraction, and canal construction. Flood control levees prevent the Mississippi River from changing course so it can't seasonally nourish wetlands

The impacts of the wetland losses are vast and varied with local, national and global implications. They give rise to a wide range of criteria to consider in addressing this problem. The impacts include damage and loss to domestic energy production and energy infrastructure, inland and coastal navigation, commercial fisheries for both shellfish and finfish, tourism, hunting, fishing, and wildlife observing recreation that is among the best in the United States. Agricultural production in the region is affected directly and much of the Nation's agricultural practices could be affected indirectly by some responses to this problem. One of the more immediate concerns to local residents and businesses is the increasing risk of damage from the loss of flood and storm protection. The effects of climate change and major natural disasters are but two major uncertainties, the resolution of which could have significantly different implications for the choices of decision makers as well as the MCDA of analysts.

2.2. BARROW, ALASKA

Planners in the Anchorage District of the US Army Corps of Engineers have defined a very different coastal problem for the northernmost city in North America, Barrow (Figure 2):

> Barrow's way of life is intrinsic to its location. Erosion and coastal storms threaten the modern infrastructure essential to the social fabric that the community of Barrow provides as a regional cultural center.

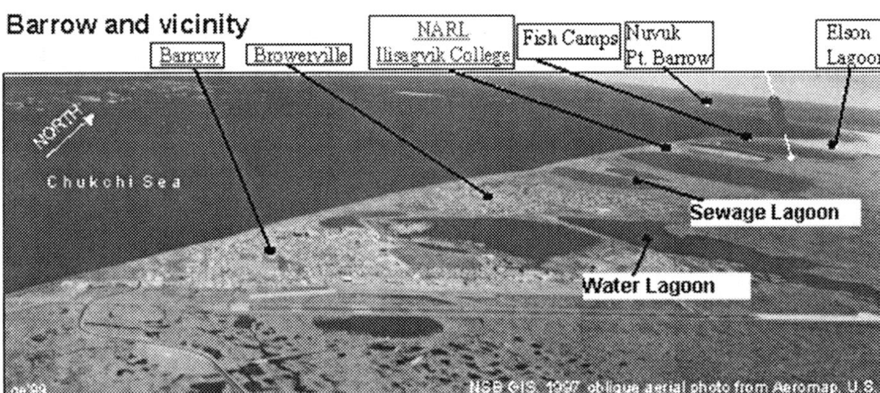

Figure 2: Vicinity of Barrow, Alaska.

Barrow's climate is an Arctic one with a daily minimum temperature below freezing an average of 324 days of the year. The population of 4,434 people (2002) is 57% Native American, 22% White, and 9% Asian living in a city that cannot be reached by land. Everything must be flown or barged into the town. The Inupiat Eskimo culture leads an essentially subsistence lifestyle. Ice cellars are common and the average harvest and consumption per resident of indigenous meats, i.e. bowhead whale, caribou, whitefish, walrus, and seal is 290 pounds annually. The North Slope Borough (NSB) government is the primary employer in this town with an economic base of oil and gas extraction, a small federal government, tourism, and some limited manufacturing.

Among the uncertainties associated with this problem is climate change, specifically the potential effects of global warming on storm regimes, and the frequency, duration and thickness of ice-cover. The fate of the town's social and economic infrastructure is equally uncertain, consequently the future of this unique North American consequence is uncertain.

The criteria that may be important to consider in any effort to solve Barrow's coastal restoration problems include:

- Future storms and erosion damages (this is a key driver of future scenarios)
- Engineering feasibility of management measures
- Environmental concerns in the coastal zone
- Archeological impacts
- Economic effects and net benefits of management measures
- Social impacts (this is another key driver of future scenarios)
- Cultural impacts on Inupiat Eskimos

3. Uncertainty

One of the emerging constants in the modern world is uncertainty. The world grows more complex. Complexity as used here, refers to such things as the size of a society, the number of its parts, the distinctiveness of those parts, the variety of specialized social roles that it incorporates, the number of distinct social personalities present, and the variety of mechanisms for organizing these into a coherent, functioning whole [10]. Our social systems grow so complex as to defy understanding. Consequently, our systems of problem solving develop greater complexity.

The world faces an increasingly rapid pace of change in almost every arena. Scientific breakthroughs make things, once impossible to conceive, commonplace. Much of this change is driven by rapid advances in technology too turbulent and rapidly changing to be predicted by human beings.

Relentless pressure on costs is now a fixture in all public decision making. We have entered a world where irreversible consequences, unlimited in time and space are now possible. Some of the wicked problems planners face can have a long latency period. For example, the ongoing efforts with landscape scale ecosystem restoration problems in the Florida Everglades, Coastal Louisiana, and the Columbia River Basin provide clear examples of problems that took decades to emerge and be recognized. The implications of the solutions being formulated may similarly take decades to be understood.

A new phenomenon of "known unawareness" has entered our lexicon. As a society we are beginning to realize that despite all we know the unknown far outweighs what is known. We have begun to suspect that there are some risks for which there may be no narrative closure, i.e., no ending by which the truth is recovered and the boundaries of the risk established.

Despite the world's rapid advances in all kinds of sciences we are increasingly dominated by public perception. When it comes to uncertainties and risks, acceptability depends on whether those who bear the losses also receive the benefits. When this is not the case, the situation is often considered unacceptable. As a result, possibility is often accorded the same significance as existence in a stakeholder's view. And this view can find its way into policy. This is in part because many things that were once considered certain and safe, and often vouched for by authorities, turned out to be deadly. The recent BSE experience in Europe and SARS experience in China provide vivid examples of this.

Responsibility in this more connected world has become less clear. Who has to prove what? What constitutes proof under conditions of uncertainty? What norms of accountability are being used? Who is responsible morally? And who is responsible for paying the costs? These questions plague planners nationally and transnationally. Consequently, many public policy decisions are subjective in nature.

All decision making and planning processes operate in this uncertain reality. Yet most of them cling stubbornly to a deterministic approach to planning and decision-making that belies the experience of business and government the world over. Coastal restoration, like other decision contexts needs a culture of uncertainty. We need decision support tools for this uncertain and subjective world. These tools include:

- Risk analysis--systematic framework for organizations to approach decision making related to risks
- MCDA—a tool useful in risk analysis or other decision support frameworks
- Planning frameworks that are realistic and useful

The latter is the focus for the remainder of this paper.

4. Traditional Planning

Traditional planning processes are generally deterministic in practice. The U.S. Army Corps of Engineers uses a six-step planning process prescribed by the Economic and Environmental Principles and Guidelines for Water and Related Land Resources Implementation Studies (P&G) summarized in the Figure 3.

A distinguishing characteristic of this and many other planning processes is the reliance on a forecast of a single most likely alternative future. The focus on a single forecast of an uncertain future is based on an understandable desire for one right answer. As a matter of practice, these forecasts are often anchored in the present. This had been a perfectly acceptable practice in the past when change was not as rapid and the social context of problems was not as complex, in other words before the world became so uncertain.

Reliance on a single forecast of the future guarantees an adversarial decision process because there are always legitimate differences in views of an uncertain future.

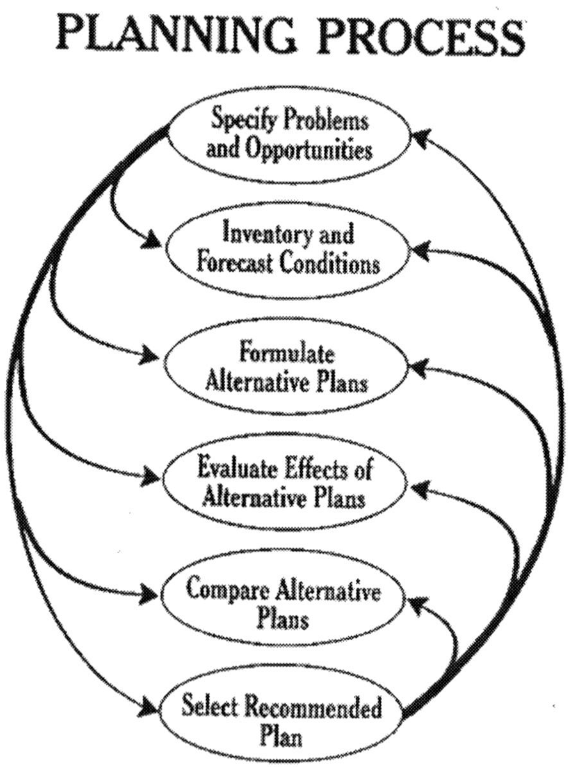

Figure 3: The six-step planning process prescribed by the P&G.

MCDA can be an integral part of the evaluation, comparison and selection steps of the planning process. The future forecast is critical to the estimation values used in the MCDA. Currently the MCDA is sometimes supplemented by a risk-based analysis of some of the decision criteria measurements. For example, it would be routine practice to use risk-based analysis to forecast expected annual storm damages in the future without any management intervention in the coastal zone and then to forecast that value again assuming specific management actions are in place and operating. A summary measurement of these risk-based values may be used as part of an MCDA analysis.

Alternatively, or in addition, the MCDA may include a sensitivity analysis of its own. But these approaches represent shades of gray in a situation that may require more black and white analysis.

4.1. FORECASTING COASTAL RESOURCES FUTURES

To provide a better context for understanding this paper's suggestions for addressing major uncertainties a closer look at a traditional planning approach is offered. There are many ways to approach an evaluation of coastal restoration management options, including gap analysis, before and after comparison, and with and without comparison. The latter is generally preferred, and is prescribed for the US water resource agencies by the P&G. The risk estimates, scientific analyses, special studies, economic and environmental analyses, opinion surveys, legal analysis and the like, upon which an evaluation of restoration measures will be based, will vary from case-to-case. But a few generic evaluation framework steps can be identified using an adverse human health risk estimate as an example below (Figure 4). These steps include the following:

- Describe the existing baseline risk condition
- Describe the most likely future condition in the absence of a change in resource management, i.e. the "without condition." In traditional planning every option is evaluated against this same without change condition, the "Future No Action" shown below. This future may exhibit an increasing, decreasing, flat or mixed trend.
- Describe the most likely future condition with a specific resource management intervention in place, i.e. the "with condition." Each intervention (A) has its own unique with condition, the "Future With Intervention A."
- Compare "with" and "without" conditions for each intervention option.
- Characterize the effects of this comparison—not all effects are equal in size, some are desirable, others are not.

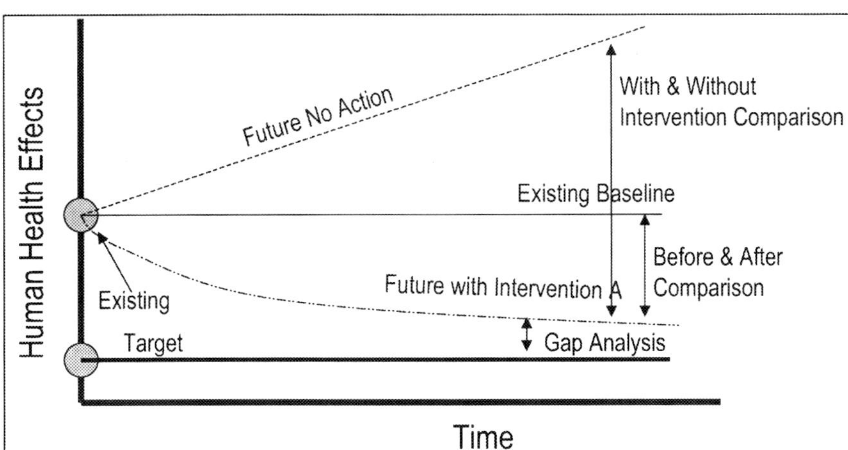

Figure 4: With and without comparison.

The National Environmental Protection Act (NEPA) process often favors a before and after comparison, while other decision processes may set a goal of achieving some target level of risk reduction and use gap analysis to gage how close different options come to achieving that target.

Using comparisons of with and without resource management interventions usually requires very sophisticated levels of analysis. But the critical flaw in this analysis in instances where there are strategic uncertainties is the reliance on a single forecast, the most likely future condition. For example, what assumptions do planners make about climate change for either the Coastal Louisiana or the Barrow, Alaska coastal restoration studies? If no climate change is assumed the evaluation of MCDA criteria in a with and without condition comparison may yield one set of values. If climate change is assumed an entirely different set of values may be obtained. The differences could be large enough to significantly influence the formulation of the best resource management action. If the assumption is wrong, the management measure's performance may be somewhere between less effective and totally ineffective. Realistic forecasts of strategic uncertainties are critical to successful MCDA in a complex coastal restoration environment.

Analysts and decision makers labor in uncertainty. A single forecast of the future will be wrong for anything but the most simplistic forecast of future conditions. Consequently, MCDA conducted in a traditional planning framework is based on what could be not what will be. In a world of many stakeholders, what could be is wide open to debate and litigation.

Coastal restoration needs a planning framework that addresses strategic uncertainties, i.e., the large and difficult uncertainties, in an explicit fashion. A wide variety of coastal problems face strategic uncertainties due to natural causes like global warming and unpredictable storm regimes including catastrophic events like the December 2004 tsunami in the Indian Ocean, or Hurricane Katrina in 2005. In addition, there are potentially strategic uncertainties due to human activities. These include a wide range of anthropogenic disturbances, changing national values and geopolitical events. Scenario planning can help analysts address strategic uncertainties.

5. Scenario Planning

One of the emerging constants in the modern world is uncertainty. There is good reason to believe that complexity and rapid as well as unpredictable change should be considered normal parts of the 21st century landscape. Faced with this reality, planners can bemoan the difficulty of decision-making or they can devise simple, effective strategies to enable themselves to cope with and even thrive in an uncertain world. Scenario planning is an effective tool for strategic decision making under uncertainty.

In Figure 5 below, circumstances in which scenario planning may be useful are suggested. When there is little uncertainty and the consequences of being wrong about these uncertainties is minor, any sort of standard decision making method will do. When there is much uncertainty and the consequences of being wrong are grave scenario planning is indicated as a viable option. A role for traditional deterministic planning remains as well as the figure shows.

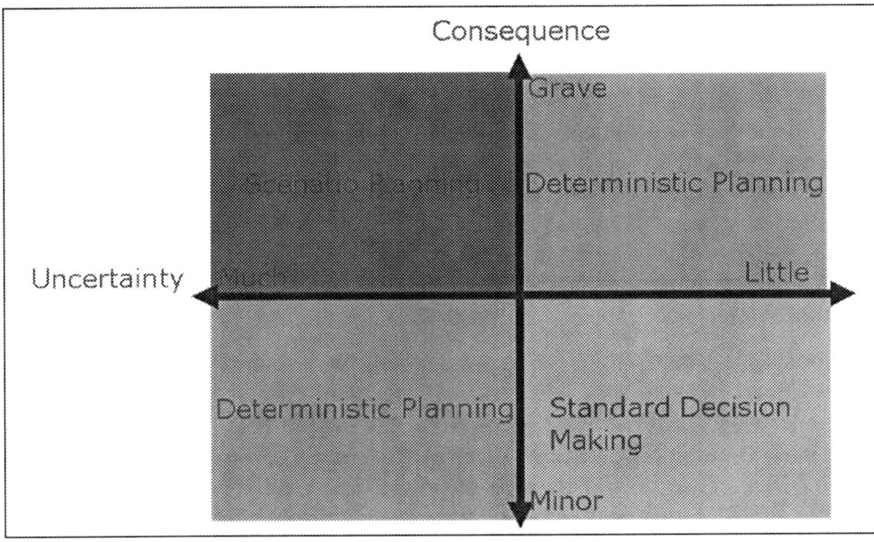

Figure 5: Opportunities for scenario planning.

Scenario planning is one technique developed in the latter part of the twentieth century for dealing with the pervasive uncertainty that confronts modern decision makers. Scenario planning is not forecasting. In fact scenario planning is rooted in the proposition that all forecasts are wrong. This approach to planning considers scenarios to be instrumental in understanding and anticipating operating environment trends. It relies on the construction of alternative scenarios as a form of sensitivity analysis performed on the unknown and unknowable future.

"Scenario" literally means an outline or synopsis of a play. The word is derived from the Italian, from *scena*, scene, that comes from the Latin *scaena* and it dates from about 1878. Herman Kahn introduced the word to its planning context, roughly a description of possible actions or events in the future, at the RAND Corporation in the 1950s. The first applications of scenarios in a planning context are thought to have been in the military strategy studies done by RAND for the U.S. government.

By the 1960's the Wharton School's H. Ozbekahn had used scenarios in an urban planning project for Paris, France. The theoretical foundations of scenario forecasting, an important component of scenario planning, were principally developed in the 1970s. Royal Dutch Shell is regularly credited with popularizing and modernizing the use of scenario planning for strategic planning in the early 1970s [11; 12]. In fact, Wack asserts it was Royal Dutch Shell that came up with the idea of scenario planning. French [5] and German [1] planners have also made early use of these methods.

The use of scenario-driven planning spread in the 1970s and by the 1980s it seems to have emerged as a distinct field of study with an extensive literature. A review of the extensive scenario planning literature [13] shows increasingly more attention was paid to the methodology of scenario planning through the 1980s. Since that time the literature has continued to grow. The emphasis in the more recent literature has been on the use of scenarios as a tool for addressing uncertainty.

There is no way to do justice to the length and breadth of the variety in scenario planning approaches in a short paper. For the purposes of this paper one approach to scenario planning is briefly summarized in the context of the with and without language developed above and in use by many planners in the US. This approach can be described as a process that develops several without project conditions rather than a single most likely alternative future without a project, as traditional planning does.

This method, developed for strategic planning by industry, recognizes strategic uncertainties in the future. Different realizations of the future could lead to quite different views about the best actions to take in the present. The uncertainties are addressed by describing different scenarios for each relevant future state of the world. Then, rather than to choose a plan based on its differences between a single without and with project conditions comparison, a plan would be evaluated against each of the possible future states of the world (scenarios, i.e., there are multiple without project conditions). The plan that performs best across all future without project conditions (or scenarios) is deemed the best plan; rather than choosing the plan that performs best if only one future state of the world is realized. Adaptive management is a modern concept that fits resource management options developed in a scenario planning context quite readily.

Scenario planning explores several alternative futures. Scenarios are: "Developed by blending data and analysis with intuition and creativity, scenario plots must 'hang together' like a well-crafted novel, stretch the imagination without going outside the bounds of believability, and consistently address issues that are critical to decision makers." [9] Although there is not a single monolithic definition of scenario analysis there are some reasonably consistent characteristics of scenario analysis in most of its forms. Some approaches to scenario construction can be found in Georgantzas and Acar [4], Audrey Schriefer [9], Clemons [2], Mercer [6], Shoemaker [8], and Ringland [7].

5.1. AN EXAMPLE

An oversimplified example of scenario creation that is representative of many of the techniques found in the literature is offered as a hypothetical illustration of the process. The example is taken from a one day effort with Barrow, Alaska planning personnel of the Anchorage District of the US Army Corps of Engineers.

After a careful identification of the problem (section 2.2) the team identified two key and uncertain drivers that will bound and shape the alternative future of Barrow and its coastal resources. These were: 1) storm damage and beach erosion regimes, and 2) social infrastructure. There were many other variables and values of interest and importance to the community and decision makers but these two defined four alternative futures as show in Figure 6.

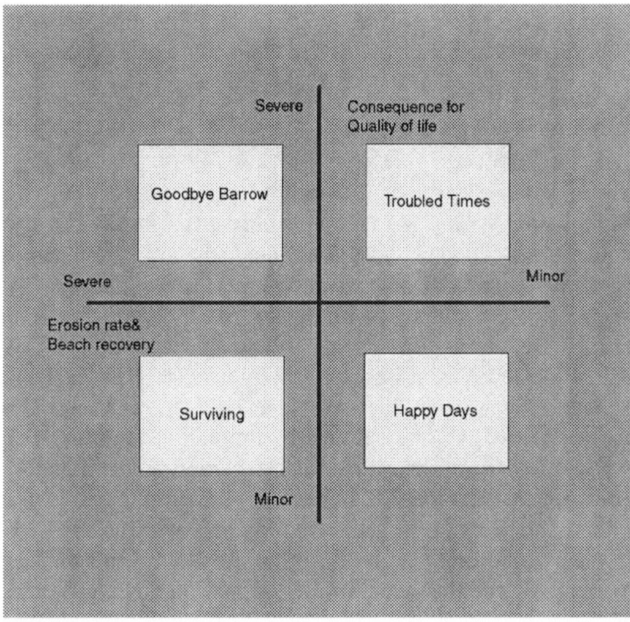

Figure 6: Four alternative futures in Barrow, Alaska.

Each scenario is described in a narrative story style that is true to the problem formulation. Scenarios (i.e., without conditions) are given a catchy name so that planners and stakeholders can begin to understand and identify with the different possible futures that could result in the absence of resource management actions. The narrative provides a clear, concise and nontechnical description of the dimensions of the coastal restoration problem that are important to decision makers and the community as they would appear in that particular future state of the world. They also provide a transparent entrée into the world of the analysis for stakeholders.

The Goodbye Barrow scenario would tell the story of severe erosion rates and beach recovery under increasing evidence of global warming that results in retreating ice cover, more frequent and more severe storms. The consequent retreat of the shoreline claims the town's road against ineffective local measures to protect it. The occasional ivu proves even more hazardous to the community in unpredictable ways. Utility interruptions begin to occur and all the people with means move. The quality of life suffers and a unique North American culture is endangered. The other three scenarios tell their own unique and different story of the future.

The traditional planning process would identify one of these futures as most likely and then all resource management options would be evaluated against it. If the future turns out to be different than the most likely forecast the efficacy of the management options is impaired. Scenario planning, by contrast, would evaluate each plan against the four scenarios. The plan with the best overall performance would be the recommended plan. Because major deviations from each scenario will have been anticipated by virtue of the scenario planning process an adaptive management component can be expected to be part of such a plan.

6. MCDA in Scenario Planning

MCDA is useful to support the evaluation of coastal restoration plans. A separate MCDA would be done for each of the four scenarios. The MCDA may include risk assessment outcomes, economic analysis, scientific measurements, quality of life indicators or measurements of any other criteria desired. Sensitivity analysis and other treatments of uncertainty within the MCDA are perfectly acceptable; however, it is worth noting that scenario analysis is used only when there are strategic reasons for doing so. This generally means there is much uncertainty with grave consequences for making the wrong decision. In such instances the strategic uncertainties addressed in the scenario planning framework may dwarf the individual tactical uncertainties in the MCDA analysis. In other words, analysts should not presume to make the MCDA any more complicated than it needs to be. The major uncertainties are being addressed in the scenario framework. The results of these analyses would provide the basis for further refinements of the resource management plans or for the selection of the best plan. A separate deterministic MCDA for each scenario in a scenario planning framework is a viable alternative to traditional planning approaches to coastal restoration problems.

7. References

1. Brauers, J., Weber, M. 1988. "A New Method of Scenario Analysis for Strategic Planning," *Journal of Forecasting,* 1988, Vol.7, No.1, pp. 31-47.

2. Clemons E.K. "Using Scenario Analysis to Manage the Strategic Risks of Reengineering," *Long Range Planning,* December 1995, Vol. 28, No. 6, pp. 123-123(1).

3. Figueira, Jose, Greco, Salvatore, and Ehrgott, Matthias, *Multiple Criteria Decision Analysis: State of the Art Surveys*, Springer, 2005.

4. Georgantzas, Nicholas C., and William Acar. *Scenario-Driven Planning: Learning to Manage Strategic Uncertainty.* Westport, CT: Quorum Books, 1995.

5. Godet, M. "Scenarios and Strategic Management: Prospective ET Plantification Stratégique," 1987, (D. Green and A. Rodney Translators).

6. Mercer, D. "Scenarios Made Easy". *Long Range Planning*, August 1995, vol. 28, no. 4, pp. 7-8+81(2), Elsevier Science.

7. Ringland, Gill "Using Scenarios to Focus R&D," *Strategy and Leadership,* 2003a, Vol. 31, No.1, pp.45-55.

8. Schoemaker, P.J.H. "When and How to Use Scenario Planning: a Heuristic Approach With Illustration," *J. Forecast,* 1991, Vol. 10, pp. 549-564.

9. Schriefer, A. and D. Mercer. "Scenarios Made Easy - Getting the Most Out of Scenarios" *The Journal of Product Innovation Management*, March 1996, vol. 13, no. 2, pp. 175-176(2) Elsevier Science.

10. Tainter, Joseph A. ***Getting Down To Earth***: *Practical Applications of Ecological Economics*, Island Press, 1996; ISBN 1-55963-503-7.

11. Wack, P. "Scenarios: Uncharted Waters Ahead," *Harvard Business Review* 1985a Vol.63 No. 5, pp. 72-89.

12. Wack, P. "Scenarios: Shooting the Rapids," *Harvard Business Review* 1985b Vol. 63 No.6, pp. 139-150.

13. Yoe, C. "Scenario-Based Planning and Decision-Making: Guidelines for Use in the U.S. Army Corps of Engineers Planning Studies and Literature Review," Institute for Water Resources, 2005.

RANKING OF AVAILABLE COUNTERMEASURES USING MCDA APPLIED TO CONTAMINATED ENVIRONMENT

A. GREBENKOV, A. YAKUSHAU
Joint Institute for Power and Nuclear Research – Sosny
Minsk 220109 Belarus

Abstract

The elimination of negative consequences of some human impacts on the environment requires a process for making rational choices among different management options. The objective of this study is to develop and employ a technique, grounded in Multi-Criteria Decision Analysis, that would weight relevant parameters when making a risk-based decision based upon the most feasible and effective choice of site remediation measures or clean-up actions among different options.

1. Introduction

Eliminating the negative consequences of some human impacts on the environment requires making rational choices among many possible options. In many cases, the choice is grounded in Multi-Criteria Decision Analysis—a system of integrated multiattribute objective and subjective (expert) assessments of several alternatives. Usually, objective parameters can be formalized and transformed into a one-attribute scheme. An example would be cost-benefit analysis of site remediation strategies and least-cost evaluations applied to the ranking of available countermeasures.

The basic notion in the application of least-cost techniques to a decision-making process is very simple: an option is selected if the resulting net benefit, including mitigation of risk and minimization of cost, exceeds that of the best alternative and not otherwise. The objective of this study is to develop and employ a technique that would weight relevant economical parameters when making a risk-based decision to determine the most feasible and effective choice of site remediation measures or clean-up actions among different options.

I. Linkov et al. (eds.), Environmental Security in Harbors and Coastal Areas, 111–119.

2. Assumptions, Methods, and Models

In the case of application of this practice to ranking of available environmental protection measures, the net benefit from introduction of a given countermeasure can be expressed as *B=V–P*, where *B* is the net benefit from application of the countermeasure; *V* is the benefit, gained as a result of expected risk reduction, from mitigating the additional risk-related cost; *P* is the cost of a countermeasure. The cost of the countermeasure includes all efforts made by the society in decreasing the environmental risk to an acceptable level, and it is normally expressed in monetary terms. The countermeasure monetary evaluation is based on the conventional economic concept that takes into account actual expenses and their distribution in time. It consists of a combination of the capital cost of the countermeasure and the subsequent operation and maintenance cost. The NPV and IRR parameters allow evaluating a countermeasure from a cost-effectiveness point of view. Obviously, the countermeasure that has the largest NPV value is the most cost-effective one. Along with IRR, this can be used for comparing and rating different countermeasures and selecting the most economically attractive remediation strategy.

In order to take into account temporal scales, a discounting method is commonly used. In accordance with discounting technique, the net benefit from application of any countermeasure can be considered as net present value (NPV), which is defined as:

$$B = NPV = \sum_{i=1}^{n} \frac{V_i - C_i - O_i}{(1+\alpha)^i}, \tag{1}$$

where V_i, is annual benefit from application of countermeasure; C_i is capital cost; O_i is operation cost; α is discount rate; and n is project lifetime.

Parameter NPV allows evaluating the countermeasure from the cost-effectiveness point of view. Obviously, the countermeasure that has largest NPV value is the most cost effective. If operation and maintenance costs and benefits from the countermeasure application are assumed to be constant throughout the lifetime of the project, the calculation of the NPV can be simplified by the use of annuity factors. These are the sum of discount factors over time periods and are defined as:

$$A(n,\alpha) = \frac{(1-\alpha)^n - 1}{\alpha(1+\alpha)^n} \tag{2}$$

In this case, the expression (14) can be rewritten as:

$$NPV = (V - O) \cdot A(n,\alpha) - C \tag{3}$$

The formulas given in Equations (1) or (3) can be used for comparing and rating different countermeasures and selecting the most economically attractive one. To compare a countermeasure to be selected for promotion with other investments the parameter of internal rate of return (IRR) can be used. IRR can be found using the following equation:

$$(V - O) \cdot A(n, IRR) - C = 0 \tag{4}$$

The task of determining an evaluation of the optimal measure from a set of measures can be formalized as a minimization process applied to several functions based on a product of the matrices responsible for environmental contamination, transport of contaminants through forage chains, effectiveness of different clean-up (countermeasure) techniques designed for specified contaminant(s), cost of available technologies(s), and monetary equivalents for averted risks from exposures to specified contaminants.

The expected reduction factors D_{il} of exposure to contaminated forage as a result of application of different countermeasures to the specified route of exposure can be defined as the following matrix:

$$[D]_{il} = \begin{bmatrix} & \text{Forage 1} & .... & ..... & \text{Forage M} \\ \text{Countermeasure 1} & D_{11} & .... & ..... & D_{1M} \\ .... & .... & ..... & .... & ..... \\ .... & ..... & ..... & ...... & ..... \\ \text{Countermeasure N} & D_{N1} & ..... & ..... & D_{NM} \end{bmatrix} \tag{5}$$

The second matrix, which has to be added to the matrix product, contains weighted dimensionless coefficients responsible for contaminant transfer between different forage chains and receptors. The matrix has the following structure:

$$[W]_{li} = \begin{bmatrix} & \text{Contaminant 1} & .... & ..... & \text{Contaminant M} \\ \text{Forage 1} & W_{11} & .... & ..... & W_{1M} \\ .... & .... & ..... & .... & ..... \\ .... & ..... & ..... & ...... & ..... \\ \text{Forage N} & W_{N1} & ..... & ..... & W_{NM} \end{bmatrix} \tag{6}$$

Components of the following matrix represent the concentration of contaminants in different food products:

$$[C]_{il} = \begin{bmatrix} & \text{Forage 1} & .... & ..... & \text{Forage M} \\ \text{Contaminant 1} & C_{11} & .... & ..... & C_{1M} \\ .... & .... & ..... & .... & ..... \\ .... & ..... & ..... & ...... & ..... \\ \text{Contaminant N} & C_{N1} & ..... & ..... & C_{NM} \end{bmatrix} \tag{7}$$

The matrix $[K]$, with components representing the expected concentration of contaminant *k* incorporated in receptor tissue with forage item *j,* can be calculated as:

$$[K]=[C]\cdot\{e\}^{T} \tag{8}$$

In Equation (8), the components of vector-column $\{e\}$ contain the hazard quotient that accounts for toxic (or radiation) dose generated by a unit concentration of contaminant k inside the receptor's body. Upper index T means that vector $\{e\}$ is to be transposed to a vector-row.

The expected reduction of annual toxic dose with due account of the countermeasure to be implemented in the given site can be presented as a vector-column $\{\nabla E\}$, calculated by using the following expression:

$$\{\nabla E\}=[D]\cdot[W]\cdot[K]\cdot\{M\} \tag{9}$$

The vector-column $\{M\}$ includes components that represent annual consumption of the food product related to corresponding food chain. A decreasing extent of annual dose $\{\nabla E\}$ (or component of it) caused by each countermeasure applied to each sources of exposure can be evaluated in monetary units by multiplying to coefficient β. This coefficient is a monetary equivalent of an averted exposure dose. Thus, the economical effectiveness of implementation of countermeasures, or NPV and IRR, is calculated as:

$$\{NPV\}_k=\left(\beta\cdot\{\nabla E\}_k-\{O\}_k\right)\cdot A(n,\alpha)-\{C\}_k \tag{10}$$

$$\left(\beta\cdot\{\nabla E\}_k-\{O\}_k\right)\cdot A(n,IRR)-\{C\}_k=0 \tag{11}$$

The proposed analysis of effectiveness of different radiation protection measures does not imply all varieties of ecological, socioeconomic, and other conditions. These conditions are to be evaluated by a decision-maker on a basis of separate set of criteria and assumptions (e.g., availability of resources and materials, other possible ecological constraints, political situation, etc.). The completeness of such evaluation depends on what task, direct or inverse, is used.

The direct task implies, first, specification and ranking of the territories / objects where the countermeasures must be implemented; secondly, choice of the countermeasures, evaluation of corresponding means (finance, materials), assessment of effectiveness and analysis of alternatives; and finally, development of a rehabilitation plan.

When an inverse task is considered, experts have to choose a countermeasure (or set of countermeasures) that would optimally meet a given criterion and satisfy foreseeable constraints (financial, ecological, social, political, and so on). Whatever the case, the proposed model of cost-benefit analysis gives an opportunity to evaluate and validate the main economical aggregates that are the most crucial bases of decision making.

3. Results of Model Application

The suggested model as an integral part of complex software was used in development of the Radiological-Hygienic Passport database (RHP-database) [1]. The objective of the database was to collect, verify and file all reliable and consistent data regarding the radioecological situation, sanitary and public health conditions, and economical and societal infrastructure for contaminated settlements in the regions contaminated after Chernobyl. Another objective was to operate as an informational and supportive tool for decision-making processes. The RHP-database is designed and produced with the relevant models incorporated in a simulation block of RHP software to calculate expected dose response and the economical parameters of countermeasures.

The RHP-database is built in Microsoft Access and has the following principal structure (Figure 1):

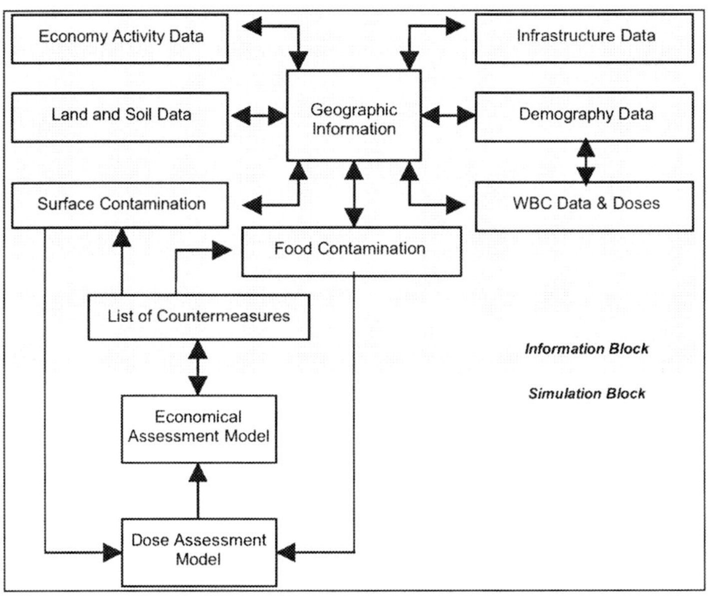

Figure 1: RHP Database structure.

The RHP-database main form is designed for opening other principal database forms, as well as for linking to our modeling block with two mathematical codes, which are responsible for assessment of annual effective doses and calculation of economic parameters of the selected countermeasures (the model described above).

From the main form (Figure 2), a user can open several major database forms using relevant buttons, namely:

1. "List of Settlements" (calls the database form containing general administrative location of all target settlements).
2. "Population" (calls the database form with demographic data).

3. “Infrastructure” (call the database form with information about industrial and communal infrastructure).
4. “Building and dwelling” (calls the database form with information about buildings of different types in the settlement location).
5. “Contamination of settlement” (calls the database form with information about contamination levels in settlement territory).
6. “Contamination of lands” (calls the database form with information about contamination levels in homesteads, pastures, hayfields, and other agricultural lands).
7. “Contamination of food” (calls the database form with information about content of radionuclides in local foodstuffs and forest food products).
8. “WBC data” (calls the database form with information about 137Cs content in human body according to the records of whole body counting).
9. “Annual effective dose” (calls the database form with information about annual effective dose recorded for the given settlement).
10. “Countermeasures applied” (calls the database form with information about radiation protection measures that have been already applied to the given settlement).

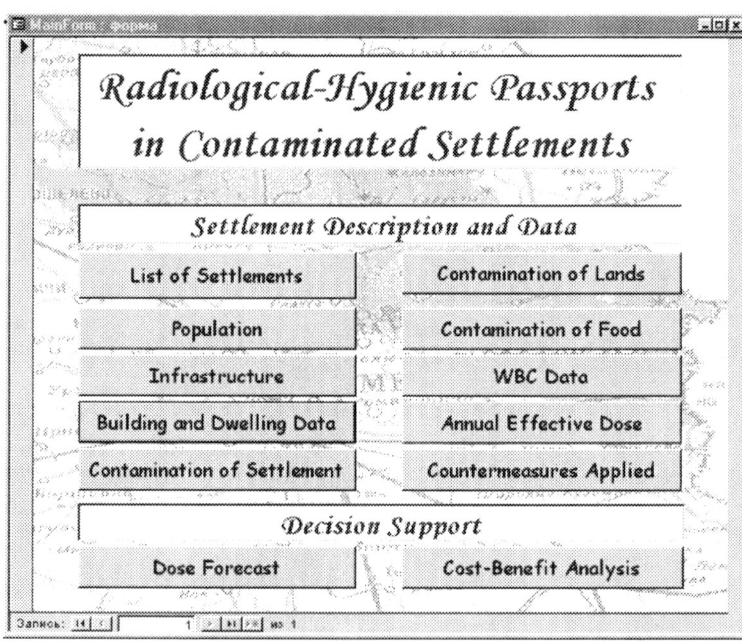

Figure 2: Database main window form.

MS Excel was used both for RHP data preparation, cross-check analysis and data entry. The Excel workbook used for data compilation concerning agricultural countermeasures and other radiation protection measures is shown in Figure 3. As it may be seen, each technology and technique, which can be chosen for the application, includes all necessary information to perform evaluation of efficiency and least-cost analysis. Here, we have five spreadsheets each describing the different clean-up technologies and remediation techniques applied to man-made surfaces, soil surfaces in urban areas, forest areas, virgin soil, and cultured soil.

The corresponding database forms above suggest first assessing a threat caused by contamination of food and evaluating the efficacy of selected countermeasure options by lowering external and internal exposures. After completion of the assessment of exposure to different contaminated food chains, the code suggests choosing a radiation protection measure from the list. Then, a user can proceed to economic analysis of the option selected by using the form shown in Figure 4. After the analysis of exposure to contaminated food chain is made, the software suggests the form shown in Figure 5 for calculation of economic parameters of different countermeasures. The form allows calculation of benefits from countermeasures, Net Present Value (NPV), and Internal Rate of Return (IRR) of realization of the specified countermeasure.

	A	B	C	D	
1	***Name of technology***	***Washing***	***High pressure water hosing***	***Replacement***	***High***
2	Tool	**Roof washer**	**High pressure turbo nozzles**	**Change of roof**	**Higi**
3	Target	Roofs	Roofs	Roofs	Roof
4	Mode of operation	Brushing and washing	Water hosing @ 150 bar	Removal and replacement	Wate
5	Design (short description of technique)	Rotating brush + rinsing water	Rotating nozzles + heated water	Set of tools	High
6	Detailed description	Read in the note-box ------>	Read in the note-box ------>	Read in the note-box ------>	Reac
7	Sufficient constraints	Non-applicable in winter	Non-applicable in winter	None	Non-.
8	Occupational exposure (short description)	External irradiation from surfaces	External irradiation from surfaces	External irradiation from surfaces	Exter
9	Inhalation/external dose ratio	0,001	0,000	0,001	0,00
10	Ingestion/external dose ratio	0,000	0,000	0,000	0,00(
11	Other impacts. Toxicity (qualitative assessment)	None unless asbestos	None unless asbestos	None unless asbestos	None
12	Other benefits (qualitative assessment)	Roof cleaned for moss and algae	Roof cleaned. Nicer appearance	New roof	Roof
13	Number of operators	1	2	3	2
14	Unit (m^2, kg, number, etc.)	m^2	m^2	m^2	m^2
15	Productivity (unit/h)	18,00	30,00	12,00	40,0(
16	Consumables per unit	Water: 10 dm^3, Gasoline: 0,25 dm^3	Water: 20 dm^3; Gasoline 0,27 dm^3	Roofing materials: 1,2 m^2	Wate
17	Man-days per unit (at 7,2 working hours per day)	0,0077	0,0093	0,0347	0,00(
18	Achievable scale of application (units/year)	129600,00	216000,00	86400,00	2880
19	Cost: Tool investment cost (Euro)	8000,00	38000,00	0,00	3600
20	Cost: Discount (Euro/year)	1600,00	7600,00	0,00	7200
21	Cost: Manpower (Euro/unit)(*)	0,094	0,113	0,425	0,08(
22	Cost: Consumables total (Euro/unit)(**)(***)	0,143	0,154	15,700	0,114
23	Cost: Overheads (Euro/unit)(****)	0,151	0,181	0,680	0,13(
24	Cost: O&M cost total (Euro/year)	50292,00	96890,40	1451962,00	9648
25	Number of man-hours exposed (h/unit)	0,06	0,07	0,25	0,05
26	Decontamination factor (DF)	2,50	3,00	1000,00	3,20
27	Exposure dose rate reduction factor (ERF)	1,10	1,15	1,20	1,16
28	Solid wastes generated (kg/unit)	0,25	0,30	12,00	0,40
29	Liquid waste generated (l/unit; dm^3/unit)	10,00	20,00	0,00	27,0(
30	Waste normalized activity (kBq/m^3 per kBq/unit)	60,0	33,3	33,3	26,5
31	Cost: Waste management (Euro/unit)(*****)	4,00	8,00	0,00	10,8(
32	(*) Cost of manpower (Euro/(man-hour)	1,70	1,70	1,70	1,70
33	(**) Price of gasoline (Euro/dm^3)	0,57	0,57	0,57	0,57
34	(***) Price of electricity (Euro/kW-h)	0,03	0,03	0,03	0,03
35	(****) Overheads as a portion of man-power (%)	160,00	160,00	160,00	160,(
36	(*****) Cost of waste management (Euro/m^3)	400,00	400,00	400,00	400,(
37					

Man-made surfaces / Soil surfaces in urban areas / Forest areas / Virgin soil / Cultured soil

Figure 3: Fragment of RHP countermeasures data in initial Excel format.

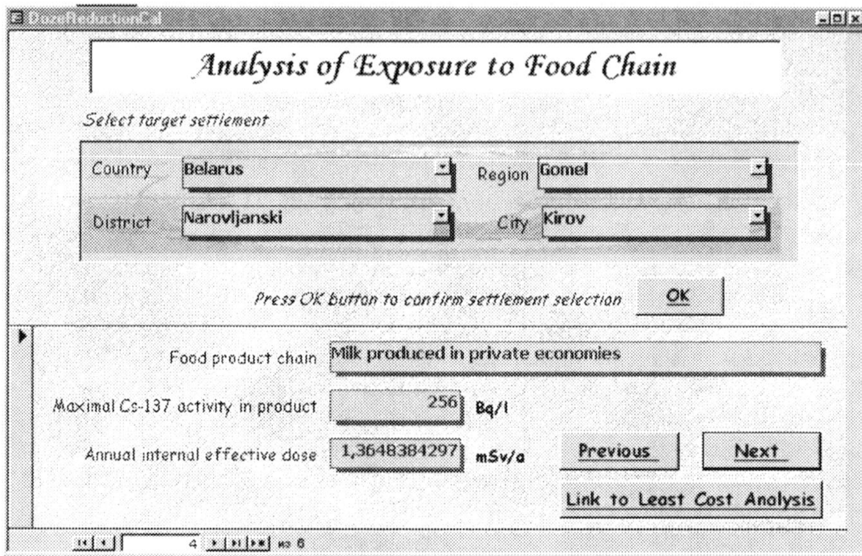

Figure 4: Form responding to "Cost-Benefit Analysis" button of the main form.

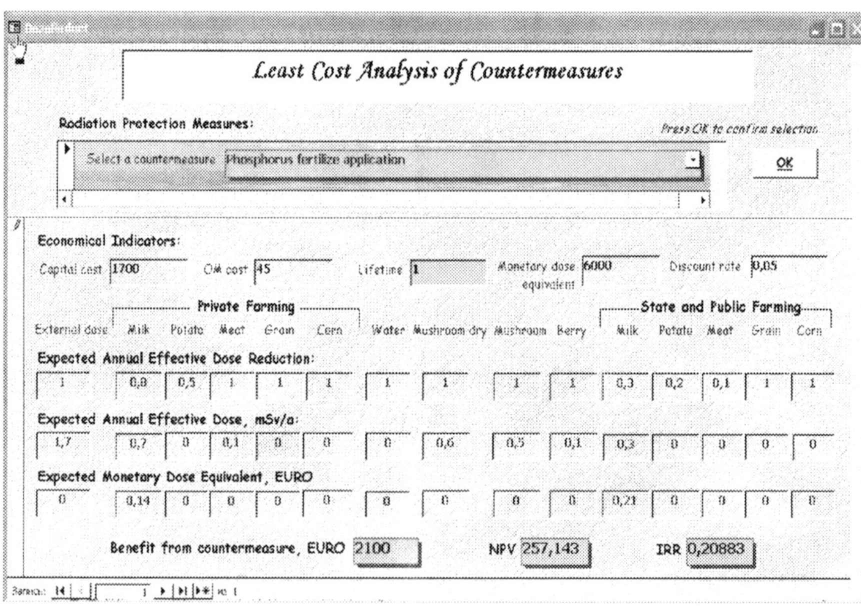

Figure 5: Form responding to "Link to Least Cost Analysis" button of the previous form.

4. Conclusion

The suggested model can find a direct task resolution implying specification and ranking of territories/objects where countermeasures should be implemented, address a choice of countermeasures, evaluate corresponding means (finance, materials), assess effectiveness and analysis of alternatives, and finally, develop a rehabilitation plan.

When an inverse task is considered, experts have to choose countermeasures that would optimally meet a given criterion and satisfy foreseeable constraints (financial, ecological, social, political, and so on). Whatever the case, the proposed model gives an opportunity to evaluate and validate the main economical aggregates that are among the most important attributes of the MCDA process.

5. Acknowledgement

This study was supported by the French-German Initiatives Programme.

6. References

1. Grebenkov, A.J. (2004). Database on Radiological Situations, Countermeasures and Medical care in Contaminated Settlements (Radiological Passports). *Proc. of International Workshop, "The French-German Initiative: Results and Their Implication for Man and Environment,," Kiev, Ukraine, October 5-6, 2004*, 19-20

ENGAGING THE GENERAL PUBLIC AND OTHER STAKEHOLDER GROUPS

J. SKEI
Norwegian Institute for Water Research (NIVA)
Gaustadalleèn 21, 0344 Oslo, Norway

Abstract

Sediment management decisions can have substantial environmental and economic consequences. Due their complexity, efforts should be made to communicate complicated technical facts, uncertainties, risk assessments, and other tools used in the decision-making process to the public and stakeholders in an easily understandable way. This paper briefly reviews the public's involvement in the decision to cap sediments contaminated with heavy metals in a fjord located on the west coast of Norway.

1. Introduction

People are here defined as society (the public), stakeholders and decision makers, constituting those being affected by a decision concerning sediment management, those being responsible for the actions taken and those who legally decide.

The objective of sediment management is to make sure that best available knowledge and technology is used to reduce the risks of environmental problems related to contaminated sediments in a cost efficient and sustainable way. Compared with end of pipe solutions, with respect to discharges to the aquatic environment, sediment remediation is complex, less predictable and very costly. Consequently, the decision making process has to be based on solid scientific knowledge, practical insight and socio-economic understanding.

As sediment management may have substantial environmental and economic consequences, the decision making process must be transparent and all relevant parties – the people - should feel an ownership to the management plans. To adopt ownership you need engagement and involvement at an early stage in the planning process.

I. Linkov et al. (eds.), Environmental Security in Harbors and Coastal Areas, 121–125.

2. The Importance of Public Participation in Decision Making

At the preliminary stage of planning, the primary stakeholder and the decision makers should play an active role. The public, however, should not be involved at this preliminary stage, which may be considered as confusing and partly chaotic. But as soon as a preliminary plan has been communicated by the primary stakeholders and the decision makers, the public should be brought into the process. The use of open meetings and public hearings, where information is given and questions may be asked, is a suitable form of engaging the public.

The involvement of stakeholders and decision makers will take place in formal meetings and in writing. The magnitude of involvement of the different parties should be formally presented in a strategy plan and accepted by all parties. This plan should contain time schedule and mile stones and any revisions should be communicated with all parties. This will clarify what is expected at any time during the planning and implementation process. This will minimize the problem of delay of the process due to unexpected complaints, mistrust and misunderstandings.

The preliminary sediment management plan should describe the environmental objectives, the risks and uncertainties, the expected timing in terms of achievement of environmental benefits and the order of actions taken. This is to avoid unrealistic expectations, which again is a source of mistrust. It should also be clearly communicated how the environmental benefits as well as failure are documented. Monitoring prior to remediation, during remediation and a long term monitoring after remediation, is of great importance and should be of interest of all parties involved.

Due to the complexity of sediment management projects, effort should be made to communicate complicated technical facts, uncertainties, risk assessments and other tools used in the decision process to the public and stakeholders in an easily, understandable way (use of simplifications, but not oversimplifications). This is a challenge, but if ownership to a management project is the objective, the parties involved will require an understanding of all potential consequences and risks, environmentally as well as economically.

Secondary benefits from sediment management projects should also be emphasized such as environmental reputation, economic benefits related to coastal zone management, increased value of land and housing in the vicinity of remediation areas, positive effects on tourism etc.

3. A Case Study of People Engagement: Capping of Contaminated Sediments in a Norwegian bay – the Process of Decision

A zinc plant situated at the head of a 200 km long fjord system on the west coast of Norway discharged waste containing heavy metals like mercury and cadmium into the Sørfjord since 1929. As a result, the sediments were heavily contaminated, particularly in a small bay close to the smelter where levels of mercury and cadmium exceeded several 100 parts per million (ppm) [1].

In 1986 the discharge of jarosite residue stopped and the waste was stored underground in caverns [2]. This led to a noticeable improvement of the levels of heavy metals in water, fish and shellfish. The decision to terminate sea disposal of jarosite waste was based on regular monitoring of water, biota, and sediments in the Sørfjord since 1979. The monitoring data clearly showed excessive contamination and the health authorities introduced warnings concerning human consumption of seafood. The level of mercury in the blood of the local fish eating population exceeded levels where health problems may occur. Consequently, there was a large local engagement with respect to remediation of the environmental situation in the area. At the same time, the local population was very much dependent on the industry with respect to employment and the level of acceptance of environmental nuisance was very high in the 70s.

Following the underground disposal of waste in 1986 (Figure 1), monitoring data clearly demonstrated the environmental improvement in the Sørfjord. The industry obtained a good reputation and the local population was optimistic with respect to the future. Environmental goals included removal of restrictions on consumption of seafood within year 2000 were formulated, but not achieved. There are still (as of 2005) restrictions regarding consumption of fish and mussels.

Figure 1: Underground disposal of solid waste (jarosite) from zinc production.

It was soon realized that contaminated sediments could act as a source of contamination for a long time. Particularly, sediments in a shallow bay close to the zinc plant. This bay was heavily influenced by residues from the zinc production and the red colored sediments were easily resuspended during stormy conditions (Figure 2).

Figure 2: Conditions near the zinc plant in the 70s.

A second round of remediation was recommended to reduce the input of trace metals to the surface water and to lower the exposure to fish and mussels. The decision was based on a research programme involving experimental work on contaminated sediments to document heavy metal fluxes, bioavailabilty and bioaccumulation (Figure 3).

It was decided to cap the sediments of the shallow bay (90.000 m^2) with a geotextile and 30 cm of sand on top. This was executed in 1992. The local population was happy about the situation and encouraged by the hope of a cleaner environment. Monitoring of the water quality clearly showed reduced levels of contamination in the surface water immediately after the capping. Unfortunately, many unexpected incidents of accidental spills at the zinc plant have during the last decade caused episodic increases in the level of contamination and at present the sandy cap of the bay has been recontaminated to a large extent.

Figure 3: Experimental work with contaminated sediments to measure fluxes of metals from sediments to water.

The lesson learned is that the capping was performed without having source control. The local people have been engaged in environmental matters for a long time, going through phases of hope and optimism and disappointment and pessimism.

4. References

1. Skei, J.M., Price, N.B., Calvert, S.E., & Holtedahl, H., 1972: The distribution of heavy metals in sediments of Sørfjord, West Norway. Water, Air & Soil Poll., 1, 452-461.

2. Skei, J.M., 1992. A review of assessment and remediation strategies for hot spot sediments. Hydrobiologia, 235/236, 629-638.

RISK AND DECISION METHODS APPLIED TO AQUATIC ECOSYSTEM MANAGEMENT

Considerations for Invasive and Endangered Species

B.S. PAYNE, A.C. MILLER, B. SUEDEL
U.S. Army Engineer Research and Development Center
Environmental Laboratory, 3909 Halls Ferry Road, Vicksburg, MS 39180-6199 USA

Abstract

Increased global trade and modern intercontinental transportation have made invasive species an increasingly prominent stressor of freshwater ecosystems. Invasive species risk assessments, which range from simple screening protocols that focus on species attributes and ecological requirements to rigorous analyses of infestation, have become an important component of environmental impact assessment. In this paper we present two recent case studies in which risk and decision methods were applied to non-toxicological environmental issues that are central to many aquatic ecosystem management programs. The first example reviews potential infestation by the zebra mussel, *Dreissena polymorpha*, on Bayou Bartholomew, located in southeast Arkansas, as a result of the proposed augmentation of low flow conditions by pumping water from the nearby and much larger Arkansas River. The second example presents a retrospective analysis of the results of transplanting fat pocketbook pearly mussels, *Potamilus capax,* from an approximately 6-km reach of a drainage ditch in eastern Arkansas. The two examples presented herein indicate the potential for improving environmental decision-making in the face of uncertainty—but in the presence of substantial information. As more rigorous attempts are made to widen and enhance applications of risk and decision methods to environmental decision-making, ecosystem management is likely to further improve.

1. Introduction

Contaminant and human health issues have been at the forefront of environmental applications of risk and decision methods. Toxicological risks were certainly the initial focus of risk assessment protocols developed in the United States [39; 54]. However, risk and decision methods have wider applicability to environmental assessment and impact analysis. Risk analysis methods are being modified to address invasive species issues (e.g., [50; 53]). Formal decision methods had their

I. Linkov et al. (eds.), Environmental Security in Harbors and Coastal Areas, 127–148.

origins in business and legal practice [12] and now are emerging in environmental management-oriented decisions [29].

The second half of the 20th century brought expansive development of information and knowledge of the biology and ecology of populations and communities in major ecosystems. While uncertainty still exists in ecology, lack of information and knowledge cannot be allowed to paralyze ecosystem restoration and management decisions. Integration of information, application of moderately comprehensive ecological models, risk analysis, and formal decision methods all offer ways to meet environmental assessment and ecosystem management needs in the face of uncertainty.

In this paper we summarize two recent examples in which risk and decision methods were applied to non-toxicological environmental issues that are central to many aquatic ecosystem management programs. The first example deals with application of risk analysis to an invasive species dispersal concern (see also [43]). The second deals with a retrospective analysis of a set of endangered species management decisions [38], showing how the use of specific decision tools can improve the decision-making process.

2. Example 1: Infestation Potential of an Invasive Species

2.1. BACKGROUND

Invasive species and habitat degradation generally are accepted as leading threats to biodiversity [13; 30; 45]. Nearly all aquatic ecosystem management programs include a component that addresses invasive species. Freshwater aquatic habitats are the most imperiled ecosystem on the planet, due to use of freshwater ecosystems for industrial and human waste removal and processing, irrigation, flood control, power generation, transportation, and drinking water. In the medium and large rivers in the United States, habitat impacts are due to dams, locks, straightened river channels, dredged navigation channels, levees, dikes, and training structures occurred with initial construction of the inland navigation and flood control system. Habitat degradation is also associated with pollution and degraded water quality, although much of this loss has been remedied or ameliorated since regulatory implementation of clean water legislation.

Increased global trade and modern intercontinental transportation have made invasive species an increasingly prominent stressor of freshwater ecosystems [30]. Invasive species risk assessments have become an important component of environmental impact assessment. These assessments range from simple screening protocols that focus on species attributes and vectors of dispersal with largely narrative presentations of species biology and ecological requirements to rigorous predictions of infestation risk.

Dispersal likelihood and the physiological and ecological requirements of a species largely determine infestation potential in a particular region. Despite dispersal, infestation level may stay low due if the species is living in only marginally suitable habitat. Conversely, invasive species tend to thrive in more suitable habitats, as predators, competitors, pathogens, and other natural control mechanisms are often lacking.

Our example focuses on how physiological and ecological preferences of a species can be used in relation to habitat characteristics to support qualitative and quantitative risk assessments. We address potential infestation by the economically and ecologically important zebra mussel, *Dreissena polymorpha*, of a small river, Bayou Bartholomew, in southeast Arkansas due to proposed augmentation of low flow by pumping of water from the nearby and much larger Arkansas River (Figure 1). This particular example demonstrates how available information on habitat conditions and species physiological ecology can be combined to assess infestation risks in a fashion that is easily communicated to various stakeholders and managers.

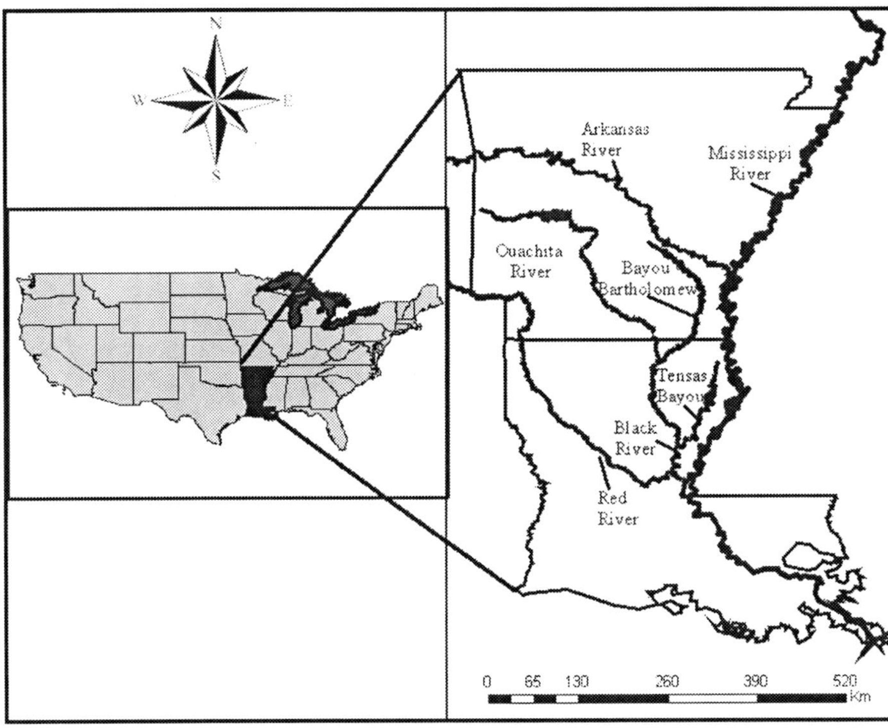

Figure 1: Map showing the locations and spatial proximity of the Arkansas River to Bayou Bartholomew and the Ouachita River.

The zebra mussel is a small freshwater bivalve native to the region of Europe surrounding the Black, Caspian, and Aral seas (see [47] and [51] for summaries of this species invasive history). Zebra mussels spread throughout much of Europe as canals, waterways, and boat traffic provided routes of dispersal that did not exist much before the Industrial Age. The zebra mussel was spread to the North American Great Lakes by ballast water exchange. This species was found in Lake St. Clair in the mid 1980s and, within a few years, spread throughout much of the inland navigation system of the United States east of the Continental Divide. Zebra mussels now occur throughout the Mississippi River from its upper reaches to the lowermost river near New Orleans, Louisiana. However, high water temperatures in summer generally prevent the relatively cool water species from thriving in the southern United States [2; 3].

2.2. PREDICTING EXPOSURE TO INVASIVE SPECIES

It is useful at first to broadly discuss a basic approach to evaluating invasive species infestation risks. Figure 2 provides an overview of the basic risk assessment protocol, tailored to invasive species. With respect to the protocol depicted in Figure 2, it is important to note that herein we deal predominately with methods of predicting exposure to invasive species. We do not address biophysical and socioeconomic effects.

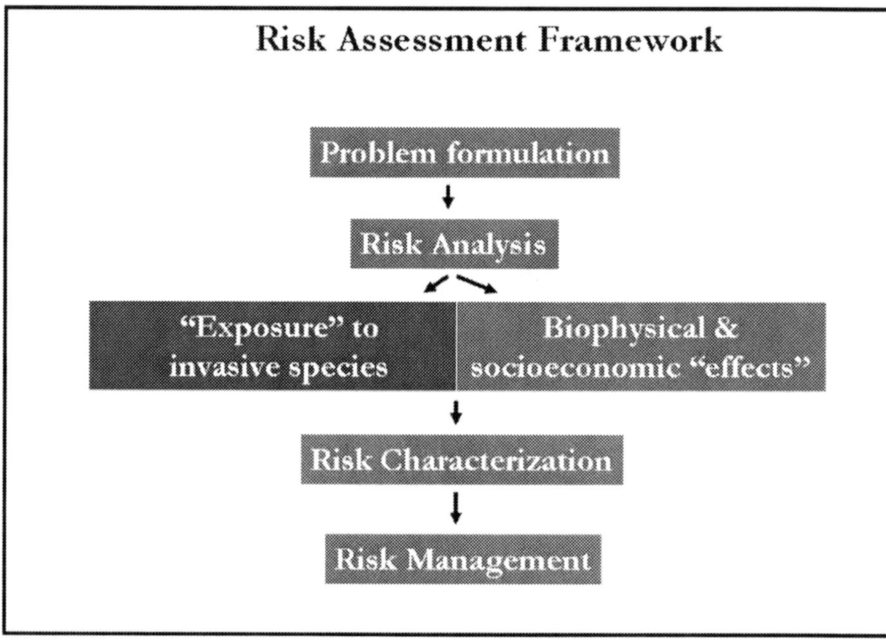

Figure 2: A simple diagrammatic representation of the risk assessment framework for invasive species; adapted from Suedel et al. 2006.

Infestation risks for most invasive species are determined by the interplay of habitat conditions and physiological tolerances. Thus, rapid and thorough compilation and analysis of existing information on the invasive species of interest is always an important early step. Different project settings and environmental concerns merit different levels of scrutiny. Thus, it is appropriate to consider risk assessment approaches that range in difficulty and cost. Experience suggests approximately three levels of risk assessment, from the least demanding to most demanding. All are based on the physiological ecology of the invasive species.

1. Level 1

   At the lowest level, useful risk guidance can be as simple as a table of habitat conditions versus tolerances of those conditions. More specifically, categories of infestation potential (e.g., low, moderate, and high) associated with a range of values for each of several critical habitat parameters. Such tables have been developed independently for a number of important invasives, or could be derived from reviews of species physiological ecology that are not aimed at a simple tabular summary. For example, Claudi and Mackie [11] did just that by compiling a simple table to guide infestation risk analysis of the zebra mussel in North America.

2. Level 2

   A higher level of detail and certainty would invoke the logic if not the accounting framework of the "habitat suitability modeling" approach developed by the U.S. Fish and Wildlife Service in the 1980s. In this approach, an attempt is made to correlate infestation potential with known habitat conditions (e.g., [26; 51]). Habitat conditions might be known from existing information, require field measurements, or some combination of both. This approach is based on organismal physiological tolerances and applied to a quantitative analysis of habitat and water quality conditions.

3. Level 3

   Ultimately, interplay habitat and organismal preferences (and potentially other factors) results in distributional patterns that are amenable to statistical analysis and can be used to build empirical models. Rigorous statistical models can be developed that allow quantification of uncertainty (e.g., [46]). A conceptually similar and rigorous approach is embodied in recent attempts to apply methods from informatics. In this approach, rule-based models of infestation potential are derived from extensive data sets on species-habitat relationships in species' native ranges [44].

All of these levels of analysis, and especially the first two, derive mostly from existing information. The final level requires paired observations of abundance and habitat conditions, usually from the invasive species home range but potential from a newly infested region. Such data sets are less certain to be available and more demanding to evaluate.

Once an invasive species has accomplished intercontinental or other wide-reaching dispersal, considerable attention is typically devoted to predicting infestation likelihood for regions and sites in reasonable proximity to those aquatic ecosystems initially infested by the recent invader. In the following paragraphs we demonstrate the use of existing habitat and tolerance data to perform an intermediate level analysis for such a prediction.

2.3. A PROPOSED INTER-BASIN WATER TRANSFER PROJECT

The inter-basin transfer of interest involves proposed pumping of water from a large river, the Arkansas River, to the headwater reach of a small river, Bayou Bartholomew, in southeast part of the State of Arkansas (Figure 1). This proposed pumping project created concern because the Arkansas River supports a zebra mussel population (although the population is seasonally suppressed in summer by high water temperature [4]). Bayou Bartholomew does not presently support zebra mussels, and the Bayou drains into the larger Ouachita River which supports many species of native unionid mussels that would be potentially harmed by zebra mussel infestations. Our purpose herein is not to analyze the project, but rather to demonstrate how information on habitat conditions and species physiological ecology can be combined to assess infestation risks in a fashion that is easily communicated to various stakeholders and managers.

2.4. THE INFESTATION RISK ANALYSIS

In brief overview, we approached the problem by: reviewing extensive published literature on the physiological ecology of *Dreissena polymorpha*, with emphasis on habitat requirements especially as they relate to the southerly distribution of this species in North America. Next we characterized the habitats in the Arkansas River, Bayou Bartholomew, and the Ouachita River with respect to critical parameters identified from the literature review. These variables included water temperature, pH, calcium, and dissolved oxygen – all of which were potentially limiting to zebra mussels in the river system under investigation. Our purpose here is not to provide a comprehensive analysis of this information (see [11] and references within), but rather to demonstrate with clear examples, how knowledge of zebra mussel physiological ecology efficiently supported risk analysis.

Habitat suitability curves for the four water quality parameters are presented in Figure 3. These are essentially similar to curves used in a prior analysis of zebra mussel infestation risk in the State of Florida [26] that, in turn, and are based mostly on published information on zebra mussel physiological ecology (e.g., [11; 47]. We do not attempt a through review of such information herein; our point is more simply that such knowledge of physiological ecology typically can be compiled for invasive species of greatest ecological and economic concern. Basic aspects of organism-habitat relationships are reasonably well known for such species, or become the focus of early investigations of pest species invasions.

Existing measurements of water quality were compiled from various source and sites and summarized in a series of figures (e.g., Figure 4). For monitored sites throughout Southeast Arkansas we compiled composite scores of habitat suitability, using a simple weighted average of habitat suitability with respect to pH, calcium, temperature, and dissolved oxygen. We included this simple regional analysis because, ultimately, concern will exist for not only Bayou Bartholomew but other stream systems in the same region.

For the three rivers of initial interest, the Arkansas River, Bayou Bartholomew, and the Ouachita River, a simple plot of pH provides the essence of our risk analysis (Figure 5). More specifically, only the Arkansas River has a pH that will generally

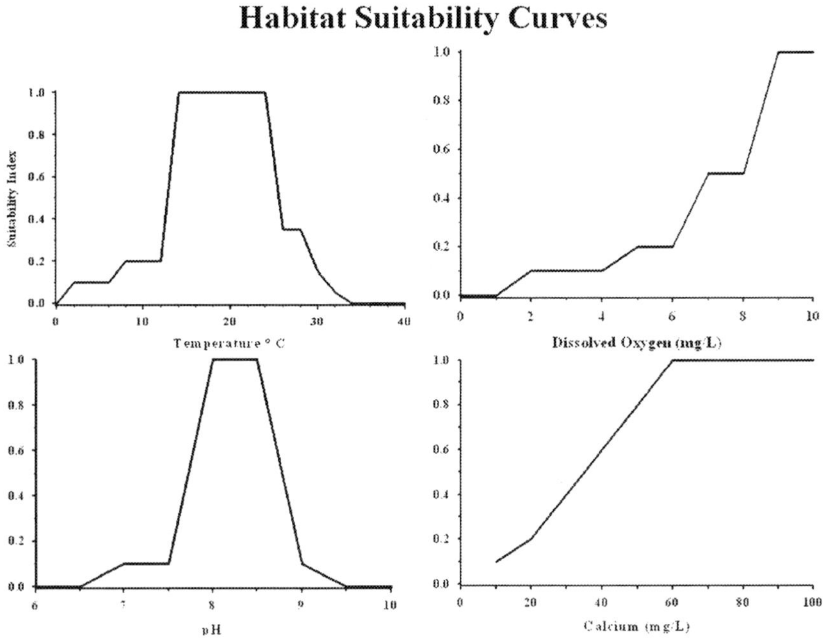

Figure 3: Habitat suitability curves showing zebra mussel preferences with respect to four often limiting water quality variables: temperature, dissolved oxygen, pH, and calcium.

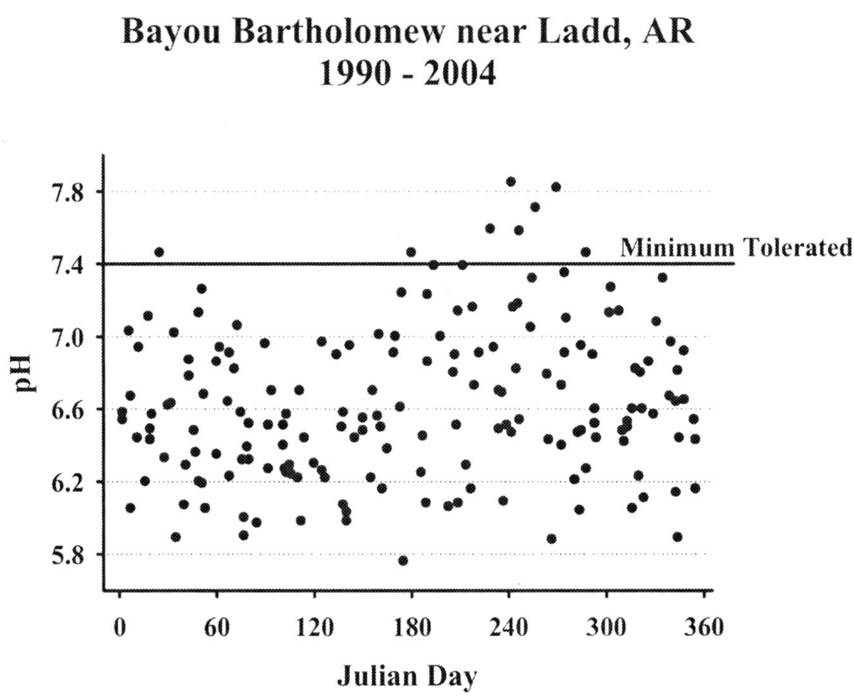

Figure 4: Example of compiled water quality data at a single site in Bayou Bartholomew compared to the minimal requirements of zebra mussel larvae.

support zebra mussels. Zebra mussels are particularly susceptible to acidic or even slightly alkaline water. During their larval stage, when their calcareous shell is first being formed, they require a pH of 7.4 or greater to survive [33]. This condition is met only in the Arkansas River. The risk of successful infestation of Bayou Bartholomew is low – only sporadically does pH equal or exceed a value of 7.4. Conditions are even more limiting in the slightly more acidic Ouachita River.

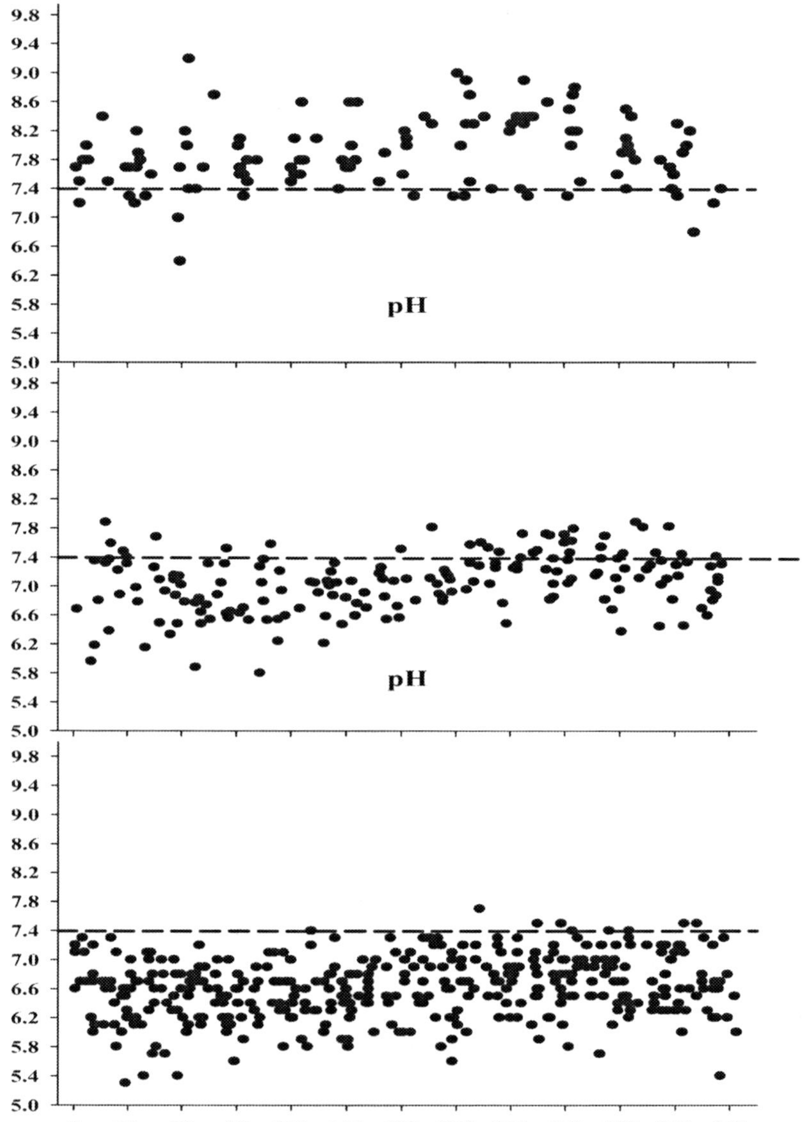

Figure 5: Summary of pH values characterizing the Arkansas River (upper plot), Bayou Bartholomew (middle plot), and the Ouachita River (lower plot). The dashed line represents the limit below which zebra mussel larvae cannot survive.

2.5. LESSONS LEARNED FROM EXAMPLE #1

Analyses of habitat conditions and physiological tolerances led to three conclusions. First, the Arkansas River near upper Bayou Bartholomew (at the location of the proposed pumping station) provided good zebra mussel habitat with respect to pH, calcium, and dissolved oxygen. However, high summer water temperatures were such that only in cool years can zebra mussels flourish in the Arkansas River [2; 3; 4]. Second, Bayou Bartholomew provides poor habitat with respect to pH and calcium ([11] and references within). Both water temperature and dissolved oxygen were slightly more suitable, but pH and calcium will be limiting for the early life stages. Third, the Quachita River offers such poor habitat with respect to pH and calcium such that it is extraordinarily unlikely that zebra mussels could establish, much less thrive, in this river.

Thus, augmentation of low flow in Bayou Bartholomew to support irrigation withdrawals by pumping of Arkansas River water is unlikely to result in colonization of the Bayou by zebra mussels. Certainly conditions for larval survival will be so poor that populations of zebra mussels are highly unlikely to be sustained in Bayou Bartholomew. Conditions in the Ouachita River, into which Bayou Bartholomew drains, are even more limiting. This particular invasive species concern should not be a major factor in environmental decision making related to this project.

3. Example #2: Retrospective Analysis of an Endangered Species Translocation

3.1. BACKGROUND

During September-October, 2002 we collected and moved more than 2,000 endangered fat pocketbook pearly mussels, *Potamilus capax,* from an approximately 6-km reach of a drainage ditch in eastern Arkansas. This translocation was aimed at protecting mussels from planned maintenance dredging and was required by a Biological Opinion prepared by the U.S. Fish and Wildlife Service. The project did not proceed as planned; we removed only about 80% of the *P. capax*. Herein, we examine mistakes made, lessons learned, and discuss procedures that might have led to a more favorable outcome. We identified three key decisions should have been thoroughly discussed prior to initiating the work: percentage of mussels to be removed, choice of recipient sites, and number of mussels to be marked and measured. Two other issues were important: the status of *P. capax* in Arkansas, and the likelihood of future dredging needs at recipient sites. Initially we felt that decision-analysis tools, used during planning, would have facilitated a better understanding of complex issues. Although they would have encouraged better discussion, it is now apparent that communication was hampered largely by the different perspectives of participants.

Native freshwater mussels (Family: Unionidae) are considered by many aquatic biologists to be the most endangered organisms in North America [58]. In 1976 twenty-four species were listed as endangered; as of January 2006 sixty-two were endangered and eight were threatened [56]. Although they reach their greatest abundance (25 to more than $100/m^2$) and species richness (20 to 30) in medium-sized to large rivers [36; 42], they are also found in ponds, lakes, and sloughs [41]. They have a unique reproductive cycle in which the newly released larval stage must undergo a two- to three-week development period on the fins or gills of a fish; hence, successful recruitment depends upon specific hosts [20; 57]. They are sedentary suspension feeders, and aside for the development period, spend their lives partially buried in substratum. Although many reasons for their endangered status have been proposed and discussed [49; 20; 34; 7; 48; 25; 58; 40; 52] large-scale alteration of free-flowing rivers in the 19th and 20th centuries to accommodate navigation was a major cause [24]. Federal agencies such as the U.S. Army Corps of Engineers often relocate endangered mussels to avoid impacts [14; 17]. Projects that could require mussel translocations include bridge construction, channel realignment, dam placement, or dredging to improve navigation or water conveyance.

The fat pocketbook pearly mussel, *Potamilus capax,* was proposed for listing on 26 September 1975 (40 FR 44392-44333), and listed as Endangered on 14 June 1976 (41 FR 24062-24067). It has a smooth, shiny, thin, and extremely globose, yellow, tan, or olive-colored shell [15]. Its range since 1970 includes the St. Francis River, Arkansas; White River, Indiana; upper Mississippi River north of St. Louis; lower Wabash River, Indiana; and lower Cumberland River, Indiana [5; 55]. However, *P. capax* is most likely to be found in slack water habitats in the St. Francis drainage, Arkansas [1; 27]. Although it can occur in sandy substratum, it typically inhabits a mixture of sand, clay, silt and sticky mud [1].

Stateline Outlet Ditch originates near the Arkansas-Missouri border west of Blytheville, Arkansas. It flows south, connects to the St. Francis River and joins the Mississippi River near Mile 672, west of Tunica, Mississippi. Near the town of Marked Tree, Arkansas, the river splits into the St. Francis Floodway to the west and the lower St. Francis River to the east. The lower reach of the St. Francis River, south of Marked Tree, is isolated from the surrounding watershed by levees, the Huxtable Pumping Plant to the south, and a pair of siphons to the north (Figure 6). Siphons are primed with a mechanical pump but they contain no turbines. Fish can go downstream into the St. Francis River through the siphons but not back up against the current.

The upper one-third of the ditch was sinuous, 25 to 40 m wide, with mostly firm, silt-sand substratum. The lower two thirds was 50 to 60 m wide and straight, and substratum consisted of flocculent mud 20 to 100 cm deep which made walking extremely difficult. The surrounding area was agricultural, although a strip of land between the ditch and the levees was vegetated with herbs, vines, silver maple, and willow. We estimated total benthic surface area at 66,500 m^2 and 170,000 m^2, in the upstream and downstream reaches, respectively. During retrieval there was no measurable water flow in the ditch.

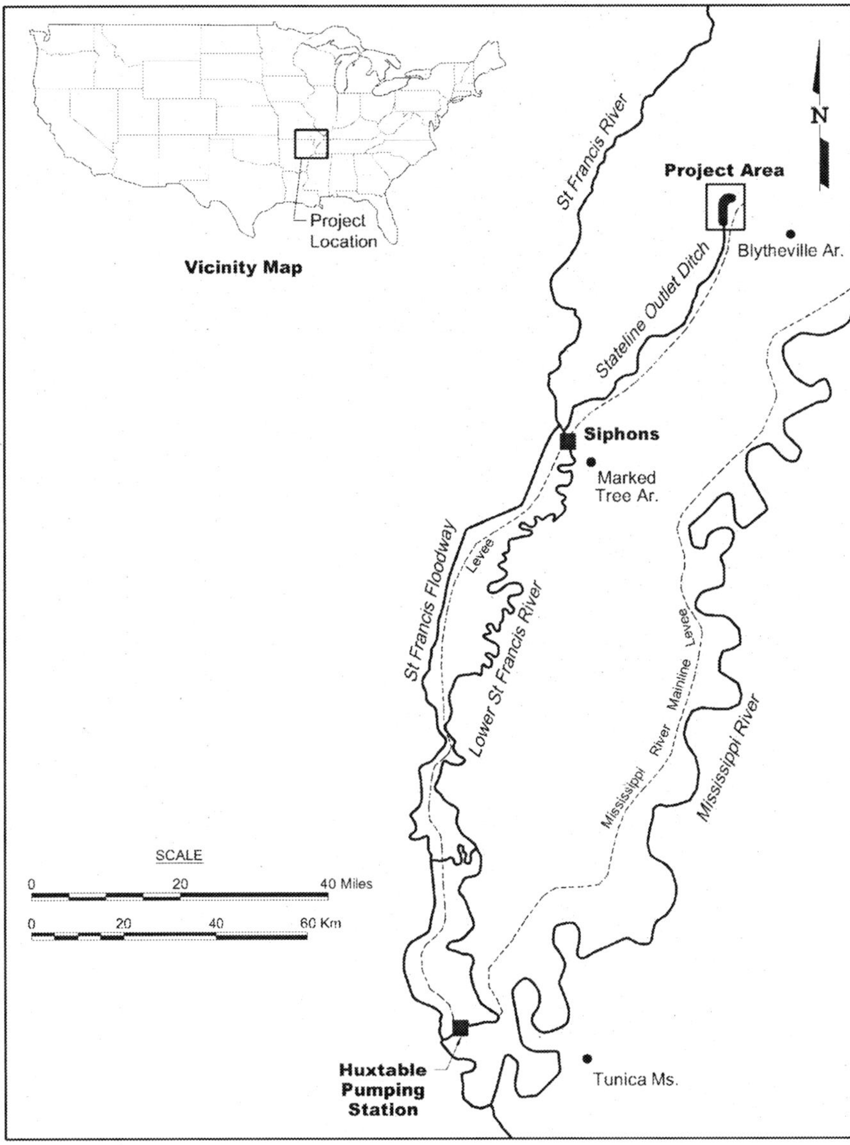

Figure 6: Location of the project area, west of Blytheville, Arkansas (left panels). Relationship of the St. Francis River, St. Francis Floodway, and levee system, located south of the project area (right panel).

3.2. IMPORTANT DECISIONS

Three important decisions affected this translocation and subsequent perceptions of its success. First was the percentage of mussels to be removed. The Biological Opinion [56] aimed at rescue of all *P. capax* in the project area, and required that all but five individuals would be moved. The Biological Opinion recognized this as potentially unrealistic, and allowed an incidental take of no more than 30 individuals that might be found in upland disposal mounds. Based on an early study [22] it was assumed that there were 3,072 *P. capax* in the project area; approximately 2,300 and 760 were in the upper and lower sections, respectively. In slight contrast to the Biological Opinion, we were asked to remove and relocate 95% (not 99%) of the mussels. Thus, although the goal of near complete removal was clear, there was ambiguity as to what could really be accomplished.

The other two important decisions concerned location of mussel recipient sites and the number that should be marked and measured. Although rationale for site selection was not discussed, presumably only those with appropriate habitat ensuring long-term survival would be suitable. The need to mark and measure mussels was related to future but unspecified growth and survival studies of translocated mussels. We were asked to mark and measure all mussels, rather than a representative subset.

There were two other important issues related to the above three questions. The first dealt with the abundance, distribution, and status of *P. capax* in Arkansas. The second concerned the likelihood of future dredging requirements at recipient sites or Stateline Outlet Ditch.

3.3. METHODS

Translocation was simple but labor intensive. Mussels were collected by wading and placed in mesh bags. They were carried to a staging area where total shell length was measured and each was engraved with an identifying number. Mussels were then packed with wet towels in coolers and transported to recipient sites and placed in substratum.

We divided the ditch into 18 reaches. Five to thirteen collectors lined up and crawled, walked, or swam, depending on water depth and the amount of mud, retrieving all live *P. capax*. Retrieval was done tactilely because of low water clarity. The area of each reach was measured and collecting time was recorded to estimate density, catch per unit effort, and depletion rate [32]. We collected mussels by hand while wading because the size of the project area (236,500 m^2) made it unreasonable to use divers equipped with scuba or surface-supplied air.

Work was not consecutive and spanned nearly two months, since collecting was restricted to low water periods. Twenty-four people participated in the 11-day project. Our inability to maintain a constant crew was partially a function of its disagreeable aspects (labor intensive, tedious, dirty, involved exposure to extremes of heat and cold, etc.). Two people left for health reasons, four commercial shell fishermen left the site with no explanation, and a commercial fisherman who had worked in other streams in this region all his life told us that this was his worst field experience.

Three relocation sites were to be used, one in a nearby ditch (# 29) and two in the St. Francis River south of Marked Tree. Ditch 29 was contiguous with the Stateline Outlet Ditch and less than 2 km away. Physical conditions in Ditch 29 (depth, water velocity, and substratum), which supported substantial numbers of *P. capax*, were virtually identical to those in Stateline Outlet Ditch. Sites on the St.

Francis River were 120 km from the project area. River flow was moderate and substratum consisted of coarse sand and silt.

3.4. LESSONS LEARNED FROM EXAMPLE #2

We had conducted a pilot test of retrieval methods in the upper reach of the ditch where an earlier study [22] indicated that most *P. capax* would be found. Results suggested that translocation was feasible and we made an estimate of the time required to remove all *P. capax*. Unfortunately, these results were misleading, since the larger and muddier lower reach was not included in the pilot study. We made a major error by not conducting more thorough test removals at various locations throughout the project area. Sufficient preliminary work should have been done to determine that the majority of the mussels were in the downstream reach and that 95% removal might not be possible. This would have provided a clearer picture of the magnitude of this translocation.

If we had examined the downstream reach in detail we would have concluded that most of the *P. capax* were located there, and they would be very difficult to remove because of deep mud. Ultimately we worked downstream reaches repeatedly without fully depleting the population. For example, in the first three passes along Reach 13 we collected 155, 39, and then 55 *P. capax*. The fact that more mussels were taken in the third versus the second pass illustrates the problem. We finally removed nearly 500 mussels from Reach 13, taking 18 on the final pass. We reworked the upper, sandier reaches three to four times and probably retrieved most *P. capax*.

We removed more than 2,000 live *P. capax* from Stateline Outlet Ditch, with the majority (78%) taken from the downstream, most-difficult-to-sample reaches. Using the depletion method of Lockwood and Schneider [32] we estimated that between 2,165 and 2,680 *P. capax* were in the project area. Therefore, we removed and relocated between 94 and 76% of the population. Translocation was stopped when it became increasingly clear that we were having difficulty removing all mussels. The following is an assessment of mistakes made and lessons learned.

We did not participate in project planning and therefore were unaware of many project details and past discussion of issues. If we had been more knowledgeable on rationale for various plans we might have been able to influence some of the decisions. A case in point is the dredging plan for Stateline Outlet Ditch, which will be discussed in more detail in the next section. We became aware of and reviewed that document after the project had been completed.

All participants (USFWS and the Memphis District) were aware of publications describing the ecology and distribution of *P. capax* [5; 10; 27; 22], the recovery plan [56], and details of previous mussel translocations [14]. Despite the fact that that everyone was aware of this literature, our restrospective analysis suggested different perspectives on four key issues. These are discussed below, with comments on how they affected decision-making and project design.

3.5. WHAT PERCENTAGE OF THE LOCAL POPULATION SHOULD BE MOVED?

In the Biological Opinion [56] it was stated that dredging would have direct and indirect effects on *P. capax*. Mussels removed by the dredge would be killed, and increased 'siltation associated with the work,' would have a deleterious effect on all others. These secondary effects would be severe enough to warrant complete (or near complete) removal.

We recently analyzed archived project specifications to determine extent of proposed dredging. Results of a hydrographic survey had been used to divide the project area into 142 sections, each 30 m (100 ft) long. Based on dredging requirements, we grouped these sections into five reaches. Half the channel in the first reach, and a 3-m strip along one bank would be affected in two downstream reaches. Two upper reaches would be completely dredged. Thus, only 50% of the project area would be affected, with less than 10% taking place in downstream reaches. The proposed action would directly affect less than 50% of the mussels, since most were in the area that was impacted least.

In the Biological Opinion [56] the recovery plan was cited [55] which stated that dredging was particularly destructive to *P. capax*. Studies by Ellis [19], Kat [28], and Brim Box and Mossa [9], were cited to bolster this statement; however, these were of mussels in general and not *P. capax* specifically. The preference of *P. capax* for sticky mud [1] suggests that this species is likely to be more tolerant of suspended sediment than species found primarily in coarse-grained substratum. Especially in reaches where only one bank was to be affected, our recommendation would have been to move mussels to the other side of the ditch. We would have suggested complete mussel removal only in two reaches that would be totally dredged.

The question of how many *P. capax* to move could have been based on genetic diversity of the population. The proportion of diversity that remains from one generation to the next is equal to $1-(1/2\ N_e)$ (N_e is the effective reproductive population) [35]. Therefore, a population of 1,000 individuals could lose 0.05%, and a population of 50 individuals could lose 1% of their diversity each year. The need to maintain genetic diversity within this population could have been addressed by estimating population size and then impacts of several removal scenarios. Since we removed approximately 2,000, the difference between total removal and 95% removal was 100 individuals, and the difference between total removal and 80% removal was 400 individuals. There must be many tens of thousands of *P. capax* in the ditches, streams, and bayous in the drainage. It is unlikely that genetic viability of *P. capax* in the drainage would be affected by these small differences.

3.6. SUITABILITY OF RELOCATION SITES

We were to use three relocation sites, one in Ditch 29, which was similar to Stateline Outlet Ditch, and two in the St. Francis River, which had silty sand substratum. We felt that *P. capax* should only be relocated in appropriate sticky mud substratum [1]. Therefore, we did not consider sites on the St. Francis River to be particularly appropriate, unless there was some overriding reason that made

Ditch 29 less desirable. Although it was not stated directly, it is likely that "range expansion" was one reason for recommending sites in St. Francis River. Although this species was already found there, moving substantial numbers to this larger area may have been perceived as having an overall benefit. Section 2.3 of the Recovery plan [55] described an attempt to augment an apparent population of *P. capax* in the Mississippi River in Missouri. Regardless of the rationale for moving *P. capax*, the prime consideration for site selection should have been chances for its long-term survival. Aside from physical conditions of habitat, the major concern should be likelihood of future developments, which would certainly require dredging. If the recipient site were dredged, *P. capax* would have to be moved a second time.

As noted above, we were unaware at the onset that the entire project area was not to be dredged. Considering the resilience of P. capax to sedimentation, we would have suggested relocating them within the project area, or slightly up or downstream of project boundaries.

These different perspectives on habitat preferences and range expansion could have been clarified during early planning with a decision tree [12] that portrayed alternatives, uncertainties, and consequences of site selection (Figure 7). Solid circles represent probability nodes which depict possible outcomes of choosing either Ditch 29 or the St. Francis River. These probabilities must add up to 100% for branches within a node and should be established beforehand. For example, if it was determined that probability of future developments requiring dredging in Ditch 29 was 80%, then the probability of not dredging must be 20%. We did not have estimates of dredging needs; however, they could have been obtained from Memphis District planners to guide site selection. Although not portrayed in Figure 7, it is likely that there would be a greater need to dredge Ditch 29 than St. Francis River. Because habitat conditions were most suitable in Ditch 29, we assess the likelihood of long-term survival in the St. Francis River as low with future developments and moderate with no future developments (Figure 7). We felt that the chances of long-time survival (future recruitment was not considered) in Ditch 29 would be moderate and high, with and without future developments.

A second consequence of site selection, range expansion, would not be accomplished (none) at Ditch 29 or in St. Francis River since *P. capax* naturally occurs there, although in low numbers. Actually, since the river is isolated from the remainder of the drainage, this alternative has no effect on overall range. Jenkinson and Ahlstedt [27] indicated that fish host for *P. capax* do not regularly go through the siphons. It is likely that the USFWS was unaware of this problem. Figure 7 illustrates how decision analysis could have been used. Portraying decisions and their consequences would have clarified issues and enhanced the likelihood of consensus, even if there was uncertainty associated with quantifying uncertainties.

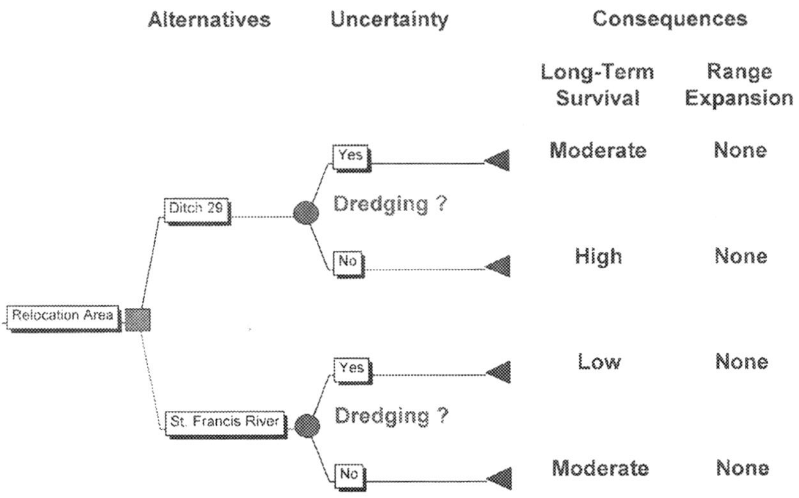

Figure 7: Decision tree illustrating alternatives and consequences of two translocation options.

3.7. MEASURING AND MARKING MUSSELS

Although not essential, most mussel translocations have a secondary goal of obtaining growth and survival data. This could have value for future projects, although in practice results can be confounded by predation, natural emigration, unexplainable mortality, and difficulty in finding translocated organisms [14]. It might seem logical to mark and measure all mussels since they had to be collected and transported anyway. However, potential logistical problems associated with holding and processing 2,000 organisms that each can weigh 300-350 grams are not trivial, especially when they are endangered, must be kept moist, and have to be carried through deep mud. Likewise there was no reason to process the entire collection; a subset of 100-200 should represent all size classes and be sufficient to estimate mortality. Finally, a sample obtained by hand-searching mud overlain by turbid water will be biased toward large organisms and length frequency histograms would underestimate recent recruitment. Unbiased samples for demographic analysis are best obtained by collecting and sieving sediments, which was done previously by Harris [22].

Using a decision tree, we judged the first three consequences of the chosen treatment scenario to vary from moderate (measure all) to low (measure a subset) to none (measure none) (Figure 8). We judged the value of measuring all or a subset as moderate, since sufficient mussels could be easily collected to obtain these data as part of another study. Regardless, it is unclear how resulting demographic or survival data would substantially contribute to the long-term success of this species. Figure 8 applies to *P. capax* and probably to most endangered species. The question for managers is simple; do the increased chances of mortality justify the need to collect such data?

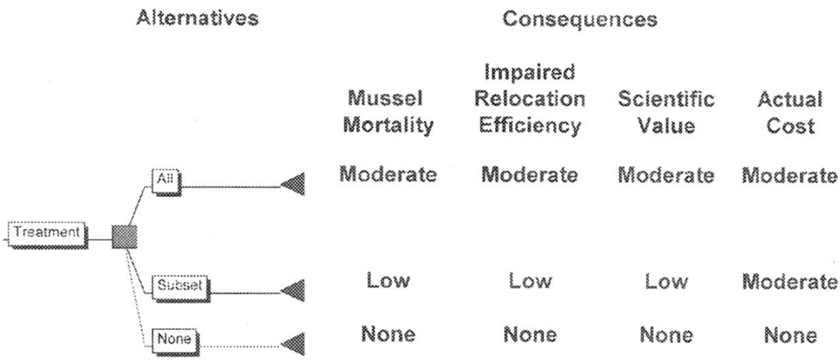

Figure 8: Decision tree illustrating alternatives and consequences of three marking and measuring options.

3.8. SHOULD *P. CAPAX* BE ENDANGERED?

Bates and Dennis [5] conducted a survey in the St. Francis drainage and reported that this species was rare and would likely soon become extinct. However, in a later, more thorough study in the same area, Clarke [10] found many hundreds and concluded that the species was not rare or spatially restricted, but common. Findings by Clarke were substantiated by subsequent investigators [1; 27]. In a review of mussels in Arkansas, Harris et al. [23] recommended that the *P. capax* be down-listed to threatened. We found several thousand *P. capax* in a 5.7 km reach of Stateline Outlet Ditch; it ranked second in abundance of 19 unionid species and comprised 20% of the fauna. The rationale for listing this species now seems questionable [37].

3.9. CONCLUSIONS FROM EXAMPLE #2

Procedural difficulties of mussel translocations to avoid impacts have been discussed by Cope and Waller [14], Losos et al. [31], Griffith et al. [21], and Parmalee and Bogan [41]. In their review of 37 projects, Cope and Waller [14] estimated that approximately 22,000 mussels died following translocations, which was likely an underestimate since mortality was reported in only 68% of the cases. Some investigators reported mortality as high as 90%. These findings alone, even if the experiences of Stateline Ditch are ignored, suggest that better assessment of risks associated with moving mussels is needed.

This retrospective analysis caused us to examine logistic problems that hindered our ability to meet project objectives. The simple decision trees developed in this retrospective analysis demonstrate how formal decision tools would have produced a better design by more clearly portraying consequences and choices. This would have focused attention on key issues, fostered better communication, and led to a common understanding of key issues. The concept of interagency cooperation through consultation is not a recent development, and was well-defined in the early 1970s [6]. It is not unreasonable to involve participants in a comprehensive decision process at the onset of an important project [8].

Difficulty in framing or making decisions was not due to incomplete information; everything needed to plan the project and assess the status of *P. capax* was available. Different perspectives obstructed meaningful communication and led to a plan that lacked foresight and was prone to errors. A major goal of the ESA is to protect ecosystems, not organisms [16; 18]. Our experiences in Stateline Outlet Ditch led us to conclude that this translocation protected neither very well. The culprit was different perspectives--which led to different conclusions--one of several reasons why decisions are hard and decision tools are useful [12].

4. Concluding Remarks

Invasive and endangered species issues, albeit central to many ecosystem management programs, are only two examples of non-toxicological problems to which risk analysis and decision methods can be applied. The two examples presented herein indicate the potential for improving environmental decision-making in the face of uncertainty - but in the presence of substantial information. As more rigorous attempts are made to widen and improve applications of risk and decision methods to environmental decision-making, the following aspects of ecosystem management are likely to be improved:

- *More systematic structuring of the decision process.* If decision makers are helped to think systematically about a problem and provided a logical framework for defining choices, better decisions are likely to result.
- *Straightforward portrayal of the consequences of decision option.* Managers and stakeholders will be better able to reach a common understanding of the relative advantages and disadvantages of management options.
- *More consistent and rational evaluations of risks and uncertainties.* This is important because behavioral and social sciences research suggest that people are inconsistent and challenged in make decisions involving risk and uncertainty – paralysis of a decision process typically will not promote better environmental management.
- *Better documentation of how a decision was reached and improved conflict resolution.* Common understanding of a complex issue and clarity of communication are made more possible as systematic approaches are taken to risk analysis and decision processes. Choices will be more defensible to various stakeholders and other interested parties.

5. Acknowledgments

These studies were funded by the U.S. Army Engineer Districts, Memphis, Tennessee and Vicksburg, Mississippi as well as the Systemwide Water Resources Program and the Aquatic Nuisance Species Research Programs of the U.S. Army Corps of Engineers Engineer Research and Development Center. The Chief of Engineers granted permission to publish this information.

6. References

1. Ahlstedt, S. A., and J. J. Jenkinson. 1991. Distribution and abundance of *Potamilus capax* and other freshwater mussels in the St. Francis River System, Arkansas and Missouri, U.S.A. *Walkerana* 5(14):225-261.

2. Aldridge, D.W., B.S. Payne, and A.C. Miller. 1995. Oxygen consumption, nitrogenous excretion, and filtration rates of *Dreissena polymorpha* at acclimation temperatures between 20 and 32°C. *Canadian Journal of Fisheries and Aquatic Sciences*, 52:1761-1768.

3. Allen, Y.C., B.A. Thompson, and C.W. Ramcharan. 1999. Growth and mortality rates of the zebra mussel, *Dreissena polymorpha*, in the lower Mississippi River. Canadian Journal of Fisheries and Aquatic Sciences, 56(5): 748-759.

4. Baker, W.P., J.N. Stoeckel, and C.J. Gagen. 2000. Heat-related zebra mussel mortality in Lake Dardanelle, Arkansas. Joint Meeting of the Arkansas, Louisiana, and Texas Chapters of the American Fisheries Society, Bossier City, Louisiana, 2000.

5. Bates, J. M, and S. D. Dennis. 1983. *Mussel (Naiad) survey – St. Francis, White, and Cache rivers, Arkansas and Missouri. Final Report.* Ecological Consultants, Inc., Ann Arbor, MI. U.S. Army Engineer District, Memphis, Memphis, TN.

6. Bean, M. 1983. *The Evolution of National Wildlife Law: Revised Edition.* Environmental Defense Fund, Prager Publications, NY.

7. Bogan, A. E. 1993. Freshwater bivalve extinction (Mollusca: Unionidae): a search for causes. *American Zoologist* 33:599-609.

8. Borsuk, M., and R. Clemen, L. Maguire, and K. Reckhow. 2001. Stakeholder values and scientific modeling in the Neuse River watershed. *Group Decisions and Negotiation* 10:355-373.

9. Brim Box, J. and J. Mossa. 1999. Sediment, land use, and freshwater mussels: prospects and problems. *Journal of the North American Benthological Society* 18:99-117.

10. Clarke, A.H. 1985. *Mussel (Naiad) study; St. Francis and White Rivers; Cross, St. Francis, and Monroe Counties, Arkansas.* U. S. Army Engineer District, Memphis, Memphis, TN.

11. Claudi, R. and G.L. Mackie. 1994. Practical *Manual for Zebra Mussel Monitoring and Control.* Lewis Publishers, Boca Raton, FL.

12. Clemen, R.T., and T. Reilly. 2001. *Making Hard Decisions with Decision Tools.* Duxbury Thomas Learning, Pacific Grove, CA.

13. Coblentz, B.E. 1990. Exotic organisms: a dilemna for conservation biology. *Conservation Biology* 4:261-265.

14. Cope,W.G., and D. L. Waller. 1995. Evaluation of freshwater mussel relocation as a conservation and management strategy. *Regulated Rivers: Research and Management* 11(2):147-155.

15. Cummings, K. S., and C.A. Mayer. 1992. *Field Guide to Freshwater Mussels of the Midwest.* Illinois Natural History Survey, Manual 5. Champaign, IL.

16. Doremus, H. 1991. Patching the Ark: Improving legal protection of biological diversity. *Ecology Law Quarterly* 18:265-333.

17. Dunn, D. L., and B. E. Sietman. 1997. Guidelines used in four geographically diverse unionid relocations. in *Conservation and management of freshwater mussels II: initiatives for the future, Proceedings of a UMRCC symposium, 16-18 October 1995, St. Louis, Missouri,* K. S. Cummings, A. C. Buchanan, C. A. Mayer, and T. J. Naimo, eds. Upper Mississippi River Conservation Committee, Rock Island, IL, 176-183.

18. Eisner, T., and J. Lubchenco, E. O. Wilson, D.S. Wilcove, and M. J. Bean. 1995. Building a scientifically sound policy for protecting endangered species. *Science* 268:1231-1232.

19. Ellis, M.M. 1936. Erosion silt as a factor in aquatic environments. *Ecology* 17:29-42.

20. Fuller, S. L. H. 1974. Clams and Mussels (Mollusca: Bivalvia). in *Pollution Ecology of Freshwater Invertebrates.* C. W. Hart Jr. and Samuel L. H. Fuller, eds., Academic Press, New York, NY.

21. Griffith, B., and J. M. Scott, J. W. Carpenter, and C. Reed. 1989. Translocation as a Species Conservation Tool: Status and Strategy. *Science* 245:477-480.

22. Harris, J. L. 2001. *Freshwater mussel survey of State Line Outlet Ditch, St. Francis River Basin, Mississippi County, Arkansas, with Population Estimate for Potamilus capax.* Report for the U.S. Army Engineer District, Memphis, Memphis, TN.

23. Harris, J. L., Rust, P. J., Christian, A. C., Posey, W. R., II., Davidson, C. L., and G. L. Harp. 1997. Revised status of rare and endangered unionacea (Mollusca: Margaritiferidae, Unionidae) in Arkansas. *Journal of the Arkansas Academy of Sciences* 51:66-89.

24. Havlik, M. E., and J. Sauer. 2000. *Native freshwater mussels in the upper Mississippi River system.* Upper Midwest Environmental Sciences Center Project Status Report 2000-04, La Crosse, WI.

25. Hayes, T. 1998. Conservation of native freshwater mussels: An overview. *Endangered Species Update* 15:108-110.

26. Hayward, D. and E.D. Estevez. 1997. *Suitability of Florida waters to invasion by the zebra mussel, Dreissena polymorpha.* Mote Marine Laboratory Technical Report No. 495.

27. Jenkinson, J. J., and S. A. Ahlstedt. 1993. A search for additional populations of *Potamilus capax* in the St. Francis and Cache River Watersheds, Arkansas and Missouri, USA. *Walkerana* 7(17/18):71-157.

28. Kat, P. W. 1982. Effects of population density and substratum type on growth and migration of Elliptio complanata (Bivalvia: Unionidae). *Malacological Review* 15:119-127.

29. Kiker, G.A., T.S. Bridges, A. Varghese, T.P. Seager, and I. Linkov. 2005. Application of multi-criteria decision analysis in environmental decision making. *Integrated Environmental Assessment and Management* 1: 95-108.

30. Lodge, D.M. 1993. Biological invasions: lessons for ecology. *Trends in Ecology and Evolution* 8:133-137.

31. Losos, E., J. Hayes, A. Phillips, D. Wilcove, and C. Alkire. 1995. Taxpayer-subsidized resource extraction harms species. *BioScience* 45(7):446-455.

32. Lockwood, R. N. and J. C. Schneider. 2000. Stream fish population estimates by mark and recapture and depletion methods. in *Manual of fisheries survey methods II: With periodic updates*. J. C. Schneider, ed. Michigan Department of Natural Resources, Fisheries Special Report 25, Ann Arbor, MI.

33. Mackie, G.L. and B.W. Kilgour. 1993. *Efficacy and Role of Alum in Removal of Zebra Mussel Veliger Larvae from Raw Water Supplies*. Third International Zebra Mussel Conference, Agenda and Abstracts, Toronto, Canada, February 1993.

34. Master, L. 1990. The imperiled status of North American aquatic animals. *Biodiversity Network News* 3:1-2, 7-8.

35. Meffe, G. K., and C. R. Carroll. 1994. *Principles of Conservation Biology*. Sinauer Associates, Inc., Sunderland, MA.

36. Miller, A. C. and B. S. Payne. 1998. Effects of disturbance on freshwater bivalve assemblages. *Regulated Rivers: Research and Management* 14:179-190.

37. Miller, A. C., and B. S. Payne. 2005. The curious care of the fat pocketbook mussel, *Potamilus capax*. *Endangered Species Update* 22(2):61-70.

38. Miller, A.C. and B.S. Payne. 2006. A retrospective analysis of a large-scale endangered species translocation project. *Environmental Practice* (in press).

39. National Academy of Sciences. 1983. *Risk Assessment in the Federal Government: Managing the Process*. National Research Council. National Academy Press. Washington, DC.

40. Neves, R. J. 1999. Conservation and commerce: Management of freshwater mussel (Bivalvia: Unionidae) resources in the United States. *Malacologia* 41(2):461-474.

41. Parmalee, P. W., and A. E. Bogan. 1998. *The Freshwater Mussels of Tennessee*. The University of Tennessee Press, Knoxville, TN.

42. Payne, B. S., and A. C. Miller. 2000. Recruitment of Fusconaia ebena (Bivalvia: Unionidae) in relation to discharge of the lower Ohio River. *American Midland Naturalist* 144:328-341.

43. Payne, B.S. and Miller, A. (2004). "A Probability Tree Applied to a Common Zebra Mussel Dispersal Issue," ERDC/TN ANSRP-04-1, U.S. Army Engineer Research and Development Center, Vicksburg, MS.

44. Peterson, A.T. and D.A. Vieglais. 2001. Predicting species invasions using ecological niche modeling: new approaches from bioinformatics attach a pressing problem. *BioScience* 51:363-371.

45. Pimentel, D., L. Lach, R. Zuniga, and D. Morrison. 2000. Environmental and economic costs of nonindigenous species in the United States. *BioScience* 50(1): 53-64.

46. Ramcharan, C.W., D.K. Padilla, and S.I. Dodsen. 1992. A multivariate model for predicting population fluctuations of *Dreissena polymorpha* in North American lakes. Canadian Journal of Fisheries and Aquatic Sciences 49(1):150-158.

47. Schloesser, D.W. and T.F. Nalepa, eds. 1993. *Zebra Mussels: Biology, Impacts, and Control.* Lewis Publishers, Boca Raton, FL.

48. Seddon, M. B., and I. J. Killeen, P. Bouchet, and A. E. Boga. 1998. Developing a strategy for molluscan conservation in the next century. *Journal of Conchology* 2:295-298.

49. Stansbery, D. H. 1971. *Rare and endangered mollusks in eastern United States.* In *Proceedings of a Symposium on rare and Endangered Mollusks (Naiads) of the U.S.,* S. E. Jorgenson, and R. E. Sharp, Eds., Bureau of Sport Fishes and Wildlife, U.S. Fish and Wildlife Service, U.S. Department of the Interior, Washington D.C.

50. Stohlgren, T.J. and J.L. Schanse. 2006. Risk analysis for biological hazards: what we need to know about invasive species. *Risk Analysis* 26(1): 163-173.

51. Strayer, D.L. 1991. Projected distribution of the zebra mussel, Dreissena polymorpha, in North America. Canadian Journal of Fisheries and Aquatic Sciences 48(8):1389-1395.

52. Strayer, D. L., J. A. Downing, W. R. Haag, T. L. King, J. B. Layzer, T. J. Newton, and S. J. Nichols. 2004. Changing perspectives on pearly mussels, North America's most imperiled animals. *BioScience* 54(5):429-439.

53. Suedel, B., T. Bridges, J. Kim, B. Payne, and A. Miller. 2006. Application of risk assessment and decision analysis to aquatic nuisance species. *Integrated Environmental Assessment and Management* (in press).

54. U.S. Environmental Protection Agency. 2004. An Examination of EPA Risk Assessment Principles and Practices. EPA/100/B-04/001. Office of the Science Advisor, U.S. Environmental Protection Agency, Washington, DC.

55. U.S. Fish and Wildlife Service. 1989. A Recovery Plan for the Fat Pocketbook Pearly Mussel *Potamilus capax* (Green 1832). U. S. Fish and Wildlife Service. Atlanta, GA.

56. U.S. Fish and Wildlife Service. 2001. Biological Opinion on effects of maintenance dredging in Stateline Outlet Ditch, Arkansas. U.S. Fish and Wildlife Service, Conway, AR.

57. Watters, G. T. 1994. *An annotated bibliography of the reproduction and propagation of the Unionidae (Primarily of North America).* Ohio Biological Survey, College of Biological Sciences, the Ohio State University, Columbus, OH.

58. Williams, J. D., M. L. Warren, Jr., K. S. Cummins, J. L. Harris, and R. J. Neves. 1993. Conservation status of freshwater mussels of the United States and Canada. *Fisheries* 18(9), 6-22.

149–160

STAKEHOLDER PREFERENCE ELICITATION

D11

R.M. COOKE
Delft Institute for Applied Mathematics
Delft University of Technology, Resources for the Future

Abstract

Arguments why stakeholder preferences cannot be modeled as utilities, multicriteria or otherwise, are reviewed. An approach to stakeholder preferences based on well known models for consumer preference in market research is proposed. Simple paired comparisons is used to represent group preferences on an affine unique scale, and regression is used to "explain" these preferences in terms of scores on a number of criteria. Using the rich body of standard regression techniques, we can analyse degree of fit, and we can deal with dependence in the "criteria". The tasks in stakeholder preference modeling can be apportioned between analysts, experts and stakeholders.

1. Introduction

Multi criteria methods are emerging in the area of risk analysis and risk management [10]. It is appropriate to recall the classical arguments why stakeholder preferences cannot be modeled as utilities, multicriteria or otherwise, in the sense of rational decision theory. An approach to stakeholder preferences based on well known models for consumer preference in market research is proposed (for a review see [5]). Simple paired comparisons is used to represent group preferences on an affine unique scale, and regression is used to "explain" these preferences in terms of scores on a number of criteria. Similar approaches to modeling valuations can be found in [12, 9, 8, 15, 11]. These approaches often use logit regression to model valuation of health states which can then be compared with "Standard Gamble" trade-off elicitation common in multi attribute utility theory [11, 19] Using the rich body of standard regression techniques, we can analyse degree of fit, and we can deal with dependence in the "criteria". The tasks in stakeholder preference modeling can be apportioned between analysts, experts and stakeholders.

I. Linkov et al. (eds.), Environmental Security in Harbors and Coastal Areas, 149–160.

2. Classical Utility Theory

Classical utility theory designates a body of techniques for evaluating outcomes that derive from the individual rational agent paradigm. The most well known of these are:

- Multi-attribute utility theory (MAUT)
- Multi-criteria decision Making (MCDM)
- Analytic Hierarchy Process (AHP)

MAUT tries to describe utility functions when the outcomes are multi dimensional (or multi attribute). *If* the preferences satisfy certain stringent independence axioms, *then* the utility on outcomes can be represented as a weighted sum of the utilities on each dimension, where the weights assigned to a dimension reflect the value of that dimension. Under these special circumstances, we can think of preferences of outcomes as

Preference (outcome) = $\sum_i$ Preference for attribute$_i$ × score of outcome on attribute$_i$.

The MAUT axioms state, roughly, that the preference for attribute$_i$ does not depend on the scores on the attributes.

MCDM is a simplification of MAUT where the independence axioms are not rigorously checked, but shorthand techniques are used to develop preferences for criteria. AHP involves cascaded sets of criteria. It is not consistent with the decision theory of a rational agent *unless* a common baseline utility (corresponding to the utility value zero) is agreed beforehand (see below).

Extending classical utility theory so as to be useful in social decision making has proven difficult, [7, 16, 1, 2], and well-meaning practitioners re-commit the same mistakes year after year, to wit:

1. ASSUMING that a group can be treated as a rational individual. The following (Condorcet voting paradox) shows why this is NOT true. Consider a population with preferences (X > Y means X is preferred to Y):
    - 1/3 of population: Beethoven > Bach > Mozart
    - 1/3 of population: Mozart > Beethoven > Bach
    - 1/3 of population: Bach > Mozart > Beethoven

    Then there is a 2/3 majority for each of the following pairwise preferences:
    - Beethoven> Bach
    - Bach > Mozart
    - Mozart > Beethoven

This is evidently intransitive. An individual with intransitive preferences could never choose, but would continually cycle through the alternatives. It has long been recognized that groups choosing by a majority rule will not in general have transitive preferences [14]. Arrow and Raynaud [2] summarize attempts to introduce axioms ensuring the transitivity of majority preference.

2. ASSUMING without verification that preference can be expressed in terms of preferences for criteria (also known as coordinates, attributes, dimensions) and scores on the criteria, as in MAUT. The following preference pattern is eminently reasonable, yet inconsistent with the MAUT axioms:

   - If unemployment is low and pollution is high, Prefer: *Close a dirty factory,* to *Keep dirty factory open*
   - If unemployment is high and pollution is low, Prefer : *Keep the dirty factory open* to *Close the dirty factory.*

   MAUT assumes that the trade-off between criteria is constant across the decision space. Preferences for criteria are measured by observing choice behavior as criteria are traded-off. Thus, subjects are asked, *If a policy A raised pollution by X but decreased unemployment by Y relative to policy B, would you choose A or B?* The above example shows that the rate at which a subject trades off may depend on values of pollution and unemployment for policy B.

3. ASSUMING without verification that the criteria scores are independent in the population of alternatives being compared. If two criteria, say CO_2 emissions and SO_2 emissions tend to favor and disfavor the same alternatives, then MCDM and MAUT would tend to introduce double counting. The extreme case of this arises if we have two cost criteria, cost in US dollars and cost in euros. Both are important but using both would clearly be double counting. A general strategy for gaming an MCDM exercise is to inject many criteria that favor your preferred alternatives (and which are therefore correlated). MAUT and MCDM have no prophylactic against correlated criteria and double counting.

4. ASSUMING that preference rations are meaningful at an individual or group level, without giving an operational definition. The presence or absence of operational definitions is illustrated as follows:

   - John prefers *Close dirty factory* to *Keep dirty factory open* $\Rightarrow$ OBSERVABLE BEHAVIOR: Vote to close
   - John's preference ratio *Close dirty factory / NOT close dirty factory = 9* $\Rightarrow$ OBSERVABLE BEHAVIOR?????

Asking for preference ratios is just like asking for temperature ratios. Ask a friend "how much hotter is boiling water than freezing water: (i) about the same (ii) hotter (iii) much hotter, or (iv) very much hotter?" Now interpret the answers as temperature ratios. Does that make sense? On the Fahrenheit scale the ratio is 212/32 = 6.625. On the Kelvin scale it is 373.15 / 273.15 = 1.366. On the centigrade scale this ratio is 100 / 0 = ∞! Yet these three scales all measure the same thing, namely temperature. Utilities, like temperatures, are measurable on an affine or interval scale. If we fix a baseline zero value, then we can compare rations of utility *differences* (of course this ratio could also be negative!). Applications of the Analytic Hierarchy Process (AHP) require participants to state their degrees of preference without establishing a baseline of zero utility. These degrees of preference are interpreted as preference ratios that have no operational meaning. Given the total lack of operational meaning for "preference degrees" the wide currency of the AHP method is cause for sober reflection, and underscores the difficulty of bringing rational methods to bear on real problems.

3. Stakeholder Preferences

Group preferences do not satisfy the axioms that enable utility theory. On the other hand, there are techniques from consumer research which yield consumer preference functions. We shall speak of *stakeholder preferences* to designate the values of a group or collective body charged with taking and/or implementing decisions. We will model these preferences with techniques from consumer preference theory.

Classical approaches [17, 4, 13, 3, 6, 18, 12] elicit pairwise preferences from experimental subjects, and "scale" the data so as to yield a preference scale which can be related to market share. Of course, "market share" is just a convenient metaphor; "group preference" or "stakeholder preference" is a better designation for present purposes. Similar approaches have been used to evaluate different health outcomes for decisions regarding treatment selection in pre-paid medical plans, and recently, in valuing health outcomes [8, 19, 15, 11].

Using standard techniques, we can analyse the degree of agreement among the stakeholders, and the degree to which the modeling hypotheses for deriving scale values fit the data. Once the scaled values have passed these statistical tests, we can use standard regression modeling to "explain" the preferences by regressing the scores on explanatory variables. A rich body of standard techniques is at our disposal to

- Analyse how well a regression model explains the scaled values.
- Derive confidence bounds for the weights of the explanatory variables.
- Analyse the correlation in the weights' sampling distributions.

For the current problems, the Thurstone pair wise comparison methods are most promising of the standard approaches. The simplest version, which will be illustrated here, makes the following assumption: after choosing the same zero and unit values on their respective utility scales, each stakeholder has a value for each alternative, and the values are independently normally distributed in the population of stakeholders with constant fixed variance. The placement of the means is determined only up to a choice of zero and choice of unit; that is these values are determined up to a positive affine transformation. This is desirable for two reasons:

- This is the same invariance structure of individual utility functions.

- This is very convenient if the scale values are to be explained by some other mode, for example a regression model (see below).

Other types of paired comparison models are discussed in section 7.

4. Implementation

The implementation involves three groups of players: an analysis team, an expert team and a set of stakeholders. Their roles are described briefly below.

4.1. ANALYSIS TEAM

The analysis team defines a set of alternatives (e.g., policies for costal defense) and a set of criteria (e.g., cost, expected number of fatalities, breach frequency, expected environmental impact, etc). A set of experts is also defined. The analysis team monitors and manages the entire process, and performs the mathematical analysis.

4.2. EXPERT TEAM

The expert team scores each alternative with respect to each of the criteria. The experts may also provide feedback on the sets of alternatives and criteria, possibly iterating the definitions.

4.3. STAKEHOLDERS

The set of consumers are given the alternatives with their scores on the criteria. Each consumer expresses his/her pair wise preference for each pair of alternatives.

5. Analysis

The analysis team analyses the consumer preferences for consistency and significance, according to standard methods. This results in an affine unique consumer preference function that assigns a preference to each alternative. The preference values are regressed on the set of criteria. This yields a coefficient B_i for criteria C_i, which optimally express the preference for each alternative as a linear function of the criteria scores. For alternative a and criteria $C_1 \ldots C_n$, we thus have:

$$PREF(a) = B_1 \times C_i(a) + \ldots B_n \times C_n(a) + \text{error}. \tag{1}$$

$C_i(a)$ is the score of alternative a on criteria C_i. Unlike multicriteria analysis, the Bs need not be positive and need not add to one; thus, they cannot be interpreted as "weights" for the criteria. However, they best explain the preference values in a lease squared sense. Standard tools are available to analyse the error and assess the degree to which the criteria scores explain the consumer preferences.

If the preferences are adequately explained by the model, the results are communicated to the problem owner. New alternatives can be evaluated using the regression model without iterating the consumer preference elicitation.

If the fit is not satisfactory, new criteria can be formulated and the regression step can be iterated. The regression model can also be extended to include interaction terms. This requires iterating the expert scoring and regression analysis, but does not require iterating the consumer elicitation. More details are illustrated in the following toy example.

6. Toy Example

We consider a toy example for modeling group preferences for autos. We want to model stakeholder preferences for autos and use this model to predict future preference behavior and drive design improvements.

The analysis team selects five autos (policies) which cover the field parsimoniously, namely, FOCUS, ASTRA, ROLLS, BMW, KA, XSRA. An expert team scores each auto on the criteria: PRICE, MONTHLY PAYMENTS, MILAGE, PASSENGER ROOM. Notice that a criteria like ROOM cannot be monotonic in value, whatever that may mean. A stakeholder will have an ideal size, and deviations above or below will be less desirable. Notice also that the criteria scores will evidently be strongly correlated.

These scores are passed to ten stakeholders, who evaluate the policies pairwise. Suppose the following preference matrix (Figure 1) emerges (the first cell entry 6.0/10 means that 6 of the 10 stakeholders preferred the FOCUS to the ASTRA).

Item	ASTRA	ROLLS	BMW	KA	XSRA
1 FOCUS	6.0/10	8.0/10	8.0/10	6.0/10	7.0/10
2 ASTRA		8.0/10	6.0/10	6.0/10	7.0/10
3 ROLLS			2.0/10	2.0/10	2.0/10
4 BMW				1.0/10	4.0/10
5 KA					6.0/10

Figure 1: Preference matrix.

The stakeholder preference summary matrix (Figure 2) shows the number of times that each stakeholder preferred each auto to some other auto. The rightmost column shows the number of circular triads in each stakeholder's paired comparisons. With 3 or more circular triads, the null hypothesis that the stakeholder in question has given his preferences at random CANNOT be rejected. The analysis team might decide to re-elicit stakeholders Rock, Cliff, Ridge and Ruby.

Stakeholder	FOCUS	ASTRA	ROLLS	BMW	KA	XSRA	Triads
Rock	4.0	4.0	1.0	1.0	2.0	3.0	4
Cliff	3.0	2.0	1.0	4.0	4.0	1.0	4
Ridge	4.0	2.0	0.0	2.0	3.0	4.0	3
Crystal	4.0	5.0	1.0	1.0	2.0	2.0	2
Jade	4.0	1.0	1.0	2.0	5.0	2.0	2
Pebble	2.0	3.0	5.0	4.0	1.0	0.0	0
Shale	3.0	2.0	0.0	1.0	5.0	4.0	0
Flint	4.0	4.0	0.0	1.0	2.0	4.0	1
Opal	4.0	4.0	0.0	1.0	4.0	2.0	1
Ruby	3.0	4.0	1.0	2.0	3.0	2.0	6
Total	35.0	31.0	10.0	19.0	31.0	24.0	

Figure 2: Stakeholder preference summary matrix.

Preference values are shown below (Figure 3). Three common models for analysing paired comparison data are shown. Thurstone C is the model used here. (For a discussion see next section.) The others are shown for the sake of comparison. For both Thurstone models a Chi square statistic tests the hypothesis that the model assumptions hold. The value 7.0349 is far from significant, thus the data do not lead to rejecting the model (the NEL [4] model would also be unrejected at 4.9667).

Item name	NEL(Bradley-Terry)	Thurstone C	Thurstone B
1. FOCUS	0.2998	0.4525	0.8546
2. ASTRA	0.2193	0.2699	0.5096
3. ROLLS	0.0416	-0.7015	-1.3273
4. BMW	0.0896	-0.2981	-0.5278
5. KA	0.2193	0.3117	0.5570
6. XSRA	0.1304	-0.0345	-0.0663
Goodness of fit :	4.9667	7.0349	
(Chi-square, DF=	10	10	

Figure 3: Preference values.

It is convenient to put the preferences from the Thurstone C model and the criteria scores in one matrix. We also add a column of ones (Figure 4). The effect of adding this column is to enable the criteria scores to reflect deviations from the mean, when the criteria scores are standardized.

	CRITERIA SCORES				
PREF	CONST	PRICE	MONTHLY	MILAGE	ROOM
0.4525	1.0000	20.0000	0.2000	14.0000	4.0000
0.2699	1.0000	25.0000	0.1800	12.0000	6.0000
-0.7015	1.0000	40.0000	0.5000	9.0000	8.0000
-0.2981	1.0000	45.0000	0.4800	8.0000	7.0000
0.3117	1.0000	12.0000	0.1200	20.0000	3.0000
-0.0345	1.0000	15.0000	0.1500	16.0000	4.0000

Figure 4: Criteria scores.

The regression analysis finds criteria scores which yield the best linear fit to the preferences, in the sense of least squares. These scores and the resulting error are shown below (Figure 5):

```
PREF = CRITERIA SCORES × REG.COEFF'S + ERROR

= CRITERIA SCORES × 1.6908  + ERROR
                    0.0409
                   -4.0369
                   -0.0403
                   -0.2126

ERROR =        0.1654
               0.0423
               0.0539
              -0.0808
               0.0577
              -0.2365
```

Figure 5: Regression analysis criteria scores.

We see that the fit is not spectacularly good. Of course the number of alternatives is small for the number of regression coefficients to be estimated. The point is to illustrate the analytical tools which can be applied. We can analyse the covariance matrix of the regression coefficients and derive confidence bounds for the coefficient estimates. The diagonal terms are the variances of the criteria, the off-diagonal terms the covariances. We see that the standard deviation of the criterion PRICE is $\sqrt{(0.0009)} = 0.03$. This large deviation reflects of course the small number of alternatives evaluated in this toy example. In this case, the problem owner would be told that the model of stakeholder preferences was not very good; we could not claim that the coefficient for price was significantly different from zero (Figure 6).

	CONST	PRICE	MONTHLY	MILEAGE	ROOM
C	1.8396	-0.0270	1.1116	-0.0758	-0.0812
Pr	-0.0270	0.0009	-0.0389	0.0012	-0.0001
Mnth	1.1116	-0.0389	2.4437	-0.0435	-0.0347
Mi	-0.0758	0.0012	-0.0435	0.0032	0.0029
RM	-0.0812	-0.0001	-0.0347	0.0029	0.0103

Figure 6: Reg coeff covariance matrix.

The off-diagonal terms indicate that the fluctuations in the criteria values are not independent (Figure 7).

REGRESSION COEFFICIENT PRODUCT MOMENT CORRELATION				
CONST	PRICE	MONTHLY	MILEAGE	ROOM
1.0000	-0.6749	0.5243	-0.9873	-0.5894
-0.6749	1.0000	-0.8426	0.6907	-0.0257
0.5243	-0.8426	1.0000	-0.4918	-0.2185
-0.9873	0.6907	-0.4918	1.0000	0.4967
-0.5894	-0.0257	-0.2185	0.4967	1.0000

Figure 7: Criteria values are not independent.

Suppose that, given scores on the criteria, the preferences were sampled from a distribution in accord with the assumptions of the regression model. Then on repeated samples, the criteria coefficients found by the least squares algorithm would fluctuate with variances on the diagonal of the covariance matrix, and the scores would be correlated as in the correlation matrix. We could obtain uncorrelated estimates of the coefficients by carefully choosing our options (cars) in such a way that the criteria scores are uncorrelated. This would be a so-called orthogonal design. Having an orthogonal design is convenient but by no means necessary.

On the Thurstone C model, each consumer's utility for the six cars is modeled as six samples from independent normal variables $X_1 \ldots X_6$ with unit standard deviation and with means given by the preference scores. The probability that X_j is most preferred is modeled as the probability that $X_j = \max(X_1 \ldots X_6)$., and this probability is the predicted market share. It can be computed by simulation, and we find (Figure 8):

CAR	PREDICTED MARKET SHARE
1. FOCUS	0.28
2. ASTRA	0.22
3. ROLLS	0.04
4. BMW	0.09
5. KA	0.23
6. XSRA	0.14

Figure 8: Thurston C model results.

Note the similarity to the Bradley-Terry scale values, which solves for market share directly from the paired comparison data (see below). We cannot extract a preference ordering from the stakeholders in this toy example, owing to intransitivities. If we could extract such an ordering, we could compare it to the predicted market shares and derive an additional check on the modeling assumptions. Pairwise comparisons are intended to deal with the volatility of consumer preference by allowing each alternative to be judged several times, in combination with all other alternatives.

7. Discussion

We briefly summarize the assumptions underlying the models Thurstone C, Thurstone B and (NEL) Bradley-Terry (there is a Thurstone A model, but it is not tractable). These are compared briefly with the logit regression approach. For a fuller discussion see Cooke [5].

Thurstone C: We assume that each stakeholder has an affine unique utility value for each alternative. We can arbitrarily choose two alternatives and scale them equal to 0 and 1. We do this for all stakeholders, using the same alternatives. The utility values for all stakeholders are now on a common scale. Each expert's values for the remaining alternatives are modeled as a sample from a random vector. We assume that the components of this vector are independent normals with unit variance and means which are solved from the paired comparison data.

Thurstone B: This is identical to Thurstone C, except that the components of the random vector of stakeholder values are not independent, but the correlation of values is constant.

The solution algorithm for either model yields values determined on an *interval scale*; that is, the values are uniquely determined when a "zero" and "unit" are fixed.

The Thurstone models make assumptions which are compatible with the theory of rational preference at the individual level, and yield preference values which are affine unique.

NEL / Bradley-Terry: NEL denotes Negative Exponential Life model. In the NEL model each stakeholder performs a thought experiment to determine which of two independent components with exponential life distributions outlives the other. The Bradley-Terry model assumes that values of alternatives are determined on a ratio scale, and that the probability that a stakeholder prefers alternative i over j is

$$V_i/(V_i+V_j) \tag{2}$$

where V_i is the value of alternative i.

Note that this ratio is invariant under linear transformations, but not under affine transformations. The computational algorithms of these two models are identical, and yield estimates of values V_i determined on a ratio scale. The Bradley-Terry solution algorithm assumes that each choice event is modeled as an independent coin-tossing experiment. The estimates of the values V_i are obtained by maximal likelihood. The Bradley-Terry assumptions would hold if each stakeholder were sampled *once* for each pairwise comparison. If each stakeholder assesses all pairs, then obviously is preferences for alternatives i and k cannot be independent of the preferences for (i , j), and (j, k).

The logit models, like the Bradley-Terry model, assume that

$$P(i > j) / P(j > i) = V_i / V_j \tag{3}$$

and find the scale values by log linear regression. The probability of choosing alternative i over j is $\exp(u_i) / (\exp(u_i) - \exp(u_j))$ where u_i is a "population utility" about which the individual utility values are distributed according to an extreme value distribution. Note the similarity with the Bradley-Terry model by putting $u_i = \ln(V_i)$.

In conclusion we remark that the stakeholder preference method sketched above has no problem dealing with intransitive group preferences. With reference to the Bach, Beethoven, Mozart example, these alternatives would all receive the same scale value. Perhaps the main virtue of the stakeholder preference approach is that it enables standard checks on model fit and model adequacy.

Finally, there are other approaches currently under development, of which one is worth mentioning here. Suppose each stakeholder in the population has a preference ranking. We could then ask, which distribution over preference rankings would produce the paired comparison data, if we sampled stakeholders randomly from the population and asked them to express their preference for each pair of alternatives? New mathematical techniques of probabilistic inversion can be used to determine the best fitting distribution over rankings.

8. References

1. Arrow, K.J. (1963) "Social choice and individual values" 2nd edition, Wiley, New York.

2. Arrow, K.J. and Raynaud, J. (1986) "Social choice and Multicriterion Decision Making", MIT press.

3. Bradley, R. (1953) " Some statistical methods in taste testing and quality evaluation" *Biometrica,* vol. 9, 22-38.

4. Bradley, R. and Terry, M. (1952) "Rank analysis of incomplete block designs *Biometrica,* vol. 39, 324-345.

5. Cooke, R.M. (1991) *Experts in Uncertainty* Oxford University Press.

6. David, H.A. (1957) The Method of Paired comparisons, Charles griffin, London, 1963.

7. French, S. (1988) *Decision Theory; an Introduction to the Mathematics of rationality* Ellis Horwood, Chichester.

8. Kind, P.(1996) "Deriving cardinal scales from ordinal preference data: the analysis of time trade-off data using pairwise judgement models", Paper presented to HESG, Brunel University.

9. Koop, G. and Poirier, D.J. (1994) "Rank-ordered logit models: an empirical analysis of Ontario voter preferences. *Journal of Applied Econometrics* 9, 369-388.

10. Linkov, I., Sahay, S. Kiker, G. Bridges, T. Belluck, D. and Meyer, A. (2005) "Multi-criteria decision analysis; comprehensive decision analysis tool for risk management of contaminated sediments."

11. McCabe, C. Brazier, J. Gilks, P. Tauchiya, A. Roberts, J. O'Hagan, A. and Stevens, K. (2004) " Estimating population cardinal health state valuation models from ordinal (rank) health state preference data" Sheffield Health Economics Group, Discussion Paper Series, Ref. 04/2.

12. McFadden, D. (1974) "Conditional logit analysis of qualitative choice behavior" in Zarembka, P (ed) *Frontiers in Econometrics*, New York Academic Press, USA, 105-142.

13. Mosteller, F. (1952) "Remarks on the method of paired comparisons: the least squares solution assuming equal standard deviations and equal correlations" *Psychometrica,* vol. 16, no. 1 3-9.

14. May, K.D. (1952) 'A set of necessary and sufficient conditions for simple majority decisions' *Econometrica* 20, 680-684.

15. Salomon, J.A. (2004) "The use of ordinal ranks in health state valuations" IHEA Conference, USA, San Francisco.

16. Savage, L.J. (1972) *Foundations of Statistics* 2end edition, Dover, New York.

17. Thurstone, L. (1927) "A law of comparative judgment" *Pschyl. Rev*. vol. 34, 273-286.

18. Torgerson, W. (1958) *Theory and Methods of Scaling*, Wiley, New York, 1958.

19. Torrance, G.W., Feeny, D.H., Furlong, W.J. Barr, R.D. Xhang, Y., and Wang, Q. (1996) "A multi-attribute utility functijon for a comprehensive health status classification system: Health Utilities Mark 2. *Medical Care* 34 (7) 702-722.

MULTIOBJECTIVE RISK/COST ANALYSIS OF ARTIFICIAL MARINE SYSTEMS USING DECISION TREES AND FUZZY EXPERT ESTIMATIONS

E. LEVNER
Holon Academic Institute of Technology
52 Golomb Street, Holon 58102 Israel

J. GANOULIS
Aristotle University of Thessaloniki
Department of Civil Engineering, Thessaloniki 54124 Greece

I. LINKOV
INTERTOX, Inc.
83 Winchester Street, Suite 1, Brookline, MA 02446 USA

Y. BENAYAHU
Tel Aviv University
Israel

Abstract

This study is devoted to the analysis of Artificial Marine Systems (AMS) and their optimisation with emphasis on their role of mitigating anthropogenic and technological threats to the environment. Historically, AMS were created and used to enhance fishing catches, protect coastal zones and maintain biodiversity. In recent years, great strides have been made in the understanding of AMS as multi-functional tools to study, monitor and even influence the global warming processes. In fact AMS can serve as global bio-filters and controllable sinks of atmospheric carbon dioxide, in the long run mitigating global warming and enhancing sustainable preservation of marine resources. A fuzzy expert system based on the Mamdani-Sugeno *modus operandi* has been designed, enabling a risk/cost/benefit analysis to be performed, environmental risk to be minimised, and the optimal structure of an AMS (location, cost, size, geometrical forms, etc.) to be found. Special attention is given to the protection of the coral reef in the Eilat-Aqaba Gulf in the Red Sea.

I. Linkov et al. (eds.), Environmental Security in Harbors and Coastal Areas, 161–174.

1. Introduction

Artificial Marine Structures (AMS) and Marine Reserves (MR) have been used for over 200 years to enhance fishing catches, protect coastal zones and maintain biodiversity. Besides these applications, they are also of scientific and educational value In recent years, great strides have been made in the understanding of AMS and MR as multi-functional tools used to study, monitor, and even influence global warming processes.

There is an increasing number of terrestrial nature reserves to protect and study areas of natural beauty and biodiversity on the Earth's surface. Similarly, long-term sea observations and monitoring using AMS and MR offer unique possibilities for studying and protecting the marine biosphere. AMS, and in particular artificial reefs (AR), may serve as convenient long-term instrumental platforms for measuring and monitoring global warming processes and anthropogenic climate changes. They may provide systematic remote measurement of variables that play a crucial role in air-sea exchange and marine environment life, like atmospheric carbon dioxide, carbon cycle, aerosols and other chemical/biological parameters. In addition, AR may serve as bio-filters capable of significantly improving seawater quality, especially in the vicinity of fish farms and sea ports. AMS, and, more widely, national and international marine reserves may serve as controllable marine sinks of atmospheric carbon dioxide, in the long run mitigating global warming and enhancing sustainable preservation of marine resources.

Using various passive and active measurement methods, the health status of ecosystems can be investigated. Changes resulting from global warming, climate change, greenhouse gas emission and other factors can be consistently monitored over the long term (years to decades). While there is no way to accurately predict individual responses to the Kyoto Protocol, long-term marine observations can be used to identify key responses as well as "perverse reactions", such as rapid degradation of coral reefs, in the next few years. Such measurements and observations may ultimately be used as input data to prognostic biogeochemical computer models and as the response of bioscience to the Kyoto Protocol.

Over 50% of AR programmes worldwide have failed due to poor site selection, inadequate planning, lack of monitoring, or because no proper management was put in place according to a set of suitability criteria [1-6]. This paper is devoted to the study of contradictory ecological effects related to the design, construction and maintenance of multi-functional AMS. Attention is given mainly to the comparative analysis of contradictory positive and negative effects related to the introduction of AMS into exploited marine ecosystems and finding optimal decisions related to the location, size, structure, maintenance and operation of AR aimed to minimise the evolved environmental risks and maximise the advantages of AMS for the regional mariculture and the environment. Special attention is devoted to the qualitative description of the role of multi-functional AMS in the Red Sea and the risk analysis of ecological benefits and rewards of large-scale AMS with respect to coral reef integrity. So far, the objective of this paper is to qualitatively and quantitatively estimate the role of AMS as bio-filters and mechanical shields for preventing and mitigating the destructive effects of global warming and seawater pollution on the health of coral reefs; the main focus being on the estimation of the impact of counter-pollution activities on coral reef health and integrity.

2. Integrated Management of AMS and Risks

AMS, and in particular AR, are man-made or natural objects placed in selected areas of the marine environment in order to promote ecosystem conservation and biodiversity maintenance and thereby improve the health and integrity of the ecological system. Any form of AMS construction will cause associated ecological and technological risks. What is of a special interest is thatAMS and especially AR may affect associated risks in two completely opposite directions, either decreasing or increasing different associated risks. In what follows, we define the spectrum of potential risks and the problems experienced by the regional environment. We focus on the following damages and risks, which have greatest impact on human health, biodiversity and the environment:

1. Risk of a loss in biodiversity, a loss of habitat and associated species.
2. Risk for the fisheries' yield.
3. Risk (danger) of beach erosion and destruction of the ecosystem equilibrium.
4. Risk (danger) for the quality of life for the human population, including aspects related to human health, water quality, recreational diving and tourism, etc.
5. Risks related to a decrease in the quality of environmental research and education.

When a large AMS is placed on the seabed the associated risks are affected in different directions. On the one hand we can observe the negative aspect of the AR structure, since.invariably there will be a loss of habitat and associated species as well as a disturbance to an equilibrium in an ecosystem, but on the other hand there is also a positive aspect, as many species will adapt and take advantage of opportunities available, and environmental factors will influence the type of species that can live in the new environment and therefore the habitat may alter, thus leading to a new, enriched ecological equilibrium.

Similarly, when considering risks to human health, on the one hand, there is a risk of water and fish contamination from toxic materials from which AMS are constructed but on the other hand AR may increase the carrying capability of the marine ecosystem and thus improve the sea water quality. . Advantages and drawbacks of AMS related to fishery and mariculture production are evident and fully recorded, at least qualitatively, in the literature (see [1-6], among many others). The risks related to beach erosion and the equilibrium of the marine ecosystem are affected in a similarly contradictory way. For any AR, an associated technological risk representing its (un)safety and technical unreliability can be estimated; the unpredictable nature of the marine environment is taken into account in these estimations. On the one hand, mistakes, inaccuracy, and ignorance in the technological risk estimations can lead to construction failures (and, unfortunately, disastrous AMS construction project failures have occurred for these reasons in the past). [1, 4] but on the other hand, AMS can considerably mitigate the risk of beach erosion and promote a (new) equilibrium of a marine ecosystem. This paper will produce an accurate risk estimate in order to identify the positive protective functions of the AMS and to find their potential effectiveness.

Our main hypothesis, which is tested and verified in the paper, is that the positive effects of AMS can be more significant than the negative ones. Moreover, AR structures can be designed to optimise the capabilities of fisheries, to increase the carrying capability of the marine ecosystem, to decrease sea water pollution and to promote sustainable development of the regional fishing industry and beach economy (tourism, surfing, education, etc.). Our study will focus on the development and aplication of two mathematical techniques:

- Decision trees [10]
- Fuzzy and interval-valued analysis and expert system assessment [7, 8, 11, 14]

The following main parameters of AMS are worth considering:

- Location - mimic the biological, geographical and economic environment in which an AMS will be placed
- Length and size of AR
- Materials and structure of AR, maximising the total environmental impact
- Geometric forms most suitable for the production and maintanance of AR

Particular interest is paid to finding optimal decisions that could benefit the economical and ecologic life of the region, including:

- Feeding commercially important species, such as fish, crabs, lobsters, weeds, etc
- Decreasing sea water pollution
- Monitoring and control of beach erosion
- Decreasing risks for human and coral health caused by seawater and beach pollution

The following advantages of AR, leading to possible mitigation of environmental risk, will be taken into account in the proposed computer models:

1. Habitat provision for fish, shell fish, lobsters, oisters, corals, marine vegetation, etc.
2. Coastal protection, including the dissipation of wave energy, reduction of beach sand losses, creation of wider beach salient, increased protection of beaches in stormy weather, etc.
3. Benefits for fishery and aquaculture

4. Benefits for human health
5. Benefits for recreational and touristic activities, including diving, surfing and sport, etc.
6. International cooperation, in particular in the Eilat-Aqaba Marine Park
7. Water protection by bio-filtering and absorbing harmful gasses from the atmosphere

3. Ecological Risks: Main Definitions

There are many definitions of environmental risk [9, 12, 18, 19]. Speaking informally, environmental risk is a quantitative measurement of ecological hazards taking into account their economic, social and related consequences. Following the U.S. Environmental Protection Agency (EPA) definition of ecological risk assessment [9, 18], we define environmental risk assessment as a quantitative appraisal of the actual or potential impact on humans, animals, plants and technological infrastructures of contaminants from a hazard.

Before *risk assessment* is explained formally and in more detail, let us define the terms *hazard* and *risk* and how we understand these definitions in the present study. *Hazard* is the potential for harm. For example, sea waters may be polluted by materials from an AMS; a hazard here is the danger for human health. Other possible types of hazards are:

- Danger of damages to fish
- Danger of damages to shell fish
- Danger for marine vegetation
- Danger for lobsters, oisters, etc.
- Danger to coral reefs and other natural resources
- Danger of beach sand losses
- Danger of beach erosion and beach pollution
- Danger of sea waves and stormy weather
- Danger for fishery and mariculture industries
- Danger for the tourism industry
- Danger for divers, surfers and other beach visitors, etc.

4. Mathematical Model

4.1. DECISION TREE

This section outlines the mathematical form of a risk minimisation model in terms of multi-criteria decision trees and fuzzy mathematical programming problems.

A *decision tree* is a flow chart or diagram representing a classification system or predictive model. The tree is structured as a sequence of simple questions, and the answers to these questions trace a path down the tree. The tree consists of nodes and arcs. The end node reached determines the final objective of classification or prediction made by the model, which can be a qualitative judgment (e.g., linguistic variables [7, 8, 11, 14]) or a numerical forecast (e.g., cost, risk, benefit, etc.). A square node of the tree represents a point at which a decision must be made, and a circular node represent situations where the outcome is uncertain. Each arc leading from a square represents a possible decision; while arcs leading from a circle represent a possible outcome and the probability of it occurring (see Figure 1).

In the multi-criteria situation, which is the subject of the present study, we shall construct several trees, each one corresponding to a single criteria, or an objective function. We will focus on two main criteria: the probability of the hazard and the amount of the expected damage.

The risk assessment procedure is based on using multiple decision trees. Each of the different decision trees is related to a single specific objective function (such as, for instance, the probability of a fatal ecological incident in the area, the impact of pollution on ecological losses and in particular the loss of natural corals, or the severity of ecological losses).

The structure of each tree reflects the following ecological "supply chain":

Type 1:

Pollution_Sources → Stressors → Protection_Targets

(this supply chain does not take counter-pollution activities into account)

Type 2:

Pollution_Sources → Pollution-Mitigation_Activities → Stressors_Mitigated→ Protection_Targets

(this supply chain takes counter-pollution activities of constant intensity into account)

Type 3:

Pollution_Sources → Pollution_Mitigation_Activities →_Strategies → Stressors_Mitigated→ Protection_Targets.

(this is a supply chain where counter-pollution activities are split into several possible strategies of different contents and intensities)

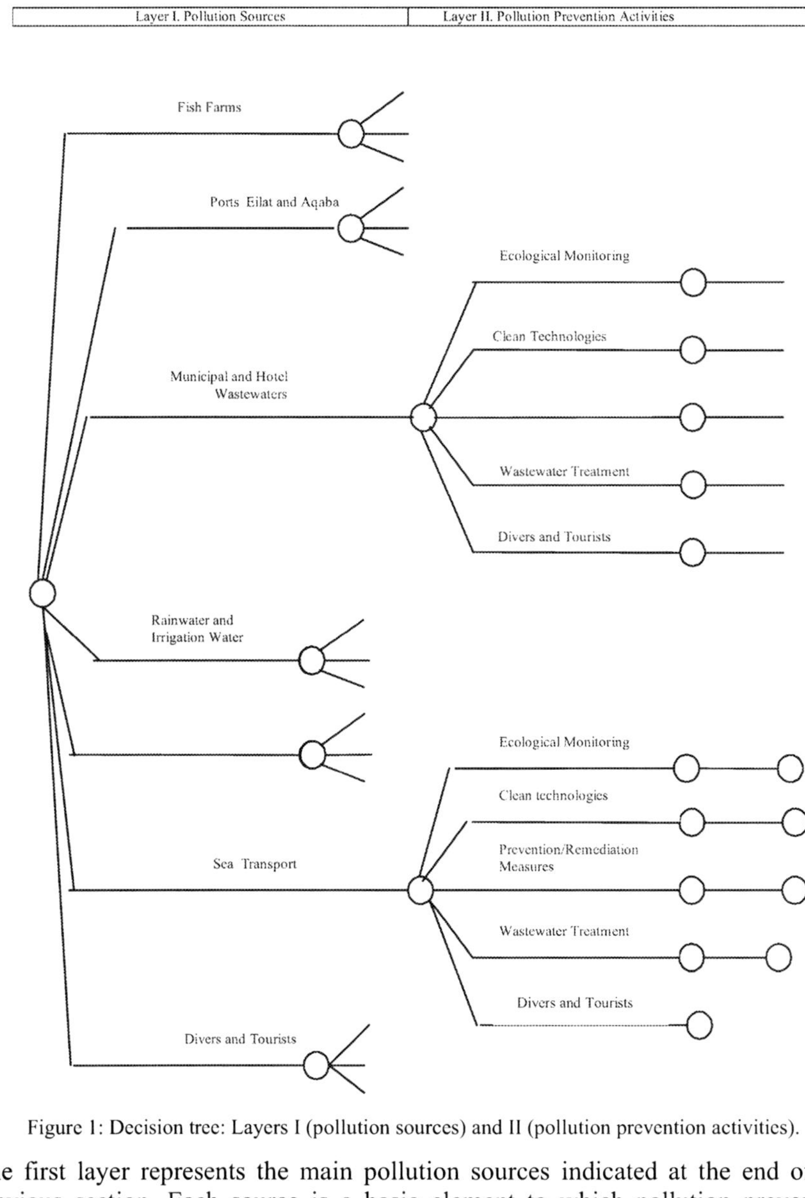

Figure 1: Decision tree: Layers I (pollution sources) and II (pollution prevention activities).

The first layer represents the main pollution sources indicated at the end of the previous section. Each source is a basic element to which pollution prevention activities are applied. The second layer (in the second tree type that includes counter-pollution activities) represents the main activities for pollution prevention and measures for risk aversion such as: regular monitoring/prevention of all types of marine pollution; introduction and development of clean technologies; introduction of water/wastewater filters and other wastewater treatment facilities; technical reconstruction and reequipment of the pollution producers, preventive and remediation measures including construction of artificial reefs, and ecological literacy education. The third level represents different classes and subclasses of ecological stressors (physical, chemical, and biological sources of damages and environmental risks) whose impact is to be mitigated by using the activities listed in the second layer when imposed upon the hazard sources listed in the first layer. The arc leading from the node i in the second layer of the tree to a node j in the third layer may be split into several possible strategies (e.g., "adventurous", "basic", and "cautious") with a set of corresponding probabilities p_{ijt} of the mitigation of a harmful stress and effects e_{ijt} of strategy i upon decreasing stress j (t = 1,2,…,T) assigned; here T is the total number of possible strategies (outcomes). Each third-level node representing a stress is assigned with a weight v_j depicting the importance or severity of the stress.

The fourth level represents different classes and subclasses of biological entities listed in Section 3 (fish, shellfish, corals, etc.) which are to be protected by mitigating harmful and toxic stressors listed in the third layer. The arc leading from the node j in the third layer to a node k in the fourth layer is assigned with a set of probabilities q_{jks} of decreasing the harmful impact of stress j upon bio-organism k (where k = 1, 2,…, K) and the positive effect f_{jks} upon organism k due to the decreasing of stress j (s=1, 2,…;S); here S is the total number of possible outcomes. Each fourth-level node representing an organism is assigned with a weight w_k depicting its size and importance in the ecological system.

The values assigned to the fourth-level nodes represent experts' evaluations of the expected results of using counter-pollution measures and strategies to decrease the stressors; these are expressed either numerically or in linguistic terms (for instance, "very strong", "strong", "medium", "weak", and "negligible"). The results of the fifth layer are the values and/or linguistic evaluations of an objective function.

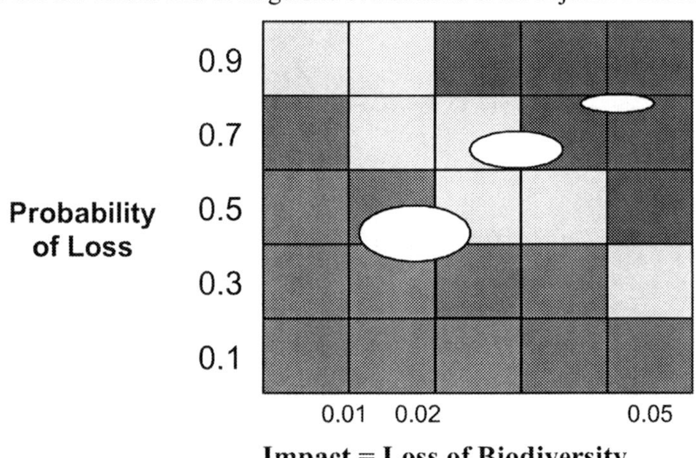

Figure 2: Qualitative risk matrix: loss of biodiversity and probability of the loss.

The final stage of the multiple-tree procedure is an integrated (quantitative and/or qualitative) estimation of the ecological risk based on the convolution of linguistic and numerical results provided by all decision trees in the risk matrix (see Figure 2).

4.2. QUANTITATIVE RISK ASSESSMENT

Ecological risk is the likelihood of harm occurring in an ecosystem and the severity of its outcome. There are many ways in which the evaluation of risks can be carried out [13]. These range from complicated numerical systems to a qualitative expert judgment of risk, such as, for example {*low, medium or high*}. This estimation depends on two criteria: the probability of the hazard and the amount of the expected damage (see Figure 3).

Most of the formal ways define risk R as the product of the *weight* w of a hazard (also called a *risk factor*) and the amount of damage, D caused by the hazard, in a monetary, material, or grade form:

$$\boldsymbol{R} = \boldsymbol{wD}. \tag{1}$$

A *risk factor* $\boldsymbol{w}$ (also called a *risk weight,* or *a risk factor number*) is the product of the likelihood (probability) $\boldsymbol{p}$ and severity $\boldsymbol{s}$ of harm arising from a hazard,

$$w = \boldsymbol{ps} \tag{2}$$

A *likelihood* rating is based on the qualitative scale shown below.

1. *Not likely.* There is really no likelihood of an accident or pollution occurring. Only under freak conditions is there a possibility of an accident or illness. All reasonable precautions have been taken. This should be the normal state of the water source or any other marine ecosystem under consideration.
2. *Possible.* If other factors were present, pollution or illness might occur, but the probability is low and the risk is minimal.
3. *Quite possible.* An accident or pollution may happen if additional factors precipitate it, but it unlikely to happen without them.
4. *Likely.*
5. *Very likely.* If the situation continues as it is, there is almost a 100% certainty that an accident or pollution will happen at least once.

Now let us establish a *severity* rating for the identified hazards using the following scale:

1. *Nil.* No risk of injury, contamination, or disease
2. *Slight.* Causing minor injury or harm.
3. *Moderate.* Causing moderate injury, harm or disease.
4. *High.* Causing death or serious injury to an individual.
5. *Very high.* Causing multiple deaths and/or widespread illnesses to the population.

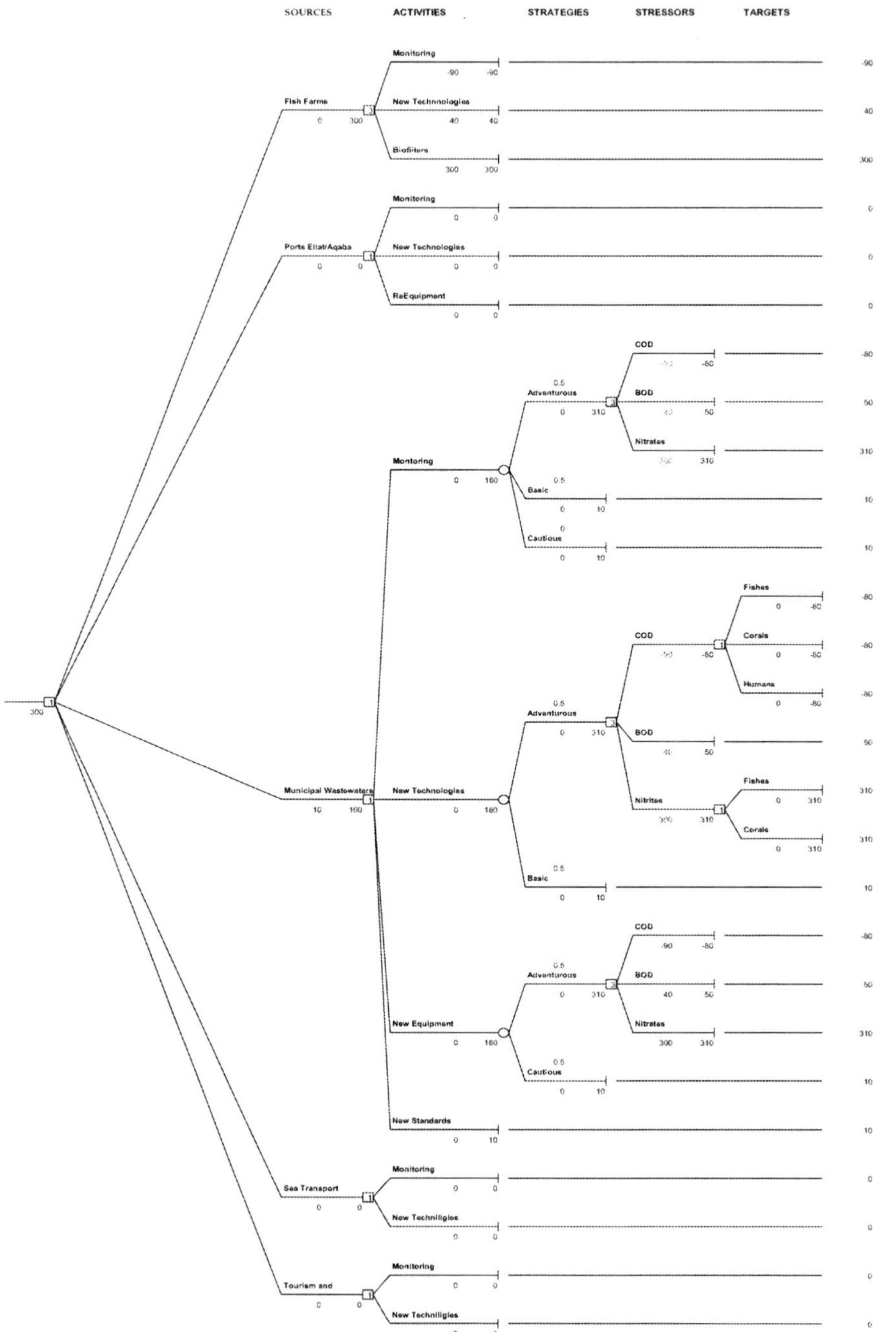

Figure 3: Decision tree for activities in the Eilat-Aquaba Gulf in the Red Sea.

A *risk factor number* is obtained by multiplying the likelihood rating by the severity rating. In the simplest case when the likelihood and severity are rated in numbers between 1 to 5, this will result in a number between 1 and 25. Such a rating enables the most serious risks to be considered first, i.e. the higher the number the higher the risk.

Risk factor numbers may be classified as follows:

- 16 – 25 *Extreme* Risk level unacceptable.
- 10 – 16 *High* Undesirable.
- 7 – 10 *Medium* May be acceptable.
- 1 – 6 *Low* May be acceptable.

The damage $\boldsymbol{D}_k$ to human or coral health caused by several toxic stressors, may be represented as

$$\boldsymbol{D}_k = \boldsymbol{v} \bullet \boldsymbol{W}^k = \Sigma_{j=1,\ldots,LxM}\, v_j(W_{jk}) W_{jk}, \qquad (3)$$

where:

L : the number of stressors;

M: the number of pollution sources;

$v_j = v_j(W_j)$: the weight for the jth stressor;

W_{jk} : the amount of harm to the kth object (i.e., protection target) caused by the jth stressor;

$\boldsymbol{W}^k = (W_{jk})$: the damage vector for the kth object caused by all stressors.

The total damage to the ecological system consisting of K objects (protection targets) is defined as follows:

$$\boldsymbol{D} = \Sigma_{k=1,\ldots,K}\, u_k\, \boldsymbol{D}_k = \Sigma_{k=1,\ldots,K}\, \Sigma_{j=1,\ldots,LxM}\, v_j(W_{jk})\, u_k W_{jk}, \qquad (4)$$

where u_k is the weight (importance of the kth target).

According to the US EPA [18], a *stressor* is any physical, chemical, or biological entity that can cause an adverse response, such as a toxic chemical.

The number M of sources, *fons et origo*, of water pollutants causing damage to human health, as well as to the health of other biological organisms, can be defined by the experts. According to [13, 15], the major producers of high-risk hazards in a typical area such as the Eilat Gulf include: spills from maritime activity and oil transport in the gulf, chemical pollutants entering the sea during transport and loading of phosphates, potash, bromides and other cargoes, microbial pollution from municipal, industrial and hotel wastewater discharge, unregulated mariculture, litter thrown from vessels or left on beaches, ballast water from vessels, contaminated storm waters and irrigation waters in winter seasons, greenhouse gasses causing airborne acid pollution, global warming causing coral bleaching and other illnesses, and anchors and divers causing physical damages to the coral reefs.

Similar to damage **D** in (4), we define the *mitigation of damage* ***MD*** caused by the counter-pollution activities, $a = 1,\ldots, A$.

$$\boldsymbol{MD} = \Sigma_{a=1,\ldots,A}\ \boldsymbol{MD}_{\mathrm{ka}} = \Sigma_{a=1,\ldots,A}\ \Sigma_{k=1,\ldots,K}\ \Sigma_{j=1,\ldots,LxM}\ v_j(W_{jk})\ u_{\mathrm{k}} W_{jka}, \quad (5)$$

where: $\boldsymbol{MD}_{\mathrm{ka}}$ is the mitigation of damage to the *k*th target caused by the *a*th counter-pollution activity;

$\underline{W}_{jka}$ is the amount of the decrease in harm to the *k*th object (i.e., a protection target) caused by the *j*th stressor as a result of the *a*th counter-pollution activity.

4.3. FUZZY EXPERT SYSTEM

4.3.1. Main Components

4.3.1.1. Inputs:

1. Pollution_Sources
2. Pollution_Mitigation_Activities
3. Strategies (or Intensity)
4. Stressors
5. Protection Targets

4.3.1.2. Outputs:

1. Probability of damage
2. Severity of damage
3. Amount of damage
4. Environmental risk decrease due to the counter-pollution activities.

4.3.2. Rules

The set of rules of the following pattern:

> If pollution mitigation activity A1 (e.g., construction of an AR biofilter) is adventurous and A2 (introduction of new dust filters at a port) is basic then stressor S1 (phosphates) will lie within the allowed standards.

4.3.3. Rules Evaluation and Aggregation

Performed according to the Mamdani-Sugeno procedure [8].

4.3.4. Defuzzification

According to the centre-of-gravity rule. Results of computations are given in Figure 3.

5. References

1. Baine, M. (2001) Artificial reefs: a review of their design, application, management and performance. *Ocean and Coastal Management* , 44: 241-259.

2. Benayahu, Y. (1985) Faunistic composition and patterns in the distribution of soft corals (Octocorallia Alconacea) along the coral reefs of Sinai Peninsula. Proc 5th Int Coral Reef Symp 6: 255-260.

3. Benayahu, Y., & Loya, Y. (1987) Long-term recruitment of soft corals (Octocorallia: Alcyonacea) on artificial substrata at Eilat (Red Sea). *Mar Ecol Prog Ser* 38: 161-167.

4. Bell, M. 2002 *Marine Artificial Reefs*. South Carolina Office of Fisheries Management, Technical Report.

5. Boersma P., J.K. Parrish, 1999. Limiting abuse: Marine protecting areas, a limited solution. *Ecol. Econ.,* 31, 287-404

6. Claudet, J., and D. Pelletier, 2004, Marine protected areas and artificial reefs: A review of the interactions between management and scientific studies, *Aquatic Living Resources,* 17, 129-138.

7. Ganoulis J. 2004) *Integrated Risk Analysis for Sustainable Water Resources Management* In: *Comparative Risk Assessment and Environmental Decision Making,* 275-285, NATO Science Series, Kluwer Academic, The Netherlands.

8. Ganoulis J. 1994 *Engineering Risk Analysis of Water Pollution: Probabilities and Fuzzy Sets*, WILEY-VCH, Weinheim, New York, Basel, Cambridge, Tokyo, 306 p.

9. Linkov I., and A.B. Ramadan (eds), *Comparative Risk Assessment and Environmental Decision Making*, Springer, Berlin, 2004.

10. Levner E., et al., 1993, *Modern Mathematical Methods of Optimisation*, Akademie Verlag, Berlin, 416 pp.

11. Levner E., A. Ptuskin and A. Friedman, *Fuzzy Sets and Systems: Theory and Applications*, Foreword by Prof. Lotfi Zadeh, Russian Academy of Sciences, CEMI Press, Moscow, 1998, 110 pp. (Russian).

12. Levner, E., and J.-M. Proth, 2003, *Strategic Risk Management of Marine Ecosystems*. A mini-course of lectures presented at the NATO Advanced Study Institute meeting "Strategic Management of Marine Ecossystems", October 1-10, 2003, Nice, France. Published in *Strategic Management of Marine Ecosystems*, Kluwer Academic Publishers, 2005).

13. Love M.S., M. Nishimoto, and D. Schroeder, 2001, *The ecological role of natural reefs and oil and gas production platforms on fishes in Southern California,* US Geological Survey, OCS Study, MMS 2001-028 (see also www.id.ucsb.edu:16080/lovelab/Report.html)

14. Marusich, L.J., 2001. *The application of fuzzy logic analysis to assessing the significance of environmental impacts: Case Studies from Mexico and Canada,* Research and DevelopmenMonograph Series, Canadian Environmental Assessment Agency, Hull, Quebec, Canada, ISBN 0-662-34457.

15. Ministry of Environment of Israel, The Environment in Israel, Jerusalem, 2002.

16. Nautilus Consultants Ltd., 2003, *Artificial Reefs in Scotland: Benefits, Costs and Risks*, Technical Report, March 2003, Edinburgh, Scotland. (Also available at http://www.nautilusconsultants.co.uk/-pdfs/Nautilus%20artificial%20reefs%20 report.pdf)

17. Szedlmayer, S., 1994 *Artificial reefs design, planning and permitting*. Aubern University Marine Extension & Research Centre.

18. USA EPA, Ecological Risk Assessment Guidance for Superfund, USA Environmental Protection Agency, PA, 1991. http://www.epa.gov/superfund/programs/risk/ecolgc.htm.

19. Whitmarsh, D, Pickering, H., 1997 Commercial exploitation of artificial reefs: economic opportunities and management imperatives. CEMARE Report.

THE *BOUCHARD-120* AND CHALK POINT SPILL RESPONSES: OBJECTIVES AND PERFORMANCE METRICS

S.P. TULER, T.P. SEAGER, R. KAY
Social and Environmental Research Institute
278 Main Street, Room 404, Greenfield, MA 01301USA

I. LINKOV
Intertox
83 Winchester Street, Suite 1, Brookline, MA 02446 USA

Abstract

Oil spills present a chronic threat to the environmental security of most major ports. While mitigation of the risk of oil spills should include prevention, major oil spills remain periodic occurrences. Consequently, spill preparedness and response are critical aspects of minimizing the damage caused by spills. Nonetheless, any major spill response engages multiple stakeholder and public groups that may have different objectives. Currently, spill managers must balance conflicts in the midst of a crisis using *ad hoc* or heuristic approaches that may be difficult to justify or communicate. Public expectations are particularly challenging to manage. In some cases, the spill response may be perceived as a failure despite the response agency's best efforts. A systematic approach to stating varied spill objectives and tracking progress may result in better management and communication and improve the credibility of spill managers. This research studies two separate spill incidents to reveal the different types of objectives held by engaged personnel and the ways that they assess the progress of the response. A total of 30 interviews are conducted and interpreted using a grounded theory approach to reveal salient objectives. Where possible, metrics relating to these objectives are elicited and the results for each spill compared. Although the quality of the study metrics is not examined in detail, we find that some spill objectives are more readily stated in terms of performance metrics than others, suggesting that spill managers may benefit from greater guidance on how to gauge progress or set goals in areas such as protection of public health and safety or mitigation of sociopolitical or economic impacts.

1. Introduction

Oil spill response is defined as encompassing all activities involved in containing and cleaning up oil spills in ways that achieve the following overarching goals [4]:

I. Linkov et al. (eds.), Environmental Security in Harbors and Coastal Areas, 175–191.

- Maintaining safety of human life.
- Stabilizing the situation to preclude it from worsening.
- Minimizing adverse environmental and socioeconomic impacts by coordinating all containment and removal activities to carry out a timely, effective response.

To achieve these goals, oil spill response efforts must include a variety of participants. They can include: federal, state, and local officials (e.g., US Coast Guard, US Environmental Protection Agency, US Department Of Energy, local fire chiefs, harbormasters, state environmental officials), the responsible party and its contractors, non-local clean-up crews hired by private contractors, environmental and community advocates at the national, state, and local levels, and community residents who have an important stake in the response, (i.e., business- and home-owners, and beach associations). Because so many different and interested parties are affected, there is a potential for conflicting ideas about how a response should be organized and implemented to achieve these goals. Different groups may have different assessments of oil spill response success because they have different objectives and some may be in conflict with others. Response strategies are always dependent on the priorities placed on protecting specific resources in the context of a particular spill. Therefore, decisions made about priorities are to a very large extent political or social as well as technical. Consequently, measuring the success of any response (for example, in accordance with the Governmental Performance Results Act of 1993) is a significant challenge. To capture potentially disparate views and to facilitate management decisions, multiple performance metrics must be employed. However, good metrics for capturing the nuances of the decision process are not always available.

This chapter addresses the question of what objectives and performance metrics are used by key stakeholders to assess two recent oil spill responses:

1. The *Bouchard-120* spill response that began on April 27th, 2003 as the tugboat *Evening Tide* ran its tanker aground and released No. 6 home heating fuel just at the entrance to Buzzards Bay, Massachusetts. (See www.buzzardsbay.org for further details.)
2. The Chalk Point spill response that began on April 7, 2000, when an intrastate pipeline that transports oil from the Potomac Electric Power Company's (PEPCO) Chalk Point electrical generating facility to residents in Prince George's County released No. 2 and No. 6 home heating fuel oil into Swanson Creek and subsequently the Patuxent River. (See www.darrp.noaa.gov/-northeast/chalk_point/index.html for further details.)

The full details of each case study are more completely described in Tuler et al. [6].

2. Oil Spill Response

Within several hours of a reported spill, an Incident Command System (ICS) incorporating federal, state, and local authorities is mobilized at the scene. Among the critical government participants in management of any major spill are representatives for the Responsible Party (RP), the Federal On-Scene Coordinator (FOSC), State On-Scene Coordinator (SOSC), and National Oceanic and Atmospheric Administration's (NOAA) Scientific Support Coordinator (SSC). The role of an SSC is to provide the FOSC scientific advice and information (such as weather and spill movement forecasts). In the initial response, the SOSC is responsible for notifying local first responders by telephone about the spill. As a coastal spill, the *Bouchard-120* fell under the jurisdiction of the USCG. By contrast, the Chalk Point spill was classified as an inland spill, over which the USEPA has jurisdiction. Other key players involved in the response and clean-up are various local, first-response officials; non-profit local advocacy groups; and contractors hired by the RP. Over a dozen government agencies may be involved in any major spill response.

In both the case study spills, failure to immediately recognize the magnitude of the spill may have complicated or delayed response efforts. In addition, strong winds and rough weather forced oil past containment booms in both cases. Consequently, oil migrated to areas that responders initially thought could be protected. In the case of the Bouchard spill, oil was "blown back" onto shores that had at first avoided oiling or had been recently cleaned.

Both response teams made extensive efforts to engage and inform the public, although these were not necessarily always perceived as successful. In the Bouchard case, summary briefings by various groups attending the incident command meetings were held several times daily initially and then twice daily once the situation was better under control. Both teams held public meetings over the course of the response, established call centers, and built or maintained websites. In the case of the Chalk Point spill, public meetings and briefings were supplemented by five issues of a newsletter that was circulated to over 27,000 area residents.

3. Research Methods

Understanding how people assess spill response efforts requires understanding of their goals and objectives. To explore these issues in the context of the study spills and responses we conducted a series of interviews and investigated published reports to gather information about:

- Roles and experiences of key responders and other interested and affected parties.
- Their concerns about spill impacts and response.
- Their views about the response.
- Their views about response performance metrics.

The interviews were designed to be semi-structured and open-ended, but they were *not* designed to systematically elicit information about relative priorities among objectives or performance metrics. Instead, at this time we were more interested in learning about the range of objectives and metrics among key participants. Initial research interviewees were identified in articles about the spill. Subsequent interviewees were selected in a snowball sample on the basis of suggestions by others. Interviewees were selected to represent different points of view, their experiences with the spill, and willingness to be interviewed.

To identify the information relevant to study objectives, we used a grounded theory approach, in which important concepts emerge *inductively* during the data analysis rather than in advance of the investigation [1, 2]. In grounded theory, data are categorized with respect to relevant similar characteristics in a process called "coding." At first, a relatively large number of categories are developed. Then, through iteration these categories are grouped into more abstract categories of conceptual relevance to the analysis; data and categories are grouped according to their relationships with each other. For example, all statements related to "reducing bird injuries" or "protecting nesting habitat" can be grouped into a category named "response should protect bird populations." This is referred to as *axial coding* in a grounded theory framework. In this way common themes among the coded objectives are identified. We then extracted all performance metrics expressed by the interviewees that were related to each of the objectives. In addition, we compiled all interviewee comments regarding appropriate uses of performance metrics for assessing oil spill response efforts and characteristics of "good" performance metrics, although we did *not* attempt to assess the quality of any of the suggested metrics either according to the interviewees own criteria for "good" metrics or other norms. (See Seager et al. [5] for a discussion of such criteria).

4. Results

We found that many different objectives for the response to the oil spill were important to our interviewees. We grouped them into endpoint, process, and resource-based metrics (following Seager et al. [5]) related to the following broad categories identified in Tables 1 through 9:

Table 1. Metrics related to protecting public health and safety.

Endpoint	Number people killed or injured. Number of mishaps during hours worked . Presence of contaminants (e.g. PAHs) in water samples? Concentrations of oil in fish tissues. Number life threatening situations to human health. Toxins in smoke plume if do in situ burning.
Process	(None identified).
Resource	Number of IRAC team members OSHA/HAZMAT trained.

Table 2. Metrics related to protection of environment.

Endpoint	Number of oiled birds, eggs or other wildlife.
	Number of miles of shoreline impacted or cleaned.
	Amount of oil or globules on shore.
	No re-oiling or residual oil causing chronic toxicity to something.
	Number of fish, birds or other wildlife killed or injured (per unit search area).
	Number of "appropriate" (not exotics) animals rehabilitated and released.
	How far sheen at surface extended out [miles]?
	How long oil stayed?
	Presence of odors of oil.
	How much grass was destroyed or acres of marsh were impacted.
	Dead and stressed organisms found (rather than estimated).
	Time to achieve background levels/concentrations of contaminant or clean-up standards and recover from clean-up related damage.
	Areas protected (e.g., by redirecting or containing oil).
	Observe water blowing over booms.
	Degree of change to beaches and sandbars from clean-up actions.
	Types of animals and vegetation present after spill cleanup.
Process	Did getting required permits delay response action?
	Time for wildlife rehabilitation efforts to set up operations.
	Rate of bird handling at rehabilitation center.
	Gallons of oil and pounds of contaminated debris recovered and disposed of.
	Accuracy of cataloguing and enumeration of findings.
	Monitoring stations established.
	Time to deploy booming and double-booming in sensitive areas.
	Immediacy of rehabilitator organization's response to call for assistance.
	Change of helicopter flight patterns in response to requests from biologists to not disturb nesting birds.
	Oil direct to sacrificial (rather than sensitive) areas.
	Oil being captured in open water before it hits the beach.
	Area covered in search and recovery.
Resource	Amount of oil containment boom deployed.
	Number of volunteers.
	Number of floating resources to pick up oil in open water oil.
	Number of bodies to manage different aspects of response, inc. SCATs.
	Is there a 'bird searcher' on each team?
	Number sandbags deployed.
	Number people on cleanup crews to deal with oiled beaches

Table 3. Metrics related to economic impacts.

Endpoint	Lost rental income, tourism dollars, property values, wages to fisher/watermen. Duration of beaches closures. Recreational opportunities that were lost that are now back to what they were? Acres and duration of shellfish areas closed or acres of closures reopened. Acres of shellfish beds lost and number of lost fishing days. Costs of laboratory work, other research studies and money spent on response. Increase in crime rate [in southern MD due to influx of people from cleanup crews]. Number of dead fish, ducks & geese.
Process	Kept track of all costs. Local municipalities reimbursed by responsible party?
Resource	(None identified).

Table 4. Metrics related to achieving legal regulatory requirements.

Endpoint	Achieve termination endpoints? Shoreline back to conditions prior to spill? No oil should come off to the touch. Evaluate response with respect to endpoints achieved. For sandy beach no visible oil, no odor of oil. For marshes no sheen. For groin (jetties between properties), riprap no sheen or no oil available when touched.
Process	Number of days until endpoints achieved. All procedures followed (e.g., NIMS).
Resource	(None identified).

Table 5. Metrics related to protection of cultural resources (Chalk Point).

Endpoint	Number of critical sites protected. Soil concentrations, smell or residual presence of oil on artifacts.
Process	Did trench digging affect sites? Experts contacted early for input about sites potentially at risk? Command responsive to requests for protection of sites? Were less destructive response actions chosen (e.g., sorbents and booms rather than burning)?
Resource	Number of GIS, hard maps, laptops and accuracy of location information. Amount of boom deployed

Table 6. Metrics related to effective and timely response.

Endpoint	Number of days company (PEPCO) shut down. Presence/absence of sheen, oil in water, tarballs, oil on shoreline, oil in sediments. Areas impacted to be cleaned up to the ecological state the environment was before the spill happened. Number of areas to be signed-off compared to total. Number of days until endpoints achieved? Is oil being contained? Patchiness of oil? Solid sheet of heavy oil. What it looks like a year later. Number of public meetings organized? Members of the public voice support? Good working relationships with all parties involved? Are response actions having the desired effect?
Process	How often beach is searched—both oiled and non-oiled? Digging holes to look for oil on shoreline? Quarterly checks to see if oil is present or not on beaches Breach of water over boom? Protection equipment put in the right place at the right time? Quality of contractor work. Number of times a orders were given but still not done. Number of public meetings, newsletters published by RP. Time it takes to get response in order. Lead resources mobilized? Number/frequency of fly-overs for real time aerial photography. Clear chain of command established? Communication to appropriate people? Accurate reporting and counting of crews in field. How quickly SCATs out in field? Are clean-up crews assigned effectively to do a good job? Is all pertinent data gathered and recorded? Data sheets available and sufficient? Local officials set up task force for spill response? Muster all forces in town? Basic training for volunteers on bird collection and rehabilitation conducted? Cleanup of impacted areas organized to be manageable and able to monitor? Access established for recovery and clean-up crews through private property? Do efforts correspond to tides? Recovery or rehabilitation of wildlife conducted? How soon have experts been called and set up triage and rehab centers?
Resource	Number of teams/people/supervisors in the field? Number of volunteers? Number of people working at one time. Hours worked. Number of monitors in field to give direction and warning to clean up crews Resources adequate for planned tasks? Cost of response. Types of skills represented on team? A 'bird searcher' on each SCAT? Pounds of sorbent material. Number of packets of baby oil for oil removal. Amount of oil containment boom.

Table 7. Metrics related to addressing public concerns, needs, and support.

Endpoint	Public reimbursements for private property losses. Number calls from public. Members of the public voice support? Public comments from critics and local residents (re: response effort, not outcomes)? Level of staining Complaints about stained rocks? Residual oil on shore (tar balls)? Was spirit of state regulations for public involvement met? Level of public trust?
Process	Relationships and trust with local officials developed? Immediacy of public meetings (number days after spill). Number of public meetings. People given examples of what was impacted and what kinds of cleanup was going on? Number of fliers and informational packets delivered door to door, visuals for media. Public provided the kinds of information it wants? Incorrect information disclosed? A timeframe for ending the cleanup established? Ongoing monitoring and addressing of issues post-spill? People have a place or someone to go to with concerns? Public receives assurances that beaches will be cleaned up to the level of their expectation? Is a forum provided to public so they can hear what's going on and give their feedback? Feelings: level of conflict/anger or happiness. Establish and keep up to date website for public information. Quality of questions from media Able to 'stay on message' during public meetings, press conferences, etc.? Unified Command accessible for public questions and comments? Amount of oil removed manually from shoreline.
Resource	Number hours agencies spent on public outreach (meetings). Number of pamphlets distributed to inform public of hazards. Number of stakeholders involved in setting clean-up standards. Number of dispatch teams arranged to reach-out to various stakeholders. Frequency of information postings on website.

Table 8. Metrics related to mitigating social nuisances (Bouchard).

Endpoint	Presence of stained rocks, oil on beach and complaints?
Process	(None identified).
Resource	(None identified).

Table 9. Metrics related to coordinated and effective response framework.

Endpoint	Clear chain of command established and Incident Command System used? Are pre-identified areas potentially being affected? Did we have information to keep Governor's office and other state Senators and Reps abreast of what was going on? Understanding of whether oil is still stored offshore re-contaminating cleaned up beaches? Is oil coming on shore several days later? Accurate accounting of volume oil spilled and on shore? Number areas cleaned as of today? Number of miles of shoreline impacted right now? Bad feelings among locals responders toward the Unified Command staff?
Process	RP is responsive? Number of hours to set-up stable incident command center. Location easily accessible? Key people became involved early? Local and federal responders notified quickly? Clear understanding of rights the state trustees have as a state agency? Is Incident Commander able to reach key people, does their phone number work, did they respond? Clear communication protocols and reliable technology working? How quickly decisions made? Are decisions correct (in hindsight)? Is there conflict or chaos in command center? Chauvinistic behavior? Cooperation? Did Unified Command resist information that did not conform to their expectations? Number and frequency of meetings, daily reports for morning meetings? Informed of meetings in advance (i.e., lead time)? Presence of watchdog to see what's going on? Modelers able to get 24 hours ahead of spill with accurate projections? Experts consulted for input on response strategies? Are there clear protocols and schedules? Plans communicated day in advance? Time it takes to implement tasks, such as boom deployment? Are players familiar with each other? Frequency of resource and personnel changes. Time taken to re-staffing response people after contractor fired. Pick-up and shipping schedule for waste generated by clean-up organized? Resources placed in the proper locations? Response organized by segments? System established to track progress? Are crews visiting hard hit areas every day and recording information in a unified way? Is all pertinent data gathered and recorded? Ability to revise objectives and activities based on monitoring effort? Accurate information obtained from the wildlife surveys and SCAT teams? Systematic, 'non-political' approach used to deploy clean-up crews. Attend to short, medium, and long term needs simultaneously? Follow 'best response' protocol and integrated command system? Coordination of volunteers performed appropriately and quickly? Clarity to all parties about stages of response effort? Equipment and personnel demobilized when no longer needed? Clear standards for sign-off established.

Resource	Number of supervisors, Spanish speaking supervisors assigned per section. Number of radios; availability of GIS and computers; phones available and working. Number of teams of trained observers walk coastline and make observations of extent and coverage area of oil Number crews trained, hours worked. Type of oil.

Interviewees in both cases wanted to protect sensitive habitats and populations of threatened and endangered species (although the methods were not designed to tell us about the relative importance or frequency within the respective case study populations). They were concerned with mitigating impacts from clean-up actions, including economic impacts related to lost recreation, tourism, fisheries, and to towns from their efforts during the response. Furthermore, they were concerned about the timely gathering and use of relevant, accurate, and credible information for decision making. Strong and flexible leadership that can learn from past experiences was important to them so that the response could be well-planned and implemented. And, they shared objectives related to addressing public concerns (e.g., providing accurate information to the public).

At the same time, there are a few interesting differences, which are reflective of the particular contexts of each spill. First, mitigating impacts to cultural resources was identified in the Chalk Point case and not the Bouchard case. This reflects the presence of significant artifacts in the region affected by the Chalk Point spill. It may also be a reflection of whom we interviewed—or more accurately, did not interview in the each case. However, it is noteworthy that no one in the Bouchard case made mention of historical or other culturally significant artifacts.

The mitigation of social nuisance impacts was not a category of objectives that emerged from our analysis of the Chalk Point interviews. However, some factors that can be related to social nuisance impacts were raised in the context of other objectives. For example, addressing the potential for an "increase in crime rate [in southern MD due to influx of people from cleanup crews]" might be considered by some to be a social nuisance, but the interviewee was clearly talking about this in the context of economic impacts.

While people in both cases expressed objectives (and performance metrics) related to establishing a coordinated and effective response framework, there were some differences in emphasis. These differences suggest the importance of the particulars of experience that inform people's views. Critical comments emerged about the integration of non-federal officials and responders into the response effort. In the Bouchard case this concern was raised about local officials and local first responders, but not in regard to state-level responders. In the Chalk Point case we found the opposite.

In the Bouchard interviews, the roles and participation of local officials and local residents were a concern. In fact, the way in which local responders along Buzzards Bay were contacted and integrated into the response was a source of contention and criticism among those we interviewed. In addition, there was a concern with how local volunteers were brought into the response effort in the Bouchard case. On the other hand, among the Chalk Point interviewees neither of these two concerns was strongly articulated; volunteers were not even mentioned in these interviews. Instead, there was a concern with whether expert input (e.g., about cultural resources) was obtained and how well federal and state agencies' were coordinated (e.g., clarity about jurisdictions). In the Chalk Point case a federal official spoke of the importance of involving local stakeholders in decisions about clean-up goals, but the metrics suggested were about one-way communication and outreach to the public (e.g., number of pamphlets). However, there are some indications that involvement of local officials would have been useful in Chalk Point in terms of understanding local nuances, such as the effect of currents that would impact oil removal efforts. There was also a concern expressed from *state* officials about their involvement in the response effort and much said about the quality of leadership and the coordination and communication among responding federal and state agencies at Chalk Point. Another difference among objectives relates to the kind of people that expressed a concern with gaining public support for the response. Only state and federal officials spoke of this as an objective. This is a very instrumental perspective that one often finds from officials involved with hazard management.

Interviewees in both cases also expressed the importance of minimizing the costs to the responsible party, although this objective might conflict with others. All objectives identified as important during the interviews are related to the protection and promotion of what people value, such as protection of critical habitats and promotion of decisions based on the best information available. *Why* they value certain things can differ; they may value certain outcomes for intrinsic reasons (e.g., value of species for their own sake) or instrumental reasons (e.g., they allow other things to be accomplished) and the relative weights given to each may also differ. Overall, a large number of performance metrics were suggested by the interviewees. However, a number of interviewees also pointed out the shortcomings of some of them. For example, one interviewee noted that the number of volunteers taking part in the response may not be a good measure because there are constraints on using volunteers (e.g., OSHA regulations) and thought that a better metric might be the number of people calling to volunteer.

However, performance metrics are best defined with respect to specific objectives, which is why we have organized the elicited performance metrics with respect to each objective in Tables 1 through 9. By coding performance metrics in the context of particular objectives it is apparent that many metrics were discussed for some objectives (e.g., mitigating ecological impacts) and few in relation to other objectives (e.g., protection of public and worker health and safety). This should not be taken to mean that those objectives associated with more performance metrics are more important than those with fewer performance metrics; this conclusion is not justified based on the evidence available. Instead, some objectives may be easier to gauge than others or there may be consensus about which metrics to use. For example, protecting health and safety can be assessed by counting OSHA-reportable injuries and fatalities and work-hours without accidents, whereas degradation of cultural resources may be difficult to capture quantitatively. Another reason may be an artifact of how we grouped objectives into larger categories. Furthermore, one might argue that many of the metrics suggested for assessing the prevention and mitigation of ecological impacts and economic impacts could also be measures for whether or not public concerns were addressed. However, they were not always expressed as such. Instead, addressing public concerns was often measured in terms of what effort was made for public outreach and involvement: How many meetings? How many leaflets distributed? Our inability to systematically identify all performance metrics with respect to each objective is a shortcoming of using interviews. This shortcoming is to be addressed by further research; we are developing methods that will ask respondents to rank-order objectives and to rank-order performance metrics for each objective.

5. Discussion

When we consider the kinds of metrics elicited in the two case studies, one observation is that the performance metrics suggested for each case differed for some objectives. For example, making the best possible decisions with limited information, uncertainties, and time pressures was identified as an important objective by many people we interviewed for the Chalk Point case. While this category of objectives was also raised in the Bouchard interviews, it seems to have generated much more attention from the Chalk Point interviewees; this may be a reflection of their perceptions of inadequacies during the Chalk Point spill response. While they found it difficult to articulate specific measurable performance metrics for them, it is nevertheless important as they view clear chain of command, strong leadership, clarity about responsibilities, and organizational jurisdictions to be closely associated with good outcomes.

Some interviewees argued that it is not always useful to define performance in terms of endpoints because they are difficult to measure in the short-term. For example, one person stated that the point was whether a comprehensive *effort* was made to recover all the wildlife. Many of the proposed metrics have to do with process, as one way of measuring "good" decision making is by the way that decisions are made. In these cases much of the attention is on measuring the quality of the effort. Endpoints are related to effectiveness, but these cannot be measured directly. For example, the "quality of decisions" is difficult to assess on the basis of decision outcomes because there can be many intervening factors that affect outcome. However, *how* a decision was made or information gathered and validated can be more appropriate. It is also apparent with objectives that are associated with the *conduct* of response effort: establishing a well-coordinated response, meeting legal requirements, and implementing an effective and timely response. On the other hand, performance metrics related to mitigating economic impacts were almost entirely related to endpoints, rather than processes or resources related. This may be a reflection of the belief that economic costs are easier to measure than impacts to ecological impacts. In fact, estimating economic impacts can be very difficult.

In some cases the endpoint metrics elicited assess indirect or interim indicators of the state of an ecological system, habitat, or population. For example, impacts to ruddy ducks were based on extrapolations of empirical field data. Similarly, some performance metrics are direct measure of economic impacts, such as lost tourism dollars and lost fisherman income, while others are indirect measures: length of shellfish bed closures, acres of shellfish bed closures, number dead ducks and geese, duration of beach closures and presence of polyaromatic hydrocarbons (PAHs) in water samples. Furthermore, endpoints suggested to evaluate the effectiveness of response were often indicators of long-term impacts. A difficulty reflected by many of the suggested metrics is that measures of effectiveness or ultimate impacts can rarely be made at the time of response because the systems are dynamic. This is a challenge in assessing progress toward goals that take a long time to realize [3]. In the case of oil spills, a desire to assess the effectiveness of the response is really about wanting to know about long-term conditions and how they differ from pre-spill conditions. To gauge long-term impacts, indirect indicators observable in the short-term were suggested.

Many of the suggested performance metrics are based on what can be counted or observed (e.g., did it occur or not occur). Metrics that are easy to measure may be more appealing from a bureaucratic perspective. How many gallons of oil and contaminated debris were removed? How many leaflets were distributed to local residents? How many birds were found dead or oiled? Were endpoints defined in plans achieved? Were state and federal standards for contaminants met? Of course, it might not be so easy to count such things accurately, claimed some of the interviewees. In fact, the focus was not always on *quantitative* metrics that some may assume can be accurately counted. For example, a state official in Massachusetts asked "was the *spirit* of state regulations for public involvement met?" which can only be answered subjectively. Problems raised by subjectivity and definitional clarity are also apparent in another example. One metric suggested was that "appropriate" response actions be used. However, the definition of what is appropriate is influenced by the type of oil spilled. For example, if the oil floats, there is less need to worry about it affecting nearby sunken ships. However, if the oil sinks, those sites may be of more concern. Appropriateness can also depend on the kind of artifacts or sites that are at risk. Chemical dispersants may not pose a risk to future research on the site when its age is already known, but they may make it difficult or impossible to use carbon dating for some sites (e.g., prehistoric sites). Similarly, effective oil removal techniques may impact the archeological integrity of a site, so what may seem appropriate with respect to one objective may be inappropriate with respect to another.

Several interviewees observed that just because something can be measured does not mean it is relevant to understanding the success of a spill response or important to many stakeholders. For example, several interviewees in both cases suggested that the amount of boom deployed could be used as a metric to assess how well a shoreline was protected from oil contamination. While it is easy to measure, it may not be a good indicator of whether the shoreline is actually impacted. As we were told by many interviewees, deployed boom was not always effective when currents or winds were strong. Another stated that "we tried to prevent oiling by putting out booming, but it's not a very accurate predictor of whether the outcome [of shoreline protection] will be achieved. Once any oil gets on a beach you still need to clean it up." Furthermore, deploying booms may have been a good decision at the time, but weather conditions can shift and cause them to fail. That is, the metric does not have a causal relationship between the state of the system and the variables that are under a decision maker's control. Similarly, an interviewee suggested that the amount of money spent—for numbers of crew, manhours, and amount of boom deployed—may be very appealing politically as a measure of performance, but questioned whether that is "doing the job to the most effective way" and suggested "there may be more cost-effective ways to look at the whole picture. Money spent is not a good measure." Nonetheless, assessment of short-term success and long-term success can be very different in natural systems [3]. Therefore, interim gauges are essential to provide more immediate feedback to decision makers. Provided the causal linkages between the interim measures (such as effort or process efficiency measures) and endpoint measures (such as recovery of wildlife populations) are well understood, the interim indicators can be an improvement on endpoint measures.

Accuracy, consistency, and reproducibility are frequently a problem—particularly with qualitative metrics. We learned that the amount of oil removed from beaches might be a relevant metric for assessing response, but some argued that the measurement is not meaningful because it is very difficult to quantify the actual amount of oil on the beach because it is mixed with sand and rocks and in sorbent material: "Especially in recovered number you are getting a mixture of oil and water…skimmers never perform as the manufacturers claim." Similarly, different people may have different ideas about what is "impacted." One way around this is training personnel so that "when they come back they say 'we've got 200 yards of shoreline heavily impacted' what one calls 'heavily impacted' is the same thing… . We try to get them to know what the shoreline types are so they can describe it in the proper terminology." However, qualitative or semi-quantitative descriptions may present a moving target and personnel continue to learn on the job, as one subject explained: "Getting the shoreline reasonably back to where it should be…it was a subjective judgment call with each section of the shoreline based on the criteria and adhering to the criteria… We refined the process as we went along over time because you have more of a sense of perspective after you've seen a bunch of different segments of the shoreline."

In several cases, objectives were posed as a "yes" or "no" question, such as "Are members of the public happy?" or "Was there trust from the public?" or "Are there good working relationships with all parties involved?" Some of the might be rephrased to suggest a more sliding scale, such as "What is the level of trust from the public?" However, spill managers would need new approaches to assessing such qualities in the midst of the response. In other cases, it maybe difficult to trace measures directly to the impact of the spill, such as number of lost fishing days suggested by a Chalk Point interviewee: are fishing days "lost" due to impacts of the oil spill or because of foul weather or other reasons? Similarly, natural mortality may be "tallied in as a result of the spill." Typically, there is a paucity of accurate baseline data on which to base comparisons or assessments. For example, "there wasn't an established criterion of how many ppm of hydrocarbons in shellfish is dangerous…There are spills here from these boats all summer long…each municipality needs to know what the baseline is and what the safe baseline is…what's allowable and what's standard?"

According to some federal and state responders, the cause-and-effect relationships between decisions and performance metrics can be very difficult to communicate to the public. This issue was raised repeatedly in regard to the impacts of clean-up efforts. Many interviewees from both case studies provided examples of when, according to them, it would have been better to cease clean-up actions because more severe impacts would result than if none or no further actions were taken. However, articulating these trade-offs to the public was difficult. There may be a disconnect between scientific metrics and measures of public perception: "We took sample after sample after sample [testing for presence of oil contaminants in fish] and everything was fine...the only problem was convincing people that everything was fine." Similarly, the public may not understand the details of response actions. In one case, "people were reporting that clean-up crews [were] being negligent, because of throwing rocks back into the water. But that was what they were instructed to do after wiping them off—and here is the rationale for doing that. After we explained this to them they were accepting of that process." Ultimately, some interviewees' feelings could be summed up succinctly by the comment made by one that "a lot of what we get to judge spills by is how the public feels about it after the response… public perception is our reality."

The causes of good or poor performance may be multiple and difficult to disentangle. In other words, while measurement of a performance metric may suggest a poor response the response might have been "good" in the context in which the decision or action was made. For example, one person suggested that a relevant metric for assessing success is the amount of oil recovered in open water before it hits the shore. However, in the Bouchard spill response, very little was picked up in the open water because of 1) weather conditions and 2) poor coordination in the early stages. While the second reason might suggest a poorly organized response, the first reason is beyond the control of the responders; it would be unfair to judge the response as poorly implemented because of harsh weather conditions that prevented a higher rate of offshore recovery. Similarly, in the Chalk Point spill there was a concern about the amount of oil recovered or removed. Several interviewees thought that *in situ* burning might have been a better response option, but the time it would take to get all the permits and necessary information in place precluded its use. The problem was not the lack of coordination, but rather other external factors limiting this option.

We found that some people assessed the response to the Bouchard and Chalk Point spills on the basis of how well preparations such as planning workshops, clean up standards, spill training, pre-spill resource procurement, staging areas permissions, contingency and emergency plans, and call lists were made prior to the response. Their concern for preparedness was related to for a) gaining public support for the response and b) ensuring effective response effort and coordination.

Furthermore, some of the interviewees assessed the quality of the response based on how well it supported the needs of damage assessment and restoration related activities. For example, several interviewees in both case studies stated that an important performance metric for the response effort was how well it developed the data necessary for damage assessment.

These findings suggest that people do not necessarily consider "response" as distinct from preparation, damage assessment, and restoration. In fact, these activities can overlap in time. Activities considered as part of the response can extend for many years after the spill event and the initial clean-up. Such is the case in the Bouchard case, where several years after the spill occurred some activities related to "response" are still being conducted.

6. Conclusions

The case studies of the *Bouchard-120* and Chalk Point spill responses suggest that spill managers and interested and affected parties can have multiple objectives. They may also use multiple measures for assessing achievement of or progress on those objectives. By and large, they were cognizant of the usefulness of assessing response efforts both during and after a spill response. Although these case studies also illustrate that objectives may not be entirely shared among interested and affected parties, in general making objectives explicit and tracking performance metrics related to those objectives may improve spill management and public communication. While we did not interview many local residents (beyond those working for a local advocacy group or local officials) for these case studies (because we intentionally selected people who were engaged with the response), a number of people spoke of their experiences with members of the public. In particular, local residents were described as wanting to know very practical information that related to impacts and response: what were clean-up schedules? what were public health effects of consuming fish contaminated (or that might be contaminated) by oil? Although some of our interviewees characterized members of the public as wanting things that were impossible or that were not reasonable, e.g., remove *all* oil, a few members of the public can be characterized as quite sophisticated in their views.

We also found that while interviewees think performance metrics are important and useful, there may be differences about *when* they should be developed and used. For example, response performance metrics were developed by the Unified Command *during* the Bouchard spill when they "sat down as a group—three weeks into the spill—because first three weeks we are just responding… . We ask the bird wildlife and rehabilitation people, what are your success factors? They said we have a goal of 20% rehabilitated. We said OK, that is the objective of the bird people. We asked the safety people, what is your objective? They said, occupational safety and health said this many hours, no mishaps. So we started tracking that number and put them on a poster so people could track them… ." The hypothesis motivating our research project is that better oil spill response can be achieved if performance metrics are defined *prior* to a spill in a systematic and collaborative process. Nonetheless, establishing metrics as part of the response process may be an improvement on management without any metrics at all.

Spill response metrics can drive which strategies are chosen and help officials gain the legitimacy and trust they hope for—if they can show that their efforts measure well on the chosen metrics. Spill response metrics can also drive learning. Most generally, a measurable understanding of the different objectives of the parties directly engaged in oil spill response and the communities impacted by the spill is likely to improve understanding, communication, response and ultimately reduce the risk of adverse impacts due to oil spills.

7. Acknowledgements

This study was funded by the Coastal Response Research Center, a partnership between the University of New Hampshire and the National Oceanic and Atmospheric Administration under NOAA Grant Number: NA04NOS4190063. Project Number: 05-983.

8. References

1. Glaser B, Strauss A. 1967. The discovery of grounded theory: Strategies for qualitative research. Chicago: Aldine.

2. Glaser B. 1992. Basics of Grounded Theory. Mill Valley, CA: Sociology Press.

3. National Research Council 2000a. International Conflict Resolution After the Cold War. Washington, DC: National Academy Press.

4. NOAA. 2001. Characteristics of Response Strategies: A Guide for Spill Response Planning in Marine Environments. A joint publication of the American Petroleum Institute, National Oceanic and Atmospheric Administration, US Coast Guard, and US Environmental Protection Agency: Seattle WA. http://response.restoration.noaa.gov/book_shelf/910_response.pdf

5. Seager TP, Satterstrom FK, Tuler SP, Kay R, Linkov I. 2006. Typological Review of Environmental Performance Metrics (with Illustrative Examples for Oil Spill Response). Integrated Environmental Assessment & Management. In press.

6. Tuler SP, Kay R, Seager TP, Linkov I. 2006. Objectives and Performance Metrics In Oil Spill Response: The Bouchard-120 And Chalk Point Spill Responses. SERI Report 06-001. Social and Environmental Research Institute: Greenfield, MA. Available at www.seri-us.org/publications.html

Part 3

Applications of Multi-Criteria Decision Analysis

COMPARATIVE ASSESSMENT OF SEVERAL MULTI-CRITERIA DECISION ANALYSIS TOOLS FOR MANAGEMENT OF CONTAMINATED SEDIMENTS

I. LINKOV
INTERTOX, Inc
83 Winchester Street, Suite 1, Brookline, MA 02446 USA

F.K. SATTERSTROM
Cambridge Environmental, Inc.
58 Charles Street, Cambridge, MA 02141 USA

B. YATSALO, A. TKACHUK
Obninsk State Technical University of Nuclear Power Engineering
IATE, Studgorodok 1, Obninsk, Kaluga region, 249030, Russia

G.A. KIKER
University of Florida
Agricultural and Biological Engineering Department, PO Box 10570, Gainesville, FL 32611-0570 USA

J. KIM, T.S. BRIDGES
U.S. Army Engineer Research and Development Center
Environmental Laboratory, 3909 Halls Ferry Rd, Vicksburg, MS 39180 USA

T.P. SEAGER, K. GARDNER
Center for Contaminated Sediments Research
University of New Hampshire, Durham, NH 03824 USA

Abstract

Over the past several decades, environmental decision-making strategies have evolved into increasingly more sophisticated, information-intensive, and complex approaches including expert judgment, cost-benefit analysis, toxicological risk assessment, comparative risk assessment, and a number of methods for incorporating public and stakeholder values. This evolution has led to an improved array of decision-making aids, including the development of Multi-Criteria Decision Analysis (MCDA) tools that offer a scientifically sound decision

I. Linkov et al. (eds.), Environmental Security in Harbors and Coastal Areas, 195–215.

analytical framework. The existence of different MCDA methods and the availability of corresponding software contribute to the possibility of practical implementation of these methods. However, even though a great deal of work has been done in justifying the theoretical foundation of these methods, real-life applications are rare. The critical attitudes of different MCDA schools toward alternative approaches may have been an obstacle in the application of MCDA. Additionally, no MCDA method is theoretically appropriate for group decision processes, and all MCDA methods and tools necessarily use significant simplifications and assumptions to rank environmental policy alternatives. Nevertheless, this paper illustrates the application of three different MCDA methods in two case studies involving management of contaminated sediments. These case studies are based on real sediment management problems experienced by the US Army Corps of Engineers and other stakeholders in the New York/New Jersey Harbor and the Cocheco Superfund Site in New Hampshire. Our analysis shows that application of three different MCDA tools points to similar management solutions, no matter which tool is applied. MCDA tools and approaches were constructively used to determine the strengths and weaknesses of each method when solving the problem.

1. Multi Criteria Decision Analysis Methods and Tools

Environmental managers must decide what they wish to achieve through environmental management and how much they are willing to pay to achieve it. Controversy arises when managers (1) have different objectives with different priorities or (2) expect different outcomes from management decisions. Those affected and involved in the decision-making must also decide what they care about, how they prioritize those concerns, and how much they are willing to pay to achieve stated objectives. There are many alternatives for the management of contaminated sediments, and there are important tradeoffs among ecological, economic, technical, and societal objectives. As an example of a tradeoff, achieving significant benefits and minimizing cost are two conflicting objectives. As a consequence, a given alternative may not take clear precedence over other alternatives with respect to every objective. This may present a dilemma to a decision-maker trying to choose a single alternative.

The common purpose of MCDA methods is to evaluate and choose among alternatives based on multiple criteria using systematic analysis that overcomes the limitations of unstructured individual or group decision-making [1, 2].

The following main categories of problems are considered on the basis of MCDA [1]:

- *Sorting* alternatives into classes/categories (e.g., “unacceptable,” “possibly acceptable,” “definitely acceptable,” etc.).
- *Screening* alternatives—eliminating those alternatives that do not appear to warrant further attention; i.e., selecting a smaller set of alternatives that (very likely) contains the ‘best’ alternative.

- *Ranking* alternatives (from "best" to "worst" according to a chosen algorithm).
- *Selecting* the "best alternative" from a given set of alternatives.
- *Designing* (searching, identifying, creating) a new action/alternative to meet goals.

Some other categories of problems— e.g. description/learning problematique (analysis of actions to gain greater understanding of what may or may not be achievable) and portfolio problematique (choice of a subset of alternatives, taking into account not only individual characteristics of each alternative, but also their positive and negative interrelations)—may also be considered with the use of MCDA approaches.

Two *key schools* within the MCDA methodologies are considered in this paper. Each is based on the specific approaches to multiple criteria analysis and methods used:

- *Value function*-based methods (see also a brief discussion below concerning the AHP method).
- *Outranking* methods.

Approaches that use value functions form the so called MAVT methods (multiattribute *value* theory). However, MAUT methods (multiattribute *utility* theory) are also often used. There are several judgements concerning the interpretation of the differences between MAVT and MAUT. For example, a value function describes a person's preference regarding different levels of an attribute under *certainty*. In contrast, a utility function is more general because it reflects the person's attitude towards taking risk (e.g., risk averse, risk seeking, risk neutral), which cannot be captured by a value function. However, Winterfeldt and Edwards [2] do not recognize the principal differences between *value* and *utility* functions and consider *utility* as "a different set of elicitation methods intended to provide consistency checks on the construction of a *value* function." Similarly, taking into account the Expected Utility Theory developed by von Neumann and Morgenstern [3], *utility theory* may be viewed as an extension/generalization of value measurement, related to the use of probabilities and expectations to deal with uncertainty [1]. Therefore, in most cases, when analyzing applied MCDA problems, authors do not distinguish MAVT and MAUT, indicating simply the implementation of MAVT/MAUT methods.

The objective of MAVT is to model and represent the decision maker's preferential system into a value function $V(\boldsymbol{a})$,

$$V(\boldsymbol{a}) = F(V_1(a_1), \ldots, V_m(a_m)); \tag{1}$$

where alternative $\boldsymbol{a}$ is presented as a vector of the evaluation criteria $\boldsymbol{a}=(a_1, \ldots, a_m)$, a_i is the assessment of alternative $\boldsymbol{a}$ according to criterion i, and $V_i(a_i)$ is the value score of the alternative reflecting its performance on criterion i (as a rule, $0 \leq V_i(a_i) \leq 100$). The most widely used form of function $F(\,)$ is an additive model:

$$V(\boldsymbol{a}) = w_1 V_1(a_1) + \ldots + w_m V_m(a_m), \tag{2}$$

$$w_i > 0\,,\ \Sigma w_i = 1, \tag{3}$$

where w_i, i=1,...,n, are the weights reflecting the relative importance of criteria (or corresponding scaling factors, [1,2]). It should be stressed, however, that for a justified implementation of the additive model (2) some requirements/axioms of MAVT should be held (among them the key ones are the *preferential independence* requirements, [1,2]). MAVT relies on the assumption that the decision-maker is rational (preferring more utility to less utility, for example), that the decision-maker has perfect knowledge, and that the decision-maker is consistent in his judgments. The goal of decision-makers in this process is to maximize the overall value $V(\boldsymbol{a})$ of alternative $\boldsymbol{a}$.

Various sophisticated methods for defining partial value functions $V_i(x)$ and assessing weights w_i have been developed both for quantitative and qualitative criteria. One of the most popular and simplest versions of MAVT is SMART (Simple Multi-Attribute Rating Technique), suggested by Edwards [4]; at present several versions of SMART are used [2,5].

One of the MAVT versions also used in this work is the PRIME method (Preference Ratios in Multiattribute Evaluation) developed by Ahti A. Salo and Raimo P. Hamalainen to assist multicriteria decision making process in case of incomplete information through the use of interval-valued ratio statements [16].

The preference elicitation consists of two phases: score elicitation and weight elicitation. For each attribute, score information is obtained through identification of the least and most preferred achievement levels, ordinal ranking of other achievement levels and elicitation of possibly interval-valued estimates for ratios of value differences (cardinal ranking). The PRIME method supports several approaches to the elicitation of attribute weights, including extension of the SWING method. The results of preference elicitation and synthesis consist of value intervals for the alternatives, weight intervals for the criteria, and dominance structures (absolute dominance and pairwise dominance) and decision rules (maximax, maximin, minimax and central values) for the comparison of alternatives.

PAIRS (Preference Assessment by Imprecise Ratio Statements) and SPAIRS (Simple PAIRS or Interval SMART/SWING), which we implement in our investigation, are also some extension of MAVT through the use of interval methods [17]. SPAIRS allows the DM to reply with intervals to the ratio questions to describe the possible imprecision in these. These intervals set constraints for the feasible weights of the criteria, and similar constraints can be set for the ratings of the alternatives. As a result, the overall values of the alternatives will also be intervals describing the possible variation in these due to allowed variation in the attribute weights and the ratings of the alternatives.

Because poor scores on some criteria can be compensated for by high scores on other criteria, MAVT is part of a group of MCDA techniques known as "compensatory" methods.

Outranking approaches imply forming an ordered relation of a given set of alternatives. Outranking methods are based on a pairwise comparison of alternatives for each criterion under consideration with subsequent integration of obtained preferences according to a chosen algorithm. Among outranking approaches, the ELECTRE family of methods developed by Roy [6] and the PROMETHEE method developed by Brans [7] are most used.

PROMETHEE is based on the performance matrix $\{z_i(\boldsymbol{a})\}$ (where $z_i(\boldsymbol{a})$ is an evaluation of alternative $\boldsymbol{a}$ against criterion i) and a chosen preference function $f_i(x)$, $0 \le f_i(x) \le 1$, with specified indifference and preference thresholds. It determines the intensity of preference for alternative $\boldsymbol{a}$ over alternative $\boldsymbol{b}$, $P_i(\boldsymbol{a},\boldsymbol{b}) = f_i(z_i(\boldsymbol{a}) - z_i(\boldsymbol{b}))$, and the preference index, $P(\boldsymbol{a},\boldsymbol{b})$,

$$P(\boldsymbol{a},\boldsymbol{b}) = \Sigma w_i \, P_i(\boldsymbol{a},\boldsymbol{b}) \tag{4}$$

where weights w_i reflect the relative importance of criteria and meet the requirements of (3). According to the features of preference functions $f_i(x)$, if $P_i(\boldsymbol{a},\boldsymbol{b})>0$, then $P_i(\boldsymbol{b},\boldsymbol{a})=0$. Preference indices are used for determination of positive outranking flow $Q^+(\boldsymbol{a})$:

$$Q^+(a) = \Sigma_b \, P(a,b) \tag{5}$$

and negative outranking flow $Q^-(\boldsymbol{a})$:

$$Q^-(\boldsymbol{a}) = \Sigma_b \, P(\boldsymbol{b},\boldsymbol{a}), \tag{6}$$

summed over all alternatives $\boldsymbol{b} \neq \boldsymbol{a}$.

According to the PROMETHEE 1 method, $\boldsymbol{a}$ outranks $\boldsymbol{b}$ if $Q^+(\boldsymbol{a}) \geq Q^+(\boldsymbol{b})$ and $Q^-(\boldsymbol{a}) \leq Q^-(\boldsymbol{b})$; $\boldsymbol{a}$ is indifferent to $\boldsymbol{b}$ if $Q^+(\boldsymbol{a}) = Q^+(\boldsymbol{b})$ and $Q^-(\boldsymbol{a}) = Q^-(\boldsymbol{b})$; $\boldsymbol{a}$ and $\boldsymbol{b}$ are incomparable if $Q^+(\boldsymbol{a}) > Q^+(\boldsymbol{b})$ and $Q^-(\boldsymbol{b}) < Q^-(\boldsymbol{a})$, or $Q^+(\boldsymbol{b}) > Q^+(\boldsymbol{a})$ and $Q^-(\boldsymbol{a}) < Q^-(\boldsymbol{b})$. Thus, PROMETHEE 1, like some other outranking methods, does not presuppose that a single best alternative can be identified, since some alternatives may be incomparable.

The PROMETHEE 2 method is based on the "net flow" criteria $Q(\boldsymbol{a})$:

$$Q(\boldsymbol{a}) = Q^+(\boldsymbol{a}) - Q^-(\boldsymbol{a}), \tag{7}$$

and it may be used for a complete ranking of alternatives (and alternative $\boldsymbol{a}$ outranks $\boldsymbol{b}$ if $Q(\boldsymbol{a}) > Q(\boldsymbol{b})$), though this approach is more disputed than PROMETHEE 1.

PROMETHEE, like other outranking methods, is considered an attractive and transparent method, although both positive and negative flows depend on the complete set of alternatives under consideration [1]. However, a drawback of outranking is that "indifference" and "preference" thresholds are essentially arbitrary and the relationship representing which alternatives outrank which depends on selection of those thresholds. One way to analyze the robustness and check consistency between thresholds is to manipulate the thresholds. Alternatively, Monte Carlo simulation can be used for exploring parameter space and deriving robust conclusions; e.g., [8].

Outranking techniques allow inferior performance on some criteria to be compensated for by superior performance on others [3-6]. They do not necessarily, however, take into account the magnitude of relative underperformance in a criterion versus the magnitude of over-performance in another criterion. Therefore, outranking models are known as "partially compensatory."

The Analytic Hierarchy Process (AHP), developed by Saaty [9], presents an integration of the additive model [1] with a distinctive determination of the decision matrix, $V_{i,a}$, and criteria weights, w_i. Within AHP, instead of defining a value function $V_i(x)$, a systematic pairwise comparison of alternatives with respect to each criterion is used based on a special ratio scale: for a given criterion, alternative $\boldsymbol{i}$ is preferred to alternative $\boldsymbol{j}$ with the strength of preference given by $a_{ij}=s$, $1 \leq s \leq 9$, correspondingly, $a_{ji}=1/s$. Then, the same procedure is implemented for n(n-1)/2 pairwise comparisons for a problem with n criteria. The obtained matrices are processed (by extracting the eigenvector corresponding to the maximum eigenvalue of the pairwise comparison matrix) and yield the values $V_{i,a}$ and weights w_i for subsequent use with the model, when preferences are aggregated across different criteria (2).

AHP may thus be considered an approach with an elicited value function (scoring) and criteria weights (weighting). However, taking into account different assumptions and approaches, proponents of AHP insist that it is not a value function method [1]. Additionally, AHP relies on the supposition that humans are more capable of making relative judgments than absolute judgments. Consequently, the rationality assumption in AHP is more relaxed than in MAVT.

We use in this work the INPRE as an interval-based AHP methods [17].

INPRE is based on preference programming that extends AHP by synthesizing interval preference statements in the hierarchy to derive intervals for the weights of alternatives.

Despite longstanding discussions on the correctness of AHP for analyzing and ranking alternatives (including specific issues such as the "rank reversal problem") [1, 10], its transparency and relatively simple pairwise judgements make it a popular decision analysis method.

2. Cocheco River and NY/NJ Harbor Case Studies

The two case studies selected for this research are representative of sediment management challenges faced by the US Army Corps of Engineers and other agencies. The Cocheco River, the first case study, is located in southeastern New Hampshire and flows toward the Gulf of Maine and the Atlantic Ocean. A section of the river, from below the dam in the center of the city of Dover to the Cocheco's confluence with the Piscataqua River, was proposed for dredging. Plans to dredge have been in the works since approximately 1996. There are many motivations for the dredging, including the economic redevelopment of Dover and the overriding goal of maintaining a navigable channel for federal navigation. Because it is a navigable waterway, the US Army Corps of Engineers (USACE) has been helping the city of Dover to coordinate the process and will be performing the dredging. There has been much debate over the need to dredge and remove sediment from the bottom of the Cocheco River. Approximately 45,000-60,000 cubic yards of sediment, some of which are contaminated with polyaromatic hydrocarbons (PAHs) and heavy metals, are planned for removal. The decision regarding what to do with the contaminated sediment was not an easy one. Regulatory constraints required secure disposal of contaminated materials (i.e., prohibiting ocean dumping). Other commonly used options (contained aquatic disposal, landfill) were not found to be useful for the site. After extensive negotiations, cement manufacture, flowable fill, wetlands restoration, and an upland disposal cell were identified as feasible alternatives for consideration (Table 1) [11].

Table 1: Alternatives under study.

Technology	Process and Hypothesis
Wetland restoration	Surrounding contaminated sediment core with clean material in new wetland cell may restore hydrologic function and ecological habitat to areas diked and/or drained.
Cement manufacture	Blending with conventional raw materials and firing in rotary kiln for manufacture of cement may destroy organic contamination; metals may be bound upon hydration of Portland cement concrete in normal construction applications.
Upland brownfield disposal cell	Dewatering, compacting, and capping on site may prevent dispersion of contaminants to the environment and allow construction of recreation space on top of cell.
Cement stabilization in flowable fill	Blending with pozzolanic material such as cement, fly ash, or blast furnace slag may bind contaminants upon hydration in normal structural applications such as trench backfilling or soil strengthening.

In the second case study, contaminated sediment management issues within the greater New York/New Jersey (NY/NJ) Harbor area are considered [12-13]. Several million cubic meters of sediment must be dredged each year to maintain navigation channels for harbor access. Due to long-term human use of the harbor area, significant contaminant concentrations have been recorded in certain areas. Additional challenges in sediment management have been created by the limitation of ocean disposal to only clean sediments and plans for deepening existing channels to allow increased access for large transport vessels. New and innovative sediment management options, along with their associated risk and decision analyses, are required for contaminated sediments within the NY/NJ harbor area and need to be systematically explored for cost-efficient risk reduction. Recognizing that the objectives of different stakeholder groups differ and often conflict, many different alternatives and criteria were considered within the NY-NJ case study (Figure 1) [14].

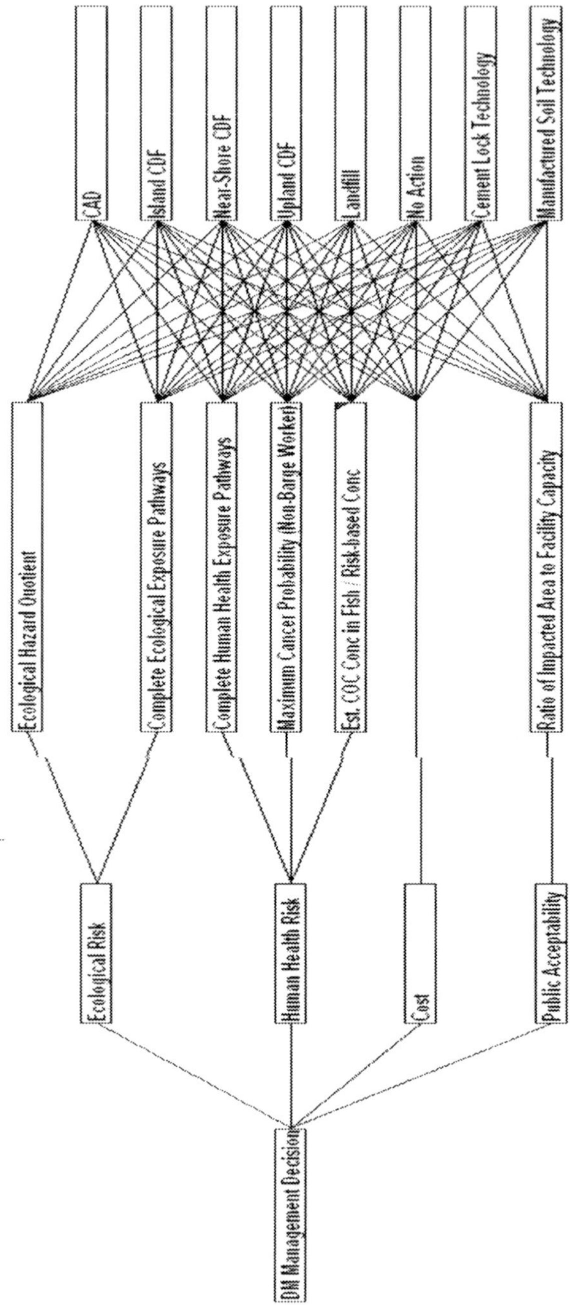

Figure 1: Value tree for NY/NJ case study: criteria and alternatives [14].

In both case studies, stakeholders in the decision making process have a wide array of concerns, some overlapping and others exclusive, about the management of contaminated sediments [11,14]. MCDA methods and tools provide a sound approach to sediment management that integrates economic and technical considerations (such as cost, human health risks, and environmental risks) with social factors (public acceptance, environmental justice, and others).

3. Methodology

To test the sensitivity of the 'optimal' management alternative to the specific MCDA method used, this project employs all three methods discussed above (MAVT, Outranking, AHP) and compares the resulting selection of a sediment management alternative for the two case studies.

The starting points for the analysis presented in this study were the performance matrices developed for the Cocheco and NY/NJ case studies. The performance matrix for the Cocheco case study (Table 2) presents an evaluation of four alternative policies using four criteria (cost, environmental quality, and impact on ecological and human health habitats) [11]. One of them, *Environmental Quality*, is a qualitative criterion. For this criterion, a three-level qualitative scale is considered: low, medium, and high. Other criteria are quantitative.

Table 2: Performance table for the Cocheco case study [11]. "No change" means 0.

Alternatives/Criteria	Cost ($/cy)	Environmental Quality	Ecological Habitat (acres)	Human Habitat (acres)
Wetlands Restoration	$75	High	10 add'l.	No change
Cement Manufacture	$30	High	No change	No change
Upland Disposal Cell	$40	Medium	No change	4 add'l.
Flowable Fill	$55	Medium	No change	No change

The performance table for the NY/NJ case study (Table 3) presents an evaluation of eight alternative policies using seven quantitative criteria [14].

Table 3: Performance table for the NY/NJ case study [14].

Criteria/ Alternatives	Confined Aquatic Disposal (CAD)	Island CDF (Confined Disposal Facilities)	Near-Shore CDF	Upland CDF	Landfill	No Action	Cement Lock Technology	Manufactured Soil Technology
Magnitude of Ecological Hazard Quotient	680.00	2100.00	900.00	900.00	0.00	5200.00	0.00	8.70
Complete Ecological Exposure Pathways	23.00	38.00	38.00	38.00	0.00	41.00	14.00	18.00
Complete Human Health Exposure Pathways	18.00	24.00	24.00	24.00	21.00	12.00	25.00	22.00
Magnitude of Maximum Cancer Risk (Non-Barge Worker)	0.03	0.09	0.04	0.04	0.30	0.20	0.02	1.00
Estimated concentration of contaminant of concern (COC) in Fish / Risk-based Conc.	28.00	92.00	38.00	38.00	0.00	220.00	0.00	0.00
Cost ($/CY)	5.00	25.00	15.00	20.00	70.00	2.00	75.00	60.00
Ratio of Impacted Area to Facility Capacity (acres/MCY)	4400.00	980.00	6500.00	6500.00	0.00	0.00	0.00	750.00

Tables 2 and 3 were transformed to fit input data formats for different software packages:

- *Decision Lab*, which employs the PROMETHEE method.
- *Expert Choice*, which uses the AHP method.
- *Criterium Decision Plus*, which implements MAVT (using the SMART approach).
- *PRIME DECISIONS* with realization of interval-based PRIME method.
- *WINPRE* (Workbench for Interactive Preference Programming) using INPRE (Interval AHP) and SPAIRS (Simple PAIRS or Interval SMART/SWING).

All the indicated packages possess wide performance capabilities, including sensitivity analysis, presentation of various output tables, and graphic user interfaces.

Experts and stakeholders were involved in structuring the MCDA problems mentioned above for the Cocheco and NY/NJ case studies, developing the performance tables, and criteria weighting [11,14]. Four theoretical stakeholder groups, and the weights which might be elicited from them, were originally elaborated within the Cocheco case study [11] (Table 4).

Table 4: Criteria weights (%), Cocheco case study [11].

Scenario/Criterion	Cost	Environmental Quality	Ecological Habitat	Human Habitat
Eco-Environmental	10	30	40	20
Human Health	10	30	30	30
Commercial	30	30	10	30
Balanced	25	25	25	25

Two of the groups (Eco-Environmental and Human Health) are considered in this work. Three scenarios for weights were proposed within the NY/NJ case study [14], and one of them (in which experts involved in the weighting process are envisioned to be affiliated with the EPA) is considered in the present investigation (Table 5).

Table 5: Mean criteria weights (%) with standard deviation, NY/NJ case study [14].

Criterion/ Scenario	EPA	USAGE	PB Meeting
Impacted Area/Capacity	7.37 (4.87)	12.46 (7.50)	10.77 (6.45)
Magnitude of Ecol HQ	20.29 (6.50)	14.72 (6.66)	16.58 (4.86)

Criterion/ Scenario	EPA	USAGE	PB Meeting
Ecological Pathways	15.36 (5.52)	12.40 (4.78)	15.87 (8.42)
Human Pathways	12.81 (5.51)	14.79 (5.54)	12.63 (4.93)
Magnitude of Cancer Risk	18.86 (4.53)	14.13 (7.86)	18.48 (6.26)
Ratio of Fish COC/Risk level	15.33 (3.89)	11.82 (6.19)	13.00 (5.30)
Cost	9.97 (4.15)	19.66 (7.38)	12.67 (6.08)

For the MCDA analysis of the Cocheco case study, the software package *Decision Lab* was originally used [11], and analysis of the NY/NJ case study was originally based on the *Criterium Decision Plus* package [14]. Implementation of other software packages for cross analysis of indicated case studies was based on performance tables (Tables 2 and 3) and corresponding criteria weights (Tables 4 and 5).

The above data and comments were taken into account for comparison of ranking orders for indicated case studies and software packages. In all the scenarios linear value functions within MAVT, and linear (for NY/NJ case study) and U-shape (for Cocheco case study) preference functions within PROMETHEE were used.

Using *PRIME DECISIONS and WINPRE* with corresponding interval-based methods we have taken into account all the criteria weights for different scenarios, Tables 4 and 5, for subsequent forming intervals for weights of the criteria.

4. Results

We discuss below only the ranking order of the alternatives for the two case studies under consideration. However, we would like to stress that ranking alternatives is not the final step but only one of the steps in the implementation of MCDA for the analysis of applied problems [1,2].

Table 6 presents an alternative ranking for the Cocheco site using different software packages. The performance matrix presented in section 3 was used in the all indicated software packages.

All MCDA software packages used within this work predict that *flowable fill* is the least attractive alternative for both stakeholder preference scenarios. *Wetland restoration* was ranked as the most attractive option by all methods for the two indicated groups of stakeholders. *Cement manufacture* and *upland capped* have ranks 2 or 3 depending on the method used (Table 6).

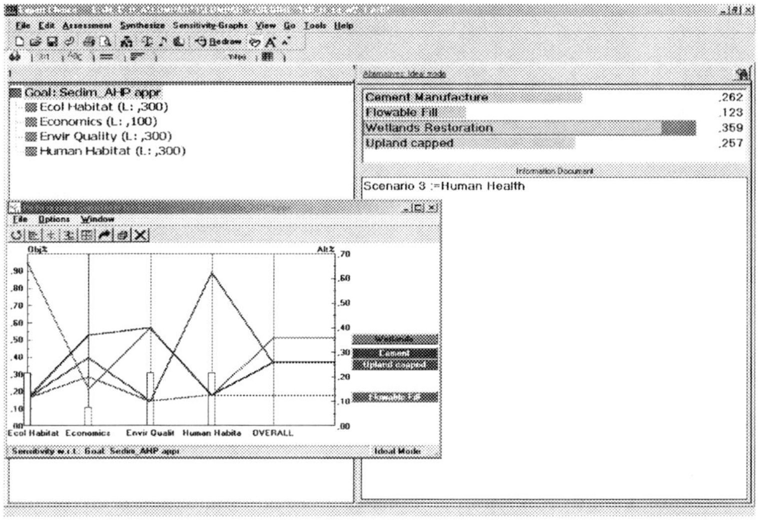

Figure 2: Case study 1: alternative ranking and sensitivity analysis, AHP/Expert Choice (human health scenario).

According to sensitivity analysis, using the PROMETHEE and Human Health scenario, increasing the weight for the Human Habitat criterion from 30% (Table 5) to 35% changes the ranking orders of the *cement manufacture* and *upland capped* alternatives, whereas doubling the weight of the cost criterion (from 10% to 20%) changes the ranks of *wetland restoration* and *cement manufacture*. Using the AHP method for the Human Health scenario of the Cocheco case study demonstrates that the difference between the *cement manufacture* and *upland capped* alternatives is negligible (Figure 2).

Table 6: Ranking alternatives for the Cocheco case study using different MCDA methods. Two criteria weighting scenarios were used: Eco-Environmentalists / Public Health.

Software & *Method*	**Alternatives**			
	Wetlands Restoration	***Cement Manufacture***	***Upland Capped***	***Flowable Fill***
ExpertChoice, *AHP*;	1/1	2/2	3/3	4/4
DecisionLab, *PROMETHEE*	1/1	2/2	3/3	4/4
CritDecPlus, *MAVT*	1/1	3/3	2/2	4/4
PRIME DECISIONS, *PRIME*	1	2	3	4
WINPRE, *INPRE*	1	3	2	4

Table 7: Ranking alternatives for NY/NJ sites using different MCDA methods.

Software & Method	Alternatives							
	CAD	**Island CDF**	**Near-Shore CDF**	**Upland CDF**	**Landfill**	**Cement Lock**	**Manufactured Soil**	**No Action**
ExpertChoice, AHP	3-4	8	6-7	6-7	1-2	2-1	3-4	5
DecisionLab, PROMETHEE 1,2	2	8	5	6	3	1	4	7
CritDecPlus, MAVT	2	7	4	5	1	3	6	8
PRIME DECISIONS, PRIME	1	6-7	4	5	2	3	7-6	8
WINPRE, SPAIRS	1	5-6	4	4	2	3	6-5	7

Implementation of MAVT/CDplus demonstrates that *wetland restoration* is the most attractive option, then *upland capped* (second) and only then *cement manufacture*, Figure 3. The Eco-Environmental scenario gives the same result (Table 6). Thus, we may state that alternative rankings for the Cocheco case study are practically the same, taking into account sensitivity analysis, independent of the method/software used.

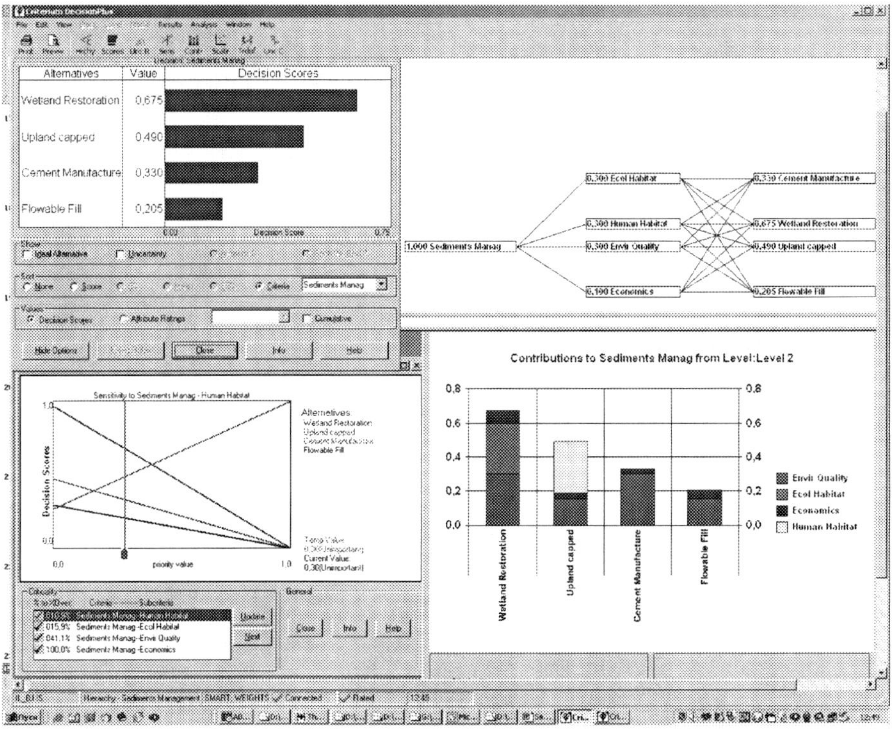

Figure 3: Case study 1: alternative ranking and sensitivity analysis, MAVT/CDplus (human health scenario).

For the NY/NJ case study, *cement lock* and *landfill* and *CAD* were ranked as top choices by all three software tools (Figure 4, Table 7). The specific order of the first three alternatives, as well as of the subsequent ones, can change due to small changes in weights, according to sensitivity analysis on the basis of methods/software used (see Figure 4 and Figure 5). That *CAD* ranks 3 or 4 within the AHP method (Table 7) is a result of uncertainty influences when transforming data from the performance table (Table 4) into the AHP scale. Differences between rank 5 and 3 are negligible (Figure 5). In addition, a "rank reversal" effect for the AHP method was observed when decreasing the number of alternatives from 8 to 5 (e.g., the rank ordering of *CAD* and *manufactured soil* changed).

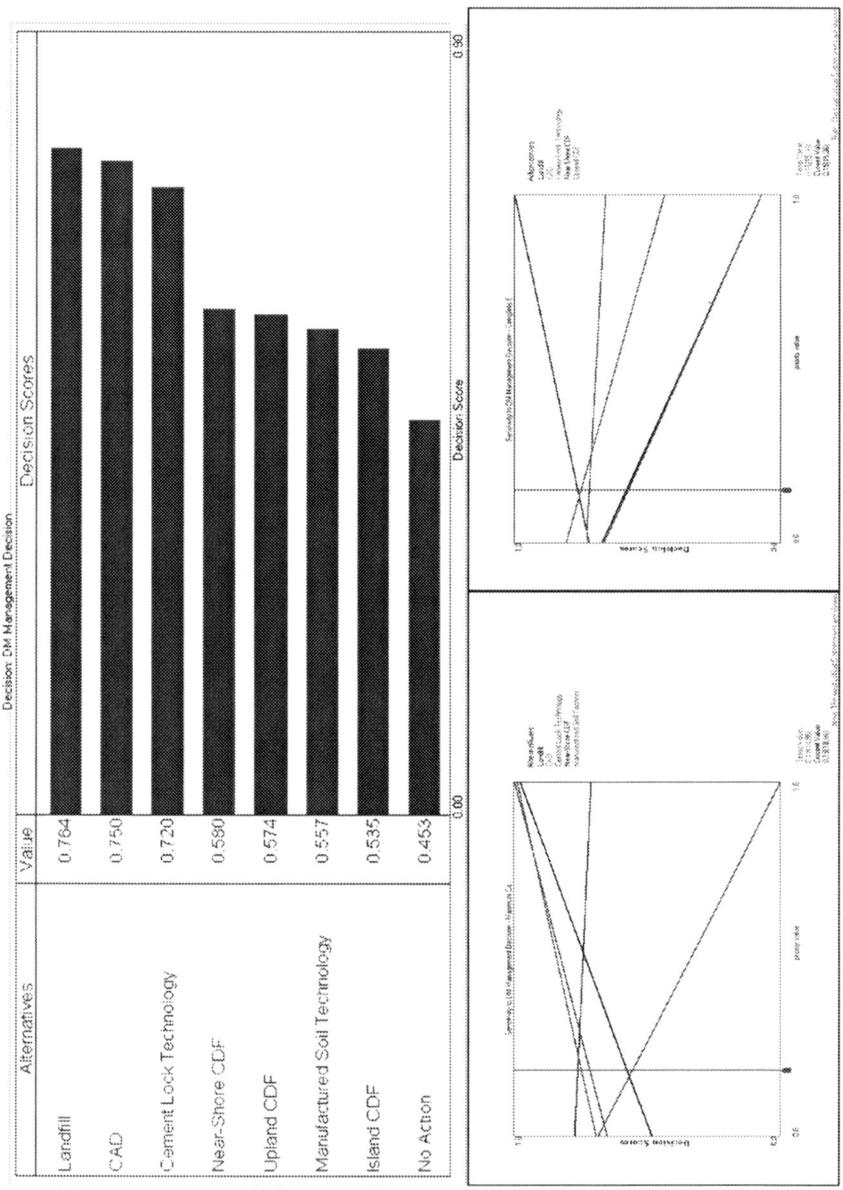

Figure 4: Case study 2: alternative ranking and sensitivity analysis, MAVT/CDplus.

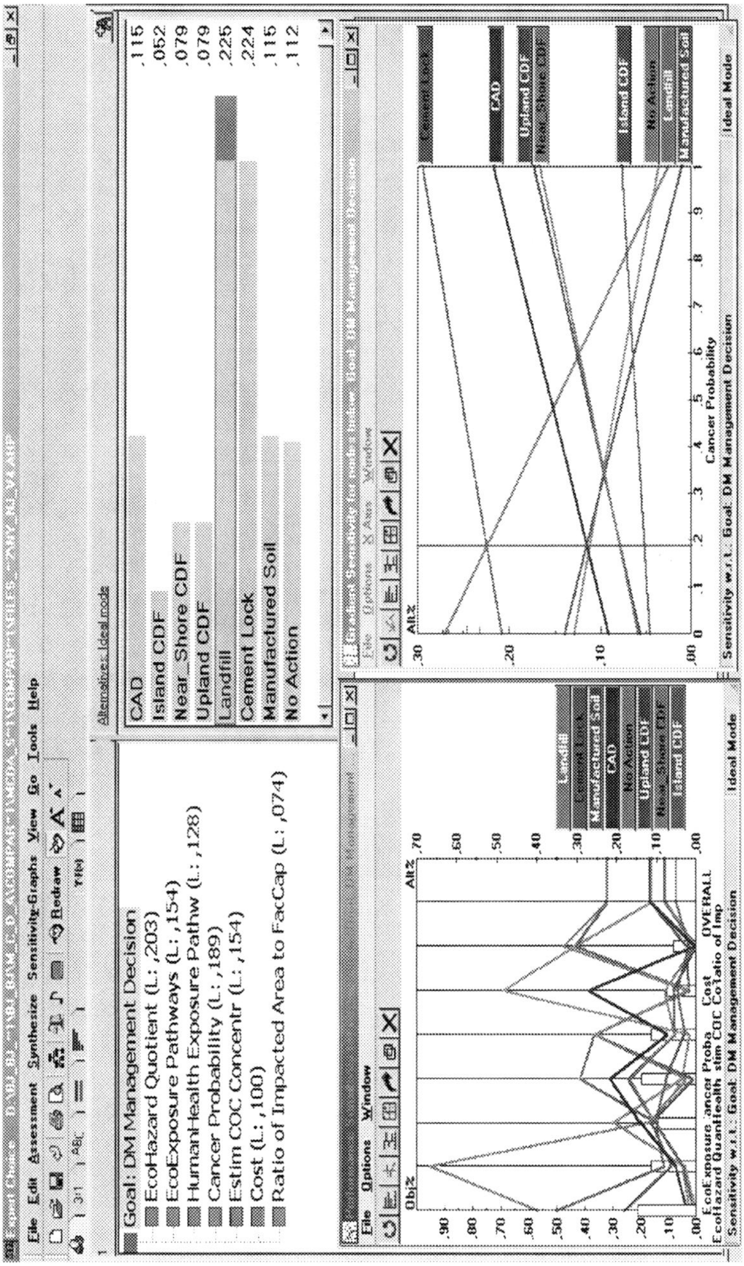

Figure 5: Case study 2: alternative ranking and sensitivity analysis, AHP/Expert Choice.

Thus, despite a relative increase in discrepancies of rank ordering for the NY/NJ case study (Table 7, as compared to Cocheco case study, Table 6), the results of such an analysis lead to the finding that *three alternatives* indicated above can be considered for further analysis as the most justified within the MCDA methodology used.

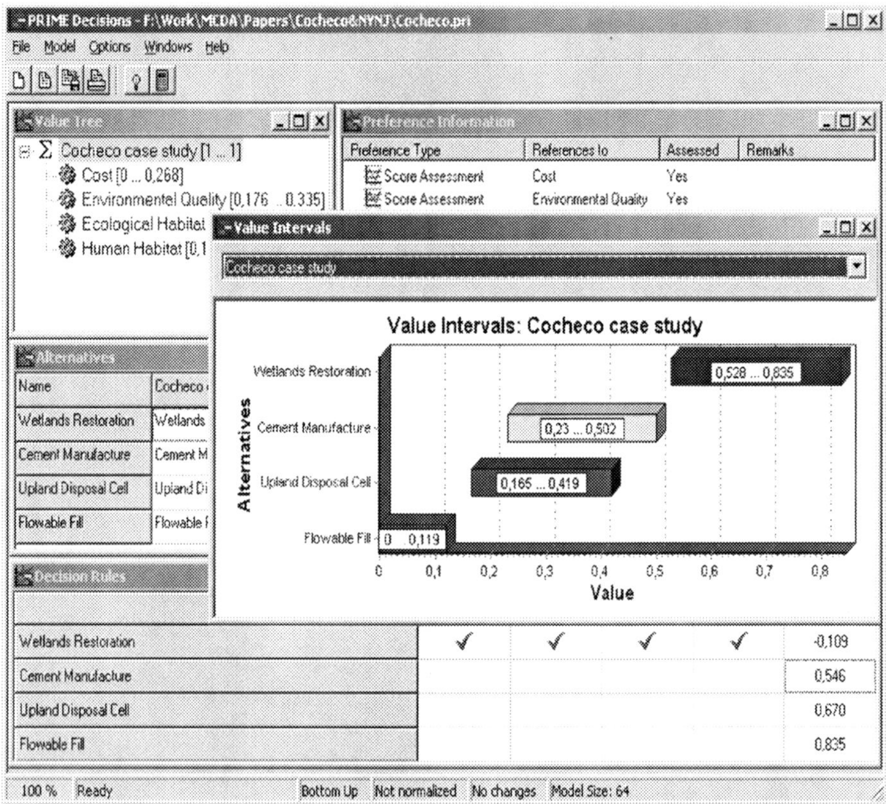

Figure 6: Case Study 1: Value intervals and decision rules, PRIME DECISIONS.

Output forms for PRIME DECISIONS with implementation of interval-based methods are presented in Figure 6. Differences in ranking alternatives for interval methods (using PRIME DECISIONS and WINPRE) in comparison with other ones for case study 2 (see Table 7), are caused by the specificity of interval methods which were applied. Such differences in the resulting ranking can be considered as uncertainty involved by using of uncertain or interval-valued statements.

5. Conclusion

Certainly, if two different groups of experts analyze a given task using the same method (e.g., outranking), we cannot state that the resulting alternative rankings will completely coincide (e.g., when groups suggest different criteria weights and/or ranking is sensitive to the preference functions or indifference/preference thresholds). And if two expert groups make their judgments under two facilitators who use different methods (e.g., MAVT and outranking), we cannot state that the ranking order should be the same even if these groups nearly coincide. In fact, if within a MAVT approach the criteria weights were elaborated as *swing weights*; i.e., are the *scaling factors* (which relate scores on one criterion to scores on other criteria), then these weights may differ from weights elicited within an outranking approach based on ranking or rating methods. In addition, implementation of value functions $V_i(x)$ and intensity of preference functions $P_i(\boldsymbol{a},\boldsymbol{b})$ developed by expert groups may also increase differences in ranking order for alternatives, which are based on the overall value function $V(\boldsymbol{a})$ and outranking flows $Q^+(\boldsymbol{a})$ and $Q^-(\boldsymbol{a})$.

Specific differences in ranking order also occur if one of the expert groups makes its judgments working with a facilitator within the AHP method while another group does the same within MAVT or outranking. Moreover, in this situation there are no well-defined and unique rules for transforming both quantitative and qualitative criteria performances from a set of data developed under MAVT/outranking into the AHP scale. Although a pairwise comparison of alternatives against a quantitative criterion is effective, an automatic transformation of pairwise ratios based on numerical values of criteria (see, e.g., Table 3) as well as from interval (0, 100) into the standard AHP value scale ($1 \le s \le 9$ and $1/9 \le 1/s \le 1$) may differ from corresponding judgments made by experts using the AHP method.

Our analysis shows that even though each MCDA method and its associated tools may use a unique theoretical background and calculation algorithms, they may be consistent in clustering environmental management alternatives. For example, the Cocheco case study shows that flowable fill is clearly the least appropriate alternative and can be safely removed from consideration. Three other alternatives may be ranked in the following order: *wetlands restoration*, and then *cement manufacture* or *upland capped.*

For the NY/NJ case study, the top three alternatives (*landfill*, *cement lock*, and *CAD*) clearly outperform the remaining five, and the ranking order is sensitive to the methods implemented and to relatively small change of weights.

The overall utility of this consideration is the ability to focus on the top few alternatives and eliminate underperforming alternatives from consideration. A comparative analysis of results on the basis of different MCDA methods (MAVT, outranking, and AHP) demonstrates similar ranking orders for alternatives within the two case studies considered. At the same time, results obtained with the use of the AHP method, though it is relatively simple and suitable for practical implementation, may raise some doubts about their validity and robustness.

Although the ranking order of alternatives may change when using different MCDA methods, a cross-platform analysis of a multicriteria problem may play an effective role in the interactive and iterative process of problem understanding and eliciting key parameters and functions of the methods being implemented, and it is also valuable for subsequent decision making. Therefore, the benefit of using more than one MCDA method in combination is an important lesson to be drawn from our results. The use of multiple methods can help managers think hard about a problem, develop a consistent set of preferences, and have confidence in their judgments. Also, although there is no such thing as a "right answer" within MCDA approaches [1], decision makers need a justified method which can be verified using other approaches (cross-validation in a broad sense).

Therefore, the cross-platform analysis presented in this paper may be useful both for screening purposes [15] and for the final choice of ranking order for alternatives, taking into account a wide range of both parameter and model uncertainties.

Certainly, there are questions concerning realization of such a cross-platform analysis; e.g.:

- Is such a cross-platform analysis, realized by analyst(s), "correct" or it should it be carried out by non-overlapping groups of experts and stakeholders?
- Can a "cross-platform" analysis give more than sensitivity/uncertainty analysis on the basis of one method?

Concerning the first question, there may be different points of view. However, we demonstrated in this investigation a positive answer regarding the second question.

The findings of this paper emphasize not the difficulties in implementation of different MCDA methods/software when analyzing multicriteria objectives, but the necessity of including MCDA specialists in solving specific practical multi-criteria problems.

6. References

1. Belton, V. and Steward, T. (2002). *Multiple Criteria Decision Analysis: An Integrated Approach*. Kluwer Academic Publishers: Boston, MA.

2. von Winterfeldt D. and Edwards W. (1986). *Decision Analysis and Behavioral Research*. Cambridge University Press, Cambridge.

3. von Neumann, J. and Morgenstern, O. (1947) *Theory of games and economic behavior*. Princeton, NJ: Princeton University Press.

4. Edwards, W. (1971). "Social utilities." *Engineering Economist*, Summer Symposium Series, 6, 119-129.

5. Edwards, W. and Barron, F.H. (1994). "SMART and SMARTER: Improved simple methods for multiattribute utility measurement." *Organizational Behavior and Human Decision Process*, 60:306-325.

6. Roy, B. (1996). *Multicriteria Methodology for Decision Aiding*. Kluwer Academic Publishers, Dordrecht.

7. Brans, J.P. and Vinckle, P. (1985). "A preference ranking organization method: the PROMETHEE method for multiple criteria decision-making." *Management Science*, 31:647-656.

8. Lahdelma, R.; Hokkanen, J.; and Salminen, P. (1998), "SMAA – Stochastic multiobjective acceptability analysis." *European Journal of Operational Research* 106, 137–143.

9. Saaty, T.L. (1980). *The Analytic Hierarchy Process*. McGraw-Hill, New York.

10. Dyer, J.S. (1990). "Remarks on the Analytic Hierarchy Process." *Management Science*, 36:249-258.

11. Rogers, S.H., Seager, T.P., and Gardner, K.H. (2004). "Combining expert judgement and stakeholder values with Promethee: a case study in contaminated sediments management. " In: *Comparative Risk Assessment and Environmental Decision Making*. I. Linkov and A. Bakr Ramadan eds. Kluwer Academic Publishers, p. 305-322.

12. Kane Driscoll, S.B.; Wickwire, W.T.; Cura, J.J.; Vorhees, D.J.; Butler, C.D.; Moore, M.; and Bridges, T. (2002). "A Comparative Screening-Level Ecological and Human Health Risk Assessment for Dredged Material Management Alternatives in New York/New Jersey Harbor." *Human and Ecological Risk Assessment* 8:3: 603-626.

13. Wakeman, T.H. and Themelis, N.J. (2001). "A basin-wide approach to dredged material in New York/New Jersey Harbor." *Journal of Hazardous Materials* 85:1-13.

14. Kiker, GA: Bridges TS: Linkov, I: Varghese, A; and Seager, T. (2005). "Application of Multi-Criteria Decision Analysis in Environmental Decision-Making. " *Integrated Environmental Assessment and Management* 1(2):1-14.

15. Hobbs, B.F. and Meier, P. (2000. *Energy decisions and the environment: a guide to the use of multicriteria methods*. Boston: Kluwer Academic Publishers.

16. Salo, A. and Hämäläinen, R.P. (2001). "Preference Ratios in Multiattribute Evaluation (PRIME) - Elicitation and Decision Procedures under Incomplete Information," *IEEE Transactions on Systems, Man, and Cybernetics - Part A: Systems and Humans*, Vol. 31, No. 6, November 2001, pp. 533-545.

17. Salo, A. and Hämäläinen, R.P.: (1992). "Preference assessment by imprecise ratio statements," *Operations Research*, Vol. 40, No. 6, November-December 1992, pp. 1053-1061.

SMAA-TRI

A Parameter Stability Analysis Method for ELECTRE TRI

T. TERVONEN, R. LAHDELMA
Centre for Management Studies, Instituto Superior Técnico, 2780-990
Porto Salvo, Portugal
Department of Information Technology, University of Turku
FIN-20520

J. ALMEIDA DIAS
Centre for Management Studies, Instituto Superior Técnico, 2780-990
Porto Salvo, Portugal

J. FIGUEIRA
Centre for Management Studies, Instituto Superior Técnico, 2780-990
Porto Salvo, Portugal

P. SALMINEN
University of Jyväskylä
School of Business and Economics, P.O. Box 35 FI-40014 Finland

Abstract

ELECTRE TRI is a multiple criteria decision aiding sorting method with a history of successful real-life applications. ELECTRE TRI requires as input several parameters, such as criteria weights, thresholds, category profiles, and lambda cutting level. We propose the SMAA-TRI method for analyzing the stability of ELECTRE TRI analysis and for deriving robust conclusions. SMAA-TRI is based on Stochastic Multicriteria Acceptability Analysis (SMAA), and it allows ELECTRE TRI to be used with imprecise, arbitrarily distributed values for weights and the lambda cutting level. The method computes for each alternative action the share of parameter values that have it assigned to different categories. We list some potential military applications. To demonstrate SMAA-TRI, we re-analyze a case study in the field of risk assessment and management.

I. Linkov et al. (eds.), Environmental Security in Harbors and Coastal Areas, 217–231.

1. Introduction

Partitioning a set of objects into groups (clusters, classes, or categories) is among the most researched areas in various disciplines. The first contributions for grouping techniques came from statistical and econometric fields in form of discriminant, logit and probit analyses [18, 41, 3, 2]. Over the last three decades, the development of grouping techniques has been based on operations research and artificial intelligence.

The groups can be defined *a priori* or *a posteriori* and they can be ordered or unordered. In the case of a priori defined ordered groups the problem is called an *ordinal classification* or *sorting problem*. In sorting problems the objects are assigned to categories based on upper and lower profiles, central objects or other norms [13].

ELECTRE TRI [45] is a sorting method that requires several parameters as input. The parameters can be divided into *preference parameters* (relative importance coefficients of criteria or weights, thresholds, and profiles) and the *technical parameter* (lambda cutting level). The weight elicitation process in general is one of the most difficult problems in MCDA, because MCDA methods are supported by mathematical models and therefore the preferences need to be expressed in mathematical terms. There are numerous weight elicitation techniques proposed for ELECTRE methods, see e.g. [30, 20, 17, 38, 33]. The different techniques may produce different values for the weights, and therefore it is advisable to perform some kind of robustness analysis when they are applied [39, 40].

Important research has been done about parameter inference and robustness analysis for ELECTRE TRI, see e.g. [34, 35, 33, 9, 10, 12, 32, 31, 36, 11].

In this paper we introduce the SMAA-TRI method that can be used for analyzing the robustness of ELECTRE TRI analysis results based on *parameter stability analysis*. Parameter stability analysis consists of analyzing the space of feasible parameters values for possible changes in the output of the method. Stability analysis allows the model to include non-deterministic parameters and it provides the DMs with more output than parameter inference. SMAA-TRI is based on Stochastic Multicriteria Acceptability Analysis (SMAA) [23, 26] that is a family of decision support methods to aid decision makers (DMs) in discrete decision making problems. The SMAA methods for the ranking problem statement (see [23, 26, 43, 24, 14, 25, 27]) are based on inverse weight space analysis that produces descriptive values characterizing the decision making problem. SMAA-TRI is the first SMAA method for the sorting problem statement.

We also demonstrate the application of SMAA-TRI by re-analyzing a case study in the field of risk assessment. There have recently been numerous military decision support applications in the field of operations research [1, 4, 5, 6, 7, 8, 19, 21, 22, 28, 37, 44]. SMAA-TRI is suitable for a large amount of military applications as well, for example:

- Recruitment. How to assign potential recruits to categories: accepted, rejected, and undecided that should be reanalyzed.

- Resource allocation. Includes various problem types, for example, classification of projects.
- Land mine detection. Assigning land zones to risk categories, a problem similar to the case study re-analyzed in this paper.

The rest of the paper is organized as follows: a comprehensive description of ELECTRE TRI is presented in Section 2. SMAA-TRI is introduced in Section 3. Section 4 contains the re-analysis of a case study. We end the paper with conclusions and avenues for future research in Section 5.

2. ELECTRE TRI Modeling

ELECTRE TRI was designed to assign a set of alternatives, objects or items (actions in general) to pre-defined and ordered categories. Each category is characterized by a lower and an upper profile. The assignment of an action to a certain category results from the comparison of the action with the profiles. The comparison is based on the credibility of the assertions "the action outranks the category profile and vice-versa". In what follows, we will assume, without any loss of generality that the scales of the criteria are ascending (therefore, all the criteria are to be maximized).

In this paper, we will use the following notation:

- $F = \{g_1, \ldots, g_j, \ldots, g_n\}$ is the set or family of *criteria*. Let $\mathcal{J}$ denote the set of criterion indices.
- $A = \{a_1, \ldots, a_i, \ldots, a_m\}$ is the set of *actions*. Let $\mathcal{I}$ denote the set of action indices.
- $C = \{C_1, \ldots, C_h, \ldots, C_k\}$ is the set of *categories* in ascending preference order (C_1 is the "worst" category). Let C denote the set of category indices.
- $B = \{b_1, \ldots, b_h, \ldots, b_{k-1}\}$ is the set of *profiles*. The profile b_h is the upper limit of category C_h and the lower limit of category C_{h+1}, for all $h \in \mathcal{B}$, where $\mathcal{B}$ is the set of profile indices.
- $w = (w_1, \ldots, w_j, \ldots, w_n)$ is the *weight vector* modeling the preferences of DMs. For the sake of simplicity, let us assume that $\Sigma_{j\in\mathcal{J}}\, w_j = 1$ (normalized weights).
- $g_j(a_i)$ is the *evaluation* of action a_i on criterion g_j for all $i \in \mathcal{I}$ and $j \in \mathcal{J}$.
- M is the evaluation matrix composed of $g_j(a_i)$ for all $i \in \mathcal{I}$ and $j \in \mathcal{J}$.
- The following comprehensive binary relations are used that allow comparing a_i and b_h:
- P is the *strict preference* relation, i.e., $a_i P b_h$ denotes the relation "a_i is strictly preferred over b_h".

- I is the *indifference* relation, i.e., $a_i I b_h$ denotes the relation "a_i is indifferent to b_h".
- Q is the *weak preference* relation, i.e., $a_i Q b_h$ denotes the relation "a_i is weakly preferred over b_h", which means hesitation between indifference and strict preference.
- R is the *incomparability* relation, i.e., $a_i R b_h$ denotes that action a_i and b_h are incomparable.
- S is the *outranking* relation, i.e., $a_i S b_h$ denotes that "a_i is at least as good as b_h".
- $\succ$ is the *preference* relation (weak or strict preference).

When the relational operator is subscripted (for example, S_j) it denotes that the relation holds with respect to the criterion indexed by the subscript.

The thresholds are denoted as follows:

- q_j is the *indifference threshold* for the criterion g_j. $q = (q_1, \ldots, q_n)$ is the vector of indifference thresholds.
- p_j is the *preference threshold* for the criterion g_j. $p = (p_1, \ldots, p_n)$ is the vector of preference thresholds.
- v_j is the *veto threshold* for the criterion g_j. $v = (v_1, \ldots, v_n)$ is the vector of veto thresholds.

These thresholds can also vary along the scale of each criterion, and in ELECTRE TRI they are always defined on profiles. In what follows we will consider variable thresholds, i.e. $q_j(g_j(b_h))$, $p_j(g_j(b_h))$, and $v_j(g_j(b_h))$.

2.1. THE CONSTRUCTION OF AN OUTRANKING RELATION

The construction of an outranking relation requires the definition of *credibility indices* for the outranking relations $a_i S b_h$ and $b_h S a_i$. Let $\rho(a_i,b_h)$ denote the credibility index of the assertion $a_i S b_h$. It is defined by using both a comprehensive concordance index, $c(a_i,b_h)$, and a discordance index $d_j(a_i,b_h)$ for each criterion $g_j \in F$ $(j \in \mathcal{J})$. The definition of $\rho(b_h,a_i)$ is similar, with the exception that the thresholds in ELECTRE TRI are always computed based on the criterion value of the profile b_h. In what follows we only exemplify the computation for the relation $a_i S b_h$.

2.1.1. *The Comprehensive Concordance Index*

The concordance index is computed by considering individually for each criterion g_j the support it provides for the assertion $a_i S_j b_h$. The partial concordance index is a fuzzy index measuring whether action a_i is at least as good as profile b_h on criterion g_j. The partial concordance indices are computed as follows, for all $j \in \mathcal{J}$, $i \in \mathcal{I}$, and $h \in \mathcal{B}$:

$$c_j(a_i,b_h)=\begin{cases}1, & \text{if } g_j(a_i)\geq g_j(b_h)-q_j(g_j(b_h)),\\ 0, & \text{if } g_j(a_i)< g_j(b_h)-p_j(g_j(b_h)),\\ \dfrac{g_j(a_i)+p_j(g_j(b_h))-g_j(b_h)}{p_j(g_j(b_h))-q_j(g_j(b_h))}, & \text{otherwise.}\end{cases} \quad (1)$$

2.1.2. *The Partial Discordance Indices*

The discordance of a criterion g_j describes the veto effect that the criterion provides against the assertion $a_iS_jb_h$. The discordance indices are computed separately for all criteria. A discordance index is also a fuzzy index, and it reaches the maximal value when criterion g_j puts its veto against the outranking relation. It is minimal when the criterion g_j is not discordant with that relation. To define the value of the discordance index on the intermediate zone a linear interpolation is used. The partial discordance indices are computed as follows, for all $j\in\mathcal{J}$, $i\in\mathcal{I}$, and $h\in\mathcal{B}$:

$$d_j(a_i,b_h)=\begin{cases}1, & \text{if } g_j(a_i)< g_j(b_h)-v_j(g_j(b_h)),\\ 0, & \text{if } g_j(a_i)\geq g_j(b_h)-p_j(g_j(b_h)),\\ \dfrac{g_j(b_h)-g_j(a_i)+p_j(g_j(b_h))}{v_j(g_j(b_h))-p_j(g_j(b_h))}, & \text{otherwise.}\end{cases} \quad (2)$$

2.1.3. *Converting a Fuzzy Relation into a Crisp One*

After determining the credibility index, the *λ-cutting level* has to be defined. The cutting level is used to transform the fuzzy outranking relation into a crisp one. It is defined as the smallest credibility index value compatible with the assertion a_iSb_h:

$$\rho(a_i,b_h)\geq\lambda\Rightarrow a_iSb_h \quad (3)$$

$$\rho(a_i,b_h)<\lambda\Rightarrow\neg a_iSb_h \quad (4)$$

$$\rho(b_h,a_i)\geq\lambda\Rightarrow b_hSa_i \quad (5)$$

$$\rho(b_h,a_i)<\lambda\Rightarrow\neg b_hSa_i \quad (6)$$

The value of λ should be in the range [0.5,1], and it describes the summation of the weights of the coalition of criteria that must support the assertion a_iSb_h.

2.2. THE EXPLOITATION PROCEDURE

The objective of the exploitation procedure is to exploit the binary relations introduced in the previous sections in order to assign actions to categories. The action a_i and the profile b_h can be compared by using the obtained relations. Based on different combinations, an action a_i can be preferred to a profile b_h ($\succ$) or vice-versa, they can be indifferent (I), or they can be incomparable (R). The fuzzy outranking relation can be decomposed into these crisp relations as follows:

$$a_iIb_h\Leftrightarrow a_iSb_h\wedge b_hSa_i \quad (7)$$

$$a_i \succ b_h \Leftrightarrow a_i S b_h \wedge \neg b_h S a_i \quad (8)$$

$$b_h \succ a_i \Leftrightarrow \neg a_i S b_h \wedge b_h S a_i \quad (9)$$

$$a_i R b_h \Leftrightarrow \neg a_i S b_h \wedge \neg b_h S ai \quad (10)$$

The sorting procedure extends two well-known procedures, the conjunctive and the disjunctive [16]. Based on these logics, there are two possible exploitation rules: the pessimistic and the optimistic. Let us consider two more "profiles", b_0 (which every action is preferred to) and b_k (which is preferred over all actions), and let Δ denote the dominance relation. The profiles must be connected with the *dominance* relation as follows:

$$b_k \,\Delta\, b_{k-1} \,\Delta\, \ldots.\, \Delta\, b_h \,\Delta\, \ldots\, \Delta\, b_1 \,\Delta\, b_0. \quad (11)$$

These are two possible rules for assigning actions to categories:

- **The pessimistic rule**: In the pessimistic rule, an action a_i is successively compared with b_k, b_{k-1}, …, until $a_i S b_{h-1}$. Then a_i is assigned to the best category C_h such that $a_i S b_{h-1}$.
- **The optimistic rule**: In the optimistic rule, an action a_i is successively compared with b_0, b_1, …, until $b_h \succ a_i$. Then a_i is assigned to the worst category C_h such that $b_h \succ a_i$.

3. SMAA-TRI Modeling

The fundamental idea of SMAA is to use Monte Carlo simulation for exploring the weight space in order to provide DMs with values characterizing the problem. The SMAA methodology has been developed for discrete stochastic MCDA problems with multiple DMs. The SMAA-2 method [26] applies inverse weight space analysis to describe for each action what kind of preferences make it the most preferred one, or place it on any particular rank. In SMAA, the criteria evaluations can be generated based on arbitrary distributions, or they can be sampled from an external source.

SMAA-TRI is developed for parameter stability analysis of ELECTRE TRI, and consists of analyzing finite spaces of arbitrarily distributed parameter values in order to describe for each action the share of parameter values that assign it to different categories. We analyze the stability of weights and the cutting level, and consider the remaining parameters to have deterministic values for easier comprehensibility. The method can easily be extended to consider non-deterministic values for thresholds. Algorithms for SMAA-TRI can be found in [42].

For analyzing ELECTRE TRI, we will denote the input for ELECTRE TRI in SMAA-TRI as follows:

1. The lambda cutting level is presented by a stochastic variable Λ with a density function $f_L(\Lambda)$ defined within the valid range [0.5,1].

2. The weights are represented by a weight distribution with a joint density function $f_W(w)$ in the feasible weight space W. Total lack of preference information is represented in "Bayesian" spirit by a uniform (constant) weight distribution in W, that is, $f_W(w) = 1/\text{vol}(W)$. The weights are non-negative and normalized: the weight space is an $n-1$ dimensional simplex in n-dimensional space:

$$W = \{w \in R^n \colon w \geq 0 \text{ and } \sum_{j=1}^{n} w_j = 1\}. \tag{12}$$

3. The data and the other parameters of ELECTRE TRI are represented by the set $T = \{M, B, q, p, v\}$. Recall that M is the criteria evaluation matrix and B is the set of profiles. These components are considered to have deterministic values.

SMAA-TRI produces category acceptability indices for all pairs of actions and categories (a_i, C_h). The category acceptability index π_i^h describes the share of possible parameter values that have an action a_i assigned to category C_h, and it is most conveniently expressed percentage-wise. It is a generalization of the rank acceptability index of SMAA-2 [26]. Let us define a *categorization function* that evaluates as the category index h to which an action ai is assigned by ELECTRE TRI:

$$h = K(i, \Lambda, \mathrm{w}, T), \tag{13}$$

and a category membership function

$$m_i^h(\lambda, w, T) = \begin{cases} 1, & \text{if } K(i, \Lambda, w, T) = h, \\ 0, & \text{otherwise}, \end{cases} \tag{14}$$

which is used for computing the category acceptability index numerically as a multidimensional integral over the finite parameter spaces as

$$\pi_i^h = \int_0^1 \int_{w \in W} f_W(w) m_i^h(\Lambda, w, T) dw d\Lambda . \tag{15}$$

The category acceptability index measures the stability of the assignment, and it can be interpreted as a fuzzy measure or a probability for membership in the category. Evidently, the category acceptability indices are within the range [0,1], where 0 indicates that the action will never be assigned to the category, and 1 indicates that it will be assigned to the category with any combination of feasible parameter values. For each action, the acceptabilities for different categories sum to unity. If the parameters are stable, the category acceptability indices for each action should be 1 for one category, and 0 for the others. In this case the assignments are said here to be robust with respect to the imprecise parameters.

The category acceptability indices provide a measure of uncertainty for the results of the sensitivity and robustness analyses as they were considered in ELECTRE TRI. While traditional way to perform sensitivity analysis in ELECTRE TRI is to consider the extremes of what can be considered possible values for the imprecise parameters [29], the category acceptability indices consider the whole space which can be determined with arbitrary joint probability distributions. Therefore, while robustness analysis for ELECTRE TRI [12] provides a result such as "depending on the parameter values, the actions is assigned either to category 2 or 3", the SMAA-TRI provides the result as "the action is assigned to category 2 with 5% of the feasible parameter values, and to category 3 with 95% of the feasible parameter values".

There are three advantages gained with the additional information:

4. The cognitive effort required in determining the extremes of the parameters considered in the sensitivity analysis is reduced, because the space can be determined to be, for example, uniformly distributed and thus small changes in the interval do not change the results dramatically.
5. Quantifying the amount of parameter values that result in "unstable" assignment determines the risk related with imprecise parameters. This will later be demonstrated in the re-analysis of the case study.
6. Weight elicitation techniques provide different weight values, and thus it seems more relevant to elicit the weights as imprecise values rather than deterministic ones (see [43]).

In addition to providing parameter stability analysis, SMAA-TRI also allows ELECTRE TRI to be applied when multiple DMs with conflicting preferences participate in the decision making process. The method allows arbitrarily distributed weights, and therefore they can be defined, for example, as intervals containing the preferences of all DMs [26]. In this case the results of the analysis (the category acceptability indices) can be used to find assignments accepted by majority of the DMs. Also the extremes of parameter combinations that assign actions to certain categories can be computed simultaneously with the parameter stability analysis.

4. Case Study: Experiments and Results

In this case study we re-analyze the recent real-world application of ELECTRE TRI in the field of risk analysis. The original analysis is presented by Merad et al. [29]. The study concentrates on France's Lorraine Region, where iron has been mined for more than a century. The underground mining tunnels have caused land subsidence, which has caused buildings to collapse. The object of this study was to partition land into zones and assign these zones into predefined risk categories in order to decide which zones need constant surveillance. We will re-analyze the assignment procedure by using the data provided in the case study.

The assignment phase consists of 10 homogenous zones (actions), $a_1, \ldots, a_{10}$ that are evaluated in terms of 10 criteria, $g_1, \ldots, g_{10}$. The criteria are not presented for brevity. There are 4 risk categories where the zones are to be assigned, Category 1 is for zones with highest risk and Category 4 for lowest. The risk categories are separated by the three profiles b_1, b_2, b_3. Performances of the zones together with profiles and thresholds are not presented for brevity. The authors used the revised

Simos' procedure by Figueira and Roy [17] to elicit the criteria weights. These (non-normalized) weights are presented in Table 1.

Table 1: Weights in the case study.

Weights	w_1	w_2	w_3	w_4	w_5	w_6	w_7	w_8	w_9	w_{10}
Non-normalized	5	1	1	1	5	1	1	20	1	10

The authors of the original case study used lambda cutting level of 0.65, but also analyzed the sensitivity of the results by altering the lambda to 0.7, 0.75, 0.8, and 0.85. In the sensitivity analysis also different profiles were applied, but the authors did not provide them in the paper. The results including the sensitivity analysis are presented in Table 2.

Table 2: Original results of the case study and sensitivity analysis.

Zones	**Results**	**Sensitivity Analysis**
a_1	Category 1	Categories 1 and 2
a_2	Category 1	Categories 1 and 2
a_3	Category 2	Stable
a_4	Category 2	Stable
a_5	Category 4	Categories 3 and 4
a_6	Category 1	Categories 1 and 2
a_7	Category 3	Categories 3 and 4
a_8	Category 4	Categories 3 and 4
a_9	Category 1	Categories 1 and 2
a_{10}	Category 4	Stable

Table 3: Weight constraints for the re-analysis.

Weights	**Lower bounds**	**Upper bounds**
w_1	3	7
w_2	0	2
w_3	0	2

Weights	Lower bounds	Upper bounds
w_4	0	2
w_5	3	7
w_6	0	2
w_7	0	2
w_8	15	25
w_9	0	2
w_{10}	7	13

We performed stability analysis to this case study with SMAA-TRI. We chose cutting level to be represented by a stochastic variable uniformly distributed in the range [0.65,0.85]. The feasible weight space was defined with constraints provided in Table III. These constraints are not probably the best constraints possible, as quantifying the imprecision should have been done along with the original case study.

SMAA-TRI was executed with 10 000 Monte Carlo iterations. The resulting category acceptability indices are presented in Table IV. Visualization of the results is important in SMAA methods, especially if there is a large amount of actions and/or criteria. Because the categories are ordered and therefore upwards inclusive, they are visualized with stacked columns in Figure 1.

The results of the re-analysis show the usefulness of SMAA-TRI. Although the stability analysis results are quite different from the ones by Merad et al. [29], SMAA-TRI provides more information. For example, compare the sensitivity analysis results for Zone 5 in Table II and the category acceptability indices for the same zone in Table IV. The original sensitivity analysis gives information that Zone 5 can be assigned to risk categories 3 or 4, and with this information the DMs (especially if they are risk-aware) should treat the zone as it would be assigned to risk category 3, which is of higher risk than category 4. But with the information provided by the category acceptability indices more informed decision can be made: regarding our imprecise and uncertain information about the parameters, we can quite safely (98% acceptability) place the zone in risk category 4.

Table 4: Category acceptability indices.

Zone	Category 1	Category 2	Category 3	Category 4
a1	100	0	0	0
a2	100	0	0	0

Zone	Category 1	Category 2	Category 3	Category 4
a3	0	100	0	0
a4	0	100	0	0
a5	0	0	2	98
a6	100	0	0	0
a7	0	34	54	12
a8	0	34	10	56
a9	100	0	0	0
a10	0	34	21	45

In this re-analysis using imprecise weights provides some interesting results. The original sensitivity analysis considered the assignment of Zone 10 stable, but by considering the weights imprecise (±30%), the assignment of the zone is quite unstable. With only 45% of the feasible parameter values the zone is placed in risk category 4, and a quite large share of the feasible values (34%) places the zone in risk category 2. If the original case study had been performed with imprecise weight values, the actions chosen based on the assignment would probably have been quite different.

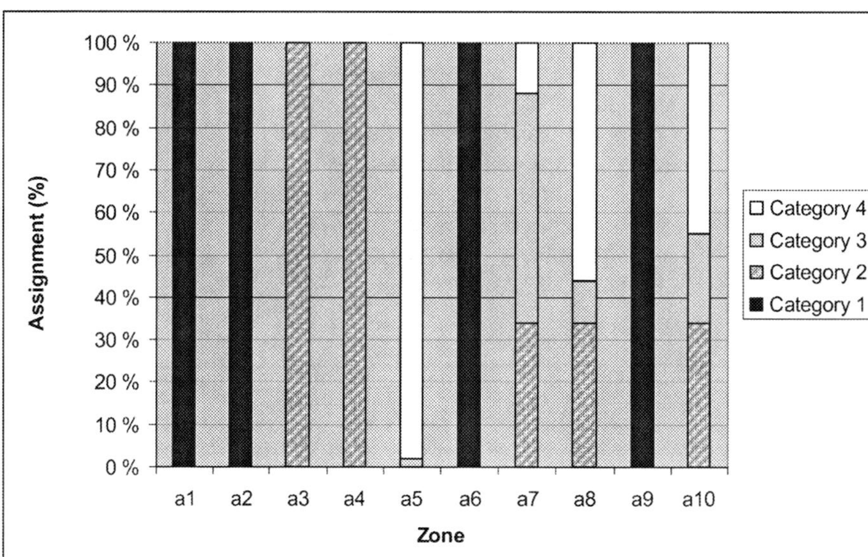

Figure 1: The category acceptability indices.

5. Conclusions and Avenues for Future Research

Defining parameter values for ELECTRE TRI model is not an easy task. Moreover, if there are multiple DMs with conflicting preferences, it might be impossible to reach consensus about weight values. With our approach the possibility to define the model by using stochastic variables "solves" these problems: the lambda cutting level can be defined with imprecise value, and the weights can be defined as intervals containing the preferences of all DMs.

In this paper we presented the SMAA-TRI method that allows ELECTRE TRI to be applied with stochastic values for lambda cutting level and weights. The SMAA-TRI analysis results in category acceptability indices for all pairs of actions and categories, and these can be used to analyze the stability of the parameters. The indices can be used also to derive robust conclusions, or if not possible, to quantify the "amount of instability" in the results induced by the imprecise parameter values.

We presented a re-analysis of the case study in which the usefulness of SMAA-TRI was demonstrated. By visualizing the category acceptability indices with stacked columns the uncertainty related with each assignment decision can be presented to the DMs in a comprehensible way. We hope that decision analysts can in the future apply SMAA-TRI for deriving robust conclusions with ELECTRE TRI both in civil and military applications. Potential military applications of SMAA-TRI are numerous, including recruitment, resource allocation, and land mine detection.

6. Acknowledgements

We thank Denis Bouyssou for enlightening comments on this paper. The work of Tommi Tervonen was supported by a grant from Turun Yliopistosäätiö and the MONET research project (POCTI/GES/37707/2001). The work of Juscelino Almeida–Dias was supported by the grant SFRH/BM/18781/2004 (Fundação para a Ciência e Tecnologia, Portugal). The work of José Figueira was partially supported by the grant SFRH/BDP/6800/2001 (Fundação para a Ciência e Tecnologia, Portugal) and the MONET research project (POCTI/GES/37707/2001). The work of Risto Lahdelma was partly supported by the Academy of Finland, decision number 215436.

7. References

1. Barnes J., Wiley V., Moore J., and Ryer D. 2004. Solving the aerial fleet refueling problem using group theoretic tabu search. *Mathematical and Computer Modelling* 39(6-8), 617–640.

2. Berkson J. 1944. Application of the logistic function to bio-assay. *Journal of American Statistical Association* 39, 357–365.

3. Bliss C. 1934. The methods of probits. *Science* 79, 38–39.

4. Chen S. 1996. Evaluating weapon systems using fuzzy arithmetic operations. *Fuzzy Sets and Systems* 77(3), 265–276.

5. Cheng C. 1996. Evaluating naval tactical missile systems by fuzzy AHP based on the grade value of membership function. *European Journal of Operational Research* 96(2), 343–350.

6. Cheng C. 1999. Evaluating weapon systems using ranking fuzzy numbers. *Fuzzy Sets and Systems* 107(1), 25–35.

7. Cheng C. and Lin Y. 2002. Evaluating the best main battle tank using fuzzy decision theory with linguistic criteria evaluation. *European Journal of Operational Research* 142(1), 174–186.

8. De Leeneer, I. and Pastijn H. 2002. Selecting land mine detection strategies by means of outranking MCDM techniques. *European Journal of Operational Research* 139(2), 327–338.

9. Dias L. and Clímaco J. 1999. On computing ELECTREs credibility indices under partial information. *Journal of Multi-Criteria Decision Analysis* 8(2), 74–92.

10. Dias, L. and Clímaco J. 2000. ELECTRE TRI for groups with imprecise information on parameter values. *Group Decision and Negotiation* 9(5), 355–377.

11. Dias L. and Mousseau V. 2004. Inferring Electres veto-related parameters from outranking examples. *European Journal of Operational Research* 170(1), 172-191.

12. Dias L., Mousseau V., Figueira J. and Clímaco J. 2002. An aggregation / disaggregation approach to obtain robust conclusions with ELECTRE TRI. *European Journal of Operational Research* 138(2), 332–348.

13. Doumpos M. and Zopounidis C. 2002. Multicriteria Decision Aid Classification Methods, Vol. 73. Kluwer Academic Publishers, Dordrecht. Preprint for NATO ARW, Thessaloniki 20-24 April, 2005.

14. Durbach I. 2005. A simulation-based test of stochastic multicriteria acceptability analysis using achievement functions. *European Journal of Operational Research* 170(3), 923-934.

15. Figueira J., Greco S., and Ehrgott M. (eds.) 2005a. Multiple Criteria Decision Analysis: State of the Art Surveys. New York: Springer Science+Business Media, Inc.

16. Figueira J., Mousseau V., and Roy B. 2005b. ELECTRE methods. Capter 4 In Figueira J., Greco S., and Ehrgott M. (eds.): Multiple Criteria Decision Analysis: State of the Art Surveys. New York: Springer Science+Business Media, Inc.

17. Figueira J. and Roy B. 2002. Determining the weights of criteria in the ELECTRE type methods with a revised Simos procedure. *European Journal of Operational Research* 139(2), 317–326.

18. Fisher R. 1936. The use of multiple measurements in taxonomic problems. *Annals of Eugenics* 7, 179–188.

19. Haapalinna I. 2003. How to allocate funds within the army. *European Journal of Operational Research* 144(1), 224–233.

20. Hokkanen, J. and Salminen P. 1997. Choosing a solid waste management system using multicriteria decision analysis. *European Journal of Operational Research* 98(1), 19–36.

21. Jaiswal N., Sangeeta Y., and Gaur S. 1995. Stochastic analysis of combat models under different termination decision rules. *European Journal of Operational Research* 83(3), 530–546.

22. Kim S., Ahn B., and Choi S. 1997. An efficient force planning system using multi-objective linear goal programming. *Computers & Operations Research* 24(6), 569–580.

23. Lahdelma R., Hokkanen J., and Salminen P. 1998. SMAA – Stochastic multiobjective acceptability analysis. *European Journal of Operational Research* 106(1), 137–143.

24. Lahdelma R., Miettinen K., and Salminen P. 2003. Ordinal criteria in stochastic multicriteria acceptability analysis (SMAA). *European Journal of Operational Research* 147(1), 117–127.

25. Lahdelma R., Miettinen K., and K. Salminen K. 2005. Reference point approach for multiple decision makers. *European Journal of Operational Research* 164(3), 785–791.

26. Lahdelma R. and Salminen P. 2001. SMAA-2: Stochastic multicriteria acceptability analysis for group decision making. *Operations Research* 49(3), 444–454.

27. Lahdelma R. and Salminen P. 2006. Stochastic multicriteria acceptability analysis using the data envelopment model. *European Journal of Operational Research* 170(1), 241-252.

28. Masud A., Metcalf P., and Hommertzheim D. 1995. A knowledge-based model management system for aircraft survivability analysis. *European Journal of Operational Research* 84(1), 47–59.

29. Merad M., Verdel T., Roy B., and Kouniali S. 2004. Use of multi-criteria decisionaids for risk zoning and management of large area subjected to mining-indiced hazards. *Tunnelling and Underground Space Technology* 19, 125–138.

30. Mousseau V. 1995. Eliciting information concerning the relative importance of criteria. In Y. Pardalos, C. Siskos, and C. Zopounidis (eds.): Advances in Multicriteria analysis. Kluwer Academic Publishers, Dordrecht, 17–43.

31. Mousseau V., Dias L., and Figueira J. 2004. Dealing with inconsistent judgements in multiple criteria sorting models. Research Report 17/2004 of The Institute of Systems Engineering and Computers (INESC-Coimbra), Coimbra, Portugal. http://www.inescc.pt.

32. Mousseau V., Figueira J., Dias L., Gomes da Silva C., and Clímaco J. 2003. Resolving inconsistencies among constraints on the parameters of an MCDA model. *European Journal of Operational Research* 147(1), 72–93.

33. Mousseau V., Figueira J., and Naux J. 2001. Using assignment examples to infer weights for ELECTRE TRI method: Some experimental results. *European Journal of Operational Research* 130(2), 263–275.

34. Mousseau V. and Slowinski R. 1998. Inferring an ELECTRE TRI model from assignment examples. *Journal of Global Optimization* 12(2), 157–174.

35. Mousseau V., Slowinski R., and Zielniewicz P. 2000. A user-oriented implementation of the ELECTRE-TRI method integrating preference elicitation support. *Computers & Operations Research* 27(7-8), 757–777.

36. Ngo The A. and Mousseau V. 2002. Using assignment examples to infer category limits for the ELECTRE TRI method. *Journal of Multi-Criteria Decision Analysis* 11, 29–43.

37. Richbourg R. and Olson W. 1996. A hybrid expert system that combines technologies to address the problem of military terrain analysis. *Expert Systems with Applications: An International Journal* 11(2), 207–225.

38. Rogers M. and Bruen M. 1998. A new system for weighting environmental criteria for use within ELECTRE III. *European Journal of Operational Research* 107(2), 552–563.

39. Roy B. 2002. Robustesse de quoi et *vis-à-vis* de quoi mais aussi robustesse pourquoi en aide à la décision ?. *Newsletter of the European Working Group on Multiple Criteria Decision Aiding*, Series 3, Number 6.

40. Roy B. 2005. A propos de robustesse en recherche opérationelle et aide à la décision, Chapt. 2. In J. Billaut, A. Moukrim and E. Sanlaville (Eds.), Flexibilité et Robustesse en Ordonnancement (forthcoming).

41. Smith C. 1947. Some examples of discrimination. *Annals of Eugenics* 13, 272–282.

42. Tervonen T., Almeida-Dias J., Figueira J., Lahdelma R., and Salminen P. 2005. SMAA-TRI: A Parameter Stability Analysis Method for ELECTRE TRI. Research Report 6/2005 of The Institute of Systems Engineering and Computers (INESCCoimbra), Coimbra, Portugal. http://www.inescc.pt.

43. Tervonen T., Figueira J., Lahdelma R., and Salminen P. 2004. An Inverse Approach for ELECTRE III. Research Report 20/2004 of The Institute of Systems Engineering and Computers (INESC-Coimbra), Coimbra, Portugal. http://www.inescc.pt.

44. Washburn A. 1994. Military Operations Research. In: S. Pollock, M. Rothkopf, and A. Barnett (eds.): Operations Research and the Public Sector, Vol. 6 of Handbooks in Operations Research and Management Science. Amsterdam: North-Holland (Elsevier Science B. V.), Chapt. 4, 67–106.

45. Yu W. 1992. Aide multicritère à la décision dans le cadre de la problématique du tri : Concepts, méthodes et applications. Ph.D. thesis, Université Paris-Dauphine.

TREATING UNCERTAIN CRITERIA AND STAKEHOLDER PREFERENCE INFORMATION IN ENVIRONMENTAL MULTI-CRITERIA DECISION PROBLEMS

R. LAHDELMA
University of Turku
Department of Information Technology, Lemminkäisenkatu 14 A, FIN-20520 TURKU Finland

P. SALMINEN
University of Jyväskylä
School of Business and Economics, P.O. Box 35, FIN-40014 Finland

Abstract

Environmental planning and decision-making processes are difficult, because they involve the evaluation of several alternatives in terms of multiple non-commensurate criteria, and a choice must be made considering the points of view of many different stakeholders with conflicting preferences. Much of the associated information can be highly imprecise or uncertain. Successful multicriteria decision aid (MCDA) methods for such problems should be able to represent explicitly not only the criteria and preference information, but also the associated uncertainty. We describe how to represent imprecise and/or uncertain criteria measurements and stakeholder preferences through probability distributions, and how to efficiently aggregate this information using stochastic simulation. As an example, we re-analyze the problem of choosing among different options for developing the Helsinki general cargo harbor. We perform the analysis by using the Stochastic Multicriteria Acceptability Analysis (SMAA-2) method with multivariate Gaussian distributions.

1. Introduction

Environmental planning and decision-making processes involve conflict analysis characterized by sociopolitical, environmental, and economic value judgments. Evaluation of several alternatives in terms of multiple criteria, results into a large body of inaccurate or uncertain data. Often the process is complicated further by the large number of decision-makers (DMs) and other stakeholders with conflicting preferences. Still, early participation of different stakeholder groups is important, because it can guarantee that all relevant points of view can be identified and

I. Linkov et al. (eds.), Environmental Security in Harbors and Coastal Areas, 233–249.

considered. Early participation also makes it possible for the different stakeholders to understand the process and easier to accept the final solution as a compromise of many conflicting goals. [16]

Multicriteria decision aiding (MCDA) methods can be used successfully in such processes. The MCDA methods define the phases of the process through their data requirements, and provide a framework for collecting, storing and processing all relevant information. This makes the decision process traceable and transparent, and makes it is possible to understand and explain why, under several conflicting preferences, a particular decision was made. The MCDA framework also makes the requirements for new information explicit, thus supporting the allocation of resources for the process.

Stochastic Multicriteria Acceptability Analysis (SMAA) methods have been developed for discrete multicriteria problems, where criteria data is uncertain or inaccurate and where it is difficult to obtain accurate or any weight information from the DMs [14]. In SMAA uncertain or inaccurate criteria measurements are represented (as in MAUT) through probability distributions. In addition, partial or missing preference (weight) information can be modeled through probability distributions. This makes it easy to model e.g. mixed ordinal and cardinal criteria and preference information.

SMAA methods are based on exploring the weight space in order to describe the preferences that would make each alternative the most preferred one, or that would give a certain rank for a specific alternative. Related ideas have been presented e.g. in [1], [3], and [2]. In the original SMAA method [10] the weight space analysis is performed based on an additive utility or value function and stochastic criteria data. The SMAA-2 method [14] generalizes the analysis to a general utility or value function, to include various kinds of preference information and to consider holistically all ranks. The SMAA-3 method [15] is based on so-called *pseudocriteria* as in the ELECTRE III decision-aid (see e.g., [4], [20], [23]). The SMAA-O method [13] extends SMAA-2 for treating mixed ordinal and cardinal criteria in a comparable manner. The multivariate Gaussian distribution was first applied in conjunction to SMAA in [12]. For applications of different SMAA methods, see e.g., [6], [7], [8], [17], [18], [19], and [22].

SMAA methods provide detailed information to describe what kinds of preferences correspond to the choice of each alternative. SMAA provides this information in the form of so-called acceptability indices that measure the variety of different preferences supporting each alternative, and central weights describing the preferences of a typical DM supporting a certain alternative. SMAA can identify good compromise alternatives that are acceptable to many stakeholders with different preferences. Such alternatives are likely to remain good solutions also in the future, subject to changing preferences, new stakeholders, and changing or more accurate criteria.

This paper is organized as follows. In Section 2 describes briefly the SMAA methods. Section 3 describes the decision problem of a new general cargo harbor for Helsinki. Section 4 re-analyzes the problem using the SMAA-2 method.

2. The SMAA-2 Method

The SMAA-2 method [14] has been developed for discrete stochastic multicriteria decision making problems with multiple DMs. SMAA-2 applies inverse weight space analysis to describe for each alternative what kind of preferences make it the most preferred one, or place it on any particular rank. The decision problem is represented as a set of m alternatives $\{x_1, x_2, \ldots, x_m\}$ to be evaluated in terms of n criteria. The DMs' preference structure is represented by a real-valued utility or value function $u(x_i,\mathbf{w})$. The value function maps the different alternatives to real values by using a weight vector $\mathbf{w}$ to quantify DMs' subjective preferences. SMAA-2 has been developed for situations where neither criteria measurements nor weights are precisely known. Uncertain or imprecise criteria are represented by a matrix of stochastic variables ξ_{ij} with joint probability density function (PDF) $f(\xi)$ in the space $X \subseteq \mathrm{R}^{m \times n}$. The SMAA-2 method does not impose any restrictions on the shape of the distribution. Criteria measurements can be represented, for example by independent uniform or normal distributions, or a multivariate Gaussian distribution to model dependent (correlated) uncertainties.

The DMs' unknown or partially known preferences are represented by a weight distribution with joint PDF $f(\mathbf{w})$ in the feasible weight space W. Total lack of preference information is represented by a uniform weight distribution in W, i.e., $f(\mathbf{w}) = 1/\mathrm{vol}(W)$. The weight space can be defined according to needs, but typically, the weights are non-negative and normalized, i.e.,

$$W = \{\mathbf{w} \in R^n \mid \mathbf{w} \geq 0 \text{ and } \sum_{j=1}^{n} w_j = 1\}. \tag{1}$$

The value function is then used to map the stochastic criteria and weight distributions into value distributions $u(\xi_i,\mathbf{w})$. Based on the value distributions, the rank of each alternative is defined as an integer from the best rank (=1) to the worst rank (=m) by means of a ranking function

$$\mathrm{rank}(i,\xi,\mathbf{w}) = 1 + \sum_{k=1}^{m} \rho(u(\xi_k,\mathbf{w}) > u(\xi_i,\mathbf{w})), \tag{2}$$

where $\rho(\text{true}) = 1$ and $\rho(\text{false}) = 0$. With a fixed realization of the stochastic weight vector $\mathbf{w}$ and stochastic criteria matrix ξ, this formula computes the rank of alternative x_k as one plus the number of alternatives x_k that are strictly better than x_i. SMAA-2 is then based on analyzing the stochastic sets of favorable rank weights

$$\mathbf{W}_i^r(\xi) = \{\mathbf{w} \in \mathbf{W} \mid \mathrm{rank}(i,\xi,\mathbf{w}) = r\}. \tag{3}$$

Any weight vector $\mathbf{w} \in \mathbf{W}_i^r(\xi)$ results in such values for different alternatives, that alternative x_i obtains rank r. Figure 1a shows the feasible weight space with missing preference information in the 3-criterion case and Figure 1b illustrates the favorable (first rank) weights with a linear value function and precise criteria measurements.

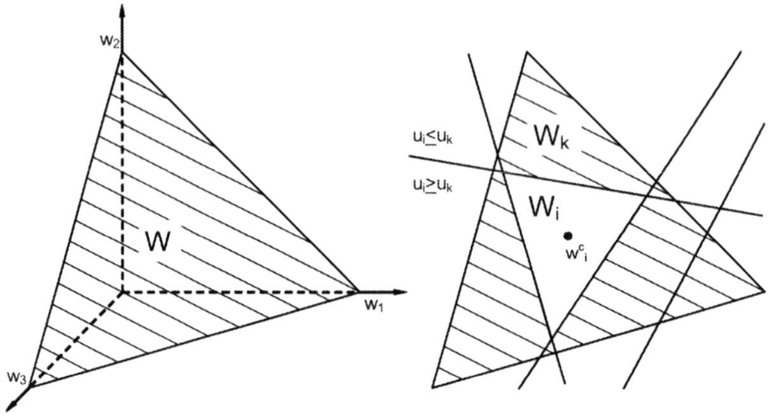

Figure 1a & b: a) Feasible weight space **W** in 3-criterion case. b) Favorable weights $\mathbf{W}_i$ for alternative.

The first descriptive measure of SMAA-2 is the *rank acceptability index* b_i^r, which measures the variety of different preferences that grant alternative x_i rank r. It is the share of all feasible weights that make the alternative acceptable for a particular rank, and it is most conveniently expressed in percent. It is computed numerically as a multidimensional integral over the criteria distributions and the favorable rank weights using

$$b_i^r = \int_X f(\xi) \int_{\mathbf{W}_i^r(\xi)} f(\mathbf{w})\, d\mathbf{w}\, d\xi \,. \tag{4}$$

The most acceptable (best) alternatives are those with high acceptabilities for the best ranks. The rank acceptability indices are in the range [0,1] where 0 indicates that the alternative will never obtain a given rank and 1 indicates that it will obtain the given rank always with any choice of weights. For comparing how different varieties of weights support each rank for each alternative, graphical examination of the rank acceptability indices is useful. Alternatives with high acceptabilities for the best ranks are taken as candidates for the most acceptable solution. On the other hand, alternatives with large acceptabilities for the worst ranks should be avoided when searching for compromises - even if they would have high acceptabilities also for fair ranks.

The first rank acceptability index is called the *acceptability index* a_i. The acceptability index is particularly interesting, because it can be used for classifying the alternatives into stochastically efficient ones (a_i>0) and inefficient or "weakly efficient" ones (a_i = 0 or near 0). The acceptability index not only identifies the efficient alternatives, but also measures the strength of the efficiency considering the uncertainty in criteria and DMs' preferences.

The *central weight vector* $\mathbf{w}_i^c$ is the expected centre of gravity (centroid) of the favorable first rank weights of an alternative. It is computed numerically as a multidimensional integral over the criteria distributions and the favorable first rank weights using

$$w_i^c = \int_X f(\xi) \int_{W_i^1(\xi)} f(w)\, w\, dw\, d\xi \Big/ a_i \,. \tag{5}$$

The central weight vector represents the preferences of a hypothetical DM supporting this alternative. Of course, the actual preferences of the DMs may be more or less incompatible with the central weight vectors. Still, presenting the central weights of different alternatives to the DMs may help them to understand how different weights correspond to different choices with the assumed preference model. This information may also aid the DMs to elicit their preferences in terms of weights.

The *confidence factor* p_i^c is the probability for an alternative to obtain the first rank when the central weight vector is chosen. It is computed as a multidimensional integral over the criteria distributions using

$$p_i^c = \int_{\xi \in \mathbf{X}: \mathbf{w}_i^c \in \mathbf{W}_i^1(\xi)} f(\xi)\, d\xi. \tag{6}$$

Confidence factors can similarly be calculated for any given weight vectors. The confidence factors measure whether the criteria data are accurate enough to discern the efficient alternatives.

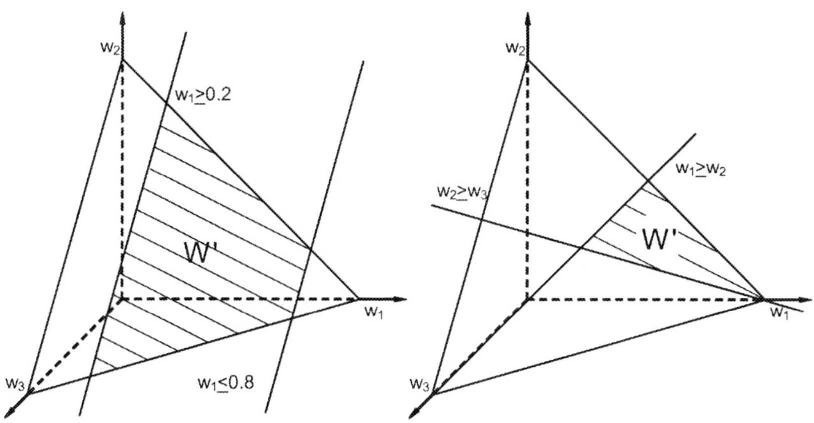

Figure 2a & b: Weight restrictions due to a) interval constraints for w_1 and b) a priority order for criteria $w_1 \geq w_2 \geq w_3$.

Specifying the weights as a probability distribution allows very flexible modeling of different kinds of preference information. For example, the DMs can specify precise weights, weight intervals, intervals for weight ratios, a priority order for the criteria, or arbitrary linear or non-linear constraints for weights. Figure 2a shows how the feasible weight space is restricted due to interval constraints for weights, and Figure 2b similarly due to a priority order for the criteria.

3. Helsinki Harbor Case

As an example we re-analyze the decision problem of constructing a new general cargo harbor in Helsinki, Finland [7]. The rapid growth of the City of Helsinki and the increase of goods traffic in its harbor had resulted in a situation where development of the existing harbor operations was not sufficiently responsive. For instance, it had been necessary to impose restrictions on the land conveyances of the harbor, which was located in the center of the city. There was no space for expansion of the existing harbor structure to raise the capacity to the level of predicted needs. Therefore, a new harbor was planned to be located in Vuosaari (see Figure 3).

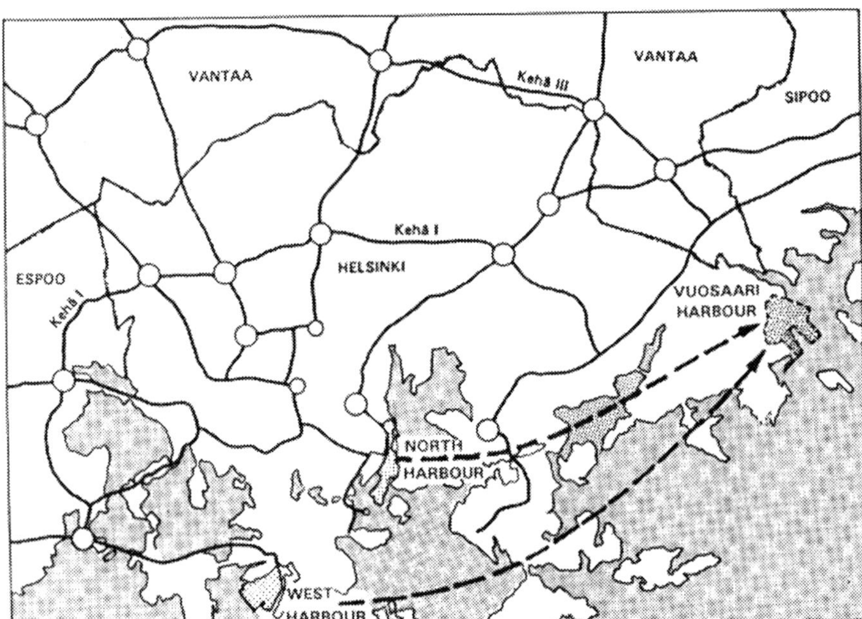

Figure 3: Transfer plans for the harbors of Helsinki.

3.1. THE ALTERNATIVES

The problem involved also the construction of navigation channels (I, II), roads (A, B, C) and railway connections (1, 2, 3, 4) leading to the harbor. As required by the Finnish Act on Environmental Impact Assessment (EIA) Procedure, a 'ZERO' alternative consisting of developing the existing harbors instead was included in the analysis. In the ZERO-alternative the City of Helsinki or its neighboring municipalities should find new locations for residential areas planned in the West harbor and North harbor districts.

The Vuosaari-alternatives entail constructing the new harbor in a present dock area, and its surroundings. Of the harbor's total surface area of 159 hectares (ha), 36 ha consist of the former dock area and 30 ha of other land area. A total of 93 hectares will consist of man-made ground in the sea. According to the plan, the harbor will be constructed in three phases. The total general cargo traffic capacity of the first and second phases will be about eight million tons. The third phase will be carried out if necessary to yield an annual capacity of 12 million tons.

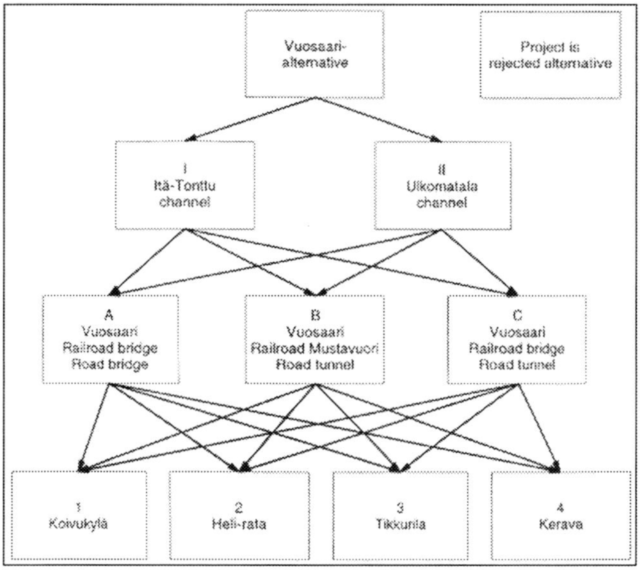

Figure 4: Project alternatives.

The draft of the present navigation channel leading to the Vuosaari dock is 7.5 meters deep, making it necessary to build a new navigation channel for the harbor. The draft of the new sea-lane will be 11 m. Road connections will be developed by extending the Kehä III-road to the harbor area, and by constructing a further collector road within the Vuosaari street network. In sections of the impact area protection of the environment is important, the roads can be conducted on bridges or into a tunnel. A new railroad connection to the main railway will be constructed. This connection can be built using a bridge across the Porvarinlahti bay, whose environmental protection is important, or Porvarinlahti can be circumvented by digging a tunnel through the Mustavuori hill. This results into three possible road/railroad combinations (A, B, C) for crossing/circumventing the Porvarinlahti bay. After this, the railroad will have four possible alternative connections, Koivukylä, Heli-rata, Tikkurila, and Kerava (1, 2, 3, 4). The original problem consisted thus of 25 alternatives as illustrated in Figure 4. Because alternatives associated with the first navigation channel (I) were excluded early in the process, $m = 13$ alternatives remained.

3.2. STAKEHOLDER PARTICIPATION

The Finnish EIA legislation requires the assessment of the direct and indirect effects, inside and outside Finnish territory, of a project or operations on human health, living conditions and amenity - soil, water, air, climate, vegetation, organisms and the interaction between them. Also included are effects on the natural diversity - community structure, buildings, landscape, townscape and the cultural heritage, as well as the utilization of natural resources. Results of the assessment are compiled into an assessment report. The coordinating authority gives a public announcement of the report with an invitation to the citizens and stakeholders to present their statements in the report. The EIA procedure is concluded when the coordination authority has given its own statement on the assessment report. The aim is that the coordination authority should take into consideration all statements received from the different groups involved. After this, the data are transferred to the actual decision-making process. In this project the DMs consisted of the members of the Helsinki City Council.

Before comparison of the actual alternatives, the following activities were performed:

1. On the basis of interviews and inquiries, the interest groups and citizens of the affected area were informed of the factors that were significant for decision-making (i.e., the proposed criteria).
2. From the above-mentioned factors, a criterion hierarchy was constructed for comparison of the alternatives. To avoid overlapping measurement, each aspect was described by only one criterion.

The impacts were divided into effects on the natural environment, people, the community structure, and utilization of natural resources. Since the project did not involve any specific utilization of natural resources (e.g., forests, ores), these effects were not included.

To investigate which factors were important for the different interest groups and residents in making their comparisons between the alternatives, eight public meetings were arranged and 11 experts and authorities were interviewed. About 250 people participated in the public meetings. The participants were also given the opportunity, by a special questionnaire, to state those factors that should definitely be taken into consideration in comparing the alternatives. 154 replies were returned.

In addition to the questionnaires presented in the public meetings, a random sample of 1800 residents of the affected area was interviewed via a survey questionnaire. This survey was done to identify resident opinions regarding the project. It thus asked people to specify those factors by which the alternatives should be assessed and compared. Another reason for this additional survey was that participation to the public meetings was biased towards those who opposed the new harbor. 496 replies were returned.

3.3. DEFINITION AND QUANTIFICATION OF THE CRITERIA

As a result of the participatory planning process, a total of n=11 criteria were identified. The criteria are described below in some detail.

3.3.1. The Sea

Impacts on the sea-area caused by the construction and use of the harbor can affect, for example, the physical and chemical properties of the water, as well as organisms such as fish and zoobenthos. The dredging and filling work of the harbor, excavation of sea sand and piling of the masses cause turbidity of water in the operation areas and their surroundings while the work is in process. Permanent disadvantageous effects will likely be caused to fishing in the neighborhood of the harbor as well as in the sea-lane district. The impact on the sea area was estimated by using a scale in which 0 = no effects, 1-2 = mild effects, 3-4 = moderate effects, and 5-6 = strong effects. This scale has traditionally been used for measuring impacts on the sea in Finland. The uncertainty of the impact assessment was estimated to remain at a level no greater than ±1, at the most, on the applied scale.

3.3.2. Ground Water

The different impact levels estimated for the ground water effects were measured on a scale from 0 to 5, where 0 means that no impact on ground water is created and 5 indicates that the operation probably affects the level and quality of ground water permanently. Assessment of the significance of the effects was not at the level of certainty since the soil and bedrock conditions of the ground transport alignment alternatives were not known in detail. The uncertainty was smallest (±0.5) in the road bridge alternatives (A) and greatest (±1 unit) in the road tunnel alternatives (B,C).

3.3.3. Emissions into Air

Harbor activities, mainly traffic, create several air pollutants, such as nitrogen and sulphur oxides, carbon monoxide, hydrocarbons and dust. It is typical of the above-mentioned emissions that they correlate strongly with each other. The nitrogen dioxide emission (ton NO_x /a) was chosen to for the comparison level to represent all emissions to the air. Predictions of the development of emissions from harbors and traffic were based on the initial concept that specific emissions will not increase, but, rather, tend to decrease. According to experts, the degree of uncertainty in the estimates can, however, be as large as ±25 %.

3.3.4. Fauna

Effects on the fauna of the region were examined on the basis of estimated change in the avifauna. The avifauna is very well known in this district and has been frequently used to indicate the ecological effects caused by changes brought about by man. Moreover, the avifauna reflects changes in other natural conditions, such as aquatic conditions, small fauna and vegetation [5], [9]. The score of the protective value of wetland avifauna in the area was used as the criterion value. Each nesting species was assigned a certain number, according to the extent of its occurrence in the natural geographic area in question and depending on how well it represents typical bird waters. The uncertainty in the resulting score was estimated as ±10 points.

3.3.5. Flora

In terms of vegetation and flora, the most valuable areas in the project's range of impact consist of the still rather unspoiled natural areas near the intended harbor. The dock-area proper, the railway yard, and the traffic lanes starting from the harbor spread over one part of the valuable vegetation areas. Effects on vegetation and flora were measured by a score of points based on the number of occurrences of vascular plants that are endangered in the province of Uusimaa and in Helsinki.

Every occurrence of each species was given points by applying the scale used by Ohenoja [21] for the valuation of mushrooms. Experts evaluated the uncertainty of the score to be ±10 %.

3.3.6. Noise

The Council of State has defined guideline values for outdoor noise levels in residential districts: 55 dB(A) during daytime and 50 dB(A) at night. The number of people residing permanently in the noise zone of daytime was chosen as an indicator of the noise effects. The noise zones caused by harbor operation and the traffic generated by the harbor were determined by measurements and model calculations. The number of residents was calculated in the noise areas on the basis of the 1994 census. Accurate calculation of the number of "noise area" residents in the original city is very difficult. Uncertainty of assessment was defined as ±500 inhabitants in the Vuosaari-alternatives, and ±100 inhabitants in the ZERO-alternative.

3.3.7. Housing and Services

The housing and services criterion refers to the possibilities provided by the different alternatives for housing and the development of the service structure in Helsinki. The construction of Vuosaari harbor creates the opportunity for building 990 000 floor square meters in West and North harbor districts and in Pasila. This would provide apartments for approximately 25 000 inhabitants. At that time, 154 500 people lived in the old city of Helsinki, and, since the number of residents is decreasing in the 2000s, without new residential areas the change offered by the Vuosaari-alternative is quite important. The change in the number of residents can also be used to describe the increased demand for services in the old city area. The change is so significant that it can prevent discontinuation of key services. The greatest uncertainties were connected with the implementation schedule and the economic situation. The uncertainty was set at ±20 %.

3.3.8. Employment

The employment criterion refers to the employment effect of the project. This was considered important because of the high unemployment level in Finland at that time. Differences between the alternatives were estimated by experts on a scale from 1 to 5, where 1 is the current situation and 5 corresponds to the employment effects when the new harbor operates at its maximum capacity (12 million tons per annum). The uncertainty was set at ±1.

3.3.9. Recreational Possibilities

Recreational possibilities refer to people's possibilities for recreation in areas specifically reserved for this purpose, or in land and water areas to which they have public right of access. The changes that would likely occur in the present situation were used as the starting point for the comparison. Construction of a new harbor means that the harbor areas of the city centre are partly released for recreational use. On the other hand, in Vuosaari, and in some other areas that are target of the construction activities, recreational use will be either totally prevented, or obstructed to various degrees. Experts estimated the significance of change linked to the different alternatives on a scale from 1 to 10 where higher values indicate better recreational possibilities. The adopted expert opinion for the uncertainty degree on the applied measure was set at ±0.4.

3.3.10. Landscape

The landscape criterion refers to the number of negative landscape effects caused by the project. On the basis of public meetings, and the resident survey, the landscape impact of container harbors was envisioned as clearly negative. Consequently, from the viewpoint of the old city, transfer of the harbor would represent a positive change, while, from the point of view of Vuosaari, rejection of the project would be positive. The landscape impact will affect those areas where the harbor activities (harbor, overland traffic connections) are visible. Comparison of the alternatives was conducted on the basis of the surface areas of the visibility zone, in square kilometers, weighted by the inverse distance to the harbor. On the basis of calculations carried out by means of map pictures, aerial photographs and observation pictures, the uncertainty degree was estimated to be ±20%.

3.3.11. Economy

The alternatives were compared by means of socio-economic benefit-cost calculations in accordance with directions on standardizing the impact assessments of traffic lane projects issued by the Ministry of Transport. A 30-year investment period with a 6% real interest rate and 30% residual value was applied. Based on sensitivity analysis of the value changes of those most critical variables (transfer value of land-areas, construction savings, construction costs, benefits of the transfer project, capacity need) the uncertainty of the economy criterion was determined to be ±0.02.

4. Criteria Measurements

Table 1 summarizes the mean criteria measurements x_{ij} and their uncertainties Δ_{ij}. The uncertainties of the criteria measurements are not all independent. For example in the Sea-criterion, the uncertainties of all alternatives associated with the same navigation channel (II) will be correlated, because the same channel will be built in all of them, and the uncertainties in the sea-effects will depend on the environment. In contrast, the uncertainty of the Sea-criterion value for the ZERO-alternative does not significantly correlate between the other alternatives, because the existing sea-channels will be used in that alternative. The following uncertainty dependencies were identified:

Table 1: Criteria measurements x_{ij} and uncertainties Δ_{ij} for the alternatives.

Alt.	Sea	GWater	Air	Fauna	Flora	Noise	Housing	Employ	Recreation	Landscape	Economy
IIA1	4	1	985	30	166	705	25000	4.5	4.2	15.1	1.75
IIA2	4	2.5	985	30	166	765	25000	4.5	4.1	15.3	1.69
IIA3	4	1.5	985	30	166	705	25000	4.5	4.3	12.7	1.75

Alt.	Sea	GWater	Air	Fauna	Flora	Noise	Housing	Employ	Recreation	Landscape	Economy
IIA4	4	1.5	985	30	166	705	25000	4.5	4.3	12.2	1.65
IIB1	4	1.5	985	35	177	705	25000	4.5	4.4	15.1	1.68
IIB2	4	2.5	985	35	177	765	25000	4.5	4.3	15.3	1.62
IIB3	4	2	985	35	177	705	25000	4.5	4.5	12.7	1.68
IIB4	4	2	985	35	177	705	25000	4.5	4.5	12.2	1.58
IIC1	4	1	985	35	166	705	25000	4.5	4.6	14.8	1.72
IIC2	4	2.5	985	35	166	765	25000	4.5	4.5	15	1.66
IIC3	4	1.5	985	35	166	705	25000	4.5	4.7	12.4	1.72
IIC4	4	2	985	35	166	705	25000	4.5	4.7	11.9	1.62
ZERO	1	0	1300	50	266	4200	0	2	1	18.8	1
Uncer-tainty	±1	±0.5, ±1, 0*	±25%	±10	±10%	±500, ±100**	±20%	±1	±0.4	±20%	±0.02

*±0.5 for bridge alternatives (A), ±1 for tunnel alternatives (B,C), and 0 for the ZERO-alternative.

**±500 for Vuosaari-alternatives and ±100 for the ZERO-alternative.

Sea Correlation between II-alternatives.

Air Correlation between Vuosaari-alternatives.

Fauna Correlation between A-alternatives and between B&C-alternatives.

Noise Correlation between 1,3&4-alternatives and between 2-alternatives.

Housing Correlation between Vuosaari-alternatives.

Employment Correlation between Vuosaari-alternatives.

5. Multicriteria Analysis by SMAA-2

In the original decision problem, the DMs refused to provide preference information. This led into developing and applying the SMAA-1 method. Uniform distributions in the given uncertainty intervals $[x_{ij}-\Delta_{ij}, x_{ij}-\Delta_{ij}]$ were used for all criteria. Uncertainty dependencies were treated by applying the same random number for each set of correlated measurements. This approach can represent only perfectly correlated and non-correlated measurements.

In this re-analysis we use the SMAA-2 method without preference information, and represent the criteria measurements, their uncertainties and uncertainty correlations by a multivariate Gaussian (normal) distribution. The mean of each criteria measurement x_{ij} is given in Table 1. The standard deviation σ_{ij} is defined so that the stated uncertainty Δ_{ij} forms the 90% confidence interval, i.e. $\sigma_{ij} = \Delta_{ij}/1.96$. Using the multivariate Gaussian distribution for criteria measurements allows specifying an arbitrary correlation (in the range [-1,1]) between any two measurements. However, here we apply the multivariate Gaussian distribution only with perfect and no correlation, which is consistent with the original analysis.

Table 2: Confidence factors and rank acceptability indices (%).

Alt	p^c	b^1	b^2	b^3	b^4	b^5	b^6	b^7	b^8	b^9	b^{10}	b^{11}	b^{12}	b^{13}
IIC3	30	18	15	12	11	9	8	7	6	5	4	3	2	1
IIC1	28	16	12	10	9	9	8	8	8	7	5	5	4	1
IIB3	17	12	11	10	10	10	9	8	8	7	6	5	4	1
IIB4	13	9	10	10	10	9	9	9	9	8	6	5	4	1
IIC4	14	8	10	12	11	11	11	10	8	8	6	4	3	1
ZERO	94	8	2	1	1	1	1	1	2	2	2	3	5	70
IIB1	12	7	9	8	8	9	9	10	9	9	9	6	5	1
IIA3	10	7	9	10	10	10	10	9	9	8	7	5	4	2
IIA4	9	6	8	9	9	10	9	9	9	8	8	7	5	2
IIA1	12	4	7	8	8	9	9	10	10	9	9	9	7	2
IIC2	4	2	3	4	5	5	6	7	8	11	12	15	15	6
IIB2	4	2	3	4	5	5	6	6	9	10	13	16	17	4
IIA2	2	1	1	2	3	3	4	5	7	9	13	18	27	9

Table 2 presents the confidence factors (p^c) and rank acceptability indices (b^r) for the alternatives. The rank acceptability indices are also shown in Figure 5. Based on the first rank acceptability index, we can see that many alternatives can be considered as the most preferred choice, depending on the preferences. Alternatives IIC3, IIC1 and IIB3 obtain the highest first rank acceptability indices, and also quite high indices for the following ranks. In the absence of preference information, these alternatives can be considered as potential choices, because they are acceptable under a large variety of different weights. Still, under suitable preferences, also the following alternatives could be chosen. Only the few last alternatives can be eliminated, because they are unlikely most preferred by any DM or stakeholder, and very unlikely most preferred by a majority. The acceptability profile of the ZERO-alternative is interesting. It receives fair acceptability for the first rank, very high acceptability (70%) for the last rank, and practically no acceptability for the intermediate ranks. This means that the Zero-alternative is probably not a very suitable compromise alternative – it divides people into those who love and those who hate it!

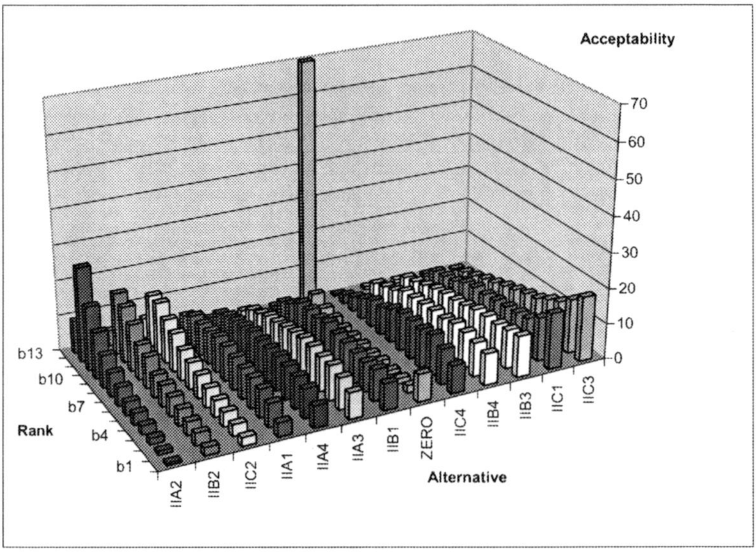

Figure 5: Rank acceptability indices (%). Alternatives are sorted by b^1.

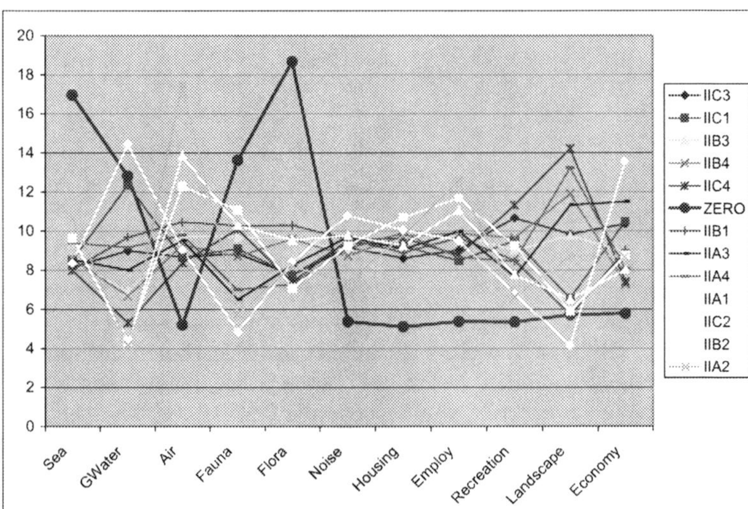

Figure 6: Central weights for alternatives.

The central weights shown in Figure 6 reveal what kind of preferences favor each alternative. We can see, for example, that the Zero-alternative would be the most preferred if much weight is assigned to the Sea, Fauna and Flora criteria, and very little weight on the Air, Noise, Housing, Employment, Recreation and Economy criteria. A DM who is uncertain about his/her preferences can use the central weights to see if his/her preferences could agree with some of the central weighs.

The relatively low confidence factors for the Vuosaari-alternatives indicate that the information is not accurate enough to reliably discriminate between them. For example, the 30% confidence factor for IIC3 indicates that even if the DMs could adhere to the central weights of that alternative, it would, considering the uncertainty in the criteria measurements, only have a 30% chance of being the most preferred one. This would be serious, if there were strong confrontations between the different Vuosaari-alternatives. However, in this case, the different Vuosaari-alternatives were rather similar – most of the DMs supported building the new harbor but did not have strong feelings about which of the top-variants was chosen. In this situation it was possible to make the decision to build a new harbor based on the current, rather imprecise information.

6. Conclusions

SMAA was initially developed for aiding DMs who may not always be willing or able to express trade-off information for criteria. Some of the non-commensurate criteria may be just too difficult for the DMs to compare, they may be unwilling to reveal their preferences, or they may anticipate that preferences may change in the future, and it is therefore not wise to fix themselves to specific weights. The inverse analysis can make providing weight information easier: instead of the question, "What are your weights?" it allows asking, "Could you adhere to these weights that favor that alternative?" However, the SMAA approach can also be used with imprecise and partial weight information provided e.g. as precise weights, as intervals, as a priority order for criteria, or as arbitrary probability distributions.

Because SMAA treats both criteria and preference information symmetrically through various probability distributions, the method can be used efficiently in processes, where initial analyses are performed rapidly with very rough criteria measurements and with less accurate or non-existent preference information. Later on, the analysis can be repeated with more precise information. The confidence factors can be used to find out if the information is accurate enough for making an informed decision, or if more accurate information is needed. The analysis can cause significant savings if costly and overly accurate measurements can be avoided, but also justifies additional measurement efforts, when necessary.

7. References

1. Bana e Costa C., 1986. A multicriteria decision-aid methodology to deal with conflicting situations on the weights. *European Journal of Operational Research* 26, 22-34.

2. Butler J., Jia J., Dyer J., 1997. Simulation techniques for the sensitivity analysis of multi-criteria decision models. *European Journal of Operational Research* 103, 531-546.

3. Charnetski J.R., Soland R.M., 1978. Multiple-attribute decision making with partial information: The comparative hypervolume criterion. *Naval Research Logistics Quarterly* 25, 279-288.

4. Figueira J., Greco S., Ehrgott M. (Eds.), 2005. Multiple Criteria Decision Analysis: State of the Art Surveys. Springer Science+Business Media, Inc., New York.

5. Furness R.W, Greenwood J.J.D & Jarvis P.J. Can birds be used to monitor the environment? In Furness, R.W & Greenwood, J.J.D. (eds.) *Birds as monitors of environmental change*. 1 - 41, Chapman & Hall, London (1993).

6. Hokkanen J., Lahdelma R., Miettinen K., Salminen P., 1998. Determining the implementation order of a general plan by using a multicriteria method. *Journal of Multi-Criteria Decision Analysis* 7(5), 273–284.

7. Hokkanen J., Lahdelma R., Salminen P., 1999. A multiple criteria decision model for analyzing and choosing among different development patterns for the Helsinki cargo harbor. *Socio-Economic Planning Sciences* 33, 1–23.

8. Hokkanen J., Lahdelma R., Salminen P., 2000. Multicriteria decision support in a technology competition for cleaning polluted soil in Helsinki. *Journal of Environmental Management* 60, 339–348.

9. Järvinen O. Conservation indices in land use planning. *Ornis Fennica* 62, 101-106 (1985).

10. Lahdelma R., Hokkanen J., Salminen P., 1998. SMAA - Stochastic multiobjective acceptability analysis. *European Journal of Operational Research* 106(1), 137-143.

11. Lahdelma R., Makkonen S., Salminen P., 2004. Treating dependent uncertainties in multi-criteria decision problems. In M.R. Rao, M.C. Puri (eds.) *Operational Research and Its Applications: Recent Trends*, Proc. APORS'2003, Vol II, New Delhi.

12. Lahdelma R., Makkonen S., Salminen P., 2006. Multivariate Gaussian criteria in SMAA. *European Journal of Operational Research*, Article in Press, Corrected Proof.

13. Lahdelma R., Miettinen K., Salminen P., 2003. Ordinal criteria in stochastic multicriteria acceptability analysis (SMAA). *European Journal of Operational Research* 147(1), 117-127.

14. Lahdelma R., Salminen P., 2001. SMAA-2: Stochastic multicriteria acceptability analysis for group decision making. *Operations Research* 49 (3), 444-454.

15. Lahdelma R., Salminen P., 2002. Pseudo-criteria versus linear utility function in Stochastic Multicriteria Acceptability Analysis. *European Journal of Operational Research* 141(2), 454-469.

16. Lahdelma R., Salminen P., Hokkanen J., 2000. Using multicriteria methods in environmental planning and management. *Environmental Management* 26(6), 595-605.

17. Lahdelma R., Salminen P., Hokkanen J., 2002. Locating a waste treatment facility by using Stochastic Multicriteria Acceptability Analysis with ordinal criteria. *European Journal of Operational Research* 142(2), 345-356.

18. Lahdelma R., Salminen P., Simonen A., Hokkanen J., 2001. Choosing a reparation method for a landfill using the SMAA-O multicriteria method. In: Multiple Criteria Decision Making in the New

Millenium, Lecture Notes in Economics and Mathematical Systems. Vol. 507. Springer-Verlag, Berlin, pp. 380–389.

19. Makkonen S., Lahdelma R., Asell A.-M., Jokinen A., 2003. Multicriteria decision support in the liberated energy market. *Journal of Multi-Criteria Decision Analysis* 12(1), 2003, 27-42.

20. Maystre L.Y., Picted J., Simos J., 1994. *Méthodes Multicritéres ELECTRE.* Presses Polytechiques et Universitaires Romandes, Lausanne, Switzerland.

21. Ohenoja, E. Keminmaan Kallinkankaan suursienistö. *Lutukka* 8, 125-129 (1992).

22. Salminen P., Hokkanen J. & Lahdelma R. Multicriteria decision analysis project on environmental problems. University of Jyväskylä, Department of Mathematics, Laboratory of Scientific Computing, Report 5/1996.

23. Vincke Ph., 1992. *Multicriteria Decision-Aid.* John Wiley & Sons, Chichester, UK.

FUZZY VS. PROBABILISTIC METHODS FOR RISK ASSESSMENT OF COASTAL AREAS

D. ANGELIDES, Y. XENIDIS
Department of Civil Engineering, Aristotle University of Thessaloniki Thessaloniki, 54124 Greece

Abstract

Coastal zones are dynamically evolving systems comprising three components, i.e. the marine, the coastal, and the land subsystem. This is a typical division of a coastal zone and the only common feature that two different coastal zones may present. Concerning other features such as landscape morphology, ecological habitats, land uses, residential development and economic activities, etc., coastal zones present a multivariate environment with various characteristics. In general, there may be three different typologies of coastal zones based on morphological criteria. However, the number of coastal zones types is extremely increased when the specific conditions met in each one of these types are taking into consideration. Several types of human settlements and habitats along with various processes of human and natural origin render an environment with many interdependencies and risks.

The purpose of this paper is to investigate the risks in relevance to human and natural processes, in all three subsystems of a coastal zone. This is an intermediate task before examining the appropriateness of two methodological approaches for risk treatment, i.e. the probability and the fuzzy theory. The two theories are presented and compared in brief to decide their applicability and effectiveness in risk assessment in the different subsystems. This paper contributes to the discussion of integrated risk assessment of coastal zones and provides clear arguments for risk analysts to select the appropriate risk assessment methodology.

1. Introduction

A common definition for coastal zones is the following: "A strip of land and sea of varying width depending on the type of the environment and its management needs" [1]. This is a generic definition that, however, only partially presents the nature of a coastal zone. For example, there is no implication about the causes of the declared variety of width of coastal zones. "A strip of land and sea" may extend

I. Linkov et al. (eds.), Environmental Security in Harbors and Coastal Areas, 251–266.

from some meters to some kilometers and there is no reference in the above definition on how this width is defined. Moreover, there is no reference about any other features that characterize a coastal zone – e.g. living organisms – but only a generic statement of "environment and management needs".

Defining the coastal zone appropriately is very important to manage, efficiently, the complex issues associated with it. Coastal zones are dynamic systems of different typologies that evolve through the interaction of the physical elements, i.e. air, water, and land, both between themselves and with natural and human processes. This definition is much more complex but also accurate concerning the real nature of a coastal zone.

An initial classification of coastal zones can be based on morphological criteria. A second set of criteria that differentiates even more the classes of coastal zones is the one consisted of natural and human processes occurring in these areas and resulting to a multivariate typology in terms of characteristics and needs. These processes do not, only, define a coastal zone's identity but, furthermore, they generate various risks. Therefore, there are certain risks associated with certain typologies of coastal areas and, of course, with various impacts on the environment and the society.

The assessment of these risks requires appropriate techniques that need to be comprehensible not only by experts of certain disciplines (e.g. engineers) but from a wider audience involving several disciplines and people living in coastal zones. The establishment of a methodology that will foster a synergy between different disciplines in coastal risks treatment is a requirement that is gradually recognized as very important in order to, successfully, assess risks and efficiently manage coastal environment and development [2].

This paper contributes to the discussion of implementing an integrated approach for coastal risks assessment in many ways. The coastal zone system is presented and analyzed in the constituent subsystems with emphasis on the processes occurring in each subsystem and their impact on the environment. This analysis reveals the interaction between the subsystems and the need to be treated uniformly in terms of risk assessment and management. Moreover, the complexity of a coastal system in reference to different typologies and natural and human processes is investigated. This is a prerequisite in order to understand the nature of risks before addressing a specific risk assessment methodology. The election of the most appropriate one is a critical issue of the whole coastal risk treatment process and should respond to all the requirements mentioned so far, i.e. appropriate for complex and interacting systems, comprehensible and simple for risk analysts emanating from different disciplines.

A comparison between probability and fuzzy theory is conducted to investigate the suitability of each approach for assessing coastal risks. These theories, although extensively discussed concerning risk analysis on other fields are not yet adequately compared regarding their applicability on coastal risk assessment. Here the advantages and disadvantages of both theories on assessing risk of coastal zones are discussed and the joined application of both approaches is suggested and argued.

2. Coastal Systems

The definition of the coastal zones suggested in the introduction refers to dynamic systems of different typologies. The standard features of these systems are the constituent subsystems as presented in Figure 1.

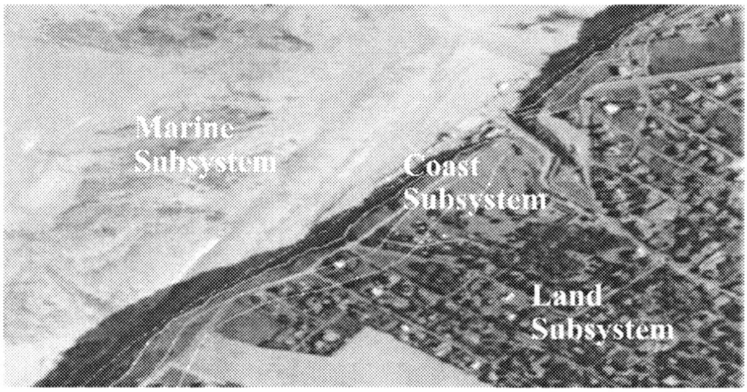

Figure 1: A schematic presentation of the coastal zones subsystems (Source of original photo: [3]).

The distinction presented in this figure between marine, coast, and land subsystems should not be considered as geographical or morphological but more as a systemic one for the reasons below:

1. The morphological constraints of each subsystem are evolving in the long term. Natural and human processes may significantly alter the area covered by each subsystem and create formations where one subsystem "enters" inside the other (e.g. alluvial deposits in the sea as extensions of the coast); therefore, the definition of each subsystem in spatial terms is not consistent from time to time and between different coasts.
2. The external limits of the marine and land subsystems are vague. Even if a formal definition of the limits between the coastal zones' subsystems could be suggested based on morphological or geographical criteria, the width of the marine and the land subsystems from the coast, would be different from case to case.

The distinction in the three subsystems should be considered in terms of basic features, important processes and respected risks identified for each zone of the coastal system. Basic features should, primarily, reflect the physical and morphological characteristics of each subsystem. Important processes should include both natural and human processes occurring in each subsystem, while respected risks represent the possible risks in the framework defined by the groups of characteristics and processes. Due to the heavily dependence between these parameters a more detailed view is required upon these characteristics of the coastal zones.

2.1. COASTAL PROCESSES

A generic classification of coastal processes should include natural and human, i.e., mainly socioeconomic processes that are linked to each other in a network of reciprocal influence. The natural resources are being exploited and eventually overexploited with the purpose of demographic and economic growth. The reckless disregard for the environmental capacity in resources leads to the degradation of the environment and the consequent loss of ability to support future growth. From this point on, the reverse course is taking place due to disturbance of the environmental equilibrium. This course involves the occurrence of intensified natural processes that jeopardize the socioeconomic status of coastal communities. Such processes are: climate changes, rises in sea level, floods, and changes in storminess, rainfall, evaporation, and freshwater unavailability. A preview on the natural and human processes that occur in the coastal systems is provided in the next subsections.

2.1.1. Natural processes

Natural processes may be of physical, chemical or biological nature. Examples of such variable natural processes and resulting impacts, as presented in [1], are:

- Coastal currents and sediment transport leading to coastal erosion or accretion
- Storm and wave conditions affecting coastal profiles
- Dispersion, degradation, adsorption and sedimentation processes affecting water and sediment quality
- Ecological succession leading to changes in habitat types and biodiversity
- Energy and material cycles affecting biological productivity.

Many of the above natural processes are affected by features of climate change, such as changes in temperature, rainfall and sea level. Climate impact may lead to:

- Higher temperatures
    - Melting glaciers and land ice
    - Thermal expansion of ocean water
- Sea level rise
- Changes in the intensity, frequency and direction of storms
- Changes in rainfall and evaporation

2.1.2. Socioeconomic processes

The impact of the overall global growth in population is expected to be such that by the year 2025 the total coastal population will be equal to the present world population [1]. In this setting, coastal zones require special attention because of the increasing pressures due to imperative demands for space and resources. Table 1 summarizes manmade pressures as identified in the EU's official information brochure on Integrated Coastal Zone Management [4].

Table 1: Manmade pressures on the coastal zone in the EU.

Problem	Impact
Increasing urbanisation	▪ Huge increase in the number of second homes built in EU coastal regions. ▪ Destruction of fragile natural habitats. ▪ Overloading of the natural environment's ability to absorb pollutants due to waste disposal systems and septic tanks of houses.
Badly planned tourist developments	▪ Huge strain on local supplies of fresh water. ▪ Inadequate facilities for disposal of solid waste. ▪ Consume of large amounts of fossil fuels for cooking, heating, vehicles, and pleasure crafts. ▪ Detrimental effect on existing local industries and on the social fabric of local communities.
Washing of inland generated pollution into the sea via streams and rivers	▪ Pollution.
Poorly conceived transport networks	▪ Pollution, overcrowding, and habitat destruction. ▪ Poor accessibility to tourist resorts and destinations.
Maritime accidents (oil slicks and chemical spills)	▪ Pollution.
Careless treatment of the natural habitats	▪ Alterations of the sea floor, beaches and shorelines. ▪ Destruction of wetlands. ▪ Dramatic reductions in fish stocks due to over-fishing. ▪ Reduction of water resources and coastal erosion.
Erosion	▪ Loss of land of ecological and economic value. ▪ Loss of property. ▪ Risk to human lives. ▪ Destruction of natural sea defences. ▪ Undermining of artificial sea defences.

This table should also include uncontrolled coastal construction, which is an important issue for many countries (e.g. Greece).

2.2. COASTAL SYSTEMS' MORPHOLOGICAL TYPOLOGIES

The processes presented in the previous section are not met in all coastal systems. Different morphologies of coastal systems provide with a different framework for each case. Human settlements and, therefore, activities are confined by spatial constraints, i.e. the coastal morphology is an important factor for the level of human presence and development of a coastal zone. Furthermore, the impact of the occurrence of two processes differs between two coasts with different morphological criteria. This means that the respective risks associated with these processes also differ. Therefore, morphology of a coastal system plays an important role that should be considered in the coastal risk analysis process. In the following subsections, a short presentation of the three main types of coastal zones identified based on morphological criteria – i.e. deltas, islands, and continental coasts – is provided.

2.2.1. Deltas

A delta is a relatively large, flat land area dissected by one or more branching rivers, which most often hosts both human settlements and unique habitats [1].

Delta plains suffer the most from natural processes such as flooding and storms, while human activities increase their vulnerability. For example, the construction of flood prevention structures reduces sedimentation and accelerates erosion and subsidence, while extraction of hydrocarbons and groundwater, uncontrolled land reclamation and inadequate land use planning jeopardizes these fragile but valuable systems [1].

2.2.2. Islands

Islands are characterized by: (a) long coastline compared to their land, (b) limited area, (c) geographic dispersion, (d) relative isolation from the continental land, (e) limited and hardly renewable natural resources. Island ecosystems may include coral reefs, sandy beaches, mangrove forests or hard, rocky cliffs. The main human activities on islands are related to fishery, agriculture and tourism. Population and economic activities are heavily concentrated in the coastal zones and this fact increases pressure and vulnerability regarding water pollution and dredging, sand and coral mining, uncontrolled coastal construction and inadequate land use planning. Natural processes with a serious impact on island subsystems are sea level rising, coastal erosion, and increased flooding.

2.2.3. Continental Coasts

Continental coastal zones present a wide range of bio-geographic features and functional uses. In most of cases continental coastal plains have an adequate elevation from the sea level that prevents from suffering due to floods. Moreover, they present a more stable shoreline compared to the respective of an island's. Although normal natural processes have a low impact on continental coastal zones, the effect of manmade pressures is extremely increased due to large population density. All kinds of human activities are taking place in continental coasts: housing, fishery, agricultural, industrial, transportation of people and goods, tourism, natural resources exploitation, and, of course, waste discharges. The respective environmental pressure from these activities may have all possible forms with the most predominant being the pollution and natural resources reduction.

3. Complexity of Coastal Systems

The analysis of coastal systems, presented above, clearly indicates that each one of these systems is prone to a different degree of complexity that results from the interaction of natural and human processes and morphological features. A hierarchical presentation of the increasing complexity of a coastal system is given in Figure 2.

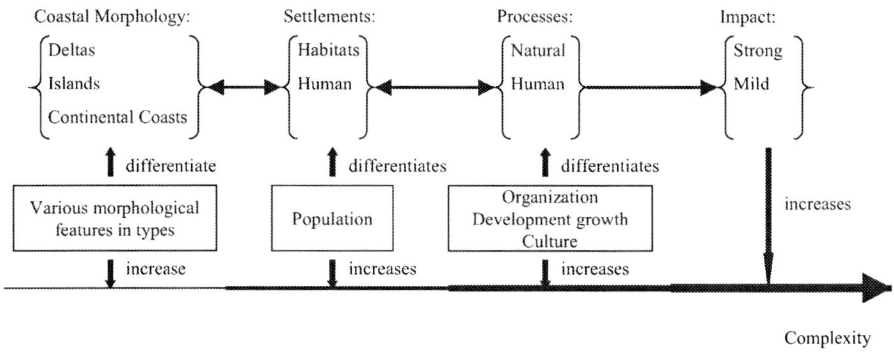

Figure 2: Parameters of increasing complexity in coastal systems.

The original parameter of complexity is the morphology of the coast. Furthermore, there may be special morphological features that differentiate significantly the degree of complexity between two coastal systems of the same morphology. A second parameter is the presence of any settlements in the coastal system. Such settlements may differ in terms of nature (i.e., human or animal settlements) and amount of population. The presence of settlements determines the nature of the processes occurring in the coastal system. Natural and human processes are also different due to the various degrees and modes of the organization of the communities in the coastal systems, the level of development they enjoy and the will for further development, the cultural background, and a numerous other parameters. Different processes have different impact on the coastal system; even if the same type of impact is expected, the severity of it is not the same. Finally, a last issue that increases complexity in a coastal system is the interaction of the parameters discussed above. The possible combinations of the different facets of each one of these parameters comprise a large amount of contingencies with uncertain result that intensifies complexity and justifies the urgent need for the adoption of an appropriate, integrated risk assessment approach for coastal systems.

4. Risk Assessment Approach

The nature and complexity of the coastal systems is a very important factor to consider prior to initiation of a risk assessment effort. The analysis conducted so far is particularized in Table 2, where physical characteristics, natural and human processes and risks are presented with reference to each coastal system. The details included in Table 2 represent the one prerequisite for a successful selection of the appropriate risk assessment methodology. The second prerequisite is the understanding of the following critical issues that, also, result from the previously presented analysis:

1. Coastal zones are not static but dynamic systems with interactions between the constituent subsystems.

2. There is a wide range of natural and human processes with varying severity of impact occurring in coastal systems of different typologies. This increases complexity and provides with a numerous number of risk scenarios to be investigated in a coastal system.

3. The nature of the above processes is dynamic and, furthermore, they are related in between through interactions and cause and effect relations (e.g. a disastrous tidal flood triggers flood defense works that in turn alter the sediment transport in the sea, etc.)

The joint consideration of all the above issues facilitates the evaluation of the available risks assessment approaches in order to adopt the most appropriate one for a coastal system. Here, the focus is on the use of probability and fuzzy theory to assess risks. The initial evaluation of these two theories and a comparison between them is required because they both represent appropriate methodology frameworks for coastal systems' risk assessment. This appropriateness lies in the basic concepts behind these theories, i.e., the stochastic or statistic nature of probabilities and the subjective judgment of possibilities of occurrence (for probability and fuzzy theory respectively). A thorough study of the information included in Table 2 and the critical issues mentioned before reveals that the nature of the coastal risk assessment could be, adequately, treated in the concepts of these two theories; therefore, this is the reason to present and compare them in brief hereafter.

4.1. BASICS OF PROBABILITY THEORY

Although probability is calculated in a standard way, its interpretation varies according to the theoretical context where it may be perceived [5]. Good [6] suggests as the most prominent probability interpretation frameworks the classical theory, the priori (or logical) theory, the relative frequency theory, and the personalistic or subjective theory. Depending on which interpretation framework is the most appropriate for the analyst, there is a different methodological approach and treatment of data. Table 3 roughly summarizes the comparison of these four popular views of probability as presented by Singpurwalla, Booker, and Bement in [5]. As it can be observed from the table, there is important criticism and strong arguments against and in favor of all four prominent probability interpretations. However, the main struggle is between the relative frequency and the subjective approaches of probability theory. While, the subjective theory is more flexible and applicable in cases of limited data, information, and knowledge, its scientific consistency is well arguable compared to the robust structure of the relative frequency theory.

Table 2: The physical characteristics, occurring processes and possible risks of coastal subsystems.

Subsystem	Marine Subsystem	Coast Subsystem	Land Subsystem
Definition	The band of sea wherein the evolution of natural and human processes has a direct impact in the tangent strip of land.	The narrow and dynamic transitional zone between the marine and land subsystems. It includes the foreshore, the beach area and natural coastal protection systems such as dunes and mangroves.	The inner part of a coastal system, which is adjacent to the coast subsystem and its width depends on the impact to the land development of the processes taking place in the marine subsystem.
Physical Characteristics	▪ Water depth ▪ Water and sediment movement ▪ Seabed composition ▪ Waves and tide ▪ Marine habitats	▪ Typical coastal profiles ▪ Hydraulic regime (storm surge water-levels) ▪ Wind and wave climate ▪ Coastal habitats, e.g. dunes and mangroves	▪ Topography ▪ Soil types ▪ Aquifer structure ▪ Groundwater resources and salinity ▪ Surface water resources ▪ Land-related habitats, e.g. wetlands
Natural Processes	▪ Coastal currents and sediment transports ▪ Storm and wave conditions affecting coastal profiles ▪ Ecological succession leading to changes in habitat types and biodiversity ▪ Energy and material cycles affecting biological productivity ▪ Higher temperatures ▪ Thermal expansion of ocean water ▪ Sea level rise ▪ Changes in rainfall and evaporation	▪ Storm and wave conditions affecting coastal profiles ▪ Dispersion, degradation, adsorption and sedimentation processes affecting water and sediment quality ▪ Ecological succession leading to changes in habitat types and biodiversity ▪ Energy and material cycles affecting biological productivity ▪ Higher temperatures ▪ Changes in rainfall and evaporation	▪ Ecological succession leading to changes in habitat types and biodiversity ▪ Energy and material cycles affecting biological productivity ▪ Higher temperatures ▪ Changes in rainfall and evaporation

Subsystem	Marine Subsystem	Coast Subsystem	Land Subsystem
Human Processes	▪ Fishery ▪ Exploitation of oil and gas reserves ▪ Navigation ▪ Tourism and recreation ▪ Waste discharges	▪ Sand extraction ▪ Water extraction from dune aquifers ▪ Exploitation of wood resources (mangroves), ▪ Aquaculture ▪ Human settlement ▪ Tourism and recreation ▪ Land reclamation ▪ Port development ▪ Transportation of people and goods ▪ Industrial activities	▪ Use of land resources ▪ Agriculture and aquaculture ▪ Human settlement ▪ Land reclamation ▪ Infrastructure facilities ▪ Irrigation ▪ Hydropower ▪ Industrial exploitation
Risks	▪ Disturbance and destruction of marine habitats due to fishing, mining, diving, anchoring, dredging and dumping ▪ Depletion of fish stocks ▪ Pollution due to oil spills ▪ Deterioration of coastal water quality due to waste discharges ▪ Sediment transports due to breakwaters and other marine works	▪ Disturbance and destruction of coastal habitats by mining, water extraction, woodcutting, settlement and infrastructure development ▪ Degradation and loss of beaches by accelerated coastal erosion ▪ Degradation of natural flood protection by mining and woodcutting ▪ Spatial conflicts for land uses ▪ Deterioration of coastal water quality ▪ Alteration of the landscape due to flood defense structures	▪ Destruction of land habitats ▪ Irrigation ▪ Flooding ▪ Shortage of freshwater resources ▪ Salinization of freshwater resources ▪ Deterioration of groundwater and surface water quality

Considering the use of probabilities for assessing risks in coastal systems it becomes evident that the theoretical context of probability should include the following aspects: (a) stochastic behavior, (b) appropriateness to deal with limited data, (c) incorporation of all appropriate information, (d) appropriateness to deal with one-of-a-kind events. For these reasons the subjective probability theory is elected as the most appropriate probability context to focus on for coastal risk management. An additional reason lies in the fact that subjective probability theory is the closest to the notions adopted in the fuzzy theory context [5]; hence it is the most appropriate probability facet to use for comparison.

4.1.1. Subjective Probability Theory and Risk Assessment

Subjective probabilities are based on stochastic activities, random, discrete or continuous variables, and probability distribution (density) functions. Probability distribution functions are used to represent the physical problem and are generated through data elaboration or use of uncertain information. In complicated situations, stochastic modeling, which introduces subjective probabilities and experts' judgments, is used to incorporate uncertainties and validate complex assumptions, parameters, and probability density functions to assess risk [7]. Equation 1 provides the simpler risk definition in the probabilistic framework.

$$p_F = P\,(L > R) = \int_0^{\infty} \left(\int_0^{\lambda} f_{LR}(\lambda, r)\, dr \right) d\lambda \tag{1}$$

In equation 1 the probability of failure (load exceeds resistance), p_F, depends on the joint density probability function, f_{LR} (λ, r), where L and R, are random variables for loads and resistances of the system respectively. Due to great difficulty in defining the $f_{LR}(\lambda, r)$ a usual simplification is to assume statistical independency between load and resistance [8].

4.2. BASICS OF FUZZY THEORY

Fuzzy theory is the framework for dealing with imprecision and vagueness of linguistic information or situations where there is no availability of data for statistical inference. The most important feature in fuzzy logic is the unclear boundaries of sets and the membership of elements in these sets. Fuzzy theory provides with the mathematical background to model intermediate areas between the binary "is" - "is not".

Fuzzy theory is not presenting variations in terms of definition or basic notions. The critical measure is the membership function, which represents numerically the degree of an element's belonging to a set. The whole mathematical context is an extension of the crisp set mathematical structure enriched with additional relations associated with the fuzzy theory framework.

Table 3: A comparison between the prominent probability theories.

	The Classical Theory	The Relative Frequency Theory	The Personalistic or Subjective Theory
Definition	The ratio of favorable cases to the number of equipossible or equally likely cases	Probability is a property of a collective, i.e., a long sequence of observations for which there is sufficient reason to believe that the relative frequency of an observed attribute will tend to a limit if the observations are indefinitely continued	Probability is a degree of belief of a given person (behaving normatively) at a given time that is measurable and conforms to other beliefs in certain ways.
Basic notions of theory	▪ There is no such thing as chance ▪ Probability is a measure of an individual's partial knowledge	▪ Probability is a measure of an empirical, objective, and physical fact of the external world, independent of human attitudes, opinions, models, and simulations ▪ Probability is never relative to evidence or opinion but determined by observations on the nature of the real world ▪ All probabilities can only be known *a posteriori*, i.e., only through observation ▪ A prerequisite to calculate probability is the existence (actual or conceived) of repetitive events	Subjectiveness in defining probability
Criticism	▪ Equipossible cases are also equiprobable ▪ There is difficulty in dividing up alternatives ▪ There are conceptual difficulties in the context of rare and unobservable events ▪ Involves personal judgment about the equally likely nature of events	▪ Collectives are difficult to construct in real life ▪ It is impossible to prove physically a relative frequency probability ▪ Rare and one-of-kind events do not have probabilities	▪ Personalistic probabilities are not consistent with scientific inference ▪ Declared probabilities may not reflect true beliefs ▪ There is no coherence in personalistic probabilities in complex real situations ▪ There is no ensuring that two persons with identical background information will declare identical probabilities ▪ Subjective probabilities are not following the dictates of the standard calculus of probability

4.2.1. Fuzzy Theory and Risk Assessment

Risk assessment based on fuzzy theory presupposes that the risk parameters are considered by means of fuzzy logic. Therefore, if R and L are fuzzy numbers representing resistance and load on a system, then a risk measure of failure for this system is given in Equation 2 [7]:

$$M = R - L \tag{2}$$

In this equation M may take both negative and positive and values indicating respectively the degree of failure or non-failure of the system. Figures 3 and 4 provide an example of the basic fuzzy risk assessment process.

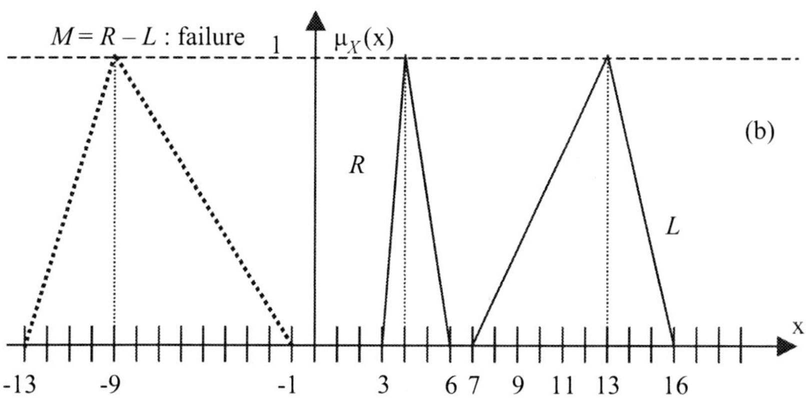

Figure 3: Fuzzy risk assessments of a system: Total Failure [7].

In Figure 3 the extreme contingency of total failure is presented following a fuzzy subtraction between the values of the minimum limit of resistance to the maximum limit of load (and the opposite for the second limit of the dashed triangle) of the triangular shape fuzzy numbers that express resistance and load [7].

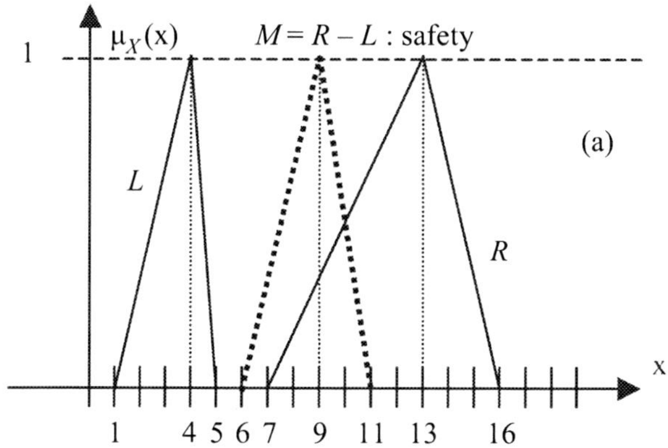

Figure 4: Fuzzy risk assessments of a system: Total Safety [7].

In Figure 4 the extreme contingency of total safety is presented following a fuzzy subtraction between the values of the respective limits of the triangular shape fuzzy numbers that express resistance and load [7]. The subtraction operation is based on the h-level intervals (a-cuts elsewhere), which are not explained further due to space limitations.

5. Comparison Between Probability and Fuzzy Theory Regarding Coastal Risk Assessment

The decision for the most appropriate coastal risk assessment approach lies in the understanding of the nature of the risks and the availability of consistent data associated with them. Tables 1 and 2 present a clear view of these risks in association with coastal subsystems and processes occurring there. An observation of these risks reveals the following:

1. Inside a subsystem, some risks maybe mostly quantitative in nature (e.g. depletion of fish stocks), while others can be mostly qualitative (e.g. alteration of the landscape due to flood defense structures). There are also risks where both quantitative and qualitative measurements are necessary (e.g. deterioration of coastal water quality due to waste discharges). This observation is critical because the predominant nature of the risk (quantitative vs. qualitative) implies or not the use of subjective judgment in the risk assessment process. Probability theory fits more with quantifiable risks (e.g. flooding) while fuzzy theory conceptualizes more efficiently qualitative risks (e.g. deterioration of coastal water quality). Even though quality is measurable (therefore quantifiable) by using several indicators, it also depends on human perceptions and intuition and therefore, fuzzy theory is most appropriate to address subjective parameters in risk assessment [9], [10], [11].
2. Where available data are consistent, credible and statistically admissible, a probability density function is the most persuasive tool to assess future risks. Natural processes, especially in the marine subsystem, are repetitive and periodical in nature (e.g. wave conditions, natural erosion, etc.) and their interaction with human interventions (e.g. breakwaters, land reclamation) can provide with data that through statistical analysis render probabilities of risk occurrence. In cases, however, where data are inconsistent (i.e., scarce or imprecise), fuzzy theory is most appropriate for risk assessment. As Suresh et al. indicate [12]: "*Fuzzy methods may be the only resort when little quantitative information is available regarding fluctuations in the parameters*".
3. Human processes with a severe impact on the coastal environment are not prone to probabilistic forecasting. Although, there are statistical data such as demographic growth, rate of development, rate of urbanization, etc., which can be used for probabilistic risk assessment, these reflect past conditions that are highly doubtful to have the same impact in the future. Therefore, the use of probabilities that are generated based on past incidents and conditions is well questionable. This is a generic issue concerning probabilities, i.e. what is the

credibility of an assessment that is based on past data and refers to future conditions where the situation can be completely different? An example is the sand extraction in the coast subsystems. In this case a period in the past of increased construction activities would provide information of large amounts of extracted sand and would generate a high probability of repetition of the phenomenon in the future, even though future construction activities could be very limited.

4. The coastal system is very complicated. There are significant interactions between human and natural processes with different impact on each coastal subsystem and different cause and effect relations among them (e.g., land uses in the land subsystem affect decisively the economic activities in the coast and the marine subsystems). Therefore, a probabilistic approach should introduce many conditional probabilities in a dynamic framework full of uncertainties. The most complex is a system the less appropriate is the use of probabilities. On the other hand fuzzy theory facilitates modeling of complexity because of its nature that provides with a simpler platform of understanding a complex system behavior.

6. Conclusions and Future Work

The comparison between the two approaches for coastal risk assessment is very revealing. The complexity of the coastal system due to interactions between human and natural processes in its marine, coast, and land subsystems does not justify the dominance of the probability-based approach for risk assessment. This dominance may be the result of focus on the marine subsystem, where the repetitive and periodic nature of natural processes, along with the gradual human interventions, allow for the collection of data that can be statistically treated and generate probabilities. However, the other two subsystems—equally important in an integrated risk assessment approach—have features that inherently render probability-based approaches ineffective. These features are imprecise and scarce qualitative data as well as non-repetitive processes necessitating the incorporation of subjective judgment and intuition to assess risks and plan for the proper mitigation measures. Human-driven processes are difficult to model with probabilities; yet they generate the most dangerous risks in the coastal environment. Therefore, a proper treatment should incorporate fuzzy theory tools to introduce complexity and uncertain outcome of human-driven processes. The integration of fuzzy and probabilistic risk assessment outputs is difficult because they have different logical backgrounds and conceptualizations. Moreover, an integrated risk assessment should incorporate the dynamic interactions between the risks in different subsystems, and therefore incorporate, in some way, results from fuzzy and probability analysis into a single output. Developing an appropriate integrated approach for coastal risk assessment calls for confronting these limitations and the concurrent use of probabilistic and fuzzy measures.

7. References

1. Dutch Coastal Zone Management Center, "NetCoast: A Guide to Integrated Coastal Zone Management", www.netcoast.nl, Coastlearn EU funded project, 2001.

2. Local Government Association's Special Interest Group on Coastal Issues, "Local Authority Coastal Risk Management: Information Pack", http://www.coastalsig.lga.gov.uk/info_pack.htm, 2004.

3. EUROSION project, Living with coastal erosion in Europe: Sediment and space for sustainability – Executive summary, DG Environment, European Commission, May, 2004, pp. 1-19.

4. European Commission, E.U. focus on coastal zones, Office for Official Publications of the European Communities, Luxembourg, 2001, pp. 1-29.

5. Ross, T.J., Booker, J.M. and Parkinson, W.J., eds. (2002) Fuzzy Logic and Probability Applications: Bridging the Gap, ASA-SIAM Series on Statistics and Applied Probability, SIAM, Philadelphia, ASA, Alexandria, VA.

6. Good, I.J. (1965) The Estimation of Probabilities, MIT Press, Cambridge, MA

7. Ganoulis, J.G. (1994) Engineering Risk Analysis of Water Pollution: Probabilities and Fuzzy Sets, VCH Verlagsgesellschaft mbH, Weinheim, Germany.

8. Freudenthal, A.M., Garrelts, J.M., and Shinozuka, M. (1966) The Analysis of Structural Safety, Journal of the Structural Division, 92, 267-325.

9. Kuchta, D. (2001) Use of Fuzzy Numbers in Project Risk (criticality) Assessment", International Journal of Project Management, 19, 305-310.

10. Lam, K.C., So, A.T.P., Hu, T., Ng, T., Yuen, R.K.K., Lo, S.M., Cheung, S.O. and Yang, H. (2001) An Integration of the Fuzzy Reasoning Technique and the Fuzzy Optimization Method in Construction Project Management Decision-Making, International Journal of Project Management, 19, 63-76.

11. de Ru, W.G. and Eloff, J.H.P. (1996) Risk Analysis Modeling with the Use of Fuzzy Logic, Computers & Security, 15, 239-248.

12. Suresh, P.V., Babar, A.K. and Venkat Raj, V. (1996) Uncertainty in Fault-tree Analysis: A Fuzzy Approach, Fuzzy Sets and Systems, 83, 135-141.

DECISION SUPPORT SYSTEMS FOR THE MANAGEMENT OF CONTAMINATED SITES

A Multi-Criteria Approach

S. GIOVE, P. AGOSTINI
Venice Research Consortium (CVR)
Viale della Liberta 5-12, Marghera-Venice, Italy

A. CRITTO, E. SEMENZIN, A. MARCOMINI
Centre Interdepartmental IDEAS - University Ca' Foscari
Calle Larga S. Marta 2137, 30123 Venice, Italy

Abstract

The management and rehabilitation of contaminated sites is a complex process encompassing environmental, technological and socioeconomic aspects. In this paper, two decision support systems (DSSs) are briefly reviewed. DESYRE (DEcision Support sYstem for REhabilitation of contaminated sites) provides an integrated platform for the management of complex information, including consideration of actual risk, selection of best technological options for site remediation, socioeconomic drivers, time, and costs. ERA-MANIA aims to improve ecological risk assessment and support expert decision making for assessment and management of contaminated sites. Both systems demonstrate the value of Multi Criteria Decision Analysis (MCDA).

1. MCDA for Environmental Problems

Environmental decision problems are usually characterized by a high level of complexity. In such a context, Multi-Criteria Decision Analysis (MCDA) represents an important and crucial step [16]. MCDA consists of one or more procedure to assist the decision maker(s) (DM) during the phases of the decision process, and takes into account possible sources of uncertainty and/or different utility functions. Sometimes the problem is expressed in the form of a decision table connecting benefit or cost criteria and alternatives. Despite the Multiple Objective Decision Making (MODM) in the MCDA problems, only a finite number of criteria and alternatives are considered. After having eliminated all the dominated alternatives (if any), the problem consists of selecting the *best* alternatives (optimal choice problem), or ranking all the alternatives (ranking problem). Moreover, we could consider both stochastic and deterministic approaches, but in what follows only the deterministic approach will be considered.

I. Linkov et al. (eds.), Environmental Security in Harbors and Coastal Areas, 267–273.

The literature about MCDA problems and methods is very rich, and cannot be discussed here. Even now, some theoretical items are still not deeply analyzed, and some confusion exists. MCDA methods are usually classified as MAUT (Multi Attribute Utility Theory), outranking (like ELECTRE), and interactive methods [25]. Another distinction regards *compensatory* and *non-compensatory* methods; in the former case interaction among attributes is possible, as in the Choquet integral [11]. A lot of MCDA methods are available at the actual state of the art, but a complete scenario is impossible to compress in a few pages.

In the context of outranking approaches we quote the PROMETHEE and ELECTRE algorithms. Other diffuse approaches in real-world applications include the family of OWA operators, fuzzy ranking methods (sometimes using inference rules), and a lot of other approaches. The most popular MAUT method is the Weighted Average aggregation operators (WA) method. In the WA method, all the criteria values are multiplied by a *weight*, a real number defined offline. Even if some mathematical methods exist to assess the values of the weights, the meaning of the weights is today much debated. A popular interpretation considers the weight as a relative *importance* coefficient (as greater the weight is, as more important its contribution), but this idea is quite controversial. In fact, starting from the Economic Theory, some authors interpret the weights as *substitution rate*. Regardless, the debate about the weight's meaning is not only a pure theoretical or linguistic problem, but is strictly connected with the method used to compute their value. Moreover, even when a *simple* method can be applied to obtain such values from the DM's opinion, particular care must be taken: otherwise the method can lead to paradoxical consequences. This can be the case of a incorrect use of the AHP methodology (and even with SMART and other methods), if the substitution rate among the criteria is not taken into account. The *local* weight ratio is nothing other than a relative measure of two alternatives with respect to some criterion. But if the *global* weights measure the relative importance of the criteria, really no information is given about the (relative) importance of the *best*, alternative, for instance (or any other one) for one criterion, and the best alternative for another one. Such information cannot be obtained from the global weight assigned offline. The substitution rates are case-dependent, and should be assessed comparing the criteria in the *current* case. Of course, such a problem does not arise in the *absolute* version of AHP [17]. Only a few authors have pointed to this important feature, see [21], [26], but most users continue to use AHP without concern about this. Another undesired item in the original additive AHP was the rank reversal phenomenon not appearing in the multiplicative version [13].

Nevertheless, given its simplicity, the WA method is very often used, not only for environmental problems. This is due to its simplicity and to the ease of interpretation. Moreover, a user-friendly hierarchical structure can be easily defined and used, partitioning the criteria into subcriteria and so on, generating a decision tree which can structure and simplify the complex original nature of the problem. Conversely, the WA method is characterized by a serious drawback, since no interaction among the attributes is possible. In fact, the *preferential independence* axiom is required. Other methods do not require such characterization—for instance, the OWA operators [27, 28], and the Choquet integral [11]—but they are more complex to understand and to apply. The Choquet integral, in particular, despite its appealing theoretical features and properties, requires a lot of parameters to be identified, and its use can be restricted to decision problems characterized by fewer criteria.

Other approaches are based on *data mining* method, where the *knowledge* of the decision structure can be directly obtained by the data. We quote the local approximation algorithms (based on *kernel* functions), the neural nets, some clustering algorithms, the decision tree (CART, ID3, etc.), etc.; see among other ones [4]. They are statistical methods, that can be applied also to solve MCDA problems. Nevertheless, their use is limited to the cases where a sufficiently great number of input-output collected data exist, or if simulated data can be easily obtained. This is difficult in the environmental application, and very often it is quite hard to create a sufficiently wide set of simulated data that the DM should analyze. A real MCDA system implies an user-friendly tool that requires the minimum (but necessary) information and data from the DM. Otherwise, the probability of erroneous answers strongly increases. Moreover, the user-friendly characteristic requires a less complex system, with an inferential motor that can be easily understood by the user(s).

Another important item regards the participative nature of the decision. Is the process characterized by only a single DM or by a team? In the latter case, where a group of Experts or DM are involved, we speak about *Group Decision Theory*. In this case the MCDA algorithm has to take into account suitable *consensus measures*, showing how much the group of decision makers agree or disagree about the alternative ranking; see for instance [5]. Very often, conflicting targets exist in a real environmental problem, given the different utility function of the different stakeholders involved in the decision process. If a consensus is not reached, or one DM disagrees with all the other ones, the System Manager has to be advised, and, if necessary, activates a feedback loop to reach a satisfying degree of consensus. The measurement and the management of the Group consensus is quite important, and needs to be carefully designed by analysts.

In the WA approach the consensus measure is based on some norm-based distance functions, easily computable and explainable to the users. All the above mentioned reasons convinced us to use the WA aggregation approach[1], in conjunction with a suitable value function to convert the criteria value on a common scale. In such a way, the WA method is no more completely compensative. This is important for environmental applications because toxic pollution, introduced by a new industry, cannot be compensated by economic development.

Thus most of the MCDA applications developed by us are based on the WA approach with a value function for the criteria. Other aggregation operators may be implemented, aiding the DM with a more complete choice of aggregation tools.

---

[1] Even if the preferential independence axiom is to be *a priori* assumed.

2. Application of MCDA in Two Decision Support Systems for the Management Of Contaminated Sites

The management and rehabilitation of contaminated sites is a complex process encompassing environmental, technological, and socioeconomic aspects. In order to facilitate this complex decision process, several attempts have been made to codify specialist expertise into decision support systems (DSSs) [2]. In the proposed DSS, Multi Criteria Decision Analysis (MCDA) plays a key-role to help both experts and decision makers. In some sense, the MCDA tool is the core of the DSS; its importance in the overall decision process is universally accepted, especially for environmental decision problems [15].

In order to manage the complex decision-making process for the rehabilitation of large contaminated sites, two DSSs called DESYRE and ERA-MANIA will be briefly described, emphasizing the role played by MCDA.

2.1. THE DESYRE DECISION SUPPORT SYSTEM

The GIS-based software DESYRE (DEcision Support sYstem for REhabilitation of contaminated sites) provides an integrated platform for the management of a heterogeneous volume of information, encompassing the consideration of the actual risk, the selection of the best technological options for site remediation, the socioeconomic drivers or constraints at the site of interest, and the time and cost perspective. Moreover, it allows active participation of both experts and stakeholders. The system provides a user-friendly and easily accessible tool which guides the user during the whole application, divided into five modules: four analytical (socioeconomic, characterization, pre- and post- risk assessment, technological) and a decision one.

In the DESYRE framework, the MCDA is applied in the definition of the pool of suitable technologies and in the comparison of alternative scenarios. During the technologies selection, a score is assigned to each technology on the basis of key-criteria, including cost, development time, efficiency (or performance), reliability, flexibility, public acceptability, and so on. The AHP method is used in this phase to weight the different criteria, whose values are then evaluated by a pool of experts. In the decision module, during the scenarios construction and comparison, the MCDA is used to derive the values of the different options. In fact the analytical steps previously performed allow to create different indices (socioeconomic, risk, technological, cost, time and environmental impact indices); these indices can be aggregated into a final index for each scenario and used to compare and rank the remediation options. The aggregation of indices is performed by experts through WA methodologies and decision makers can adjust the weight of each index according to their preferences.

2.2. THE ERA-MANIA DECISION SUPPORT SYSTEM

The ERA-MANIA DSS is a decision support system that aims at improving the Ecological Risk Assessment procedure (ERA) and supporting the expert/decision maker in the assessment and management of contaminated sites. It was developed according to the Weight of Evidence [3] and the TRIAD [19] approaches, where the results provided by a set of Lines of Evidence (LoEs) are gathered and compared to support the assessment and evaluation of the ecosystem impairment caused by the stressor(s) of concern. The developed DSS consists of two modules: "Comparative Test Tables" and "Integrated Ecological Risk Indexes". The former aims at comparing the different tests or LoEs belonging to three investigated experimental areas (chemistry/bioavailability, ecology and ecotoxicology) to guide the expert/decision maker in the choice of the most suitable set of tests to be applied to the case study. Both numerical and non-numerical criteria need to be considered and, using an ad-hoc data entry system, all those are converted into a common numerical scale. Two categories of actors are required. The first category, the System Experts (SE), assign the criteria and other parameter weights offline (the objective criteria), while the second category, the Data Experts (DE), insert numerical judgments for the non-numerical criteria (subjective criteria) based on their own experience. In both cases, the Direct Assignment method (DA) is applied and the final score is obtained by the subsequent application of the WA operator. Group Decision Theory (GDT) methods are applied to measure the consensus among the weights and the other numerical data inserted by the SE and the DE. The system provides some checks to detect the presence of a too-low consensus or if one expert disagrees with most others about some data. Thus a feedback data entry loop can be provided.

Finally, the "Integrated Ecological Risk Indexes" module provides qualitative and quantitative tools that allow assessment of the terrestrial ecosystem impairment (i.e., the impairment to the biodiversity and functional diversity of the terrestrial ecosystem) by integrating the heterogeneous information obtained by the LoEs application. The test results are initially converted into a common scale using a value function, whose parameters are previously defined offline by the DE, and then aggregated using the GM operator.

3. Acknowledgements

The DESYRE project was founded by the Italian Ministry for University and Scientific Research, and based on a research collaboration among the Venice Research Consortium (CVR), the Ca' Foscari University of Venice, the National Research Council (CNR) and the Thetis S.p.a. ERA-MANIA project was founded by the the Italian Government Commissary for the rehabilitation of the Bormida Valley (Italy), and based on a research collaboration among the Ca' Foscari University of Venice, the Italian Environmental Protection Agency (APAT), the Venice Research Consortium (CVR) and the Dutch Institute of Public Health and Environment (RIVM).

4. References

1. ASTM-American Society for Testing and Materials, Standard Provisional Guide for Risk-Based Corrective Action, Final report, PS 104-98, ASTM, Philadelphia, 1998.

2. Bardos, R.P., Mariotti, C., Marot, F., Sullivan, T., Framework for Decision Support used in Contaminated Land Management in Europe and North America, Land Contaminated & Reclamation, 9, 149-163, 2001.

3. Burton, Jr., G.A., Chapman, P.M., Smith E.P., Weight-of-Evidence approaches for assessing ecosystem impairment. Human and Ecological Risk Assessment 8: 1657-1673, 2002.

4. Bishop, C., M., Neural networks for pattern recognition, Clarendon Press, Oxford, 1995.

5. Carlsson, C., Ehrenber,g D., Eklund, P., Fedrizz,i M., Gustafsson, P., Lindholm, P., Merkuryeva G., Riissanen T., Ventre A.G.S., Consensus in distribuited soft environments, European Journal of Operational Research, 61, 165-185, 1992.

6. Chen S.C., Hwang C.L., Fuzzy multiple attribute decision making, Springer-Verlag, 1991.

7. Chen, C.T., Extensions of the TOPSIS for group decision-making under fuzzy environment, Fuzzy Sets and Systems, 114, 1-9, 2000.

8. Chen, S.J., Hwang C.-L., Hwang F.P., Fuzzy multiple attribute decision making, Springer-Verlag, Berlin, 1992.

9. Eastman, J.R., Kyem, P.A.K., Toledano, J., Jin, W., GIS and Decision Making, UNITAR Explorations in GIS Technology - Vol.4, United Nations Institute for Training and Research, Geneva, 1993.

10. Golden B.L., et al., The analytic hierarchy process, applications and studies, Springer-Verlag, 1989.

11. Grabisch M., The application of fuzzy integrals in multicriteria decision making, Fuzzy Sets and Systems, 92, 1997, 167-189.

12. Isaaks, E.H., Srivastava, R.M., An introduction to applied geostatistics, Oxford University Press, New York, 1989.

13. Lootsma F.A., Fuzzy logic for planning and decision making, Kluwer Academic Publishers, 1997.

14. Lootsma F.A., Multi-criteria decision analysis via ratio and difference judgement, Kluwer Academic Publishers, 2000.

15. Munda G., Social multi-criteria evaluation: methodological foundations and operational consequences, European Journal of Operational Research, 158, 662-677, 2004.

16. Munda, G., Fuzzy information in multicriteria environmental evaluation models, Report EUR 15602EN of Joint Research Center of European Commission, 1994.

17. Ramanathan R., A note on the use of the analytic hierarchy process for environmental impact assessment, Journal of Environmental Management, 63, 27-35, 2001.

18. Ramanathan, R., Ganesh, L.S., Group preference aggregation methods employed in AHP: an evaluation and an intrinsic process for deriving members's weightages, European Journal of Operational Research, 79, 249-265, 1994.

19. RIVM - Rutgers M, JH Faber, JF Postma, H Eijsackers, 2000. Site-specific ecological risk: a basic approach to function-specific assessment of soil pollution. The Netherland Integrated Soil Research Programme Report, Vol.28, RIVM, Aquasense, Alterra, NL.

20. Saaty, T.L., The analytic hierarchy process, McGraw Hill, New York, 1980.

21. Schoner B., Wedley W.C., Choo E.U., A unified approach to AHP with linking pins, European Journal of Operational Research, 64, 364-392, 1993.

22. Triantaphyllou E., Multi-criteria decision making methods: a comparative study, Kluwer Academic Publishers, 1999.

23. US-EPA, Risk assessment guidance for superfund Vol 1, Human health evaluation manual, Final Report, EPA/540//1-89/002, Washington DC, 1989.

24. Van Den Honert, R.C., Lootsma F.A., Group preference aggregation in the multiplicative AHP. The model of the group decision process and Pareto optimality, European Journal of Operational Research, 96, 363-370, 1996.

25. Vincke, P., Multicriteria decision-aid, Wiley & Sons, Baffins Lane, Chichester, England, 1992.

26. Wedley W.C., Choo E.U., Schoner B., Magnitude adjustment for AHP benefit/cost ratios, European Journal of Operational Research, 133, 342-351, 2001.

27. Yager R.R., et al., The ordered weighted averaging operators, theory and applications, Kluwer Academic Publishers, 1997.

28. Yager, R.R., On ordered weighted averaging aggregation operators in multicriteria decision making, IEEE Trans. Systems Man Cybernet., 18, 183-190, 1988.

VISUALIZATION OF PARETO FRONTIER IN ENVIRONMENTAL DECISION MAKING

A.V. LOTOV
State University – Higher School of Economics; Department of Mathematics for Economists; Lomonosov Moscow State University Faculty of Computational Mathematics and Cybernetics; Russian Academy of Sciences
Dorodnicyn Computing Centre, 40 Vavilova str., 119991 Moscow, Russia

Abstract

This paper describes environmental applications of a multi-objective decision aid method based on interactive visualization of the Pareto frontier. The method provides an opportunity to explore the Pareto frontier, that is, the criterion tradeoffs, for three and more criteria. The knowledge about criterion tradeoffs helps stakeholders to better understand the problem, specify the preferred combination of the criterion values (the goal), and compute the associated decision alternatives. The method is introduced on the basis of an example related to coastal water pollution. Then, real-life applications of the method are described in short. Possible applications of this method in the case of risk and uncertainty are discussed. A new Web version of the method can be used for informing lay stakeholders about environmental risks.

1. Introduction

The need for a holistic view of the problem and the involvement of ordinary people (lay stakeholders) in the decision process are two of the most important features of the modern environmental decision process. For example, the European Water Framework Directive [1] that specifies guidelines for the integrated river basin management requires implementation of the holistic view of the problems and improvement of public participation in the river basin management planning. A holistic view can be supported by the application of multi-criteria decision aid (MCDA) methods. The requirement of public participation means, however, that these methods must be understandable for lay stakeholders. Moreover, application of these methods on the Internet for educating lay stakeholders seems inevitable.

This paper is devoted to a new MCDA technique that satisfies these requirements. The technique, known as Interactive Decision Maps, provides interactive and

I. Linkov et al. (eds.), Environmental Security in Harbors and Coastal Areas, 275–292.

animated visualization of the Pareto frontier in a form that proved to be understandable for computer-literate people. It was implemented on the Web and used in real-life problems.

To discuss the main features of the Pareto frontier methods and, in particular, of the Interactive Decision Maps techniques, we start by classifying the MCDA methods according to the role of the decision maker (DM). The current form of such a classification is provided in [2]. Four groups of methods are considered (see Figure 1):

1. No preference methods
2. *a priori* preference methods
3. Interactive methods
4. *a posteriori* preference (Pareto frontier) methods

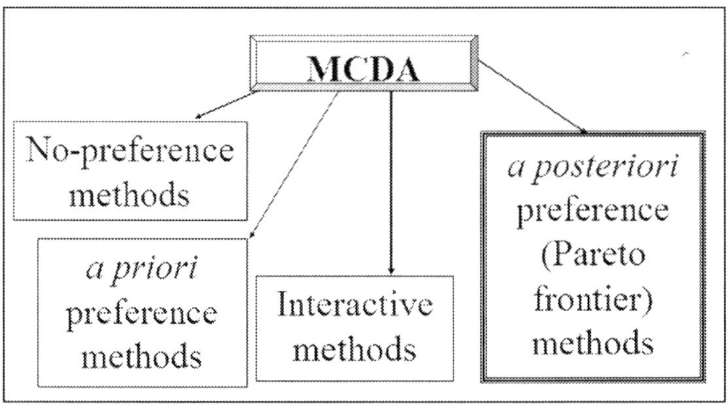

Figure 1: Classification of the MCDA techniques according to the role of the decision maker.

In the *no preference* methods, where the DM's preferences are not taken into account, the multi-objective problem is somehow converted into a single optimization problem and solved (see, for example, [2]). The DM (if exists) may either accept or reject the solution. Such methods can be used in situations where the DM is absent.

The *a priori preference* methods are based on modeling the DM's preferences before a particular variety of feasible decision alternatives is considered. Utility functions are often used for modeling the preferences. A brilliant Multi-Attribute Utility Theory (MAUT) was developed in this field [3]. However, to apply MAUT, one needs to satisfy rigorous assumptions. Moreover, the theoretically sound preference modeling techniques are based on boring procedures, during which the DM has to compare multiple pairs of criterion points. For this reason, the scope of real-life application of the MAUT-based methods is fairly narrow. Instead, heuristic simplified preference-based procedures are used, such as weighting the criteria, the AHP method, outranking approaches, goal programming techniques, etc.

One can pose a general question, whether the DM has got preferences before a particular problem is considered. In the *interactive* methods [2] that combine a step-by-step exploration of the variety of feasible decision alternatives with a step-by-step development of a preference model, one does not need to answer this question. However, the *interactive* methods use in general the comparison of a large number of criterion points, too. Therefore, they meet the same difficulties as the *a priori preference* methods.

It is extremely important that both the *a priori preference* and *interactive* methods require consistent behavior of the DM. Modern psychology has proved experimentally that human beings are often not consistent in preference-related procedures (see, for example, [4,5]). The situation is usually met in the case of weighting and other techniques that make use of linear compensations [6]. To explain the human behavior, modern psychology applies the concept of mental models, which help human beings to anticipate the future and make their decisions. It is assumed [7] that a mental model comprises several levels (Figure 2):

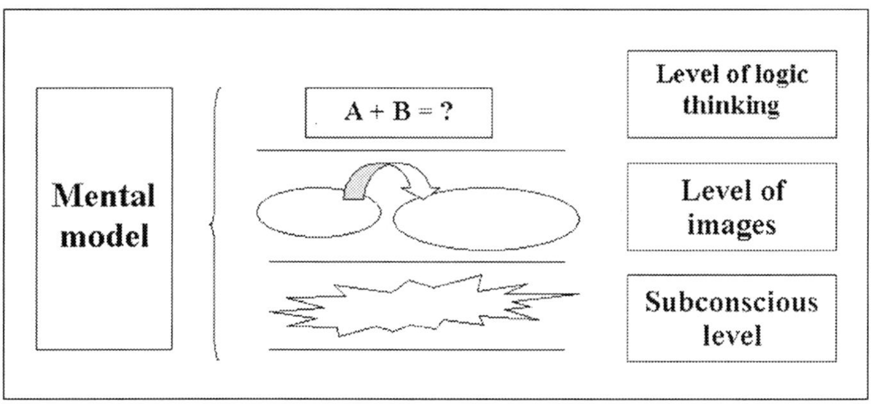

Figure 2: Levels of a mental model.

1. The upper, rational (logical) level is based on logical inference;
2. The second level includes images, relations of which (in contrast to the upper level) may be not consistent; and
3. The third level contains vague subconscious relations.

All the levels interact, and processes of coordinating them are permanently under way, especially in sleep. However, certain discordance between the levels is a natural feature of human thinking processes. Mental decision processes incorporate activities of all the levels, and it is very complicated to determine, which level is responsible for the result of the comparison of decision alternatives. One has to take into account that the imaginary and subconscious elements of the decision process are usually camouflaged by a logical inference that is used to justify a decision after it has been made. Therefore, inconsistent answers are met fairly often in the preference modeling procedures.

Goal programming based on the single-shot identification of a goal (introduced in [8]) can be considered as an example of the preference modeling approach that is psychologically simple for a human being. For this reason, the approach has found broad real-life applications. However, the users meet certain difficulties in the process of setting the goal without knowing the feasibility frontiers and the criterion tradeoffs.

In this paper we concentrate on the *a posteriori preference* methods that are gradually obtaining a broader recognition. For example, a large number of engineers have discovered and started to develop such methods during the last decade [9]. In the *a posteriori preference* methods, the basic role is played by the Pareto frontier, which is comprised of the nondominated criterion points, that is, such points that cannot be improved in one criterion without worsening at least the value of one of the remaining criteria. Approximating the Pareto frontier is carried out before the DM is involved into the decision process. Then, the DM is informed on the Pareto frontier. Note that the DM does not need to answer questions on his/her preferences. Often, a free search among the nondominated criterion points is applied. In the last case, the DM loosely obtains knowledge on feasible criterion values. Then, he/she can express his/her preferences by identifying a preferred point of the Pareto frontier.

The *a posteriori preference* methods were initiated in 1950s by S. Gass and T. Saaty [10]. The form of presentation of the Pareto frontier to the DM plays a crucial role in the methods of this group. Most of the methods provide information on the Pareto frontier in the symbolic form of a list of nondominated criterion points. If the number of criteria is greater than two, it is extremely complicated to assess such information.

Computer visualization has substantially improved the situation. Visualization of information—i.e., transforming symbolic data into geometric figures—can support human beings in better understanding the problems. Successful applications of computer-based visualization techniques are met permanently. Visualization can be used in the framework of the Pareto frontier methods, too. Visualization of the Pareto frontier for supporting environmental decision making for two decision criteria was introduced in [11]. This paper is devoted to a new technique for visualization of the Pareto frontier for three and larger number of criteria. By supporting the DM's identification of the goal point we shift the single-shot goal programming from the *a priori* group of MCDA methods to the *a posteriori* group of methods.

In what follows, we introduce the Interactive Decision Maps (IDM) method for visualization of the Pareto frontier and discuss its application in environmental decision problems. In Section 2 the mathematical formalization of IDM is provided. Section 3 is devoted to an example problem of ocean waste disposal. In Section 4 the IDM-based Reasonable Goals Method (RGM) is introduced and its environmental applications are described. Section 5 is devoted to real-life applications of RGM/IDM in Web. In Section 6 the application of RGM/IDM for supporting decision making under risk and uncertainty is discussed. Finally, we show that the Web version of the technique can be used for informing lay stakeholders of possible risks.

2. Mathematical Formulation of the Interactive Decision Maps

We assume that the feasible decisions x belong to a compact set X of the linear metric space R^n and that the vector of criterion values y is a point of the linear metric space R^m. The criterion vector y is assumed to be given by the vector function f: $R^n \rightarrow R^m$. Then, the feasible set of criterion vectors is given by $Y=f(X)$, that is,

$$Y = \{y \in R^m : y = f(x),\ x \in X\}. \tag{1}$$

Let us assume that the decrease of the values of the criteria is desirable. Then, a point $y'' \in R^m$ is preferred to a point $y' \in R^m$ ($y'' \succ y'$, that is, y'' dominates y') if $y'' \leq y'$ and $y'' \neq y'$. In this case, the Pareto (nondominated) frontier $P(Y)$ of Y is defined as

$$P(Y) = \{y' \in Y : \{y'' \in Y : y'' \leq y', y' \neq y''\} = \varnothing\} \tag{2}$$

In visualization of the Pareto frontier, an important role plays the Edgeworth-Pareto Hull (EPH) of the feasible criterion set, which is defined as

$$Y_p = Y + R_+^m = \{y^* \in R^m : y^* = y_1 + y_2, y_1 \in Y, y_2 \in R_+^m\}, \tag{3}$$

where R_+^m is the nonnegative cone of R^m. In Figure 3 the frontier of the set $Y=f(X)$ is given by the solid line. One can easily identify the points of the Pareto frontier, which is given by the bold line, by using the following rule: the cone of points $y'' : y'' \succ y'$ does not intersect with the set Y_p for a nondominated criterion point y'. The frontier of the set Y_p is given by the bold and dashed lines. Importantly, $P(Y_p) = P(Y)$. In other words, the set Y_p is the largest set that has the same Pareto frontier as the set Y. It includes, along with the feasible criterion points, all criterion points dominated by the feasible points. By proceeding to Y_p, we get rid of the dominated frontier of Y, which usually make the display too complicated.

In contrast to various techniques for approximating the Pareto frontier developed since 1955, we approximate the Pareto frontier for its subsequent visualization. The first ideas known now as IDM were introduced about 30 years ago [12]. Its concepts, methods and real-life environmental applications are summarized in the book [13]. The main feature of IDM consists of the direct approximation of the EPH. An approximation of the EPH is used for fast interactive visualization of the Pareto frontier as the frontiers of two-criterion slices (cross-sections) of the EPH.

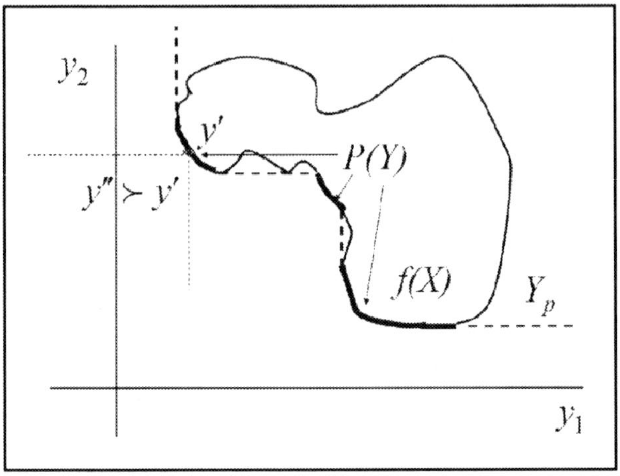

Figure 3: Illustration of $Y=f(X)$, $P(Y)$ and Y_p.

A two-criterion slice of Y_p passing through a point $y^* \in R^m$ is defined as follows. Let us consider a pair of criteria, say u and v. Let z^* be the values of the remaining criteria in the point y^*. Then, a slice of the set Y_p, which passes through the point y^* and is related to the pair (u, v), is defined as (we do not care about the order of the criteria)

$$G(Y_p, z^*) = \{(u, v) : (u, v, z^*) \in Y_p\}. \qquad (4)$$

The slices of the Pareto frontier are displayed as frontiers of the slices of the EPH. A collection of such frontiers, for which the value of only one of the remaining criteria can change, constitutes a decision map. Therefore, a decision map can provide information on criterion tradeoffs between two criteria, depending on the values of the third criterion. The influence of other criteria is displayed in the framework of the IDM technique by animation. Since the EPH has already been approximated, modern computers require only a few seconds for computing and displaying hundreds of decision maps. Due to it, animation is possible by successive demonstration of decision maps associated with monotonically changing the values of the fourth criterion.

Methods for approximating the EPH used in the framework of IDM depend on the model. In the linear case, they are based on the combination of the single criterion optimization with the Fourier convolution of the linear inequality systems [13], and the EPH is approximated by a polyhedral set. In the nonlinear case, the random search and the statistical evaluation of approximation quality are combined with the local optimization, and the EPH is approximated by a collection of domination cones with vertices located in points, which are close to the Pareto frontier. A detailed description of the approximating algorithms is given in the book [13].

Applications of IDM include national economic planning, environmental planning, water management, national energy planning, machinery design, etc [13]. We provide an example of its application in the problem related to pollution abatement in a sea bight.

3. Case Study: Ocean Waste Management Decisions

In this section we illustrate the application of IDM with an ocean waste disposal example, requiring difficult decisions concerning cost and resulting pollution. We reconsider the old problem of choosing sewage sludge disposal sites in the New York Bight [14].

In the 1980s, contamination of the New York Bight (Figure 4) was a concern of the US Environmental Protection Agency (EPA). In 1985, the EPA ordered New York City and the remaining users of the inner Bight region to begin shifting their dumping operations to the 106-mile site. However, the decision turned out to be too expensive. For this reason, allocation of sludge between three different disposal sites was proposed: a 12-mile site, a 60-mile site, and a 106-mile site. The problem was modeled by T.M. Leschine (University of Washington) and W.A. Verdini (Arizona State University). They developed a multi-layer stochastic pollution transport model that helped to evaluate the pollution at several important monitoring stations inshore and offshore for a given waste dumping at the three possible disposal sites. On the basis of the stochastic model, an influence matrix was constructed [14] to reexamine the EPA decision in a way that permits simultaneous multiple-site dumping.

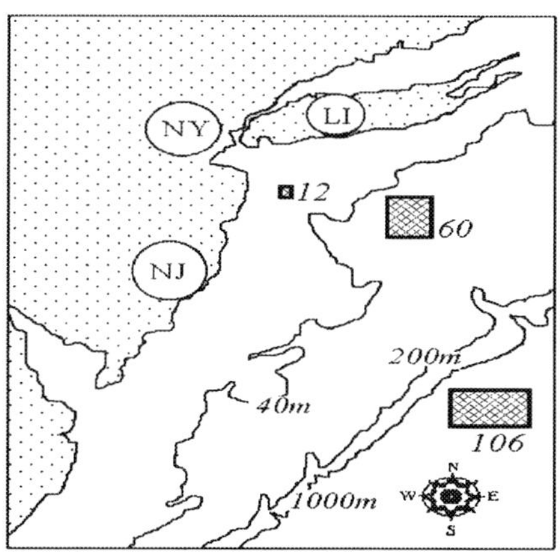

Figure 4: Map of the New York Bight region.

In the model, all sludge was assumed to be produced in New York City (52%), New Jersey (41%), and Long Island (7%). Production of sludge was assumed to be constant from year to year. Two types of vessels were used for the transportation of the sludge: towed barges and self-propelled barges. The decision variables included the number of self-propelled and towed barge trips from sources to possible sites. Constraints related to the ocean's assimilative capacity and to the annual dumping capacity of barges were taken into account.

In [14], three decision criteria were used to evaluate different sludge disposal strategies:

- Total cost of sludge disposal operation (in millions of US$)

- Pollution concentration at inshore monitoring station (percentage)
- Pollution concentration at offshore monitoring station (percentage)

This problem was reexamined once again in [15] using IDM. We provide first a particular slice and the decision map drawn by hand. Then we provide a gray copy of the decision map produced by IDM.

To begin with, let us restrict the total cost by some value, say $15 million. Then all feasible values of inshore pollution and of offshore pollution are given by the variety of feasible criterion vectors on the criterion plane (see Figure 5 where the variety is shaded).

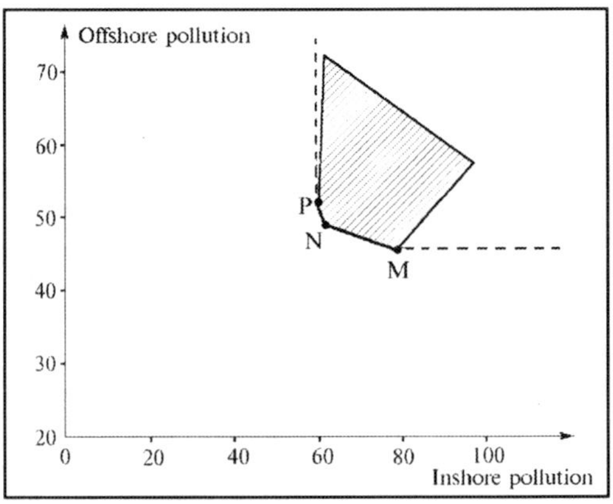

Figure 5: The variety of feasible criterion vectors and its EPH, which frontier is depicted by the dashed line.

Since it is preferable to decrease both inshore and offshore pollution, we are interested in its southwestern frontier [P, M], which is the Pareto frontier. The frontier of the EPH is depicted by a dashed line. As usual, the variety of feasible criterion vectors and its EPH have the same Pareto frontier. Note that in the neighborhood of point ***M***, a small decrement in the offshore pollution requires a substantial increment in the inshore pollution. On the contrary, in the neighborhood of point ***P***, just a small rise in the inshore pollution (say, the movement from ***P*** to ***N***) results in a sharp decrement in offshore pollution. One can easily understand how the offshore pollution is transformed into the inshore pollution if efficient strategies are used. In other words, the criterion tradeoff for inshore and offshore pollution is displayed in a clear way in Figure 5.

In Figure 6, a decision map is provided. It is a collection of two-dimensional slices of the EPH for the three criteria, where several tradeoff curves for inshore and offshore pollution are related to different constraints imposed on values of total cost. These values are given directly in Figure 6.

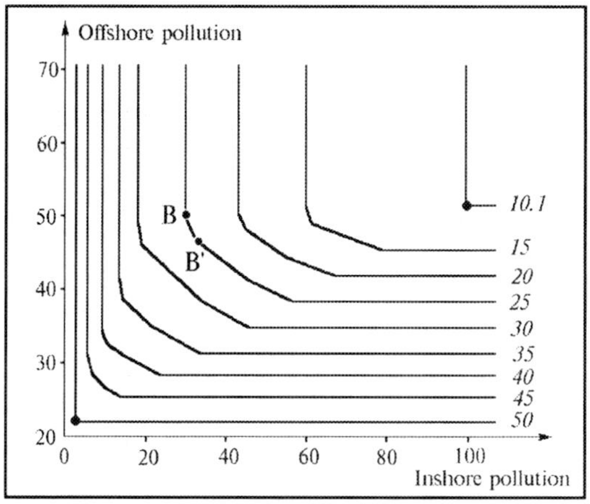

Figure 6: A decision map for several fixed costs.

The decision map helps to understand the influence of an increment in the total cost for the improvement of the environment (i.e., a reduction in the inshore and/or the offshore pollution). The costs are constrained between $10.1 million and $50 million. The tradeoff curves in Figure 6 are drawn with bold lines. They have the following important feature: there is a conflict between inshore and offshore pollution, except for the $10.1 million and $50 million frontiers, which consist of just one point. The criterion tradeoff changes while it is moving along the tradeoff curves, and also depends on cost.

Now let us consider a gray copy of the color computer display (Figure 7) that informs the user of relations among five criteria (two additional criteria are introduced – the number of self-propelled barge trips to the 106-mile site from the New York City and the number of towed barge trips from the New Jersey).

To display such information, decision maps with scroll-bars are used. The decision map displays tradeoff information for the first three criteria (in the gray copy, cost is given by shades of gray, pollution is given on axes), and the scroll-bars provide information about two additional criteria: their sliders inform on constraints imposed on the associated criteria. In Figure 7, the sliders describe constraints on the number of barge trips the to the 106-mile site from the New York City and from the New Jersey. The user can move the sliders manually, changing by this one of the constraints imposed on the number of trips. Automatic monotonic movement of a sliders results in the animation of the decision maps. Using these tools, the user can explore the influence of the fourth and fifth criteria on the tradeoffs among cost and pollution. In addition, snapshots of the animations can be displayed in the form of a decision map matrix. However, this topic is beyond the scope of our paper (see [13]).

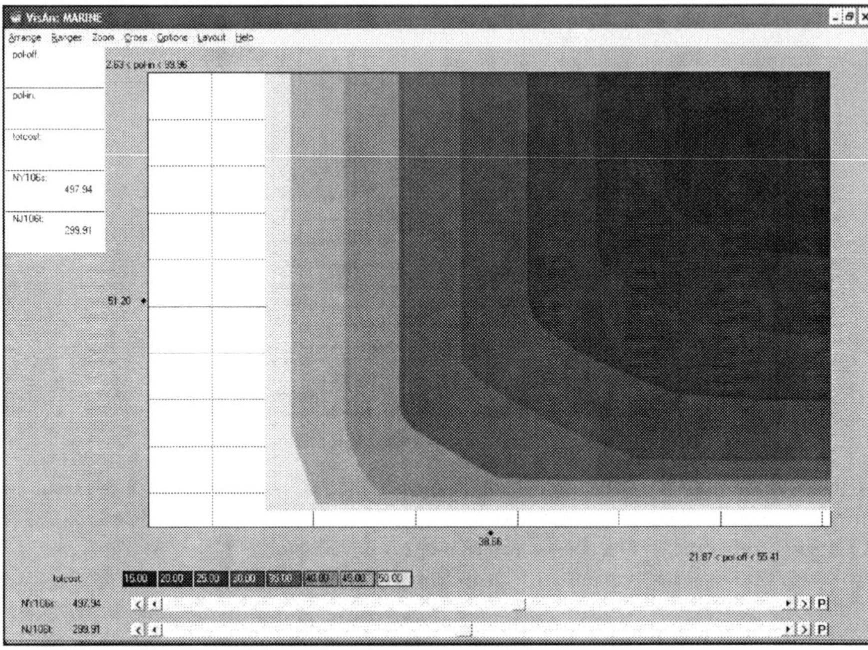

Figure 7: Gray copy of a decision map with scroll-bars that specify the constraints on barge trips.

Usually we advise users to restrict themselves to five criteria. Indeed, psychological studies prove that seven is the maximum number of objects a normal human being can deal with. Since the criterion tradeoffs are fairly complicated, restricting to five criteria seems to be natural for human beings. However, since the users apply our technique without our control, they often violate this recommendation and manage to study up to nine criteria simultaneously. Technically, the software provides such an opportunity (additional scroll-bars are displayed for any reasonable number of criteria).

The user can identify a preferred feasible combination of criterion values (the goal) at one of the frontiers by the computer mouse click. By this way, his/her preferences are expressed. Since the identified goal is feasible, the computer finds a decision alternative, which provides achievement of the goal. Note that, in contrast to the other goal methods, the user identifies the goal on the basis of the tradeoff information. It is very important that the goal is feasible. Due to it the problem of infeasibility of the goal, which is usually met in other goal methods, vanishes here. For this reason, IDM can be considered as the tool for visualization of the goal identification procedure. This IDM application is called the Feasible Goals Method. It can be considered the new visualization-based form of the goal approach.

The study of the sludge dumping is an example application of FGM/IDM. Real-life environmental applications of the technique for water quality planning on the request of the Russian Federal Program "Revival of the Volga River" are described in [13] and [16]. These studies use an interface between IDM and GIS.

Note that the sludge dumping problem is linear, and so the EPH is convex. However, the FGM/IDM can be applied in the case of nonconvex models and criteria, too. Such applications are described in [13].

4. RGM/IDM Method

As was shown in Section 3, FGM/IDM is used for studying the Pareto frontier and selecting a preferred decision using linear mathematical models. In contrast, RGM/IDM, introduced in 1990s (see, for example, [17]), is aimed at screening-aimed visualization of decision alternatives given in the form of decision matrices that may contain a large number of alternatives. Several numerical attributes that contain performance indicators must be specified as the selection criteria. Due to it, the alternatives are associated with the points in the criterion space. Then, the EPH of the convex hull of the criterion points is approximated and IDM is applied to support the identification of the goal at the Pareto frontier. However, this time the goal is not feasible, but only reasonable (that is, close to feasible criterion points). Therefore, this method is known as the Reasonable Goals Method.

Let us consider the mathematical formulation of RGM/IDM. We consider a decision matrix (table) that contains N rows (alternatives) and several columns, any of which is related to an attribute. Let us assume that m attributes are specified to be the selection criteria. Then, each alternative can be associated to a point of the m-dimensional linear criterion space R^m. Criterion values for the alternative number j are described by the criterion point y^j, which coordinates are $y_1{}^j$,..., $y_m{}^j$. Since the decision matrix contains N alternatives, we obtain N feasible criterion points y^1, y^2, ..., y^N. RGM/IDM is based on enveloping the feasible criterion points, i.e. on considering the convex hull of them denoted by YC and exploration of its Pareto frontier by approximating the EPH of YC denoted by YC*. An approximation of the set YC* has the same form as the EPH considered in the previous section, so it can be explored with the help of IDM as in the previous section.

Let y* be the goal identified by the user. In contrast to FGM/IDM, now the goal is usually not feasible. For this reason, screening of the database is applied, that is, selecting a relatively small number of feasible criterion points, which are close to the identified goal, and providing the associated rows to the user. Various concepts of proximity can be applied in this case. Details of the screening process are discussed in Chapter 4 of the book [13].

Let us consider several environmental applications of RGM/IDM. One of the studies was carried out on the request of Russian Ministry for Agriculture. In the framework of this study, RGM/IDM was used in the decision support system for water quality planning in a small region. The DSS was adapted to a small region in the basin of the Oka River, which is the second largest tributary of the Volga Rive. Eight sources of wastewater discharge were specified in the region. A large, but finite number of decision alternatives were prepared by environmental engineers. These alternatives (namely, 390,625 alternatives) were given as rows of a decision matrix, which columns contained decision information that included the cost of the alternatives as well as resulting pollution concentrations.

Three pollutants were taken into account: nitrates, phosphates, and BOD. Conventional units were used to measure pollutant concentration, in the framework of which concentration equals to one in the case the environmental requirements are satisfied precisely. The maximal concentrations (in the river) of these three pollutants were used as the screening criteria. The fourth criterion was the cost of the project measured in millions of US$. The studied problem was fairly artificial since the environmental engineers failed to collect data concerning the most important sources of pollution discharge. However, the study gives an idea how the decision maps may look.

For example, Figure 8 contains a gray copy of a decision map that describes all 390, 625 described by four criteria. Cost is given in horizontal axes, BOD (p3) is given in vertical axes, and nitrate concentration (p1) is given by shading. It turned out that phosphate concentration does not influence the decision map. For this reason it was excluded from the study. The cross represents a potential goal point. By its fixation, the user can get 11 decision alternatives as it is said in the decision map. Detailed information on the study is given in [13] and [18]. The selected alternatives can be displayed in geographic maps.

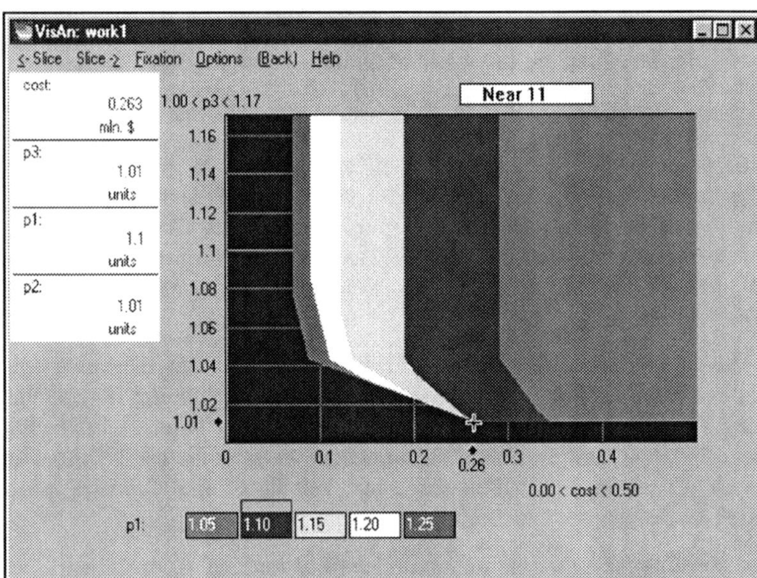

Figure 8: A gray copy of the color decision map for about 400 K decision alternatives.

Another environmental application of RGM/IDM is related to national energy planning at the Israeli Ministry of National Infrastructures (see [13] and [19]). A number of alternative strategies for long-term development of Israeli electricity production were created. All the strategies were considered as the plans for reduction of air pollution level, too. The following attributes were considered: percentage of carbon dioxide discharge reduction, percentage of nitrogen discharge reduction, additional total abatement cost, marginal abatement cost, and average cost of electricity. RGM/IDM was used for exploration of the Pareto frontier and selecting a small number of interesting strategies for subsequent final decision making.

Other applications of RGM/IDM are possible, too [13]. They include selecting from large lists of environmental, technical, financial, medical, personal, and other decision alternatives. Several of them are interfaced with GIS.

5. Web Applications of RGM/IDM

Experimental application of IDM on the Web began in 1996. A refined version based on Java was started in 2000 in the form of a web application server [13, 20]. The web application server implements multi-tier architecture and consists of the calculation server, web server application, and graphical presentation layer. The structure of user interaction with the web application server is given in Figure 9.

The calculation server is actually an executable module coded in C++. It processes given table data and approximates the set YC*. It can be compiled and executed on any platform. The Pareto frontier visualization window is a Java applet executed inside the user's browser. MS Internet Explorer, v. 4.0 or higher may be used to display it. The web application is coded in Java and JSP and serves several interfacing purposes: it helps user to prepare a table with alternatives, invokes the calculation server to process it, transmits the applet with calculated data, and handles the goal point identified by the user to generate selected alternatives. The web application can be executed on any web server that supports JSP and Java servlets. The web application server is located at http://www.ccas.ru/mmes/-mmeda/rgdb/index.htm.

The user has first to input the table to be explored and submit the query. Then, the server envelops the criterion points and sends the Java applet along with the set YC* to the computer of the user. The animated decision maps look like the decision maps for a stand-alone computer, and so there is no need to discuss them specially. Then, the applet transmits the goal identified by the user to the server, and the server returns the selected rows to the user.

The first real-life application of the Web application server is related to supporting remote negotiations and decision making in regional water management in the Werra River project, Germany. The research is funded by German Federal Ministry of Education and Research. Several hundreds of alternative strategies were elaborated that describe the environmental development of the Werra River basin during five years. The users are water engineers who use computers in their offices in the city of Eisenach, Germany, and the calculation server is located in Berlin at WASY, GmBH. The engineers have to select a strategy by using IDM, Web GIS, produced by ESRI and other decision support techniques (see [21] for details). It is planned that ordinary people from the Werra River basin will get an opportunity to explore the totality of the prepared alternatives and to individually select the one they prefer most through the Internet.

It is clear that a wide range of Web-served applications can be considered. If needed, they can interfaced with various Web GISs.

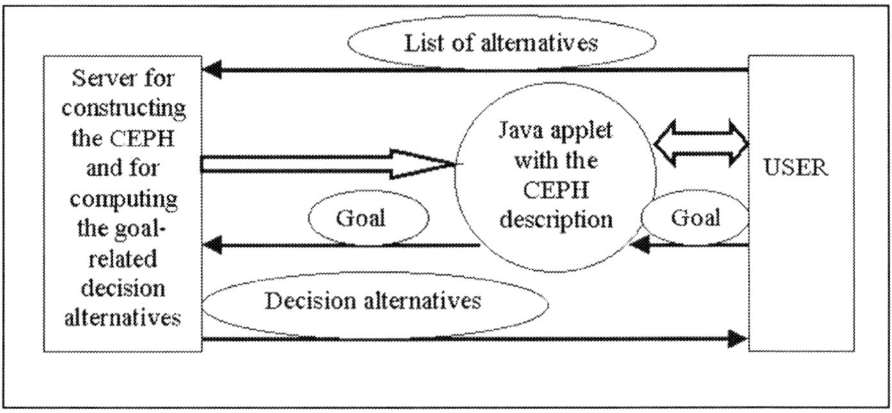

Figure 9: Scheme of user interaction with the demo Web application server.

6. Pareto Frontier Visualization in Decision Making with Risk

RGM/IDM can be used in the case of decision making under risk, that is, with stochastic information. In this case, the decision matrix must describe all performance indicators needed for decision support. For example, mathematical expectations and probabilities of important events may be included. Let us imagine a decision matrix that describes a finite number of alternative designs of a dam on a river that helps to avoid , at least partially, the negative consequences of floods. Assume that any alternative is described by three performance indicators that are used as the decision criteria: expected losses because of flood plus annual construction and maintenance cost, in million US$); the probability of a certain high losses (in percents); and the probability of the catastrophic losses (in 0.01 of percent). It is assumed that the meanings of high and catastrophic losses were specified by experts. Surely, it is desirable to minimize the values of all three criteria.

Figure 10 contains a gray copy of the color decision map.

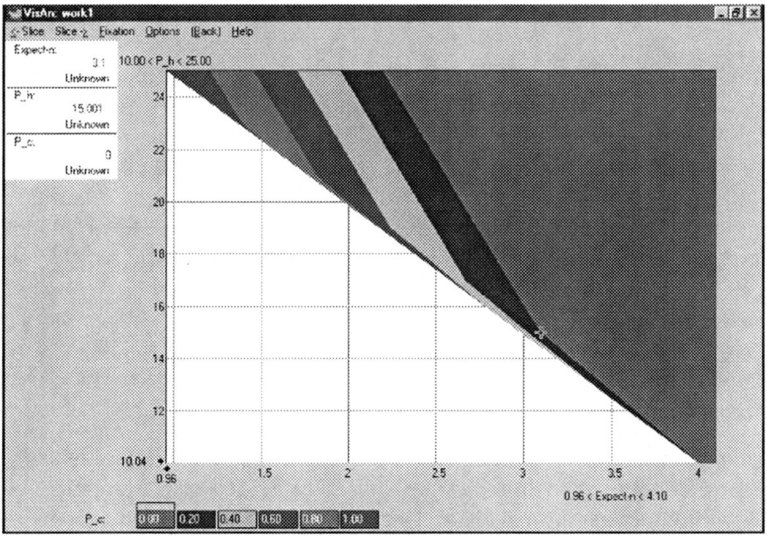

Figure 10: Decision map for decision support under risk.

The values of expected losses are given on the horizontal axis, and values of probability of high losses, denoted by P_h, are given on the vertical axis. The Pareto frontiers among expectation and P_h for several constraints imposed on the probability of catastrophic losses denoted by P_c are given (southwest frontiers). The relation between shading (color) and the values of P_c are given in the palette under the picture. One can see that probability of the catastrophic losses P_c changes from 0 to 1 in steps of 0.2. One can easily note an important feature of the variety of alternatives under study—if the probability of high losses P_h is less than about 15%, then all Pareto frontiers are very close. This means that, for these values of P_h, increasing P_c from 0 to 1 requires a very small increment of expectation of losses and costs. Further, for low values of P_h, the Pareto frontiers display linear dependence of P_h on expectation of losses. Therefore, the decision maker inevitably has to make a difficult tradeoff among P_h and expectation. For this reason, the point marked by the cross may be preferable and may be identified as the reasonable goal. The precise position of the cross is given in the tablet: expected losses equal to \$3.1 million, $P\_h = 15\%$, $P\_c = 0$. The computer finds a small number of alternatives that are close to the goal.

It is important that the subjective preferences and experience of the DM are used in selecting the goal. Using decision maps, the DM can inform other stakeholders and general public about the reasons why he/she prefers a certain decision. In turn, lay stakeholders can use the same decision maps to criticize the decision maker.

Let us mention that RGM/IDM can be applied in the case of uncertain information (by enveloping criterion boxes instead of points) and in the case of uncertain futures (robust decision making). However, these topics are beyond the scope of the paper.

7. Conclusion

RGM/IDM supports exploring large varieties of alternatives (decision matrices that include costs, benefits, risks, etc.) in a simple graphic form. Web application of RGM/IDM can help multiple lay stakeholders use the Internet to study feasibility frontiers and criterion tradeoffs. As a result, individual specification of risks by lay stakeholders is possible. In this way, they can select preferred feasible decision alternatives. Therefore, IDM can be used in the framework of a democratic paradigm of environmental decision making, which is based on the desire to involve lay stakeholders in the process of decision making. It is clear that lay stakeholders usually have minimal knowledge of how to solve environmental problems. This knowledge gap may be extremely dangerous. IDM-based web tools can help lay stakeholders study environmental problems and base their actions on a better understanding of the issues.

Long-time systematic application of IDM in the computer laboratory works for university students and sporadic application of IDM-based computer games by people without university education (including schoolchildren) makes us hope that the software can be used by any computer-literate lay stakeholder in the process of preparing political actions related to the final choice.

8. Acknowledgments

Writing of this paper was financially supported by the Russian Foundation for Basic Research (project no. 04-01-00662), by the program for State Support of Leading Russian Scientific Schools (project no. NSc-1843.2003.1), by the Program no. 16 for Fundamental Research of Russian Academy of Sciences (project no. 2) and by the Program no. 3 for Fundamental Research of the Department for Mathematical Sciences of Russian Academy of Sciences. Author is grateful to the anonymous reviewer who carefully read the paper and corrected typos and vague statements.

9. References

1. Directive 2000/60/EC of the European Parliament and of the Council of 23 October 2000.

2. Miettinen, K.M. (1999) Nonlinear Multiobjective Optimization, Kluwer Academic Publishers, Boston.

3. Keeney, R. L. and H. Raiffa (1976) Decisions with Multiple Objectives: Preferences and Value Tradeoffs, Wiley, New York.

4. Tversky, A. (1969) Intransitivity of preferences, Psychological Review, 76, 31-48.

5. Johnson, E. and Schkade, D. (1985) Bias in Utility Measurement, Management Science, 36, 406-424.

6. Borcherding K., Schmeer S., Weber M. (1993) Biases in Multiattribute Weight Elicitation, in J.P. Caverny, M. Bar-Hilel, F.N. Barron, H. Jungermann (eds.) Contributions to Decision Research, North Holland, pp. 3-28.

7. Lomov, B.F. (1999) Problems and strategies of psychological studies, Nauka Publishing House, Moscow (in Russian).

8. Charnes, A. and Cooper, W.W. (1961) Management Models and Industrial Applications of Linear Programming, vol. 1, Wiley, New York.

9. Evolutionary Multiobjective Optimization Web page, Maintained by C.A. C. Coello, <http://www.lania.mx/~ccoello/EMOO/>

10. Gass, S., and Saaty, T. (1955) The computational algorithm for the parametric objective function, Naval Research Logistics Quarterly, 2, 39-51.

11. Cohon, J. (1978) Multiobjective Programming and Planning, John Wiley, New York.

12. Lotov, A.V. (1975) Exploration of economic systems with the help of reachable sets. In Proc. Int. Conf. on the Modeling of Economic Processes (Erevan, 1974), Computing Center of the USSR Academy of Sciences, Moscow, pp. 132-137 (in Russian).

13. Lotov A.V., Bushenkov V.A., and Kamenev G.K. (2004) Interactive Decision Maps. Approximation and Visualization of the Pareto Frontgier. Kluwer Academic Publishers: Boston.

14. Leschine, T.M., Wallenius, H., and Verdini, W.A. (1992) Interactive Multi-objective Analysis and Assimilative Capacity-based Ocean Disposal Decisions, European Journal on Operational Research, 56(2).

15. Lotov, A., Chernykh, O., Bushenkov, V., Wallenius, H., and Wallenius, J. (1998) Interactive Decision Maps, with an Example Illustrating Ocean Waste Management Decisions, in: T. J. Stewart and R.C. van den Honert (eds.) Trends in Multicriteria Decision Making, Springer, Lecture Notes in Economic and Mathematical Systems, 465, 313-323.

16. Lotov, A.V., Bourmistrova, L.V., Efremov, R.V., Bushenkov, V.A., Buber, A.L., and Brainin, N.A. (2005) Experience of Model Integration and Pareto Frontier Visualization in the Search for Preferable Water Quality Strategies, Environmental Modelling and Software, 20(2), 243-260.

17. Lotov, A., Bushenkov, V., Chernov, A., Gusev, D. and Kamenev, G. (1997) INTERNET, GIS, and Interactive Decision Maps, in: J. of Geographical Information and Decision Analysis, v.1(2), 119-143, <http://www.geodec.org/gida_2.htm>.

18. Bourmistrova, L., Efremov, R., Lotov, A. (2002) A Visual Decision Making Support Technique and its Application in Water Resources Management Systems, Journal of Computer and System Science Int., 41, 759-769.

19. Soloveichik, D., Ben-Aderet, N., Grinman, M., and Lotov, A. (2002) Multi-objective Optimization and Marginal Abatement Cost in the Electricity Sector – an Israeli Case Study, European Journal on Operational Research, 140, 571-583.

20. Lotov, A.V., Kistanov, A.A., Zaitsev, A.D. (2004) Visualization-based Data Mining Tool and its Web application, in: Y. Shi, W. Xu, and Z. Chen (eds.) Data Mining and Knowledge Management, Chinese Academy of Sciences Symposium CASDMKM (Beijing, 2004), Springer-Verlag, Berlin, pp. 1-10.

21. Dietrich, J., Schumann, A.H., Lotov, A.V. (2004) Workflow Oriented Participatory Decision Support for Integrated River Basin Planning, in: IFAC Workshop on Modelling and Control for Participatory Planning and Managing (Venice 2004), <http://www.elet.polimi.it/IFAC_TC_Environment/venice2004/programme/>.

22. Lotov, A.V. (2003) Internet Tools for Supporting Lay Stakeholders in the Framework of the Democratic Paradigm of Environmental Decision Making, Journal of Multi-Criteria Decision Analysis, 12, 145-142.

SPATIALLY-EXPLICIT POPULATION MODELS WITH COMPLEX DECISIONS

Fish, Cattle, and Decision Analysis

R.A. GOODWIN
U.S. Army Engineer Research and Development Center
Environmental Laboratory, 333 SW 1st Ave, Portland, OR 97208 USA

V. PANDEY, G.A. KIKER
University of Florida
Agricultural and Biological Engineering Department, Frazier Rogers Hall, P.O. Box 110570, Gainesville, FL 32611-0570 USA

J.B. KIM
U.S. Army Engineer Research and Development Center
Environmental Laboratory, 3909 Halls Ferry Rd, Vicksburg, MS 39180 USA

Abstract

Many approaches exist for modeling the response of animals to environmental condition and change. Regardless of the model selected, uncertainty is a major component in the modeling of complex physical-biological relationships. Structured methods exist for handling uncertainty in these modeling studies, and can facilitate decision-making among stakeholders with differing values. We describe two different approaches for modeling population response to environmental pattern. Then, we propose a simple means for incorporating uncertainty into the modeling process using structured and transparent means. First, a model formula is selected and applied with a structured uncertainty analysis during parameterization. Second, Monte Carlo simulation is applied to propagate the uncertainties in the model outputs induced by the uncertain inputs. Finally, multi-criteria decision analysis (MCDA) is applied to prioritize model forecasts (i.e., of the likely input conditions) according to perceived value, relevance, accuracy, and uncertainty. The structure discussed is simple and can be modified in many ways to meet the demands of a particular study. This paper provides (1) a brief look at alternatives for modeling animal populations and (2) how these types of models can be applied within a structured and transparent framework for handling uncertainty that saves time, money, and effort.

I. Linkov et al. (eds.), Environmental Security in Harbors and Coastal Areas, 293–306.

1. Introduction

Modeling is a critical tool for understanding complex ecosystem function. Modeling helps synthesize information in the effort to gain insight into system structure, processes, pattern, and likely response to alternative management strategies. The complexity needed in an ecological model depends on the range of scales of important system dynamics under study and whether the model is to be used as a screening tool or as a method for decoding process dynamics at a first principles level. This chapter briefly describes two models for modeling the movement dynamics of individuals and how output from these models can be used within a structured decision analysis for management decision-support. This chapter is divided into four sections: (1) a brief review of spatially explicit modeling concepts, (2) description of an integrated frameworks approach for decoding and forecasting 3-D fish movement behavior response to hydrodynamic and water quality pattern, (3) an integrated cattle/hydrological/nutrient model for estimating non-point source pollution, and (4) a simple decision-analytical approach for incorporating uncertainty and prior belief networks into the decision-making process that, inevitably, modeling is meant to support.

2. Overview of Spatially-Explicit Modeling Concepts

Spatially-explicit models are increasingly being used to understand and forecast animal movement and spatially important population dynamics [11]. Models range in complexity from describing the number of patch habitats occupied by a population to complex individual/agent-based models describing spatial habitat occupancy at the individual level. Meta-population models describe the movements of clusters of individuals between habitat patches that are either isolated from one another or have limited exchange of individuals [2]. Meta-population models can be spatially structured so they incorporate information about habitat relationships and the characteristics of the landscape in which the meta-population exists [3].

In an individual-based model, the behavior, growth, reproduction, and other important characteristics of each individual are tracked through time. These models provide ecologists with an effective way to explore the mechanisms through which population and ecosystem ecology arises from how individuals interact with each other and their environment. Individual-based models can take the form of agent-based models when individuals and characteristics of the environment are treated as interacting agents [40]. Multi-agent systems (MAS) are influenced by computer sciences and social sciences [7] and give more prominence to the decision-making process of the agents and to the social organization in which these agents are embedded. Ferber [13] has defined a multi-agent system being composed of the environment, objects, agents, and relations and operations. MAS has been effectively used in a variety of cases such as modeling sheep's spatial memory [10], prediction of duck population response to anthropogenic cases [26], and forecasting the effects of alternative water management scenarios in South Florida on the long-term populations of white-tailed deer and Florida panther [1].

3. Modeling Impact of Hydropower Dams on Fish Movement Behavior

Sustainable water resource management requires an understanding of the relationship between the physicochemical environment and habitat utilization by target aquatic biota. This understanding is the critical foundation upon which forecasting tools are developed that accurately forecast how habitats (and ultimately the aquatic populations they support) change in response to alternative water resource management plans. Two related questions must be answered to develop this understanding: 1) what are the hydrodynamic, water quality, and other cues used by aquatic biota to move through the habitat mosaic and 2) what are the criteria they use to either select or reject one specific habitat of the many they encounter. Unfortunately, relatively little is known about these cues and dynamics, which impact the success of management actions.

Development of simple animal movement models is confounded by the different theoretical approaches used in the analysis of animal movement and aggregation: Eulerian (biomass flux), Lagrangian (individual movement), and agents (discrete rules) simulation [33]. Model power and simplicity are achieved by coupling the theoretical treatments in a manner that maximizes the utility and minimizes the liability of each approach [29]. The Numerical Fish Surrogate uses a Eulerian-Lagrangian-agent method (ELAM, [17]) for mechanistically decoding and forecasting movement patterns of individual fish responding to abiotic stimuli. An ELAM model is an individual-based model (IBM) coupling a (1) Eulerian framework to govern the physical, hydrodynamic, and water quality domains, (2) Lagrangian framework to govern the sensory perception and movement trajectories of individual fish, and (3) agent framework to govern the behavior decisions of individuals.

In general, animals have a multitude of time-varying streams of information and must select between numerous behaviors. The Lagrangian particle-tracking algorithm, supplemented with a game theoretic agent-based, event-driven foraging behavior algorithm [5], tracks individual movements within a hydraulic and/or water quality model. Abiotic information at the location of the individual is transformed into streams of information, systematically organized, and then evaluated as motivations, or utilities. A mathematical hierarchy is developed representing the integration of information from various sensory modalities that may take varying precedence during the changing phases of a behavioral sequence [38; 31]. In each increment of time, using the cues on the presence or absence of the agents characterized by stimuli being above or below a threshold change level (Figure 1), the fish tracks the expected utility of each behavior and elicits the behavior with the maximum expected utility. The expected utility (U_i) from behavior B_i depends on the behavior's maximum or intrinsic utility (u_i) times the probability (P_i) of obtaining the utility, minus the bioenergetic cost (C_i) of the behavior as:

$$U_i(t) = P_i(t) \cdot u_i - C_i(t) \quad (1)$$

The probability of obtaining the utility depends on the previous probability and whether or not the fish encounters the agent in increment *t*-1 to *t* and is expressed as:

$$P_i(t) = (\ 1 - m_i\ ) \cdot e_i(t) + m_i \cdot P_i(t\text{-}1) \tag{2}$$

where m_i is a memory coefficient weighting the present event and past probability $P_i(t\text{-}1)$, and $e_i(t)$ is a Boolean measure equal to unity if the stimulus change threshold is exceeded in increment *t*-1 to *t* and zero otherwise. Perceived change in the strength of a stimulus may be treated using log, e.g., $I(t)/I_a(t)$ in Figure 1, linear, e.g., $d(t) - d_a(t)$ in Figure 1, or other means for calculating change. Subscript a indicates the strength of the stimulus to which the individual is already acclimatized.

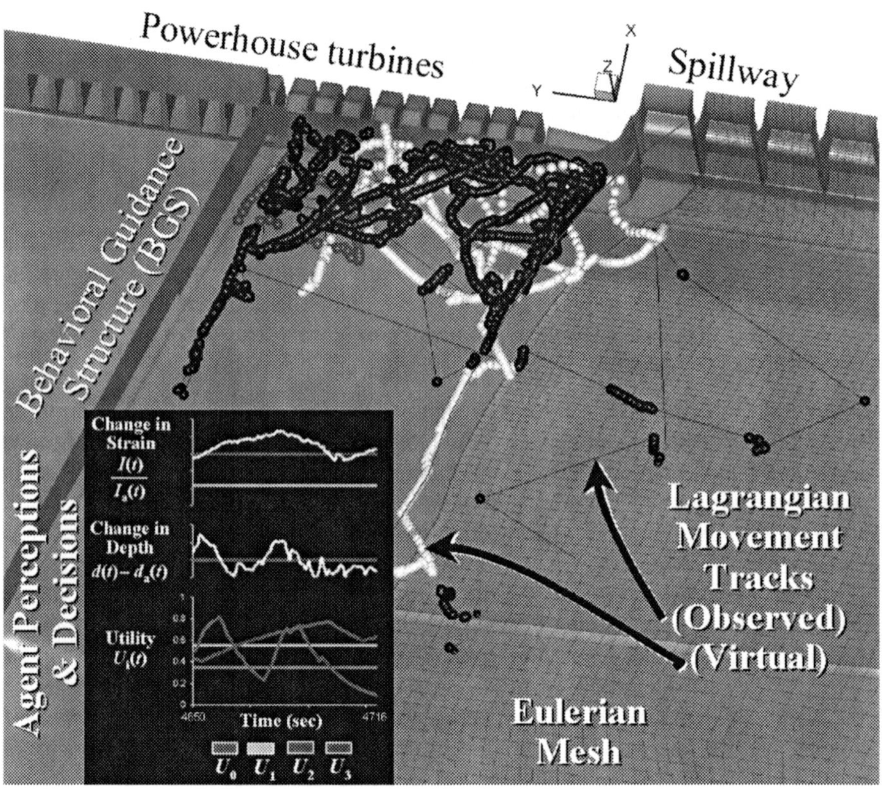

Figure 1: Illustration of virtual (yellow/red) and observed (black) juvenile salmon (Lagrangian) movement tracks at Lower Granite Dam. BGS is a suspended steel wall in the forebay. CFD model (Eulerian) mesh highlighted for a portion of the modeled physical domain. Plot of fish perceptions and decisions in black box for time t=4650 sec to 4716 sec corresponds to red portion of the yellow virtual track. Perceived change in each stimulus (agent) illustrated with the white line. Thresholds illustrated in upper two plots as solid horizontal colored lines. Utility for each agent-behavior couplet is calculated based on Boolean events triggered when the perceived change in a stimulus exceeds the corresponding threshold thought to induce a response behavior. Observed fish tracking data from Cash et al. [8].

The Numerical Fish Surrogate has been applied to determine the impact of (1) pump-storage and release operations on the movement of blueback herring responding to water quality and hydrodynamic patterns [16; 30] and (2) hydraulic structures to effectively attract and collect outmigrating juvenile salmon for bypass around hydropower dams [17]. For both applications, model output consists of individual fish tracks, which can be grouped, converted, or rescaled so as to be of the same form as observed fish data. In the blueback herring model, virtual gillnets and boat-mounted hydroacoustics were used to sample the virtual fish population as it moved within the 2-D laterally-averaged hydrodynamic and water quality model CE-QUAL-W2. Virtual fish data were compared to fish distributions as measured by actual gillnets and hydroacoustic instrumentation deployed on the reservoir. In the juvenile salmon model, the number of virtual fish passing into each turbine unit, spillbay, and bypass structure was counted as they moved within a 3-D computational fluid dynamics (CFD) model of the hydropower dam forebay. Virtual fish passage proportions were compared to the proportions of actual fish measured by instrumentation deployed at the dam, and virtual fish tracks were compared to tracks of acoustically-tagged juvenile salmon [8] (Figure 1). In both cases, the modeler could adjust behavior rules until the movements of virtual fish approximated observed fish. Once accuracy of the model was deemed sufficient, managers could assess the impact of alternative management strategies by running the fish model using forecasted hydrodynamic and water quality conditions.

4. Modeling Cattle for Estimating Non-Point Source Pollution

Interaction of animals with their surrounding environment is a complex aspect of ecology. These interactions often produce complex movement patterns that pose a challenge to ecological modelers. Herbivore dynamics have been modeled using classical predator-prey relationships [32] and energy balance relationships that account for the balance between energy required for herbivore body maintenance and the amount gathered by foraging [20; 12]. A number of models have been developed for describing animal responses to environmental inputs, including GRAZE [34], SPUR2 [18], and SAVANNA [9]. The SAVANNA ecosystem model is a spatially-explicit, process-oriented modeling system developed to simulate ecosystems occupied by ungulate herbivores. The model is composed of several submodels, which describe various processes and vary in complexity. The herbivory submodel simulates forage intake by diet selection, forage abundance, and forage quality. An energy balance submodel simulates body weight of the mean animal of each species based on differences between energy intake and energy spent. Stafford Smith [39] described a detailed mechanistic model in which they added a behavioral submodel to simulate the ecology of an arid zone sheep paddock in pastoral areas of South Australia.

High-density animal operations are especially of interest as it can potentially be a cause of concern with regards to its impact on the environment [6]. Pastureland and dairies can become an important source of diffuse or nonpoint source pollution if adequate practices are not implemented or in cases when livestock is allowed to approach or enter surface waters. In regions such as south Florida where cattle-ranching and dairy-farming are important agricultural activities there is concern of increases in nutrient loadings from these agricultural lands. Phosphorus loading from rangelands and its subsequent movement into the drainage waters (Lake Okeechobee) is a major environmental concern [4]. The primary source of phosphorus has been non-point source agricultural runoff, particularly from beef cattle ranching and dairy farming, the two primary land uses in the Lake Okeechobee watershed [15]. Best Management Practices (BMPs) can reduce nutrient losses in drainage waters to an environmentally acceptable level, while simultaneously maintaining an economically viable farming operation for the grower [6]. Regulatory agencies such as the South Florida Water Management District (SFWMD) are currently assessing the efficiencies of water quality BMPs such as fencing, nutrient management, and water retention in wetlands to reduce phosphorus loads to Lake Okeechobee from cow-calf operations in the Okeechobee basin [41].

Since cattle defecation is of major concern, it is evident that in order to develop a complete understanding of the animal-plant-soil system in a ranch system, spatiotemporal dynamics of grazing cattle must be understood. This information will aid in developing a comprehensive understanding of ecological interactions. At the MacArthur Agro-ecology Research Center (MAERC), Buck Island Ranch (BIR) in Lake Placid, Florida, a 4,168-ha full-scale commercial cattle ranch, Global Positioning Services (GPS) collars were placed on grazing cattle to track their movements. Movement behavior patterns of individual cattle in the pasture domain at BIR (Figure 2) are decoded so the impact of different management practices on cattle movement and, therefore, non-point source pollution and water quality can be evaluated.

Modeling of such movement is particularly challenging as movement patters are highly complex and span all across the domain (Figure 2). Development of animal movement sub-model is purposed which will represent cattle movement and behavior. This model will be an add-on component, which in combination with existing ACRU2000 agrohydrological model will provide a more integrated picture of the animal-plant-soil system. The ACRU agrohydrological modeling system [37] is a multi-purpose and multi-level integrated physical conceptual model that can simulate streamflow, total evaporation, and land cover/management and abstraction impacts on water resources at a daily time step. The original ACRU model, developed in FORTRAN, was restructured entirely with an object oriented programming language (Java) and the version was re-named to ACRU2000 [23]. The advent of ACRU2000 made the model more compliant with spatial hydrological aspects and addition of newer modules. Research efforts into development of ACRU2000 have made it compatible to the unique Florida hydrology i.e. high water table, Flatwood soils (sandy), flat topography and plenty of isolated wetlands [25; 42]. The model will consist of a movement submodel coupled to ACRU2000 [23], an object-oriented agrohydrological model that simulates stream flow, total evaporation, land cover/management, and abstraction impacts on water resources.

Even though ACRU has been extensively tested within the BIR system, it is expected not be free from uncertainty in its predictions. Predictive uncertainty from mathematical models can be problematic for stakeholders who rely on model-based decision support. The following section discusses uncertainty in model predictions and its use within structured decision analysis.

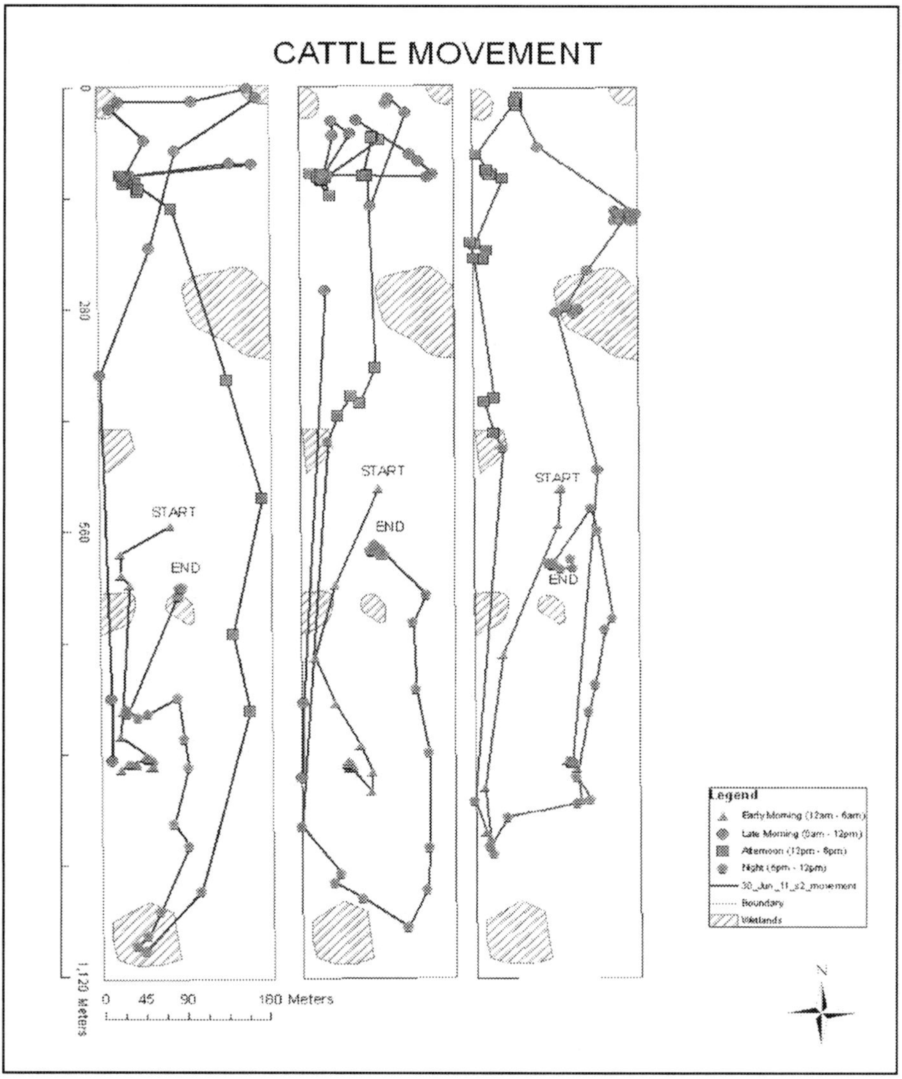

Figure 2: Map of three GPS-collared cattle tracks during one day at the BIR experimental pasture.

5. Handling Uncertainty in Population Modeling

Uncertainty is part of modeling populations and is introduced by, among other sources, (a) conflicting patterns in and alternative sources of observed data and (b) alternative model formulations and parameterization. A structured means is needed to handle the four distinctive types of uncertainty in modeling efforts [36]:

- Temporal – uncertainty of future or past states,
- Structural – uncertainty due to complexity,
- Metrical – uncertainty in measurement, and
- Translational – uncertainty in explaining results.

Decision-support modeling is best undertaken with a structured approach for handling uncertainty [28; 35; 27].

Uncertainty can be incorporated into the modeling process using a structured and transparent means as follows:

1. Selection of model parameters and application of a structured uncertainty analysis for parameterization
2. Propagation to calculate the uncertainties in the model outputs induced by the uncertain inputs
3. Application of MCDA to rank model forecasts of management impact based on perceived value, relevance, accuracy, and uncertainty

5.1. STEP 1

Since the mathematical model is nonlinear, it is important to explicitly consider uncertainties in model inputs and parameters. Mathematical models may include hundreds of parameters. It is not practical to consider uncertainties in each and every one. We must decide which parameters are the most important to the overall model's output and what uncertainties should be propagated.

The simplest way to screen for important uncertain parameters is a first order analysis [28]. Uncertainty importance is defined as a product of a normalized parameter's sensitivity and its normalized uncertainty (coefficient of variation).

The following equation shows a dimensionless index of uncertainty importance for parameter θ [28].

$$UI(\theta) = [(\mathrm{d}x/\mathrm{d}\theta)(\mu_\Theta/\mu_x)]\,[\sigma_\Theta/\mu_\Theta] \tag{3}$$

where μ_Θ= nominal value (*e.g.*, mean value) for parameter θ,

σ_Θ = standard deviation for parameter θ, and

x = model output. The most interesting model variable, x can be chosen for this test, or several can be considered.

The first squared bracketed term on the right-hand-side represents a normalized sensitivity (or "elasticity" the percent change in outcome x induced by a 1 percent change in model input θ) and the second squared bracketed term represents normalized uncertainty (or "coefficient of variation") of the parameter. *UI* can be used to rank the model parameters.

The uncertainty importance method is a local approach that considers output sensitivity at a nominal value θ of the input. If the model is highly nonlinear and there is a large uncertainty in θ, this procedure may distort the relative importance of different uncertainties, as the index may change considerably if different values of θ are considered. An alternative approach is the tornado diagram. The latter approach allows us to compute the effect on the outputs by changing each parameter from its lowest value to highest value while other parameters are fixed at nominal value.

5.2. STEP 2

Once the important parameters are selected, the question is: how can we use assumed distributions of input parameters $\underline{\Theta}$ to estimate the distribution of model outcomes (*i.e.*, uncertainty propagation)? A quantitative risk analysis method, the Monte Carlo method, is often used to approximate model output distributions [14]. The Monte Carlo simulation is used to propagate these uncertainties through model to derive a discrete distribution for the model outcomes. Monte Carlo simulation draws a value at random from the distribution for each input, then one for each input defines a scenario. The entire process is repeated many times producing independent scenarios with corresponding output values. These output values constitute a random sample induced by the input distributions. Because of sample error, the Monte Carlo method is only an approximation to the exact distribution of outcomes. Therefore the accuracy of the method can be improved by increasing the number of independent samples [28]. Alternatively, during the Monte Carlo process, the Latin Hypercube Sampling (LHS) Method in variance reduction techniques is used to reduce the standard error of the mean (*i.e.*, the standard deviation of the sampling distribution of the mean), improving computation efficiency [21]. A description of these various techniques is beyond the scope of this brief chapter.

5.3. STEP 3

Multi-criteria decision analysis (MCDA) provides a means for scientists and managers to collaboratively prioritize, a priori, observed data available for different species, life-stage, season, and environment, all of which have varying levels of relevance, accuracy, and uncertainty with regard to the target issue. Collaborative decision-making ensures that stakeholders are aware of the tradeoffs made in using specific data to develop the decision-support model.

MCDA uses utility functions to develop a scalar index of performance rather than monetary measures. MCDA has been increasingly recommended for integrating ecological, social, and economic objectives in order to evaluate alternatives. It has two major roles [19]):

- To provide information on *trade-offs* by displaying the relative performance of alternative strategies. This allows decision makers to understand the relative advantages and disadvantages of each alternative.
- To help decision makers systematically apply their *values* to the management problem in such a way as to rationally and efficiently document their process and recommendations for a preferred alternative.

The steps of a MCDA are (1) to identify the fundamental objectives (criteria) and alternatives; (2) to quantify the impact of the alternatives on the stated fundamental objectives to be achieved; (3) to examine trade-offs; (4) to elicit and apply the value judgments that result in a ranking of alternatives; and (5) to amalgamate step integrates all the values, weights, and criteria into an overall score that allows the alternatives to be ranked in a manner that reflects the objectives and values of the decision-makers [24].

The value judgments consist of two parts; value scaling and weighting judgment. First, value scaling is the creation of a utility function for each criterion. These functions describe a person's preferences regarding different levels of each criterion. The functions translate the physical criterion into a measure of value, and are scaled between 0 and 1, representing the worst and best values, respectively. Utility functions can take various shapes: linear, nonlinear, or a stepwise shape reflecting how an individual's degree of satisfaction changes as the score for a particular criterion changes.

Second, weighting represents differences in value attached to different criteria. Essentially, the weights assigned to each criterion represent the rate at which people are willing to trade off portions of the criterion range among the criteria. Therefore, the relative importance of criteria and weights should be determined by considering the full range of possible performance of each alternative in terms of each criterion. There are several approaches to eliciting the weighting structure for a decision model, including direct weighting, the analytic hierarchy process (AHP), the swing method, and the trade-off method. A description of these various techniques is beyond the scope of this brief summary of MCDA.

While we present these elements in the form of a sequential list, iterations among these steps are necessary. As a part of this process it is critical to determine who the stakeholders and participants in the decision process are, since the MCDA process depends on an assessment of their beliefs and preferences in order to establish the objectives to be achieved, the alternatives to be examined, and the weights that reflect the participants' priorities among these objectives.

One important thing is that the uncertainties are important and can change the decisions due to human's perception about risk. Decision-makers and stakeholders commonly have different attitudes toward risk and uncertainty as reflected in their views about risk outcomes and the distribution of those outcomes. Decision-makers are often observed to be risk averse because they want to achieve their objectives with more certainty. For a given expected value of predicted outcomes from a model, people generally prefer lower variance. As a result, the expected performance for a management alternative being considered by a given decision is determined by the combined effect of the expected value of the alternative and its variance [22].

The key benefits of MCDA are its emphasis on the importance of the values of decision-makers and stakeholders in the course of establishing criteria, the explicit incorporation of these values into the decision-making process, and the ability the method offers to evaluate the contribution of specific values and criteria to overall ranks or decisions through a form of sensitivity analysis Thus, MCDA can improve the quality of decisions that involve multiple criteria by making a decision more explicit, rational, and efficient.

6. Discussion

Myriad approaches exist for tackling the issue of individual and population movement behavior response to environmental condition and perturbation. While uncertainty is a major component in any forecast of complex system dynamics, structured methods for approaching and describing uncertainty facilitate decision-making among stakeholders with differing values. We have attempted to describe two different approaches to modeling population dynamics and then how information from these models can be incorporated within a structured framework for handling uncertainty.

The structure we proposed is simple and can be modified in a variety of different ways. Forging consensus in complex environmental problems represents a considerable challenge due to the fact that people may have different and sometimes irreconcilable views, different priorities, different objectives, and different beliefs about outcomes. MCDA and uncertainty analysis provide the means to identify the reasons and causes for disagreements among parties that hinder cooperation and negotiation. Through interviews and group discussions within the MCDA process, the stakeholders and decision makers can facilitate a reexamination of their values, reflect on their implications, and resolve inconsistencies. We believe that the use of decision analysis provides a formal structure within which stakeholders and decision makers can think more consistently about their values, communicate them to each other, document the process, and explicitly consider important uncertainties that are currently disregarded or treated simplistically.

Uncertainty will inevitably be addressed in variety of different forms depending on the focus of the study. Time, effort, and money are used most efficiently when a structured and transparent means is employed for handling uncertainty in a decision-support modeling effort.

7. References

1. Abbott, C.A., Berry, M.W., Comiskey, E.J., Gross, L.J., and Luh, H.K. (1995) Computational Models of White-Tailed Deer in the Florida Everglades. UTK-CS. http://www.cs.utk.edu/~berry/-deer/deer.html

2. Akçakaya, H.R., Burgman, M.A., and Ginzburg, L.R. (1999) Applied Population Ecology: Principles and Computer Exercises using RAMAS EcoLab 2.0. Sinauer Associates, Sunderland, MA.

3. Akçakaya, H.R. (2001) Linking population-level risk assessment with landscape and habitat models. *The Science of the Total Environment* 274:283-291.

4. Allen, L.H., Ruddell, J.M., Ritter, G.J., Davis, F.E., and Yates, P. (1982) Land use effects on Taylor Creek water quality. *Proceedings of Specialty Conference on Environmentally Sound Water and Soil Management*, ASCE, Orlando, Florida, 20-23 July, 67-77.

5. Anderson, J.J. (2002) An agent-based event driven foraging model. *Natural Resource Modeling* 15(1):55-82.

6. Bottcher, A.B., Tremwel, T.K., and Campbell, K.L. (1995) Best management practices for water quality improvement in the Lake Okeechobee watershed. *Ecological Engineering* 5:341-356.

7. Bousquet, F. and Page, C.L. (2004). Multi-agent simulations and ecosystem management: a review. *Ecological Modelling* 176:313-332.

8. Cash, K.M., Adams, N.S., Hatton, T.W., Jones, E.C., and Rondorf, D.W. (2002) Three-Dimensional Fish Tracking to Evaluate the Operation of the Lower Granite Surface Bypass Collector and Behavioral Guidance Structure during 2000. Final Report prepared by the U.S. Geological Survey Columbia River Research Laboratory for the U.S. Army Corps of Engineers, Walla Walla District, Walla Walla, Washington.

9. Coughenour, M.B. (1993) Savanna-Landscape and Regional Ecosystem Model. Natural Resource Ecology Laboratory, Colorado State University, Fort Collins, Colorado.

10. Dumont, B. and Hill, D. (2001) Multi-agent simulation of group foraging in sheep: effects of spatial memory, conspecific attraction and plot size. *Ecological Modelling* 14:201-215.

11. Dunning, J.B., Stewart, D.J., Danielson, D.J., Noon, B.R., Root, T.L., Lamberson, R.H., and Stevens E.E. (1995) Spatially explicit population models: current forms and future uses. *Ecological Applications* 5(1):3-11.

12. FAO (Food and Agriculture Organization) (1991) Guidelines: Land Evaluation for Extensive Grazing. *FAO Soils Bulletin* 58. FAO, Rome.

13. Ferber, J. (1999) Multi-Agent Systems: An Introduction to Distributed Artificial Intelligence. Addison-Wesley, Reading, MA.

14. Fishman, G. S. (1996) Monte Carlo: Concepts, Algorithms, and Applications: Springer Series in Operations Research. New York, Springer-Verlag.

15. Flaig, E.G., and Reddy, K.R. (1995) Fate of phosphorus in the Lake Okeechobee watershed, Florida, USA: overview and recommendations. *Ecological Engineering* 5:127-142.

16. Goodwin, R.A., Nestler, J.M., Loucks, D.P., and Chapman, R.S. (2001) Simulating mobile populations in aquatic ecosystems. *Journal of Water Resources Planning and Management* 127(6):386-393.

17. Goodwin, R.A., Nestler, J.M., Anderson, J.J., Weber, L.J., and Loucks, D.P. (2006) Forecasting 3-D fish movement behavior using a Eulerian-Lagrangian-agent method (ELAM). *Ecological Modelling* 192:197-223.

18. Hanson, J.D., Baker, B.B., and Bourbon, R.M. (1992) SPUR II Model Description and User Guide: GPSR Technical Report No. 1. USDA-ARS, Great Plains Systems Research Unit, Ft. Collins, Colorado.

19. Hobbs, B. F., and Meier, P. (2000) Energy Decisions and the Environment: A Guide to the Use of Multicriteria Methods: International Series in Operations Research & Management Science, Vol. 28. Boston, Kluwer Academic Publishers.

20. Hobbs, N.T., and Swift, D.M. (1985) Estimates of habitat carrying capacity incorporating explicit nutritional constraints. *Journal of Wildlife Management* 49(3):814-822.

21. Iman, R. L., and Helton, J. C. (1988) An Investigation of Uncertainty and Sensitivity Analysis Techniques for Computer-Models. *Risk Analysis* 8:71-90.

22. Keeney, R. L., and Raiffa, H. (1976) Decisions with Multiple Objectives: Preferences and Value Tradeoffs. New York, Wiley.

23. Kiker, G.A., and Clark, D.J. (2001) The development of a Java based, object-oriented modeling system for simulation of southern African hydrology. ASAE Paper No. 01-2030.

24. Kim, J. B., Hobbs, B. F., and Koonce, J. F. (2003). Multicriteria Bayesian Analysis of Lower Trophic Level Uncertainties and Value of Research in Lake Erie. *Human and Ecological Risk Assessment* 9:1023-1057.

25. Martinez C.J. (2006) Object-oriented hydrologic and water-quality model for high-water-table environments. Ph.D. Dissertation. Gainesville, FL: University of Florida.

26. Mathevet, R., Bousquet, F., Page, C.L., and Antona, M. (2003) Agent-based simulations of interactions between duck population, farming decisions and leasing of hunting rights in the Camargue (Southern France). *Ecological Modelling* 165:107-126.

27. Morgan, M.G., and Dowlatabadi, H. (1996) Learning from integrated assessment of climate change. *Climatic Change* 34:337-368.

28. Morgan, M.G., and Henrion, M. (1990) Uncertainty: A Guide to Dealing With Uncertainty in Quantitative Risk and Policy Analysis, Cambridge University Press, New York, New York.

29. Nestler, J.M., Goodwin, R.A., and Loucks, D.P. (2005) Coupling of engineering and biological models for ecosystem analysis. *Journal of Water Resources Planning and Management* 131(2):101-109.

30. Nestler, J.M., Goodwin, R.A., Cole, T., Degan, D., and Dennerline, D. (2002) Simulating movement patterns of blueback herring in a stratified southern impoundment. *Transactions of the American Fisheries Society* 131:55-69.

31. New, J.G., Fewkes, L.A., and Khan, A.N. (2001) Strike feeding behavior in the muskellunge, *Esox masquinongy*: contributions of the lateral line and visual sensory systems. *Journal of Experimental Biology* 204:1207-1221.

32. Noy-Meir, I. (1975) Stability of grazing systems: an application of predator-prey graphs. *Journal of Ecology* 65:459-481.

33. Parrish, J.K., and Edelstein-Keshet, L. (1999) Complexity, pattern, and evolutionary trade-offs in animal aggregation. *Science* 284:99-101.

34. Parsch, L.D., and Loewer, O.J. (1995) GRAZE Beef-Forage Simulation Model: User's Guide.

35. Reckhow, K.H. (1994) Importance of scientific uncertainty in decision-making. *Environmental Management* 18:161-166.

36. Rowe, W.D. (1994) Understanding uncertainty. *Risk Analysis* 14(5):743-750.

37. Schulze RE. (1995) Hydrology and Agrohydrology: A text to accompany the *ACRU 3.00* Agrohydrological Modelling System. WRC Report No. TT69/95. ACRU Report No.43.

38. Sogard, S.M. and Olla, B.L. (1993) Effects of light, thermoclines and predator presence on vertical distribution and behavioral interactions of juvenile walleye pollock, *Theragra chalcogramma Pallas*. *Journal of Exp. Mar. Biol. Ecol.* 167:179-195.

39. Stafford Smith, D.M. (1988) Modeling: Three approaches to predicting how herbivore impact is distributed in rangelands. New Mexico. Agricultural Experiment Station Regular Research Report: 628:1-56.

40. Topping, C.J., Hansen, T.S., Jensen, T.S., Jepsen, J.U., Nikolajsen, F., and Odderskaer, P. (2003) ALMaSS, an agent-based model for animals in temperate European landscapes. *Ecological Modelling* 167:65-82.

41. UF-IFAS (2002) Demonstration of water quality best management practices for beef cattle ranching in the Lake Okeechobee Basin. University Of Florida, Institute of Food and Agricultural Sciences.

42. Yang, L. (2006) Coupled simulation modeling of flatwoods hydrology, nutrients and vegetation dynamics. Ph.D. Dissertation. Gainesville, FL: University of Florida.

Part 4

Applications of Risk Assessment

RISK ASSESSMENT IN DETECTION AND PREVENTION OF TERRORIST ATTACKS IN HARBORS AND COASTAL AREAS

M.K. ZAIDI
Radiological and Environmental Sciences Laboratory
1955 Fremont Avenue, Idaho Falls, Id 83401 USA

Abstract

The terrorists are seeking ways to attack the supply line of important goods and may cause damage to the environment, food, and the neighboring sea life. To prevent such an attack, the developing countries are doing major work in research and development of systems to detect such chemicals/explosives, which may be used by the terrorist groups to attack the developed nations. Risk assessment in detecting and preventing terrorist attacks is used to determine the accident sequences that lead to system failures, to remove weak links of the system, and also to help those who regulate the shipping and port establishments.

1. Introduction

> *There will never be a 100 percent guarantee of security for our people, the economy, and our society. We must resist the urge to seek total security— it is not achievable and drains our attention from those things that can be accomplished.*

James S. Gilmore III is the former governor of Virginia and was in office when the Pentagon was attacked on Sept. 11, 2001. Those involved in such acts of terrorism have a different culture, languages, and mindset.

The purpose of Risk Assessment (RA) in detecting and preventing terrorist attacks is to determine the accident sequences that lead to system failures, to remove weak links of the system, and also to help those who regulate the shipping and port establishments [1]. We plan to discuss the safety and security of specific maritime assets and strategic target points in the face of potential terrorist attacks. Protective and responsive measures which may have to be taken to reduce the risk and mitigate the consequences of these attacks. The attacks may be very diverse in nature and unpredictable. The attack may come from the air, sea, or land and may include a variety of weapons, explosives, etc. Their target may be populated areas, bridges, shipper, tanker, cruise ships, and waterways. Preparing for all these unknowns calls for a risk management approach.

I. Linkov et al. (eds.), Environmental Security in Harbors and Coastal Areas, 309–316.

RA is a systematic, analytical process to consider the likelihood that a threat will harm an asset or individuals and to identify actions that to reduce the risk and mitigate the consequences of an attack or event. Risk Management (RM) principles acknowledge that while risk generally cannot be eliminated, enhancing protection from unknown or potential threats can reduce it.

An effective RM approach includes a threat assessment, a vulnerability assessment, and a criticality assessment. A threat assessment identifies and evaluates threats based on various factors, including capability and intentions as well as the potential impact of an event. It will never be known whether every threat or event has been identified and complete information about the threats identified are available. The other two approaches—vulnerability and criticality—are essential for being better prepared against threats. A Vulnerability Assessment (VA) is a process that identifies weaknesses that may be exploited and suggests options to eliminate or mitigate those weaknesses. A Criticality Assessment (CA) is a process to systematically identify and evaluate an organization's assets based on a variety of factors, including the importance of its mission or function, where people are at risk. CAs are important because they provide a basis for prioritizing which assets require higher or special protection. The national Command Control Communication and Intelligence (C3I) system will provide help to those who are in need.

Acts of terrorism have increased greatly since the terrorist attack of September 2001. They are associated with illicit trafficking of explosives, narcotics, chemical weapons, hazardous chemicals, radioactive materials, and human beings. The terrorist would like to strike vulnerable civilian or military targets in nontraditional ways. They don't want to have a direct confrontation with any superpower or their military forces as they are confident that they will never win so these nations face a diffuse threat. Their aim is to inflict heavy human, environmental and economic losses and create fear, chaos, and harassment in people's minds. Civilian maritime assets are considered attractive targets as world trade heavily depends on maritime transportation of energy and other goods. It is easy to attack a cruise ship, an oil tanker, liquid gas carrier, or nuclear waste ship passing through narrow waterways and straits including those with bridges across them or important ports and harbors especially those adjacent to densely populated areas.

The world's total movement in containers is estimated to be about 75 million 20-foot equivalent units (TEU). The oil transported worldwide is over 5 M tons per day. The oil passes through narrow straits and pipelines which are target points due to their potential for closure. The disruption of oil supplies through these target points could have a significant impact, at an international level, on world oil prices, stability and security, and locally on environment, economy, security, and peace. A single typical ship can carry up to 6000 TEU. Although the container has become indispensable to world commerce, it has also turned out to be a convenient way to smuggle drugs, contraband and illegal immigrants. About 10 million containers enter the USA alone each year and only about 2% of these are being inspected. The most general shipping routes are shown in Figure 1 [2].

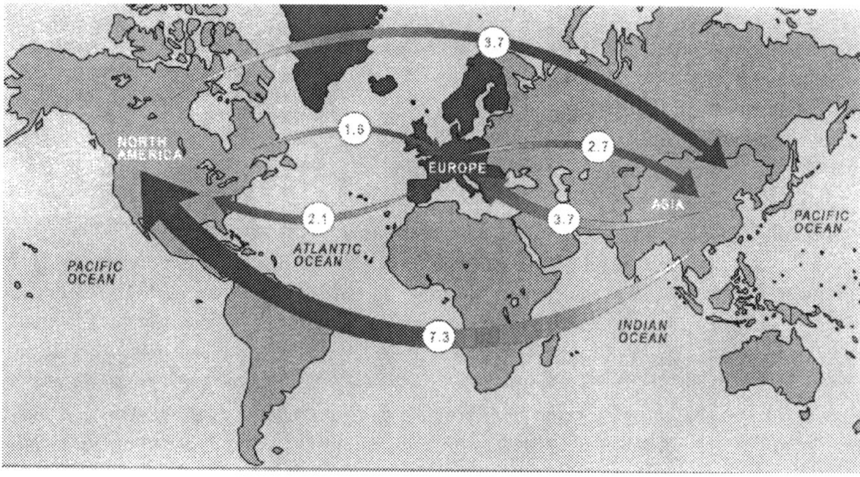

Figure 1: Arrows show the most general shipping routes [2].

To face this significant smuggling risk, different sized portal monitors are being used to screen containers for radioactive material and other substances. This paper will discuss different types of inspection systems. Modern cargo inspection systems are noninvasive imaging systems. The systems use penetrating radiation scanning (gamma and x-rays) to detect objects. They provide no information on the nature of the object. A new technique, neutron activation analysis, detects and identifies small volumes of threat materials such as explosives and drugs hidden in large containers. The neutrons produce a reaction in the object which allows the elemental composition to be identified, so as to tell whether the material is a harmless object, or an explosive, drugs, etc. The project is being developed at Rudjer Boskovic Institute, Zagreb, Croatia, and the Institute of Physics, Padua, Italy [3].

The explosive detection system, developed locally at the US DOE operated Idaho National Laboratory (INL) and some other national laboratories in the United States, is a nonintrusive, noncontact inspection technique that can—within seconds—identify explosives hidden in a small cargo truck or similar vehicle. It can scan the chemical makeup of a truck's load to quickly determine whether it contains explosives or drugs. The systems have been designed for reliable operation by technicians with minimal training. Other developing countries are not behind—they are actively working in developing such modalities according to their resources, need, and some support from developing countries. Still in many underdeveloped countries where labor is cheap, they do all such checking by hand and get good results. In developing countries where the labor is not available or is very costly and they have modern technologies and resources available, they are looking for scientific means to address these harsh challenges.

1.1. EXPLOSIVE DETECTION SYSTEM

The same technology is being developed to detect illicit trafficking of radioactive material at the Idaho National Laboratory (INL) located in Idaho, USA. INL has designed and manufactured an Explosive Detection System. It is a simple, nonintrusive, noncontact inspection technique that can within 300 seconds identify explosives hidden in a small cargo truck or similar vehicle. The system uses neutron activation technology in a portal-type configuration. Trucks or automobiles entering a building parking garage, military base, or embassy grounds would be required to stop within the system's inspection zone. After the driver exits the vehicle, the process would begin. Using a technique called Pulsed Thermal Neutron Analysis, high-energy neutron output causes nuclear excitation of materials within the vehicle. The patent-pending system uses detectors to identify elements within the targeted cargo that indicate the presence of explosives. The whole process takes about five minutes and—like an X-ray taken in a dental office—it leaves no lasting radiation effects on the inspected truck, cargo or facility in which the system is used. The system is quick, inexpensive, and reliable, due in part to its simple, yet robust design that incorporates few moving parts that can break down. The system also monitors the health of its individual components, which allows an operator to quickly pinpoint potential problems and make adjustments. The graphical user interfaces—what the operator sees on the computer screen—are easy to understand and eliminate potential for ambiguous interpretation. The Explosive Detection System is designed for reliable operation by a technician and requires minimum training [4].

1.2. CONCEALED WEAPONS DETECTOR

For small, concealed weapons, the INL-designed concealed weapons detector was installed at a local courthouse in 1998. Over the years it has stopped scores of weapons from entering the courtrooms and the building. The next-generation INEEL concealed weapons detector is now commercialized and marketed as SecureScan 2000 by View Systems, Inc. The sensors detected a student trying to sneak a razor blade in his mouth in a local school. The system is a passive device that senses disturbances in the eaerth's ambient magnetic field—disturbances such as those caused by a weapon passing through the aperture of the portal. The scientists have further enhanced to the process to increase system sensitivity, reduce false alarms, and recognize evolving weaponry. The latest detector is sophisticated enough to discriminate between threat and no threat items such as keys and coins. It can identify threat items such as a box cutter or a razor blade [5]. They are very well suited for offices at the seaport and other terminals.

1.3. AIRBORNE BOOST-PHASE INTERCEPTORS (ABIS)

A single robotic flying dragonfly could easily detect enemy troops or an ambush waiting over the next hill or around the next corner [6]. ABI carrying 90-kg Kinetic-kill vehicle (KKV) may be used to challenge intercontinental ballistic missiles (ICBMs). INL is also developing hand held instruments—a gamma-ray spectrometer with improved sensitivity and resolution to detect low level of radioactivity due to fallout. Unmanned Vehicle program is another emerging and rapidly expanding program with a focus on supporting a wide variety of defense and national security needs. The current program is used by various defense agencies in development of a wide spectrum of capabilities, including the next-generation counter-mine robots.

Figure 2: The new INEEL weapons detector.

2. Promoting Nuclear Security

The IAEA's nuclear security plan of action has three main areas of focus: Prevention, Detection, and Response.

- **Prevention**. This involves preventing any illicit or nonpeaceful use of nuclear or other radioactive materials — including acts of terrorism. This includes identifying and reducing risks by:
  - Physical protection of nuclear materials in use, storage, or transport.
  - Physical protection of related nuclear facilities.
  - Return of high energy uranium (HEU) from research reactors to countries of origin, and converting those facilities to use low enriched uranium (LEU).

    This often involves partnering—with the IAEA, with the host country, with a donor like the US government or the Global Threat Reduction Initiative (GTRI) operated by the US Department of Energy, and with the country to which the HEU will be returned [7].

At a time of growing concern over threats of terrorism, the security of nuclear and radioactive material is an urgent and serious issue [8]. In August 2002, a coordinated transport of fresh fuel was conducted from the research reactor at the Vinca Institute of Nuclear Sciences, in Yugoslavia, to Russia, the country of origin. The removal of 1.8 tons of uranium enriched to a level 2.6 percent, another 6.6 pounds of low-enriched uranium and approximately 1000 highly radioactive sources had been moved to safe place in the USA in June 2004 from an IAEA safe vault in Baghdad, Iraq [9]. Very recently, six kilograms of HEU were safely returned to Russia Federation from Czech Republic [7].

- Interim protection measures at these vulnerable locations (e.g. research reactors).
- Control of radioactive material—the recovery of lost or "orphaned" sources was done in Georgia with international cooperation. They had conducted surveys and searched to recover and secure highly radioactive sources .
- Management and control of materials through effective Source Selection Advisory Council (*SSAC)* and adherence to the *Code of Conduct* on Radioactive Sources.
- The IAEA is continuing to help former Soviet Union (FSU) states with *Design Basis Threat (DBT) assessments*, which are vital for the identification, categorization and security of radioactive sources, as well as risks to nuclear facilities [10].

- **Detection**. This involves putting systems in place to help countries detect any illicit activity:
  - Includes border patrols, better equipment at border crossings.
  - Training of customs officials.
  - Increased cooperation between law enforcement officials.
  - IAEA maintains the *Illicit Trafficking Database (ITDB)*, contains information on 576 confirmed cases of illicit trafficking in nuclear and other radioactive materials as of April 2004. The ITDB has recorded 29 such cases so far in 2004 and 60 cases occurred in 2003 [11].
- **Response**. This involves providing rapid assistance to governments, on request, to respond to emergencies:
  - To date, most incidents have involved helping with the recovery and securing of radioactive sources.
  - Also put plans in place for responding to acts of sabotage or acts of terrorism involving nuclear or radioactive materials [12].

  What you should do: Leave the immediate area on foot, do not panic, do not take public or private transportation such as buses, subways or cars because if radioactive material (RAM) was involved, they may contaminate cars or the public transport, go inside the nearest building and stay inside as it will save your exposure to any RAM from a DB that may be on dust at the scene, remove clothes as soon as possible, place them in a plastic bag, and seal it. Removing clothes will remove most of the contamination caused by external exposure to RAM. Saving the contaminated clothes will allow testing. Take a shower or wash. Washing will reduce the amount of RA contamination on the body and will effectively reduce total exposure, be on the lookout for information—use a radio, TV or phone. After emergency personnel can assess the scene and the damage, they will tell you if radiation was involved [13].

3. Conclusions

Risk assessment in detecting and preventing terrorist attacks should be a routine practice to determine the accident sequences that lead to system failures, to remove weak links of the system, and also to help those who regulate the shipping and port establishments. To prevent a terrorist attack, the developing countries are doing major work in research and development of new systems to detect chemicals and explosives that may be used by terrorist groups to attack the developed nations. The terrorist have no cities that can be bombed nor they are focused on self-preservation so the nuclear deterrent is ineffective against terrorist groups, as they are constantly shifting targets and modes of attack demands a more cooperative and flexible international response. *There will never be a 100 percent guarantee of security for our people, the economy, and our society. We must resist the urge to seek total security—it is not achievable and drains our attention from those things that can be accomplished.* With fear, we avoid challenge and with confidence we conquer them.

4. References

1. McCormide NJ. 1981. Reliability and Risk Assessment, Academic Press, P-383.

2. NATO News. (2004). Science for Peace—New Projects Launched. V 66 (1), 5

3. Valkovic V. et.al.(2004). Inspection of Shipping Containers. R.Phys.Ch. 71(3-4).897-98.

4. Idaho State Journal. Nuclear Site's Security. 6/17/2004

5. Inside INEEL. Summer 2004. Page 6

6. Myhra, D. (2004). Bugging the Enemy! Homeland Defense 2 (5). 22-23

7. Lopatto, J. (2005). Highly enriched uranium repatriated from Czeck. HPJ 88 (3), 272-73

8. Steinhaeusler F. (2003). Protecting the Sources. IAEA Bulletin 45(1). 17-20

9. Idaho State Journal, July 8, 2004

10. 6 Code of Federal Register 10CFR 73.1 (a)

11. IAEA. (2003). Incidents of Illicitly Trafficking Nuclear and Radioactive Materials.

12. Miller K.(2004). Health Physics Considerations in Medical Radiation Emergencies. Operational Rad. Safety. 87, Suppl.1. S19-24

13. http://www.iaea.org/NewsCenter/PressReleases/2002/prn0209.html.

COMPLEX HEALTH RISK ASSESSMENT AND ANALYSIS FROM EXPOSURE TO IONIZING RADIATION, CHEMICAL CONTAMINANTS AND OTHER SOURCES OF HARM

V. DEMIN
RRC "Kurchatov Institute"
Kurchatov Sq. 1, 123182 Moscow, Russia

Abstract

Needs in practical use of health risk (HR) assessment and analysis in Russia began to rise at the end of the 80s. They came from decision making on the radiation and social protection of population on the territories that suffered from Chernobyl and other radiation accidents, nuclear weapon tests, etc. The current ecological and public health protection regulation concerning the development of hazardous industries, especially the fuel-power complex, appeal also to HR analysis. In recent years, the growing need for HR assessment has been observed in the activity connected with utilization of nuclear submarines and rehabilitation of the corresponding contaminated territories. According to the preliminary results of HR assessment, chemical contamination can produce a higher impact on population and occupational workers' health than ionizing radiation (IR).

RRC "Kurchatov Institute" has been involved in the development of HR assessment tools (methodology, computer modules, and data bases) and the regulatory aspects of the use of HR analysis in decision making, case studies, etc. One of the directions of this development is to produce a common and transparent basis for HR assessment of different sources of risk and the basis for complex decision making.

A number of factors create the complexity of HR analysis and decision making. Series of different risk indices are needed in HR assessment and their aggregation continues to be an intractable problem. HRs competition principally makes nonlinear any decision-making task. The necessity for economic evaluation of population health damage in the decision making adds an additional dimension.

Current different approaches in risk assessment and establishing safety standards, developed for IR, chemicals, and other sources of harm, are analyzed. Some recommendations are given to produce a common approach. A specific individual risk index $\mathfrak{R}$ has been proposed for safety decision making: establishing safety standards and other levels of protective actions, comparison of various sources of risk, etc. The index $\mathfrak{R}$ is defined as the partial mathematical expectation of lost

I. Linkov et al. (eds.), Environmental Security in Harbors and Coastal Areas, 317–327.

years of healthy life due to exposure during a year to a risk source considered. The more concrete determinations of this index for different risk sources derived from the common definition of $\mathfrak{R}$ are given. Generic safety standards (GSS) for the public and occupational workers have been suggested in terms of this index. Secondary specific safety standards have been derived from GSS for IR and a number of other risk sources including environmental chemical pollutants. Other general and derived levels for decision making have also been proposed. Recommendations are given on methods and criteria for comparison of various sources of risk. Some examples of risk comparison are demonstrated in the frame of different comparison tasks.

1. Introduction

Decision making in environmental projects can be complex and seemingly intractable, principally due to the inherent existence of tradeoffs between sociopolitical, environmental, and economic factors. Health risk (HR) assessment gives part of the data for this decision making. But HR assessment itself can be the case of multi-dimension and multi-criteria decision analysis.

Needs in practical use of HR assessment and analysis in Russia began to rise at the end of the 80s. They came from decision making on the radiation and social protection of population on the territories that suffered from Chernobyl and other radiation accidents, nuclear weapon tests, etc. The current ecological and public health protection regulation concerning the development of hazardous industries, especially the fuel-power complex, appeal also to HR analysis. In last years the growing need in HR assessment has been observed in the activity connected with utilization of nuclear submarines and rehabilitation (U&R) of the corresponding contaminated territories. There are two different harbors and coastal areas in Russia, where such activity has taken place: the Northwest (Archangelsk and Murmansk regions) and Far East regions. This activity began many years ago. But political changes in the USSR and Russia and subsequent economic difficulties prevented the proper and timely performance of U&R activity. This circumstance complicated the U&R problems. Russia has longstanding environmental problems in the Northwest and Far East regions, which create national and international concern. Now U&R work is carried out in Russia with the technical assistance of the EU and other countries. In HR assessment for decision support of U&R works, radiation as well as nonradiation protection aspects should be considered.

According to the preliminary results of HR assessment in the frame of U&R activity, chemical contamination can produce a higher impact on population and occupational workers health than ionizing radiation (IR).

A number of factors create the complexity of HR analysis and decision making: different risk sources, different approaches and series of different risk indices used for risk assessment, etc. Different risk indices are needed in HR assessment and their integration continues to be an intractable problem. HRs competition principally makes nonlinear any decision-making task. The necessity for economic evaluation of population health damage in decision making adds an additional dimension and level of complexity. In some important cases the complexity stems partly from the imperfectness of the HR assessment methods.

To meet practical needs in HR analysis, the development of HR assessment means and regulation began in the frame of a few research state programs. RRC "Kurchatov Institute" has been involved in these R & D. The directions of this activity are: development of HR assessment tools (methodology, computer modules, and data bases), regulatory aspects of the use of risk analysis in decision making, case studies, etc. Both radiation and nonradiation HR, including risk from chemical contamination of the environment, have been considered. The tool includes methodology and a computer module for HR assessment. The module is called BARD: bank of data for assessment of risk. The tool has been continuously developed using new achievements and responding to practical demands [1-5]. One of the objectives and directions of this development is to produce a common, unified, and transparent basis of HR analysis for different sources of risk. Application areas of the tool are:

- Assessment of the radiological and nonradiological consequences of radiation accidents and nuclear weapons tests.
- Assessment of the radiation risk from any source of radiation exposure (natural, medical and technogenic).
- HR assessment and comparison for different energy systems.
- Analysis of the public health in terms of risk and health-demographic indices.
- Assessment of HR from other technogenic sources (if input data are available, e.g. risk from chemical air pollution).

The paper is based on some results of R & D mentioned above.

2. Common Methodical Basis

In the past, much attention was paid to radiation risk. Now, due to achievements in epidemiological studies and methodology development, HR assessment can be used for chemical contamination of the environment and other sources of harm. The aggregation or comparison of HR assessment data for decision making is possible primarily when using a common basis for HR risk assessment. This common basis should include:

1. General method of HR assessment.
2. ERF (for a risk source considered) produced in the form of age-cause-specific mortality or morbidity rate.
3. Country or regional health-demographic data (HDD).

Only all three parts allow the calculation of any necessary set of specific or general HR indices in all possible dependencies and to make some aggregating or averaging.

Point 1 is well known in the methodology of risk assessment. HDD can be easily received from a state statistic department or from WHO database. All difficult problems are connected with point 2.

Health effects of ionizing radiation and many chemical contaminants are nonspecific. ERF can be obtained only by carrying out epidemiological studies. As a rule, such studies for low exposures are very complicated, long-term, and expensive. In many cases they have a limited statistical power.

An important factor in HR quantification is how to transfer site-specific risks across populations. In other words, how risk coefficients estimated for one population apply to another population with different HDD. For IR and some limited number of chemical pollutants, ERFs were obtained on the basis of results of epidemiological studies and were presented in the form mentioned above (point 2).

The most part of these ERFs are of multiplicative risk mode dependence: excess incidence rate of health effects is proportional to baseline rates ("spontaneous" rates). In this case, the transfer across populations is made not of ERF total, but only relative risk coefficients. If such dependence is not only a methodical approach and it reflects the medical-biological nature of formation of the health effect, it can be a good argument for the validity of the transfer. If risk assessment is made for some specific population and the results are presented in an averaged form it is not generally valid to transfer these results to another population.

The multi-criteria approach in decision making with HR assessment data is partly inevitable and partly comes from methodical imperfectness or lack of knowledge. In HR assessment from different sources of harm we have deal with a variety of health effects (different diseases and mortality with different damage expressed in LLE) and correspondingly with sets of risk indices. In decision making, an aggregation of HR assessment data should be done to the necessary and possible extent. Two methods of aggregation have been studied in HR assessment research. The first one is producing common aggregated risk index and then monetary evaluation of its unit. In literature there are proposals of such values as DALY (disability-adjusted life years) or QALY (quality adjusted life years)—see, e.g. [8]—which are candidates for the common aggregating index. Another way consists in using all specific HR risk indices and separate monetary evaluation of all their units. The integration is made in the final stage of risk assessment: the monetary evaluation of human health damage.

A good example of the aggregated HR index is given by ICRP with its effective dose concept [6]: one radiation risk index integrates different health effects (fatal, nonfatal cancers, and hereditary effects). But this index is specific only for the radiation risk assessment. One should note that the first way might in some aspects contradict the economic damage evaluation due to not quite properly taking the time factor into account.

Usually the second way is used in HR assessment. Now the newest draft regulation document on the monetary evaluation of HR has been elaborated in Russia in this way.

3. Special Risk Index and Decision-Making Levels

Up to now there is no common approach in establishment of health protective levels for different sources of risk. To protect population or personnel from dangerous industry activities, the risk limits are usually established in the terms of individual mortality rate r. It is obvious that such limits can't be properly used for the risk sources with delayed health effects (ionizing radiation, impact of some chemicals especially with carcinogenic or hereditary effects, etc.).

In radiation protection the generalized risk index R_E is constructed and used to produce such quantity as the effective dose for establishing safety standards [6]. It is not so simple to use this index for comparison of HR for ionizing radiation and other risk sources and to compare radiation and nonradiation safety standards. Now the use of HR assessment in establishing safety standards for some chemical contaminants is in progress. What risk index could be used in this case?

Cases of death from different sources of risk can differ considerably by the damage expressed in the loss of life expectancy (LLE). For chemical and radiation carcinogenesis there is a relatively large latent time interval T_l (between an exposure and appearance of a cancer): for solid cancers the minimal value of T_l is equal to 5-10 years, averaged one – 40 – 50 years. Due to this the averaged LLE ($L^{l.c.}_{av.}$) for one lethal radiogenic cancer is considerably less than averaged LLE ($L^{i.d.}_{av.}$) due to immediate death from an accident or adverse effect. Analogous situation is for the chemical carcinogenesis.

In the accordance with the ICRP estimation [6] the global values of these quantities are equal to

$$L^{l.c.}_{av.} = 15 \text{ years}, \quad L^{i.d.}_{av.} = 35 \text{ years}. \tag{1}$$

Values of LLE (specific or averaged) from sources of risk can be estimated for the population of any country or region using a computer code like BARD. The risk coefficient of the effective dose g_E [6] (the globally averaged specific LLE for ionizing radiation), which includes all health effects (cancer mortality and morbidity, hereditary effects), is equal to

$$g_E = \begin{cases} 1 \; man{\cdot}yr\,/man{\cdot}Sv & \text{for public,} \\ 0.8 \; man{\cdot}yr\,/man{\cdot}Sv & \text{for professional exposure.} \end{cases} \tag{2}$$

After analyzing the aspects of establishment of safety standards, decision making levels, the risk indices and ways of their integration for different risk sources, especially ICRP recommendation [6], the proposal was made concerning the special risk index $\mathfrak{R}$ for the risk standardization and comparison [1,3,5]. $\mathfrak{R}$ is quantitatively defined as LLE caused by the annual exposure to the risk source considered:

$$\mathfrak{R} = \text{annual exposure} * \text{damage (LLE) from the exposure unit.} \quad (3)$$

For different risk sources, the exposure can be measured with different quantities. For immediate death due to an accident or an adverse effect, the exposure is expressed in terms of the probability of death. In this concrete case

$$\mathfrak{R} = r \cdot g_r . \quad (4)$$

Here r is the individual death rate (in other words, annual probability of death), g_r is LLE due to a death in the year considered.

For ionizing radiation dose D (absorbed, equivalent, or effective – in the dependence of the type of radiation and an application area) is used as a measure of exposure. For this case $\mathfrak{R}$ is expressed as

$$\mathfrak{R} = d_{i.r.} \cdot g_{i.r.} , \quad (5)$$

where $d_{i.r.}$ is the ionizing radiation dose rate, $g_{i.r.}$ is LLE from the dose unit.

As a rule, ERFs for chemical contaminants are referred to exposure ε_{ch} - the time integral of a concentration of a contaminant in the air or water. The dimension of ε_{ch} is [yr*μg/m^3]. The expression (4) for $\mathfrak{R}$ should be rewritten in the following concrete form:

$$\mathfrak{R} = \, e_{ch} \cdot g_{ch} , \quad (6)$$

e_{ch} is the exposure rate (the annual exposure); its dimension is [(yr*μg/m^3)/yr]; g_{ch} is LLE from the exposure unit.

The meaning of $\mathfrak{R}$ can be seen from Eqs. (3) – (6). It is relative LLE: LLE in years referred to 1 year under the exposure to a risk source. The dimension of this value is [yr(LLE)/yr]. In a statistical sense $\mathfrak{R}$ is conditionally the share of the year which is lost due to the exposure to a risk source during this year. In this sense $\mathfrak{R}$ can be called *the relative damage*. Really lifetime years are lost after the exposure. Taking into account this comment, $\mathfrak{R}$ can be in some conditional sense considered a dimensionless quantity.

The definitions (3) – (6) can be also generalized for morbidity.

In establishing the health protection standards and levels it is reasonable to use the averaged values *in* Eqs (4) – (6) as it was done, e.g., in the radiation protection for radiation risk coefficients [6]. In average through age $g_r = L^{i.d.}_{av.} = 35$ years, see Eq. (1). In Eq. (5) the effective dose and its coefficients can be used:

$$\mathfrak{R} = d_E \cdot g_E , \quad (7)$$

where d_E is effective dose rate (annual effective dose), g_E is well-known LLE risk coefficient for the ionizing radiation, see Eq. (2).

From the literature one can find some important examples for the chemical airborne pollutants [7,9]:

$$g_{ch.} = \begin{cases} 0.0006 \text{ yr } / \text{ yr} \cdot \mu\text{g/m}^3 \text{ for } PM_{2.5} \text{ and EU population ,} \\ 0.0008 \text{ yr } / \text{ yr} \cdot \mu\text{g/m}^3 \text{ for } PM_{2.5} \text{ and Russia population ,} \\ 0.002 \text{ yr } / \text{ yr} \cdot \mu\text{g/m}^3 \text{ for As and global population .} \end{cases} \tag{8}$$

Cancer is considered a health effect of air pollutant As. For $PM_{2.5}$, health effects are diseases of respiratory and circulatory systems and possible death from them.

Using risk index $\Re$ the universal safety standards (risk limits $\Re_n$) can be established. To have them equivalent to radiation safety basic standards [6]

$$d_{E,n} = \begin{cases} 1 \; mSv/yr \text{ for public,} \\ \\ 20 \; mSv/yr \text{ for personnel.} \end{cases} \tag{9}$$

and taking into account that some precaution was laid in these ICRP effective dose limits, the following universal risk limits are proposed for consideration by the specialists in health risk analysis and members of regulatory bodies (see Eqs (7), (2) and (9)):

$$\Re_n = \begin{cases} 0.0007 \quad \text{for public,} \\ \\ 0.01 \quad \text{for personnel.} \end{cases} \tag{10}$$

Based on these general risk limits (safety standards), derivative limits for different risk sources can be obtained. For risk sources with nondelayed health effects (accident and adverse effects) the derivative risk limits in the terms of death risk rate r are equal ($r_n = \Re_n / g_r$, $g_r = L^{i.d.}_{av.}$, see Eqs (10), (1), (4))

$$r_n = \begin{cases} 2.0 \cdot 10^{-5} / \text{ yr} \quad \text{for public,} \\ \\ 2.9 \cdot 10^{-4} / \text{ yr} \quad \text{for personnel.} \end{cases} \tag{11}$$

For the chemical air pollutants considered above the derivative health protection standards in the term of the annually averaged air concentration c_n should be ($c_n = \Re_n / g_{ch}$, see Eqs (8) and (10)):

$$c_n = \begin{cases} 1.2\ \mu g/m^3 & \text{for } PM_{2.5} \text{ and EU population,} \\ 0.9\ \mu g/m^3 & \text{for } PM_{2.5} \text{ and Russia population,} \\ 0.4\ \mu g/m^3 & \text{for As and global population.} \end{cases} \quad (12)$$

It is also simple to calculate the derivative chemical safety standards for personnel in the terms of maximum permissible concentration c_n from the definitions (8) and (10).

"De minimus" level $\Re_{d.m.}$ is proposed to be established as $\Re_{d.m.} = 10^{-5}$. Accordingly the secondary "de minimus" levels should be $r_{d.m.} = 3*10^{-7}$ /year (risk sources with nondelayed health effects) and 20 μSv/year (ionizing radiation).

One should note the effect of competition of different risks. A man dies only once. This manifests itself in the fact that the total lifetime risk is equal to unity. A change in one of the death causes automatically leads to a change (renormalization) in the lifetime risk indices of other sources in action, even though they are statistically independent. These risk indices, like many others, are not additive. Annual risks considered mutually compete with themselves, decreasing each other. However, for impacts on the level of health protection standards, the effect of risk competition can be ignored. Moreover, this ignorance gives an element of precaution (a reserve of reliability) in the establishing the safety standards, although it is rather insignificant.

4. Risk Comparisons

There are several different tasks for risk comparison, which can be met in papers published. The following ones have been met more often:

1. Comparison of the individual risk rate from different risk sources, e.g., risk from transports, nature disasters, risk from hazardous industrial plants, etc.
2. Comparison of safety standards for risk sources, approved on the basis of different approaches, e.g., radiation and chemical protection standards.
3. Comparison of risk sources by the total risk (damage), e.g., different power production systems.
4. Comparison of population health risks from different diseases or any other sources of risk for different regions of Russia, different countries, etc.

The proper risk indices have not always been chosen for such comparisons; e.g., individual death rate (annual probability of death) often used is not valid for radiation exposure: annual exposure in a current year can cause a radiological cancer few decades later, but annual probability of death in a current year can be caused by an exposure during previous years. As it was explained above, the number of deaths can be an improper damage index for comparison of different sources. In such a comparison, LLE is more suitable. In ExternE project comparisons, LLE is usually used [10].

Here is an example of the individual risk comparison (task 1) is given. It is known that in Russia the probability of death caused by lightning blows and motor-transport accidents is on average equal to 10^{-7}/yr and $2\cdot10^{-4}$/yr, respectively. Annual effective dose outside the protection zone of a nuclear power plant (NPP) is not higher than 0.01 mSv/yr. Then the individual risk from these sources of risk in the terms of quantity $\mathfrak{R}$, proposed above, is equal to

$$\mathfrak{R} = \begin{cases} 4\cdot10^{-6} & \text{- lightning blows;} \\ 7\cdot10^{-3} & \text{- motor-transport accidents;} \\ \leq 10^{-5} & \text{- living near the NPP protection zone .} \end{cases} \quad (13)$$

Let one consider the comparison of health risk, e.g., from spontaneous fatal cancers for France and Russia (task 4). As it follows from data available values of $\mu(a)$ - age-cause-specific cancer mortality rate – are rather close to each other in France and Russia: the differences are not higher than 20 – 30%. What HR index can be properly used for the comparison ?

Let it be the lifetime risk R . The calculation (using BARD) gives the following results: the value of R for Russia in 1.5 – 2 times less (in dependence upon the cancer type), than for France. The same results would be obtained for radiological cancers and with other averaged or lifetime risk indices. The reason for such results is the following: the life expectancy in Russia is considerably less then in France (for males the difference is more than 10 years). In other words this fact can be explained by risk competition. If some risk factors begin to increase (in Russia they are the circulatory system diseases and mortality from accidents and adverse effects), then value of R for the risk source considered will decrease under the condition that its $\mu(a)$ is not changed.

If the comparison is made in the terms of the standardized annual mortality m_s (averaged through the standard age distribution), then rather close results are obtained for France and Russia: for males of France and Russia $m_s \approx 300$ (per 10^5 people), for females $m_s \approx 150$ (Russia) and 170 (France). In the calculation with BARD the European age distribution standard, adopted by WHO, is used.

It should be clear that the last comparison is more valid than the previous one. Quantity m_s is the one of the principal indices of the health – demographic analysis. But this quantity is not also a universal index for comparisons. It has its own limitations. Its value depends upon the choice of the standard.

The shortcoming of the HR indices connected with the effect of risk competition can have an influence upon the possibility of finding the optimal protection against the complex of risk factors [1].

Several other demonstrative examples of risk comparison were presented in [3].

5. Conclusions

One of the objectives and directions of our research activity is to improve decision making by producing a common, unified basis for complex health risk assessment of different sources of harm. Along this direction, the special risk index $\mathfrak{R}$ is proposed for safety decision making. General safety standards for public and occupational workers are suggested in terms of this index. Secondary safety standards have been derived for ionizing radiation and a number of other risk sources. Using this and some other risk indices in different tasks of risk comparison is analyzed. The averaged values of risk indices are usually taken in establishing decision-making levels. In particular, the coefficients of the effective dose, which are used in the definition of the effective dose itself and the radiation safety standards, are based on the globally averaged lifetime risk indices. These averaged values are also used in the secondary safety standards proposed in the paper. However, in the consideration of issues of risk comparisons it was noted that the averaged and lifetime risk indices do not always properly reflect the real impact of the risk sources due to the effects of risk competition. It is necessary to critically analyze the use of averaged and lifetime risk indices in the different safety decision levels adopted. The proposal for a special standard population has been elaborated to overcome this problem. Some ideas along this line can be found in [1].

6. References

1. Demin, V.F. (1999). Methodological Aspects of Health Risk Assessment. *Atomnaya Energiya 86* (1): 46-63.

2. Demin, V.F. (2000). Health risk assessment and management: some directions of R & D and practical use in Russia. Proceedings of the intern. conf. «PSAM 5 – Probabilistic Safety Assessment and Management", 27.11 - 1.12.00, Osaka, Japan, ed. by S.Kondo and K.Funita, Universal Academy Press, Tokyo, Japan, V. 1, 281–287.

3. Demin, V.F., Golikov, V.Ya., Ivanov, E.V., et al. (2002). Standardisation of different risks. *Hygiene & Sanitary* (6):30-35.

4. Demin, V.F., Krylov D.A. (2004). Estimation of positive ecological consequences of "gas pause" in the heat and electricity generation of Russia. *Energonadzor & Energoeffectiveness* (2):37-45.

5. Demin V.F. (2004). Common basis for health risk assessment and safety decision making concerning exposure to ionizing radiation and other sources of harm In the proceed. of the international conference ECORAD-2004 (The scientific basis for environment protection against radioactivity), 6 – 10 September 2004, Aix-en-Provence, France; IRSN, France, p. 01/06.

6. ICRP, Recommendations of the Commission - 1990 (1991), ICRP Publication 60, *Ann. ICRP, 21*.

7. Leksell, I., Rabl, A. (2001). Air Pollution and Mortality: Quantification and Valuation of Years of Life Lost. *Risk Analysis 21*(2):35 – 47.

8. Murray, C.J.L. (1995). Weighted estimation of morbidity: a conceptual basis for the method of assessment of DALY (disability-adjusted life years), Harvard Center for Population and Development Studies, Cambridge, USA.

9. Novikov, S.M., Rumyantsev, G.I., Zoldakova, Z.I., et al. (1998) Problems of cancer risk assessment foe chemical contaminants of the environment. *Hygiene & Sanitation* (1):12-19.

10. Externalities of Fuel Cycles: "ExternE Project". (1994 – 1998). Volumes 1 - 8, European Commission, DG XII.

ROLE OF RISK ASSESSMENT IN ENVIRONMENTAL SECURITY PLANNING AND DECISION-MAKING

R.J. WENNING
ENVIRON International Corporation
6001 Shellmound Street, Ste 700, Emeryville, CA, 94608 USA

S. DELLA SALA
Autorita Portuale di Venezia (Venice Port Authority)
Zattere 1401, 30123 Venice, Italy

V. MAGAR
ENVIRON International Corporation
123 North Wacker Street, Ste 250, Chicago, IL, 60606 USA

Abstract

This chapter explores the role of risk assessment in environmental security planning and decision-making at commercial shipping ports and harbors. Environmental risk assessment is a discipline that has matured over the past nearly 50 years, evolving from assessment of chemical exposures on human health and wildlife to comparative analysis of the net risks and benefits to the environment associated with the implementation of different remedy alternatives at contaminated land and sediment sites. In recent years, partly due to the events of 11 September 2001, escalating tensions in poor and under-developed countries, and increasing scarcity of natural resources, several countries and international organizations have raised concerns about environmental security. For commercial and industrial shipping ports, which are often located in heavily populated urban areas and sensitive coastal environments in both developed and developing countries, the application of risk assessment methods is an important first step towards protecting critical industrial, environmental and utility infrastructure and understanding environmental response and prevention requirements and capabilities. In the context of environmental risk assessment, there are at least three security challenges that must be overcome. First, quantitative prediction is needed, with a high degree of confidence, of the range of possible damages and potential threats posed to both human health and the environment. The nature of this work itself poses a security challenge because of the potential sensitivity of the information that must be compiled and evaluated. Second, the information and risk predictions generated by environmental risk assessment to describe the outcome of different possible disaster events and environmental scenarios must be realistic and

I. Linkov et al. (eds.), Environmental Security in Harbors and Coastal Areas, 329–341.

plausible. And, third, the results of an environmental risk analysis must inform decision-makers of different disaster prevention/response action plans, their associated capital investments, and net impact on risk. Ultimately, the risk assessment process can provide port authorities and local or national government authorities the appropriate tools to prioritize prevention/response actions must be to minimize or prevent scenarios most likely to adversely impact the environment, cause injuries or fatalities to port workers and residents in surrounding communities, and result in short or long-term economic impacts.

1. Introduction

Quantitative environmental risk assessment has evolved considerably over the past 50 years from a predominantly retrospective tool for analysis of exposures to chemicals in the workplace, radioisotope effects on human health, and assessment of potential threats to wildlife and human health posed by chemical releases [13; 25] to include vulnerability analysis of critical infrastructure and comparative analysis of environmental risks posed by different environmental cleanup options at contaminated land and sediment sites [5; 22; 20; 33]. At present, risk assessment is used increasingly in the U.S. and other countries to evaluate the life cycle of new chemical substances and to weigh the net risks, costs and benefits to the environment associated with different environmental management strategies [21; 16; 19]. The evolution of risk assessment as a discipline coincides with our rapidly expanding knowledge of chemical environmental fate, ecology, geochemistry, human and wildlife toxicology and exposure modeling [24].

More recently, the application of quantitative environmental risk assessment has expanded to include environmental security. Applications of risk assessment include analysis of environmental disaster scenarios at chemical manufacturing plants and in heavily populated urban areas, coastal ports, and harbors where commercial shipping activities, power plants and chemical or petroleum transfer and storage facilities may be vulnerable to natural disasters, industrial accidents or terrorism [11; 35]. The different methods proposed to identify and rank vulnerabilities to critical infrastructure are based on risk assessment [1]. Radiological weapons and bioterrorism have also renewed interest in risk assessment as a means to evaluate and respond to public health threats [17; 26]. The U.S. Environmental Protection Agency requires a risk management plan and "Consequence Analysis" at industrial facilities to describe "worst case" and "expected" environmental release scenarios as part of the permitting process [32]. Similarly, the United Kingdom requires submittal of "Safety Cases", which are intended to demonstrate the level of risk associated with offshore oil and gas production facilities [9].

This chapter explores the emerging use of environmental risk assessment in environmental security, including its application to disaster prevention planning and its usefulness to decision-making when faced with weighing the risks, costs and benefits of different or competing security needs and environmental threats.

2. Defining Environmental Security

At present, several definitions of environmental security have been proposed without a clear consensus of its meaning [23; 2; 3]. In the context of human health and ecological risk assessment, environmental security arises from activities that protect, prevent and anticipate events that may harm the environment (including its natural and biological resources), local inhabitants, and government at any scale. The level of security achieved at a coastal port and harbor is proportional to the level of investment in programs and infrastructure that protect industrial plants and municipal facilities from intrusion and undesirable threats; prevent the exploitation of identified vulnerabilities at such facilities that are capable of disrupting operations and services, thereby endangering local inhabitants and the environment; and are capable of responding to threats and incidents when they occur.

A broader definition of environmental security that is not addressed herein and deserves recognition concerns the challenges to human populations posed by the depletion or limitation of critical resources and the conflicts that may arise to acquire, replace or defend the affected resources. In this context, environmental security revolves around the central idea that environmental problems – in particular, resource scarcity and environmental degradation – may lead to violent conflict between and among states and societies [7]. Environmental concerns such as ozone depletion, chemical pollution, loss of arable land and water, and food scarcity are increasingly part of international political discourse and policy initiatives focused on improvement of the standard of living for peoples in poor and under-developed countries [6; 4; 30]. There is little debate that these are, indeed, environmental security challenges that will grow in importance over the coming decades.

Setting aside the question of whether adopting a broad or narrow view of environmental security is appropriate, five common elements of environmental security identified by Glenn et al. [8] are relevant regardless of either perspective. Environmental security entails considerations of (1) public safety from environmental dangers caused by natural or man-made processes due to ignorance, accident, mismanagement or design; (2) amelioration of natural resource scarcity; (3) maintenance of a healthy environment; (4) amelioration of environmental degradation; and (5) prevention of social disorder and conflict (i.e., promotion of social stability).

These five elements of environmental security are applicable to industrial ship ports and commercial harbors. Environmental security considerations for ports and harbors focus on: protection, prevention and response planning activities that preserve the functions of industrial and municipal tenants; and protection, prevention and response planning activities that maintain or preserve the environment, public health and safety from the potentially adverse effects posed by waste streams and catastrophic events (e.g., accidents, spills releases and terrorism). Actions or events that jeopardize environmental security and disrupt services provided by port facilities and other critical infrastructure may also elicit indirect, secondary impacts that result in environmental contamination and impose immediate and/or long-term damages on ecological resources and to public health. It is evident that chemical or material releases (either accidental or purposeful, in the context of terrorism) to the environment and the environmental alterations caused by such releases can result in actual or perceived human health and ecological risks. Additional security issues may arise in the event of conflicts between different stakeholder groups due to different perceptions of the significance of the environmental perturbation or potential threat.

3. Defining Risk

According to Belluck et al [2], there is an unavoidable convergence of what previously seemed disparate concepts - critical infrastructure, environmental security, risk assessment and risk management - and a need for tools that explicitly identify risks and provide a basis for making decisions that blend the relevant environmental, political, social and economic considerations.

In the context of environmental security at shipping ports and in coastal environments, risk is broadly defined as the possibility, or threat, of damage or destruction of shipping facilities, supporting infrastructure, the environment or the local population. Consistent with the different types of risks identified by Rowe [27], there are four important types of security risks that should be considered as part of an environmental security assessment process: (1) mission or function risks; (2) asset risks; (3) human health risks; and (4) ecological risks. Mission (or function) risks are the vulnerabilities that exist and have the potential to prevent a facility or organization from accomplishing its stated purpose. Asset risks are the vulnerabilities that exist and have the potential to damage critical infrastructure and other tangible physical assets such as buildings and equipment. Human health risks are well-defined in both the scientific and regulatory literature and reflect the vulnerabilities that have the potential to affect the health and safety of the local population and workers employed at shipping facility. Ecological risks are the vulnerabilities that exist and have the potential to degrade water, sediment, soil, air and biological (i.e., non-human) resources. In each case, the security risks posed by a threat may be acute and short-lived, or chronic, imposing longer-term affects on facilities, infrastructure, human health, and the environment.

The U.S. Department of Homeland Security has characterized the risk assessment paradigm for addressing environmental security [31]. Risk assessment precedes decision-making on environmental security and the management of potential threats. According to USDHS [31], risk is mathematically expressed as:

$$\text{Risk [R]} = \text{Consequences [C]} \times \text{Likelihood [L]}$$

where, likelihood [L] is further defined in terms of a specific vulnerability [V] that is exploited by a specific threat [T]. The likelihood of a breach in security is a conditional probability and can be expressed mathematically as:

$$\text{Likelihood [L]} = \text{probability of Threat [T]} \times \text{probability of Vulnerability [V]}$$

A threat is defined as any indication, circumstance, or event with the potential to cause loss or damage to shipping facilities, supporting infrastructure or the local population. Vulnerability is a weakness that can be exploited to gain access to a facility or critical infrastructure. For example, vulnerabilities at ports and harbors might include, but are not limited to, building characteristics, petroleum storage tanks, power generating plants, waste storage and removal equipment and security practices and procedures.

In this context, risk is defined more fully as the product of consequences (or impacts) caused by loss or damage to a valued asset and the likelihood (or probability) that the asset may be damaged or destroyed by a particular man-made (e.g., terrorism, over-fishing or exploitation of a different natural resource) or naturally occurring event (e.g., earthquake, global warming or extreme weather event). To determine the relative degree of risk, the probability, or likelihood, of occurrence of the undesirable event (either man-made or naturally occurring) must be estimated. The assets and their vulnerabilities judged using the USDHS [31] paradigm to be at greatest risk become the basis for deciding where and how to focus protection and prevention measures.

It is important to keep in mind, however, that defining acceptable levels of risk are rarely determined successfully by a mathematical formula. Risk levels will vary with time, circumstances, social and political values, and management attitudes toward risk to the human safety and the environment. The managers or owners of critical infrastructure must ultimately decide what constitutes an acceptable level of risk based on a blending of qualitative considerations and quantifiable criteria. Several recent advancements in decision analysis provide new paradigms for making judgments made regarding impact, threat, and vulnerability that help determine risk priorities [18].

4. Determining Environmental Vulnerabilities and Risks

The terrorist attacks of 11 September 2001 prompted the U.S. and several countries and international organizations to renew their focus on security issues, primarily the protection of critical infrastructure and measures to safeguard the general population. The U.S. Congress, for example, enacted legislation that required the Department of Homeland Security to analyze vulnerabilities and to suggest security enhancements for "critical infrastructure." The two key legislative actions that emerged following this mandate, the Public Health Security and Bioterrorism Preparedness and Response Act of 2002 (P.L. 107-188) and the Maritime Transportation Security Act (MTSA, P.L. 107-295), require vulnerability assessments, security plans, and incident response plans for chemical facilities that supply drinking water or are located in ports. The MTSA, in particular, calls for numerous wide-ranging and narrowly focused studies of the shipping industry, emphasizing U.S. ports and ships entering U.S. ports.

The events of 11 September 2001 also prompted a concerted international effort to address security concerns at industrial shipping ports. Members of the International Maritime Organization [14] adopted a series of resolutions and guidelines comprising a comprehensive security regime for international shipping, culminating in the International Ship and Port Facility Security Code (ISPS). The ISPS specifies a series of security measures that were adopted in July 2004 by international shipping companies and ports of Contracting Governments [15]. Under ISPS, performing a security and threat assessment was the first required step towards protecting critical industrial, environmental and utility infrastructure and understanding environmental response and prevention requirements and capabilities [15]. In the context of ISPS requirements for the international ports and maritime communities, three important steps to performing an assessment of environmental security risks included: (1) identification of plausible threats; (2) determining the potential impact of plausible threats; and (3) balancing the potential impacts of plausible threats with the implementation of safeguards such as prevention measures and response plans.

At approximately the same time, the American Bureau of Shipping (ABS) developed several recommendations on the approach to performing risk assessments relevant to the evaluation of maritime activities and shipping ports [36]. The risk assessment process (Figure 1) could be approached either qualitatively or quantitatively. ABS [36] identified several analysis techniques and models that could aid in conducting risk assessments (Figure 2). According to ABS [36], the key to successful risk analysis was choosing the right method (or combination of methods) that best fit the needs and mission of each port or maritime facility. Although Figure 2 lists each method only under its most common step, to avoid repetition, some methods could be used for more than one step in the risk assessment process.

The ABS [36] approach applied an environmental risk assessment paradigm to environmental security and disaster decision-making and prevention planning comparable to the U.S. paradigm to evaluate the human health and ecological consequences of chemical contamination. Consistent with the U.S. risk assessment paradigm, the ABS [36] approach recommended a four-step process:

1. Hazard Identification
2. Frequency Assessment
3. Consequence Assessment
4. Risk Evaluation

The value of risk assessment in performing an environmental security assessment at port and maritime facilities is evident in the ABS paradigm. The elements of a risk assessment are briefly described in each step of the ABS [36] four-step process.

FIGURE 1
The Risk Assessment Process

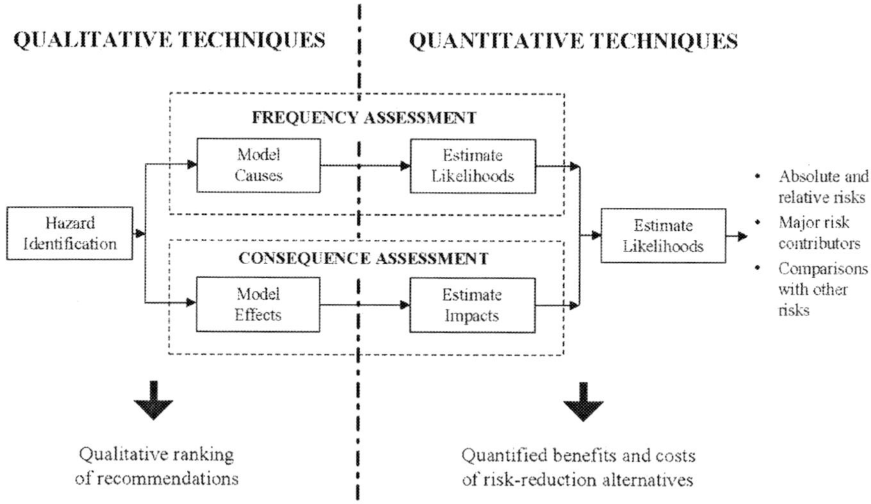

FIGURE 2
Overview of Risk Assessment Methods

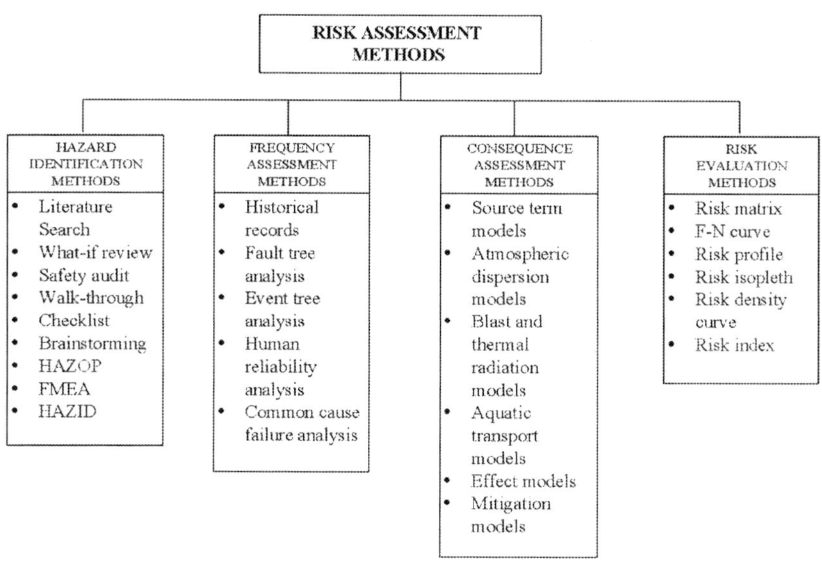

Figures 1 and 2: Risk assessment process and methods.

4.1. HAZARD IDENTIFICATION

The first step toward identifying potential hazards involves the development of a baseline understanding of current facility functions and surrounding environmental (air, soil, surface water, sediment and ecology) conditions, and includes consideration of disaster prevention and response measures already in place. This step is accomplished through various detailed facility infrastructure and environmental surveys. The information developed from these surveys is used in the subsequent steps to evaluate and prioritize the seriousness of different environmental disaster scenarios. In some cases, after establishing and understanding baseline facility and environmental conditions and identifying potential hazards, qualitative methods of assessing frequency and consequence are sufficient to proceed directly to the risk evaluation step. In other cases, a more detailed quantitative analysis is required.

4.2. FREQUENCY ASSESSMENT

The next step in the risk assessment process is to estimate the frequency at which either man-made or naturally-occurring disasters may occur. Risk assessment models can help estimate an incident frequency (e.g., frequency and severity of naturally occurring events, human error, or mechanical or structural failure) using historical records and readily-available databases (e.g., Bureau of Labor Statistics, NOAA storm, severity, and duration statistics; and U.S.G.S. earthquake data).

4.3. CONSEQUENCE ASSESSMENT

The third step involves the development of operations scenarios describing the range of possible disaster events and potential environmental outcomes associated each event. Consequence assessment typically involves detailed environmental fate and exposure models to predict spatial and temporal scales of the consequences of natural or man-made disasters. An appropriately applied model, coupled with an uncertainty analysis of the factors used in the model, may be used to evaluate appropriate response actions or preventative measures and the probabilities for their success. Several environmental fate models such as atmospheric dispersion models, blast and thermal radiation models and hydrodynamic models are reviewed in the literature and may be useful for this effort [10].

The use of environmental fate and exposure models in risk assessment typically involves four activities:

1. Characterize the source of materials or substances potentially released as a consequence of the man-made or natural hazards
2. Measure or estimate (using site-specific models, correlations to similar situations or observations reported elsewhere) the release and transport of materials to the environment and surrounding human population as a consequence of the event
3. Identify the effects of materials or substances on the environment and surrounding human population

4. Quantify the health, safety, environmental and/or economic impacts of the release

Considerable efforts have been invested researching the toxicological effects of exposure to toxic substances on human health and wildlife; several empirical databases are available describing both acute and chronic effects of exposures on humans and animals [28; 12; 29; 34]. Engineering databases are also available describing the effects of fires and explosions on structures and equipment (see e.g., www.nist.gov/srd/), and large, sophisticated experiments have been performed to validate computer algorithms for predicting groundwater, surface water, and atmospheric dispersion of toxic substances from both point releases and non-point sources. These and other resources are helpful for predicting the consequences of unforeseen spill events or accidents.

The result from the consequence assessment step is an estimate of the statistically expected exposure of the target population to the hazard of interest and the safety/health effects related to that level of exposure. Consequences are usually stated in the expected number of injuries or casualties in the surrounding human population or, in some cases, exposure to certain levels of toxic materials or substances released to water or air. Consequences also may be stated in terms of ecological damages such as the reduction in animal populations, loss of ecological habitat and impairment of soil, sediment, water, and air quality. In both cases, the estimates derived from predictive environmental fate and exposure models customarily account for either average or worst-case meteorological conditions and population distribution and may even include mitigating factors, such as evacuation and sheltering.

Consequence estimates will typically have large uncertainties owing to the absence of detailed site-specific information and the inherent unknowns with regard to the disaster or accident scenario itself. The uncertainties are typically associated with several individual assumptions embedded in the algorithms used to develop the environmental fate and exposure models and can be accounted for through the use of statistical methods such as Monte Carlo analysis. It is not uncommon for estimates to vary by several orders of magnitude. The level of uncertainty reflects basic uncertainties such as chemical/physical properties, differences in average versus time-dependent meteorological conditions, and differences in acute and chronic toxicological or ecological responses to toxic exposures. The outcomes are statistically based to give some frequency of occurrence, and a probability of impact. Because government environmental regulatory agencies often operate under (or are driven by) worst-case scenarios, it is necessary to highlight both likely and worst-case outcomes. The likely outcomes of consequence analyses, which are generally more realistic and less dramatic and improbable, are more useful in communications with the general public about the probabilities and consequences of a disaster or accident scenario.

4.4. RISK EVALUATION

The fourth and final step in the risk assessment process involves the evaluation of the relative risks associated with different disaster scenarios. Information is generated to identify, prioritize, select, and implement response actions and prevention planning. Several qualitative and quantitative methods of risk evaluation are available (Figure 2).

At present, advanced assessment tools such as multi-criteria decision analysis (MCDA) and comparative risk analysis (CRA) are emerging as integral components of both risk evaluation and decision-making. A review of MCDA and CRA methods is available elsewhere [18]. Briefly, the approach combines the probabilities associated with different threats (either man-made or natural), the various consequences of the threats, and the mission, asset, human health and ecological risks posed by the threats. The approach provides information on the value of different response actions and preventative measures, and the probabilities for successful mitigation or prevention of threats posed by man-made events (e.g., terrorism) and natural disasters (e.g., extreme weather events). In the absence of this approach, planning the best possible responses to contain a man-made or natural disaster, mitigating the environmental effects and, more importantly, preventing events from occurring will likely continue to challenge decision makers involved in environmental security issues. This is due to the quantitative and qualitative information developed from environmental modeling and risk assessment, cost and benefit analyses, opinion polls and other data-generating methods that contribute to ranking different environmental security threats and associated decision-making alternatives. MCDA and CRA also have the additional advantage of visualizing tradeoffs among multiple, conflicting criteria and quantifying the uncertainties necessary for comparison of multiple response actions and preventative measures.

5. Risk Assessment Challenges and Path Forward

In the context of environmental risk assessment, there are at least three challenges in environmental security and disaster decision-making and prevention planning. The first security challenge that must be addressed includes quantitative prediction with a high degree of confidence of the range of possible environmental damages and the potential threat posed to both human health and ecology. The nature of this work itself poses a security challenge because of the potential sensitivity of the information that must be compiled and evaluated.

The second, and perhaps equally important, security challenge is the use of the information and environmental risk assessment to describe the outcome of different possible disaster events and environmental scenarios. A closely related security challenge addresses the choice of risk assessment tools, including air, hydrodynamic and ecological models and the assumptions that populate the different models, and costs required to undertake an appropriately detailed and realistic risk analysis and investments necessary to reduce uncertainties in the model.

A third security challenge involves the use of the results of comprehensive environmental risk analysis to inform decision-makers of different disaster prevention/response action plans, the capital investments associated with each action plan, and the expected outcome of the plan as compared to the status quo. This information will allow decision-makers to prioritize capital investments to minimize or prevent adverse impacts to the environment, injuries or fatalities to port workers and residents in surrounding communities, and socioeconomic impacts.

The path forward clearly entails proposing a widely-acceptable framework for evaluating environmental security in both developed and developing countries, perhaps under the auspices of the United Nations and other international organizations. At present, there is not a single method accepted by regulatory agencies and risk experts for performing an assessment of environmental security risks at industrial and other non-military facilities, including commercial shipping ports and harbors. For example, a preliminary evaluation of vulnerability assessment methods conducted by the U.S. Department of Homeland Security found 24 different risk assessment methods [31]. In some cases, risk assessment methods were relevant to only a single critical infrastructure (e.g., water supply), while other methods were more broadly applicable to several types of critical infrastructure. Given the broad range of critical infrastructure (e.g., power generating facilities, port facilities, government centers, etc.), it is probably necessary to develop more than one security assessment framework that incorporate subjective semi-quantitative methods or numerical models that generate results based on unique assumptions and other specific considerations. By using our standardized assessment and audit approach, we can help organizations uncover and understand potential risks to their people, facilities, local communities, transportation systems, the environment, and other physical and intangible assets.

6. References

1. Apostolakis, G.E., Lemon, D.M. 2005. A screening methodology for the identification and ranking of infrastructure vulnerabilities due to terrorism. Risk Anal. 25(2):361-376.

2. Belluck, D.A., Hull, R.N., Benjamin, S.L., Linkov I. 2005a. Environmental security, critical infrastructure and risk assessment: Definitions and current trends. In: Linkov, I., Kiker, G., eds. Environmental Security and Risk Assessment. Kluwer Academic Publishers (this volume).

3. Belluck, D.A., Hull, R.N., Benjamin, S.L., Alcorn, J., Linkov, I. 2005b. Are Standard Risk Acceptability Criteria Applicable to Critical Infrastructure Based on Environmental Security Needs? In: Linkov, I., Kiker, G., eds. Environmental Security and Risk Assessment. Kluwer Academic Publishers (this volume).

4. Dalby, S. 2002. Environmental Security. Minneapolis: University of Minnesota Press.

5. Efroymson, R.A., Nicolette, J.P., and Suter G.W. 2004. A framework for net environmental benefit analysis for remediation or restoration of contaminated sites. Environ Manage. 34(3):315-331.

6. Gleditsch, N.P. 1998. Armed conflict and the environment: a critique of literature. Journal of Peace Research, 35/3 May: 381-400.

7. Gleditsch, N.P. 1997. Conflict and the Environment. Dordrecht: Kluwer.

8. Glenn, J.C., Gordon, T.J., Perelet, R. 1998. Defining Environmental Security: Implications for the U.S. Army. Editor: Molly Landholm. AEPI-IFP-1298

9. *Health and Safety Executive (HSE). 1992.* Regulation 4 of the Offshore Installations (Safety Case) Regulations 1992 (SI 1992/2885) (SCR) United Kingdom. http://www.og.dti.gov.uk/regulation/-guidance/reg_offshore/app10.htm. Accessed 25 July 2005.

10. Hedden, K.F. 1984. Multimedia fate and transport models: an overview. J Toxicol Clin Toxicol 21(1-2):65-95.

11. Hendrickson, R.G. 2005. Terrorist chemical releases: assessment of medical risk and implications for emergency preparadness. Hum Ecol Risk Assess. 11:487-499.

12. Heurgren-Carlstrom, G., Malmberg, E. 2003. Online information resources of toxicology in Sweden. Toxicology. 21;190(1-2):63-73.

13. Hopper, L.D., Oehme, F.W. 1989. Chemical risk assessment: a review. Vet Hum Toxicol. 31(6):543-554.

14. International Maritime Organization (IMO). 2004. IMO's global programme on maritime and port security - the work continues. http://www.imo.org/Newsroom/mainframe.asp?topic_id=848&doc_id= 3656. Accessed on 22 July 2005.

15. International Maritime Organization (IMO). 2002. International Ship and Port Facility Security Code (ISPS). Conference of Contracting Governments to the International Convention for the Safety of Life at Sea, 1974: 9 - 13 December 2002. http://www.imo.org/Newsroom/mainframe.asp? topic_id=583&doc_id=2689Accessed 25 July 2005.

16. Johnson, B.B., Chess, C. 2003. How reassuring are risk comparisons to pollution standards and emission limits? Risk Anal. 23(5):999-1007.

17. Karam, P.A. 2005. Radiological terrorism. Hum Ecol Risk Assess. 11:501-523.

18. Kiker, G.A., Bridges, T.S., Varghese, A., Seager, T.P., Linkov, I. 2005. Application of multicriteria decision analysis in environmental decision making. Integ Environ Assess Manage. 1(2):95-109.

19. Knapp, K.C., Weinberg, M., Howitt, R., Posnikoff, J.F. 2003. Water transfers, agriculture, and groundwater management: a dynamic economic analysis. J Environ Manage. 67(4):291-301.

20. Latora, V., Marchiori, M. 2005. Vulnerability and protection of infrastructure networks. Phys Rev E Stat Nonlin Soft Matter Phys. 71(1 Pt 2):015103. Epub - January 20.

21. Lin, B.L., Tokai, A., Nakanishi, J. 2005. Approaches for establishing predicted-no-effect concentrations for population-level ecological risk assessment in the context of chemical substances management. Environ Sci Technol. 39(13):4833-4840.

22. Leung, M., Lambert, J.H., Mosenthal, A. 2004. A risk-based approach to setting priorities in protecting bridges against terrorist attacks. Risk Anal. 24(4):963-984.

23. McNeil, F. 2000. Making Sense of Environmental Security. North-South Agenda. Paper. Thirty-Nine. http://www.miami.edu/nsc/publications/pub-ap-pdf/39AP.pdf

24. Monosson, E. 2005. Chemical mixtures: considering the evolution of toxicology and chemical assessment. Environ Health Perspect. 113(4):383-390.

25. Paustenbach, D.J. 1989. Important recent advances in the practice of health risk assessment: implications for the 1990s. Regul Toxicol Pharmacol. 10(3):204-243.

26. Pavlin, J.A. 2005 Medical surveillance for biological terrorism agents. Hum Ecol Risk Assess. 11:525-537.

27. Rowe, W.D. 1988. An Anatomy of Risk. Robert E. Krieger Publishing Co., Malabar, FL.

28. Russom, C.L. 2002. Mining environmental toxicology information: web resources. Toxicology.173(1-2):75-88.

29. Schwela, D. and Hakkinen, P.J. 2004. Human exposure assessment resources on the World Wide Web. Toxicology. 198(1-3):169-176.

30. Swatuk, L.A. 2004. Environmental security in practice: Transboundary natural resources management in Southern Africa. Presentation in Section 31 of the Pan-European Conference on International Relations, The Hague, 9-11 September.

31. United States Department of Homeland Security (USDHS). 2003. Vulnerability Assessment Methodologies Report. Phase I Final Report. Office for Domestic Preparedness, Washington, D.C. July.

32. United States Environmental Protection Agency (USEPA). 1996. Accidental Release Prevention Requirements: Risk Management Programs Under the Clean Air Act, Section 112(r)(7); List of Regulated Substances and Thresholds for Accidental Release Prevention, Stay of Effectiveness; and Accidental Release Prevention Requirements: Risk Management Programs Under Section 112(r)(7) of the Clean Air Act as Amended, Guidelines; Final Rules and Notice. Federal Register 61(120):31667-31720. 20 June.

33. Wenning, R.J., Sorensen, M., Magar, V. 2005. Importance of implementation and residual risk analyses in sediment remediation. Integ Environ Assess Manage. 2(1):in press.

34. Wexler, P. 2004. The U.S. National Library of Medicine's Toxicology and Environmental Health Information Program. Toxicology. 20:198(1-3):161-168.

35. Zaidi, M. 2005. Risk assessment in detecting and preventing of terrorist attacks in harbors and coastal areas. In: Linkov, I., Kiker, G., eds. Environmental Security and Risk Assessment. Kluwer Academic Publishers (this volume).

36. American Shipping Bureau (ABS). 2000. Guidance Notes on Risk Assessment Applications for the Marine and Offshore Oil and Gas Industries. *Houston, TX: ABS Publications.*

HUMIC SUBSTANCES AS A NATURAL FACTOR LOWERING ECOLOGICAL RISK IN ESTUARIES

Z. FILIP, K. DEMNEROVA
Department of Biochemistry and Microbiology
Institute of Chemical Technology Prague, Technicka 3-5, 166 28 Prague, Czech Republic

Abstract

Humic substances are considered to be a necessary component of terrestrial and aquatic ecosystems and the most persistent part of the decomposing organic matter. In marine environments about 90% of the dissolved carbon, as well as a major part of the sedimentary organic carbon, is humic matter. In the salt marsh estuaries, in addition to the humic matter brought about by rivers, a predominant autochthonous vascular plant *Spartina alterniflora* serves as a source of humic substances. Estuarine humic substances consist of C, N, and O, containing functional groups that can interact with inorganic elements and organic compounds. In this way humic substances apparently affect the bioavailability and toxicity of different pollutants. Although they could be used and structurally transformed by natural assemblages of microorganisms in some way, a considerable part of estuarine humic substances persist, apparently because of their adsorption on mineral components of estuarine sediments such as clay minerals. Due to their specific characteristics, including natural longevity in ecosystems, humic substances represent a natural factor apparently lowering ecological risk in salt marsh estuaries.

1. Introduction

In the last decades coastal areas have attracted still more attention worldwide because of human overpopulation, extensive industrial and commercial activities, and consequently increasing risk of environmental pollution. Due to this development, some coastal areas have lost their long-time attributed natural characteristic as an eternal "ultimate sink" for any kind of discharged material. On the contrary, they have been recognized as sensitive areas that require care, thorough management, and sometimes even remedial treatments, due to manmade contamination by hazardous substances. Linkov et al. [15] presented a complex framework for managing contaminated aquatic sediments, based on multi-criteria decision analysis and including ecological health, human health, public acceptance,

I. Linkov et al. (eds.), Environmental Security in Harbors and Coastal Areas, 343–353.

and costs as four basic factors. The authors divided the actual efforts to be applied to: (i) Remediation of contaminated sediments and aquatic ecosystems (whenever necessary); (ii) Reduction of contaminants introduced into aquatic ecosystems (a basic requirement); and (iii) Optimum use of water and coastal resources (a preventive measure). Broad international cooperation appears imperative to realize this complex strategy in a broad scale.

In spite of growing anthropogenic pressure, sedimenting shorelines exist in temperate latitudes that show rather stable natural environments. Salt marshes at the coastline of Georgia in the southeastern USA may serve as an example. They receive input from five differently polluted rivers, which drain some 89,500 km^2 of extensively used uplands, and bring vast amounts of both natural and anthropogenic organic and inorganic matter. Nevertheless, these salt marshes maintain a high rate of primary plant production; approximately 1400 g C m^{-2} y^{-1}, of which about 85 % is incorporated into stands of a smooth cordgrass *Spartina alterniflora.* Thus, a recent sediment of the marshes is likely to contain a high percentage of organic matter derived from the cordgrass growing there. Humic substances usually represent the major components of the organic matter in marine sediments [16, 24], and they are also known largely to react with inorganic and organic pollutants [9, 22, 23, 18, 25]. We attempted to elucidate their origin in a salt marsh, and also their ability to react with some metals and organics that are known as pollutants. In this report we summarize our experimental data obtained and published individually some years ago [1, 2, 5, 6, 12], in order to indicate the important role of humic substances as a factor affecting quality and functions of salt marsh sediment as a part of the marine environment.

2. Comparison of Humic Ssubstances from Salt Marsh Sediment and *S. alterniflora* Plant Material

Sediment was sampled from a depth of 0-15 cm in the salt marshes of Sapelo Island (Georgia), and after removing coarse particles (> 2 mm) it was dried and finely ground. Samples of live (fresh) and standing-dead leaves and culms of *S. alterniflora* were collected from plants of intermediate-height (0.5 m) growing at the same site. Plants were treated similarly to sediment; i.e., dried at 60°C and finely milled. Portions of the sediment and plant material were extracted under N_2 with an alkali (0.1 *M* NaOH + 0.1 *M* $Na_4P_2O_7$ mixture 1:1) for 24 h. Details can be found elsewhere [12]. Obviously, sediment produced the highest amount of humic substances (Table 1); the yield ratio being approximately 1:2:15 for the fresh plant material, the dead plant material, and sediment on an ash-free basis. Based on an elemental composition of original samples (not shown) it was calculated that 0.64% of the C of the fresh *S. alterniflora* was extracted as humic substances, and in the dead *S. alterniflora* and in the salt marsh sediment this part increased to 1.42% and 9.38 % of C, respectively.

Table 1: The yield and elemental composition of humic substances from sediment and *S. alterniflora*..

Sample	Humic substance (g dry wt./100 g)	Element (ash-free %) C	H	N	O	Ash (%)
Sediment	2.17	40.4	4.5	2.5	52.2	21.9
S. alterniflora						
Fresh	0.56	46.0	5.9	5.8	42.3	20.2
Dead	1.09	40.4	5.6	3.6	33.8	4.7

Humic substances extracted with alkali from salt marsh sediment and the fresh or dead biomass of *S. alterniflora* contained several biologically as well as environmentally important trace elements. Some metals bind more strongly with humic material than with other components of the plants or sediment (Table 2).

Table 2: Concentration of some elements in sediment, fresh and dead *S. alterniflora*, and in the derived humic substances (µg g-1 dry matter).

Element	Sediment whole material	Sediment humic substance	Fresh *S. alterniflora* whole material	Fresh *S. alterniflora* humic substance	Dead *S. alterniflora* whole material	Dead *S. alterniflora* humic substance
Cd	< 1	1.3	0.1	0.2	< 0.2	0.2
Cr	42.0	48.0	0.3	0.8	2.5	1.5
Cu	26.0	30.0	0.5	33.0	8.0	6.0
Mo	5.0	61.0	0.3	5.0	0.3	3.0
Sn	1.3	0.9	0.03	0.7	0.1	0.3
Pb	15.0	12.0	0.2	0.1	0.8	4.0
Zn	44.0	25.0	12.0	48.0	20.0	15.0

The humic substances have been found to consist of 50-70% rather single-structured organics such as phenols, monosaccharides and amino acids [7]. According to Huljev [13], however, the stability constants for metal-humic complexes are one or two orders of magnitude higher than those for metal-amino acids, -carbohydrates, or –phenol complexes. This could also apply for Cu, Pb, and Zn in our study, although there is no consistent behavior through the samples. Raspor et al. [20] also reported the contents of aliphatics less pronounced in the interactions of trace metals with humic substances isolated from estuarine sediments. For some metals (Cd, Cr, Pb) detected in our humic substances there was an increasing concentration in the order fresh plants < dead plants < sediment, and this reflects the same trend in the whole material. In the experiments by Kerndorff and Schnitzer [14] the sorption of trace metals tended to increase with rise in pH, decrease in metal concentration and increase in concentration of humic acids in the equilibrating solution. There were indications of competition for active sites (COOH and phenolic OH groups) on the humic acids between the different metals.

The FTIR spectra (Figure 1) reveal an insight into structural similarities and differences of the humic substances.

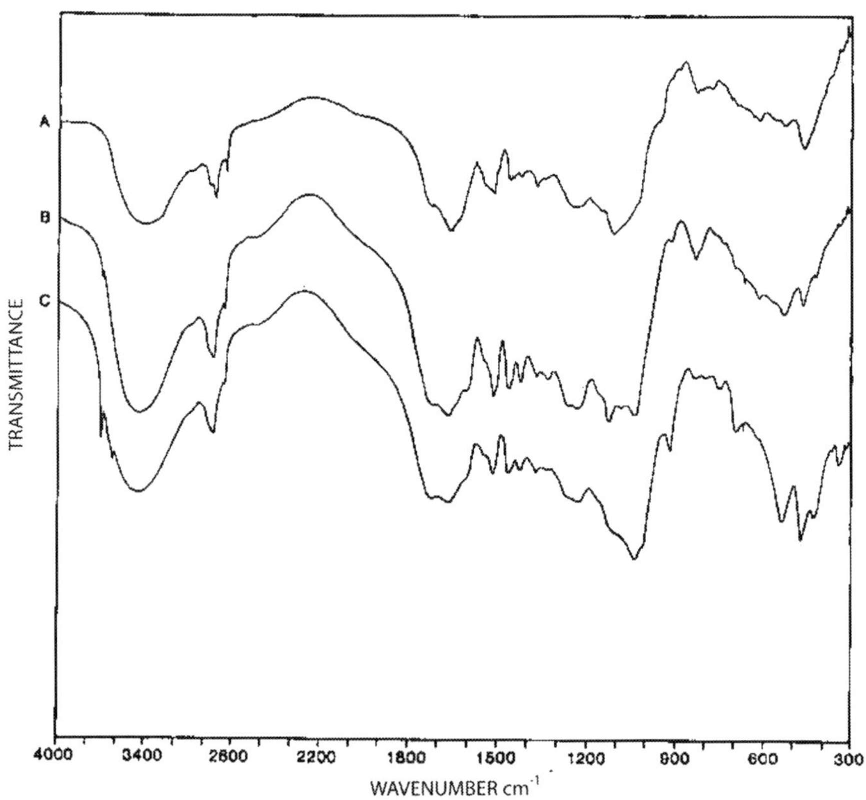

Figure 1 : FTIR spectra of humic substances extracted from (A) fresh *S. alterniflora*, (B) dead *S. alterniflora*, (C) salt marsh sediment.

Strong bands are evident near 3400 and 2900 cm^{-1} and correspond mainly to H-bonded OH and aliphatic C-H groups. The band at 2859 cm^{-1}, which also reflects C-H stretching from both methyl ($-CH_3$) and methylene ($C-H_2$) groups in aliphatic compounds, becomes weaker in the order fresh plant > dead plant > sediment. The sediment humic substances show an additional OH-double band at 3698 and 3620 cm^{-1} which is probably related to clay moieties. A ($C-H_2$) groups in aliphatic compounds, becomes weaker in the order fresh plant > dead plant > sediment. The sediment humic substances show an additional OH-double band at 3698 and 3620 cm^{-1} which is probably related to clay moieties. A shoulder at 1725 cm^{-1} indicates the presence of carbonyl and free carboxyl groups. A broad band at 1665-1660 cm^{-1} which is characteristic of the amide linkages of proteins is also somewhat diminished in the order fresh plant > dead plant > sediment-related humic substances. Peaks which could be the C-O stretching of carbohydrates appear at 1037 and 1034 cm^{-1} for the dead plant and sediment humic substances respectively. In the sediment humic substances, however, Si-O vibrations of clays may also participate in absorption at the same wavelength range. Several bands in the region 913-429 cm^{-1} and a high ash content suggest that clay minerals and other silicates are present.

3. Metals and Organics Binding Capacity in Estuarine Humic Substances

Humic substances are known to bind numerous metals and affect their bioavailability and toxicity as well as geochemical transport and fate [26, 23]. In our experiments [3], we concentrated on copper binding with estuarine humic substances as a probe by which differences could be followed in binding with various competing elements such as Ca^{2+}, Mg^{2+}, Al^{3+}, and Fe^{3+}. These elements can be found in most aquatic environments. FTIR spectroscopy was used to characterize the isolated humic matter-metal complexes and attempt to deduce types of binding sites for specific metals. The copper binding capacity (CuBC) ranged from 0.171±0.047 to 0.239±0.026 μg atm Cu^{2+} mg^{-1} humic matter (Table 3), and demonstrated an increase in the order of humic substances from fresh plant < dead plant < sediment. Aluminium (III) and iron (III) had a greater effect on blocking the CuBC sites in humic substances than did the divalent alkaline earths; up to 28% of the CuBC sites in sediment humic substances were blocked by these elements.

Table 3: Copper binding capacity (CuBC) of salt marsh humic substances, and influence of competing ions.

	Humic substances		
	Sediment	**Dead *S. Alterniflora***	**Fresh *S. alterniflora***
CuBC	0.239±0.026	0.190±0.009	0.171±0.047
Competing ions[2]			
Ca^{2+}	100	100	92
Mg^{2+}	100	93	83
Al^{3+}	85	58	78
Fe^{3+}	72	57	78

[1]CuBC reported as μg atm Cu^{2+} mg^{-1} humic matter corrected for ash content

[2]Reported as percent of original CuBC remaining

There are structural differences between protonated (H^+), and metal (Cu^{2+}) complexed humic substances (Figure 2). The FTIR spectrum of protonated humic substance (2,A) show pronounced peaks at 1716 cm^{-1} and 1221 cm^{-1}, which are typically assigned to the C═O and C—O stretching in protonated carboxylic acids. Complexation with Cu^{2+} did not result in removing but shifting the peak of C═O to 1724 cm^{-1}. However, the peak at 1221 cm^{-1} disappeared. New peaks appeared in the copper-humic complex spectrum at 1539 cm^{-1}, 1265 cm^{-1}, and 1230 cm^{-1}. The peak at 1539 cm^{-1} is assigned to mono-substituted amides, while the doublet at 1265 and 1230 cm^{-1} has been assigned to symmetrical bending of coordinated N—H groups with Cu^{2+} in the N-bonded biuret complex.

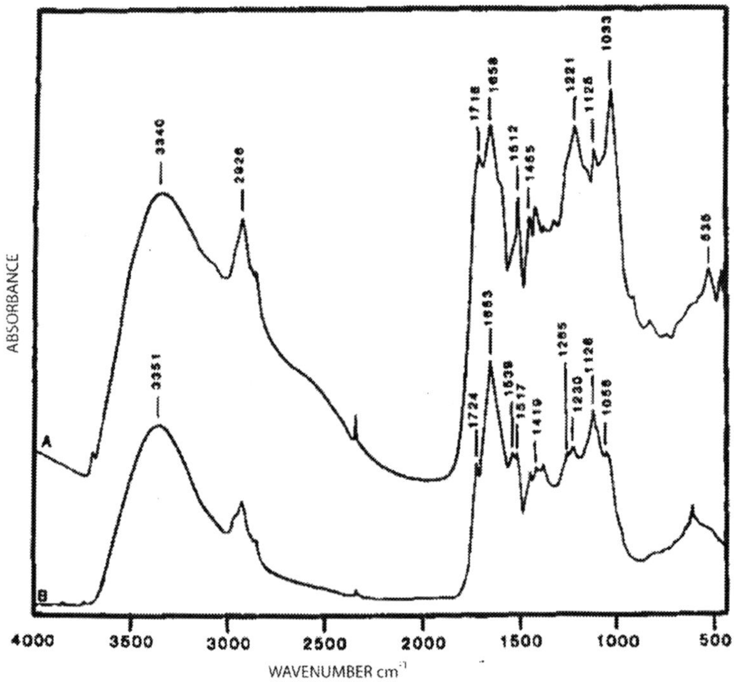

Figure 2 : FTIR spectra of humic substances extracted from salt marsh sediment (A) protonated; (B) complexed with Cu^2.

Thus, there appeared to be FTIR spectral evidence for partial copper complexation through nitrogen groups in this humic-copper complex. The FTIR spectra of the Ca^{2+}, Mg^{2+}, Al^{3+}, and Fe^{3+} complexes with humic substances (not shown in a figure) had no distinguishable peak in the 1715-1725 cm^{-1} region nor was there a significant peak at 1539 cm^{-1}. Furthermore, the region between 1265 and 1230 cm^{-1} also lacked an indication of coordinated N—H groups. Thus, it appeared that these ions bind to humic substances rather through carboxylic functional groups.

Besides oxygen and nitrogen containing functional groups that can interact with inorganic elements, humic substances are relatively large molecules which contain considerable aliphatic and aromatic organic carbon groups. These structural units can interact with different organic molecules, including hydrophobic ones such as polycyclic aromatic hydrocarbons (PAH) through electrostatic or van der Waals bonding to form organic-organic Complexes [17]. PAHs are ubiquitous in nature, and as carcinogens they represent a health risk. In our experiments [1] we used anthracene, phenanthrene, and pyrene. Despite the increase in ring size from anthracene to pyrene, the three PAH compounds demonstrated almost identical values of a binding constant (K_{oc}) with sediment humic substance; i.e., 4.97 (anthracene), 5.00 (phenantrene), and 5.03 (pyrene). Using a benzo(a)pyrene (BaP) we could demonstrate a decrease of the octanol-water partitioning coefficient ($\log_{10}K_{ow}$) from ~6.0-6.5 to ~ 4.2 in complexes with salt marsh-related humic substances which indicates that the presence of humic substances enhances the solubility of BaP in aqueous solution [5]. It has been found that metabolites from PAHs react with functional groups of humic substances by condensation processes to form stable chemical bonds; i.e., the bound moieties may be stabilized against degradation [21]. However, our studies have shown that sedimentary fulvic and humic acids could be affected by exposure to natural sunlight. The effects result in cleavage and loss of smaller organic compounds which consequently may affect the binding capacity of remaining humic substances towards pollutants [4].

4. Complexation of Estuarine Humic Substances with Minerals

River waters transport large quantities of sedimentary materials into the estuaries, about 30% of which are clay minerals, particularly montmorillonite and kaolinite. In addition, these tidally influenced water bodies also receive an important influx of clays from shoreline erosion and bottom scouring. The bottom muds of Sapelo Sound, for example, contain a clay fraction composed of 30-45% kaolinite and 50-60% expandable clays [19]. Different southeastern USA salt marsh sediments were found to consist of 91% silt and clay fraction; others to contain up to 82% quartz sand [8]. In our experiments, we investigated adsorption and transformation of salt marsh-related humic substances by quartz and clay minerals [10]. The amounts of adsorbed humic substances did not differ substantially for the individual humic preparations (Table 4). All of them were adsorbed from 9.4% to 11.8% on quartz, from 80.2% to 84.6% on kaolinite, and 79.9% to 89.9% on montmorillonite. Since the amounts adsorbed on kaolinite and montmorillonite were quite similar, very little interlayer adsorption apparently occurred on montmorillonite.

Table 4: Adsorption of estuarine humic substances (in mg TOC^{1} l^{-1}) on quartz and clay minerals.

Humic Substances	Original TOC	Adsorption on Quartz	Adsorption on Kaolinite	Adsorption on Montmorillonite
Sediment	98.4	9.3 (10%)	83.2 (84.6%)	88.5 (98.9%)
Dead *S. Alterniflora*	89.9	9.9 (11%)	72.1 (80.2%)	71.8 (79.9%)
Fresh *S. alterniflora*	95.8	11.3 (12%)	77.3 (80.6%)	80.7 (84.2%)

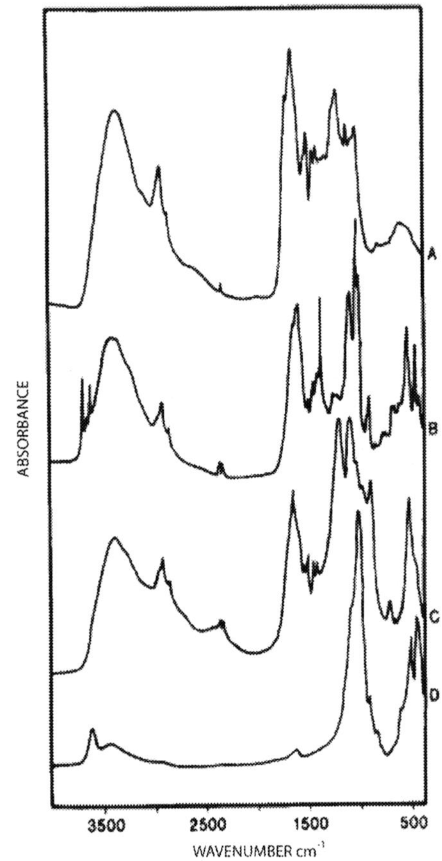

Figure 3: FTIR spectra of (A) humic substances extracted from salt marsh sediment; (B) humic substances fraction remaining in solution after adsorption on kaolinite; (C) humic substances fraction re-extracted by 0.1 M NaOH + 0.1 M Na4P2O7 from kaolinite; (D) humic substances fraction re-extracted as (C) from montmorillonite.

An attempt was made to remove the adsorbed humic substances from the complexes with quartz and clay minerally by acidic or alkalic extraction. When using 0.1 *N* H_2SO_4, up to 20.2% of humic substances bound on quartz but only up to 6.7% of those bound on montmorillonite were recovered. The desorption of humic substances was more complete when using a mixture of 0.1 *M* NaOH + 0.1 *M* $Na_4P_2O_7$. A maximum yield of 42% of humic substances was extracted from quartz-humic complexes, and, surprisingly, the relative amount of sediment humic substances extracted from montmorillonite was higher (47%).

In the FTIR spectra of sediment-related humic substances before and after complexation with clay minerals (Figure 3), transformation of humic substances apparently was caused by their contact with and adsorption on clays. This indicates that aliphatic chains such as in alkanes and some primary and secondary amides with their carbonyls, and further carboxyl and OH groups are responsible for the humic-clay interactions. Simultaneously, an increase in aromacity in humic substances remaining after adsorption on montmorillonite was indicated, e.g., at 803 cm^{-1} (C—H groups in aromatics). Humic substances re-extracted from kaolinite and montmorillonite complexes contained a large proportion of the respective clays, and this could also be clearly recognized from the FTIR spectra (Figure 3 C,D). In fact, the individual preparations contained between 50% and 89% ash, which was composed of kaolinite and montmorillonite, respectively. For this reason, in humic substances extracted from the humic-montmorillonite complex, the organic structures were completely overshadowed by the respective clay mineral (Figure 3 D).

All this indicates that in presence of clay minerals the preparations of "free" humic substances were transformed in a fraction of humins which is known to be highly stable in natural environments. Nevertheless, we could also demonstrate that original humic substances obtained from salt marsh sediment can be effectively utilized by the natural microbial assemblage indigenous to the same sediment [11]. An average utilization amounted to 65% (w/w), and the humic substances served preferentially as sources of nitrogen. Simultaneously, distinct changes in elemental and structural characteristics occurred (e.g., carbon content increased by 14% and the same was true for the proportion of phenolic structures as indicated in the respective FTIR spectra). These features are characteristics of diagenetic transformations that apparently occur to sedimentary humic substances due to the activities of autochthonous microbial populations.

5. Conclusions

Humic substances originating in the *S. alterniflora* stands have been estimated to represent about 56% of C that is included in the surficial salt marsh sediment per annum [1]. These substances have significant potential as biological nutrients, as sites binding different inorganic and organic anthropogenic pollutants, and also as basic structural components of organic-mineral complexes. There is no doubt that the indicated interactions exert influences to both lower and higher organisms inhabiting salt marsh estuaries. It appears also that humic substances comprising the major organic components in recent sediments contribute to long-term ecological stability in these extended coastal environments.

6. Acknowledgement

The authors gratefully acknowledge support for this study received from the European Commission, Brussels, by approval of a Marie Curie Chair at the Institute of Chemical Technology, Prague.

7. References

1. Alberts, J.J. and Filip, Z. (1994a). Humic substances in rivers and estuaries of Georgia, USA. Trends in Chemical Geology 1: 143-162.

2. Alberts, J.J. and Filip, Z. (1994b). Effect of organic solvent pre-extraction of source substrates on elemental composition, Fourier Transform Infrared spectra and copper binding in estuarine humic and fulvic acids. In: Humic Substances in the Global Environment and Implications on Human Health. N. Senesi and T.M. Miano eds. Elsevier, p. 781-790.

3. Alberts, J.J. and Filip, Z. (1998). Metal binding in estuarine humic and fulvic acids: FTIR analysis of humic acid-metal complexes. Environmental Technology 19: 923-931.

4. Alberts, J.J., Filip, Z. and Hertkorn, N. (1997). Interaction of light with estuarine fulvic and humic acids. In: The Role of Humic Substances in the Ecosystems and in Environmental Protection. J. Drozd, S.S. Gonet, N. Senesi and J. Weber eds. PTSH –Polish Society of Humic Substances, Wroclaw, p.539-544.

5. Alberts, J.J., Filip, Z. and Leversee, G.J. (1989). Interaction of estuarine organic matter with copper and benzoe(a)pyrene. Marine Chemistry 28: 77.87.

6. Alberts, J.J., Filip, Z., Price, M.T., Williams, D.J. and Williams, M.C. (1988). Elemental composition, stable carbon isotope ratios and spectrophotometric properties of humic substances occurring in a salt marsh estuary. Organic Geochemistry 12: 455-467.

7. Alberts, J.J., Filip, Z., Price, M.T., Hedges, J.I. and Jacobsen, T.R. (1992). Cu-oxidation products, acid hydrolysable monosaccharides and amino acids of humic substances occurring in a salt marsh estuary. Organic Geochemistry 18: 181-180.

8. Bradley, P.M. and Morris, J.T. (1990). Physical characteristics of salt marsh sediments: ecological applications. Marine Ecology Program Series 61: 245-252.

9. Choi, W.-W. and Chen, K.Y. (1976). Associations of chlrorinated hydrocarbons with fine particles and humic substances in nearshore surficial sediments. Environmental Science & Technology 10 : 782-786.

10. Filip, Z. and Alberts, J.J. (1994a). Adsorption and transformation of salt marsh related humic acids by quartz and clay minerals. The Science of the Total Environment 153: 141-150.

11. Filip, Z. and Alberts, J.J. (1994b). Microbial utilization resulting in early diagenesis of salt-marsh humic acids. The Science of the Total Environment 144: 121-135.

12. Filip, Z., Alberts, J.J., Cheshire, M.V., Goodman, B.A. and Bacon, J.R. (1988). Comparison of salt marsh humic acid with humic-like substances from the indigenous plant species Spartina alterniflora (Loisel). The Science of the Total Environment 71: 157-172.

13. Huljev, D.J. (1986). Trace metals in humic acids and their hydrolysis products. Environmental Research 39: 258-264.

14. Kerndorff, H. and Schnitzer, M. (1980). Sorption of metals on humic acids. Geochimica et Cosmochimica Acta 44: 1701-1708.

15. Linkov, I., Sahay, S., Kiker, G., Bridges, T and Seager, T.P. (2005). Multi-criterias decision analysis: A framework for managing contaminated sediments. In: Strategic Management iof Marine Ecosystems. E. Levner and J.M. Proth eds. Kluwer, Amsterdam, p. 1-26.

16. Mayer, L.M. (1985). Geochemistry of humic substances in estuarine environments. In: Humic Substances in Soil, Sediment and Water. G.R. Aiken, D.M. McKnight, R.L. Wershaw and P. Mac Carthy eds. Wiley, New York, p. 211-232.

17. Merkelbach, R.C.M., Leeuwangh, P. and Deneer, J. (1993). The role of hydrophobicity in the association of polychlorinated biphenyl congeners with humic acid. The Science of the Total Environment,Supplement 1993: 409-415.

18. Myneni, S.C.B., Brown, J.T., Martinez, G.A. and Meyer-Ilse, W. (1999). Imaging humic substance macromolecular structures in water and soil. Science 286: 1335-1337.

19. Pinet, P.R. and Morgan Jr. W.P. (1979). Implications of clay provenance studies in two Georgia estuaries. Journal of Sediment and Petroleum 49: 575-580.

20. Raspor, B., Nürnberg, H.W., Valenta, P. and Branica, M. (1984). Studies in sea water and lake water on interactions of trace elements with humic substances isolated from marine and estuarine sediments. Marine Chemistry 15: 217-230.

21. Richnow, H.H., Reidt, C., Seifert, R. and Michaelis, W. (1994). Chemical cross-linking of xenobiotica and mineral constituents to humic substances derived from polluted environments. In: Humic Substances in the Global Environment and Implications on Human Health. N. Senesi and T.M. Miano eds. Elsevier, Amsterdam, p. 1043-1048.

22. Schnitzer, M. and Kerndorff, H. (1980). Effects of pollution on humic substances. Journal of Environmental Sciences and Health B15: 431-456.

23. Senesi, N. (1992). Metal-humic substance complexes in the environment. Molecular and mechanistic aspects by multiple spectroscopic approach. In: Biochemistry of Trace Metals. D.C.Adriano ed. Lewis Publ., Boca Raton, p.429-496.

24. Vanderbroucke, M., Pellet, R. and Debyser, Y. (1985). Geochemistry of humic substances in marine sediments. In: Humic Substances in Soil, Sediment and Water. G.R. Aiken, D.M. McKnight, R.L. Wershaw and P. MacCarthy eds. Wiley, New York, p. 249-273.

25. Varrault, G. and Bermond, A. (2002). Trace metal sorption ability of insolubilized humic acids. Environmental Technology 23: 421-428.

26. Weber, J.H. (1988). Binding and transport of metals by humic materials. In: Humic Substances and their Role in the Environment. F.H. Frimmel and R.F. Christman eds., Wiley, New York, p. 165-178.

RISK ASSESSMENT APPLICATIONS IN CROATIA

D. SINKA, D. SKANATA
Enconet International
Miramarska 20, 10.000 Zagreb, Croatia

Abstract

The first technological risk assessment in the Republic of Croatia was accomplished in 1994. Since then, this procedure has been applied in various fields with different goals and according to different methodologies. However, it can be stated that the total number of assessments performed so far is limited.

In the majority of cases the basic objective of technological risk assessment implementation was to meet legal obligations. According to the valid legislation of the Republic of Croatia, the risk assessments have to be made within the scope of the emergency planning, with the objective of reducing unacceptable risks in the company's manufacturing, processing, storing, or transporting hazardous substances and within the framework of environmental impact assessment (EIA). The minor part of the risk assessments has been executed on the initiative of operators of the potentially hazardous facilities, on the encouragement of local authorities or under the pressure of the local population. The results of the implementation of these assessments have been used to optimize allocations for operational safety, for needs of spatial planning, and with the objective of setting safety priorities in large industrial areas.

In the paper examples of risk assessment applications in the Republic of Croatia are given. The examples are related to the industrial zones located near major population centers, development of plans and procedures for off-site nuclear emergency preparedness, oil and gas processing and crude oil transport by pipeline. The examples include the objectives of implementation, the descriptions of applied methodology and the main results.

According to the opinion of the authors, in the Republic of Croatia the procedure of technological risk assessment has not yet been recognized as a useful decision-making support tool. Such opinion is based on the number of performed assessments as well as on their effects. Some of the main reasons for the adverse state are undoubtedly the following: lack of information, unfavorable economic situation, low level of safety culture, modest possibilities of education, and distrust of the profession.

I. Linkov et al. (eds.), Environmental Security in Harbors and Coastal Areas, 355–366.

One of the steps directed towards the improvement of the situation could be the founding of a nonprofit institution called "Risk Analysis Institute." Establishing of the Institute should facilitate knowledge transfer concerning the field of risk analysis to interested parties in the Republic of Croatia. The Institute should cover the area of technological and other kinds of risks, and among other things it should have a promotional, educational, and advisory role. The projects of the Institute would be carried out in cooperation with the economy, state institutions, and nongovernmental organizations. In accordance with the plans, the Institute could start work in 2005.

1. Introduction

The risk assessment related to the technological facilities has been performed in the Republic of Croatia for the last ten years in different fields, with different motives and in accordance with different methodologies. The basic motive for implementation of the majority of assessments is meeting legal obligations. The technological risk assessment nowadays represents a legal obligation in the following fields:

- Off-site emergency planning
- Process safety management of highly hazardous substances
- Environmental impact assessment (EIA)

Laws regulating risk assessment have been taken over for the most part from US or EU legislation or a combination of the two. At the moment, harmonization of the legislation between the Republic of Croatia and the EU is being worked on because of preparations for the full membership. This process is expected to result in stricter legal requirements; i.e., the risk assessment implementation is expected to become obligatory in some additional fields.

Aside from meeting legal obligations, risk assessments are also performed on the initiative of operators of potentially hazardous facilities, or under the pressure of local authorities or local population. The usual motives for implementation in such cases are to optimize allocations for operational safety or to set criteria for spatial planning.

2. Application Examples

Within this chapter, four short examples of risk assessment implementation are presented. The examples include the scope and main objectives of the assessment, basic information about the methodology, and the most significant results. In the first example, the first technological risk assessment performed in Croatia (at least for what the authors know) is described. The assessment was implemented within the framework of the UNEP, UNIDO, IAEA, and WHO Inter-Agency Programme, and concerns large industrial zones. In the second and third example, risk assessments performed with the objective of meeting legal obligations in the field of emergency planning and environmental impact assessment (EIA) are described. The fourth example concerns the assessment performed on the common initiative of the technological facility management and the local community with the objective of defining criteria for spatial planning.

2.1. EXAMPLE 1: SETTING SAFETY PRIORITIES IN LARGE INDUSTRIAL AREAS

The Case Study Zagreb Project was an attempt to introduce an integrated approach to assess and manage risks associated with industrial complexes in the City of Zagreb [1]. The main tasks of the project were (1) to identify and assess risks to the public due to major industrial accidents with off-site consequences in fixed installations handling, storing, or processing hazardous substances and (2) to classify those installations on a risk-priority basis.

The assessment was performed by applying Rapid Risk Assessment Methodology to obtain a preliminary, general and quantitative overview of risks from various production facilities in a particular industrial and urban area, in order to identify the priorities for further detailed risk analyses. Consequently, the results obtained can be used only for a relative comparison of risks and not at any rate for absolute assessment, or management of risks for individual facilities.

The methodology is based on the set of basic data (such as type of activities, type and quantity of hazardous substances, population density in the vicinity of facilities, meteorological data, etc.). The series of successive elimination criteria (i.e., presence of maximum quantities of hazardous substances that can participate in an accident, boundary distances from the inhabited area, and spatial range of the accident effects respectively) was applied, eliminating facilities of negligible risk from further analysis.

In order to specify risk priorities (and only for that purpose), societal risk acceptability criteria were introduced. As an upper acceptability level of societal risk of maximum 10 fatalities per accident the value of 10^{-4}/year was assumed, whereas the threshold of negligible risk under the same conditions was 10^{-6}/year. The area between these two values was considered the risk reduction area. The n-n^2 rule has been used for other values of societal risk, saying that if the number of fatalities in an accident is increased by n times, the probability of an accident should be reduced by n^2 times.

The results obtained, combined with the risk acceptance criteria (straight lines) are illustrated in an F-N diagram on Figure 1. As it is shown, altogether 9 storage facilities have fallen within the unacceptable risk area (2.5% of the total number of the facilities surveyed). Also, 17 facilities (storage and process) have been recognized as the facilities in which the level of safety has to be improved.

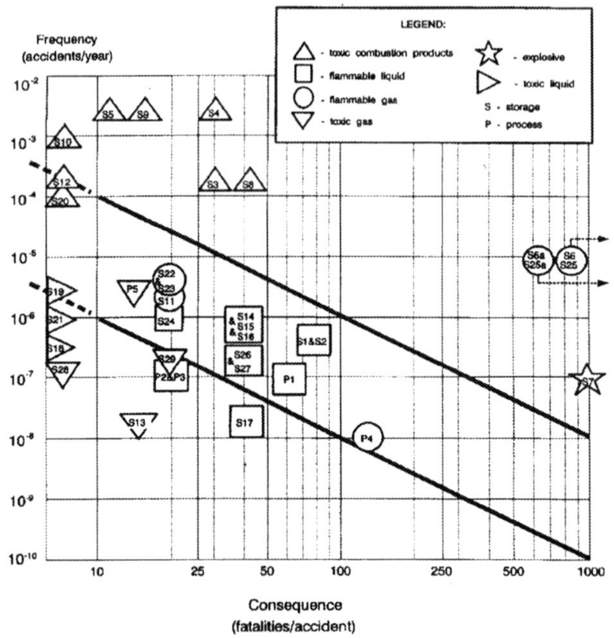

Figure 1: Graphical presentation of the Case Study Zagreb Project results.

2.2. EXAMPLE 2: NUCLEAR EMERGENCY PLANNING ON THE NATIONAL LEVEL

Efforts to improve emergency response plans to be undertaken in case of nuclear accident have recently been made in the Republic of Croatia. Special emphasis has been placed on Krsko NPP in Slovenia and Paks NPP in Hungary, these two being the plants nearest to the territory of the Republic of Croatia. However, one could ask whether planning and preparation would be conducted in the same way if, instead of distance, a more comprehensive criterion would be adopted: risk.

In order to demonstrate possibilities of using risk assessment in development of nuclear emergency preparedness, individual risk was estimated for citizens of the four biggest population centers (Zagreb, Osijek, Rijeka and Split) [2]. The assessment referred to all power reactors located at a distance up to 1,000 km that were in operation during the data collection phase. Distances between power reactors included in the assessment and the population centers in the Republic of Croatia are 40 km or more. For that reason, only nuclear accidents that can cause significant consequences at greater distance; i.e., accidents that include release of larger quantities of radioactive substances, have been taken into account.

The assessment was performed using the newly developed methodology. Assessment of consequences; i.e., the received radioactive doses, was composed of three individual analyses: source term analysis, atmospheric dispersion analysis, and exposure analysis. Source term analysis was based on the release assessment method developed by the U.S. Nuclear Regulatory Commission (NRC).

Atmospheric dispersion was analyzed using the Lagrange "puff" model. The exposure analysis was limited to the early phase exposure pathways: (1) exposure to direct radiation of the radioactive cloud, (2) inhalation of aerosols, particles and gases coming from the radioactive cloud, and (3) exposure to radiation of radionuclides deposited on the ground. Frequency assessment was based on results of the conducted probability safety analyses (PSAs) and on assessment of "probability of unfavorable direction" (probability for the radioactive cloud, formed out of the substances released from the nuclear power plant during the accident, to pass over the selected point of exposure). With assessed consequences and frequency, the risk of nuclear accident was calculated as a mathematical product of two parameters.

The results demonstrate that, when dealing with emergency preparedness development, risk assessment proves to a be very useful and widely applicable technique. If risk would be taken as basic criterion in emergency planning, the results shown in Figure 2 would directly indicate the necessary preparation level for each of the four population centers.

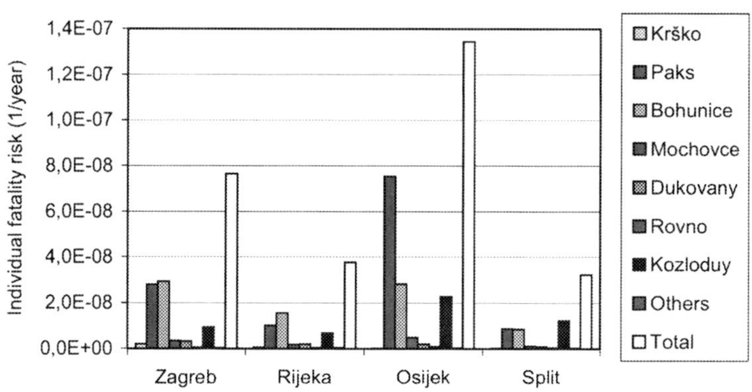

Figure 2: Contributions of individual power plants to total individual risks.

Furthermore, the assessment of risks from individual power plants (Figure 2) and power plant types (Figure 3) indicates to which facilities the greatest attention should be paid. Risks from groups of power plants formed in accordance with their respective distance from exposure location (Figure 4) shows what kind of tools for determining consequences and protective actions during a nuclear accident should be made available and used.

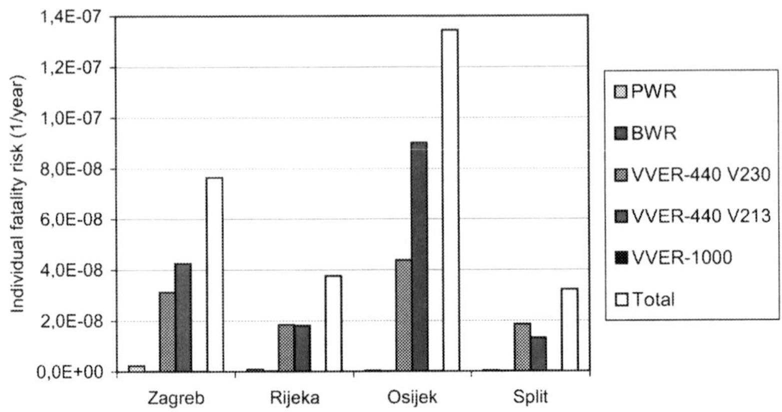

Figure 3: Contributions of particular power plant types to total individual risks.

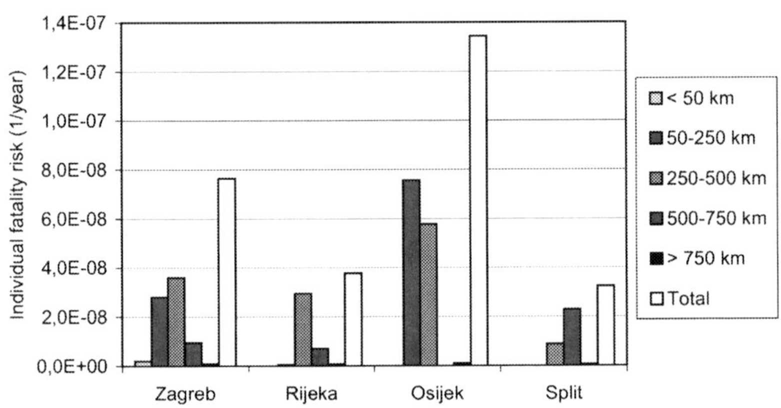

Figure 4: Contributions of groups of power plants to total individual risks.

2.3. EXAMPLE 3: CRUDE OIL PIPELINE TRANSPORT

In this example the quantitative risk assessment performed for the pipeline transport system is described [3]. The system is used to transport crude oil from the Croatian-Hungarian border to the port Omisalj in the North of the Adriatic. It consists of three terminals, two pump/relief stations, and 289 km of pipeline with the belonging block stations. The performed risk assessment is a part of the environmental impact assessment (EIA), the document made with a goal of estimating the social acceptability of the proposed technological solution.

The risk assessment for the oil transport system has been performed in accordance with eight relevant accident scenarios. They are:

1. Construction fault
2. Materials fault
3. System malfunction
4. Human error
5. External corrosion
6. Internal corrosion
7. Natural hazard
8. Third-party activity damage

The scenarios above cover all sorts of accidents that have already happened within the system and also those that have been happening in the Western European oil pipeline systems in the last 30 years. The assessment did not cover the accidents resulting from malicious acts; i.e., vandalism, sabotage, or terrorist attacks.

Frequency assessment, as well as the assessment of the consequences of accident scenarios, was based on the analysis of the data from the CONCAWE database. In this database, data related to the accidents in the Western European oil pipeline systems have been stored since 1971. The main results of the frequency assessment are shown in Figure 5.

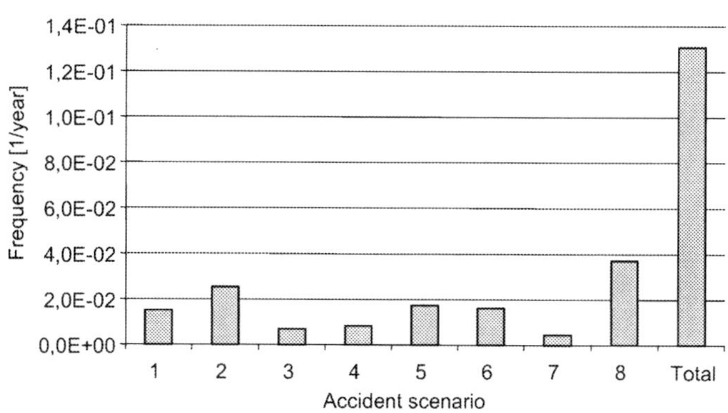

Figure 5: Frequency assessment results.

The goal of the consequence assessment was to define the magnitude of expected crude oil releases related to the realization of particular accident scenario. For this purpose, four spill categories were defined and then their relative contributions to each of the accident scenarios were assessed (see Figure 6).

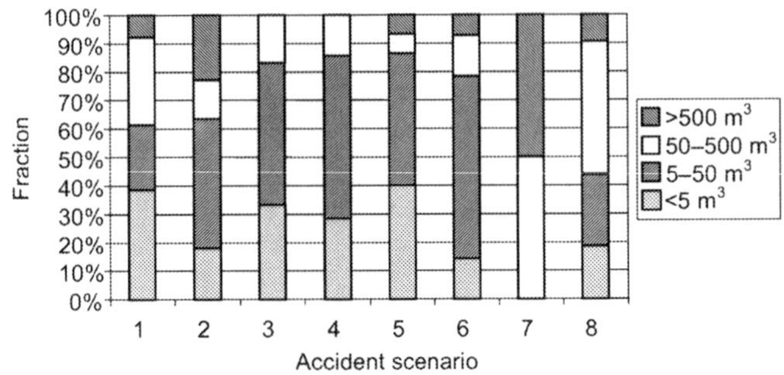

Figure 6: Consequence assessment results.

The final result of the assessment consisted of (1) health risks for the oil pipeline system workers and other persons that might find themselves in the vicinity and (2) environmental risks. The human health risks were expressed as loss of life probabilities due to a technological accident, and the environmental risks as (1) the expected average annual quantities of crude oil released into the environment, (2) the expected average annual surface pollution (agricultural, forest, and other), and (3) the probabilities of groundwater or surface water pollution. An example of the final assessment results is shown in Figure 7. The figure refers to the environmental risks expressed as the expected average quantities of crude oil to be released from the pipeline per year of operation.

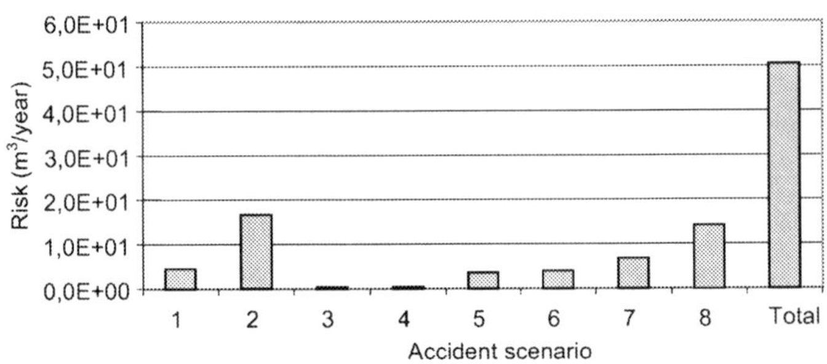

Figure 7: Risks expressed as expected average annual releases.

2.4. EXAMPLE 4: GAS REFINING WITHIN URBAN AREAS

ETAN is a plant constructed for recovery of valuable substances found in natural gas before it is distributed to consumers. It consists of process facilities, storage facilities, and truck/train loading facilities. The plant is located on the outskirts of a town of 8,000 inhabitants. So far, spatial planning related to areas surrounding ETAN has been based on a worst-case consequence assessment carried out almost 20 years ago. The main goal of the assessment performed now was to give more objective estimation of the level of risk that ETAN poses to the neighboring community [4].

ETAN plant risks are primarily related to high quantities of energy contained within various flammable substances. That energy can be released in case of fire or explosion. Hazard identification results demonstrated that only accidents related to instantaneous or continuous releases from storage facilities, followed by immediate or delayed vapor cloud fire or vapor cloud explosion, could cause serious consequences beyond the boundary fence.

Within consequence assessment, releases to the atmosphere; atmospheric dispersion; fires; and explosions of propane, butane, pentane, and condensate have been modeled. Releases and atmospheric dispersion have been analyzed using a heavy gas model included in ALOHA computer code. The code was developed by the U.S. Environmental Protection Agency (EPA). Vapor cloud ignition positions for various atmospheric transport directions have been defined by applying the method proposed by the Center for Chemical Process Safety (CCPS). Explosion effects have been estimated using TNT equivalency model developed by the U.K. Health and Safety Executive (HSE).

Frequency assessment was carried out using Event Tree Analysis (ETA). The initial data for the analysis; i.e., the data related to the components' reliability, were taken from literature published by the U.S. Federal Emergency Management Agency (FEMA). Results obtained have been compared to ETAN plant historical data.

Individual risks posed to the members of the neighboring community were presented as risk profiles (transects) and risk contour plots (see Figure 8).

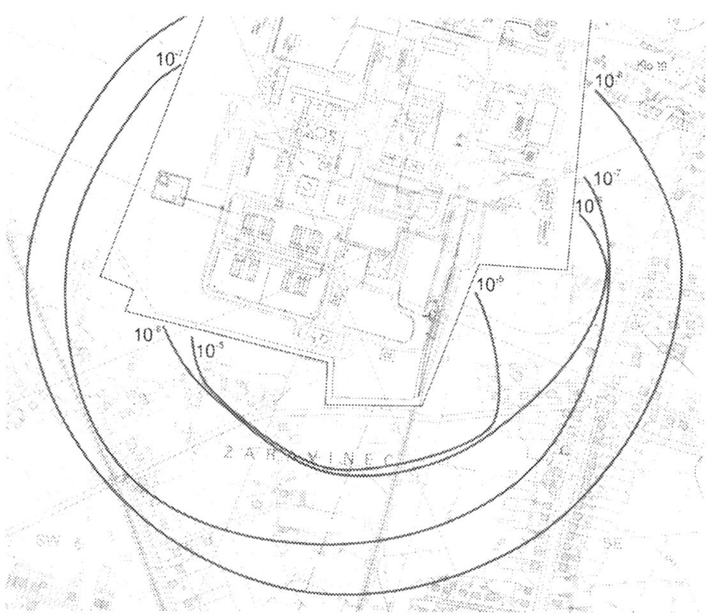

Figure 8: Individual fatality risk contours (annual risks).

Societal risks have been presented as F-N curves. In order to facilitate future spatial planning activities, acceptable risk values from various sources have been collected and compared with assessment results.

Using the ETAN plant as an example, it has been shown that quantitative risk assessment can be applied to support informed decision making related to spatial planning. The results obtained by such assessments can be especially valuable in cases where spatial planning criteria are not imposed by the law or recommended.

3. Discussion

Although it has been implemented for a longer period of time and in different areas, according to the judgment of the authors the technological risk assessment has not yet become recognized as an important decision-making aid tool in Croatia. Such an estimate is based on a relatively small total number of performed assessments and on the fact that the assessment results are often used in an unsatisfactory way. Misapplication of the results can be shown on the examples from the previous chapter. As mentioned, one of the basic goals of the Case Study Zagreb Project (Example 1) was to identify technological facilities for further detailed analysis. Although the facilities have been identified, a detailed analysis has never been carried out. The risk assessment for the ETAN plant (Example 2) has been made for the purpose of setting criteria for spatial planning. However, fact is that the building of some controversial objects in the vicinity of the plant had been started before the assessment was finished and the results presented to the local authorities.

Figure 9 shows an attempt to identify and systemize the causes of the unsatisfactory conditions of risk assessment applications in Croatia. It can be noticed that nine causes have been connected with a small number of performed assessments and four causes with modest effects; i.e., with misapplication of the assessment results. It can also be noticed that the biggest number of causes is connected with a lack of information and education, which refers to the state institutions and local government, as well as to the economy, media, nongovernmental organizations, and the public. To a significant degree, this unsatisfactory condition results from a difficult economic situation and the general level of social development.

The founding of the Risk Analysis Institute, which is now going on, is considered to be an important step towards the improvement of the condition. The Institute is going to work as a nonprofit institution and cooperate on its projects with governmental institutions, the economy, and nongovernmental organizations. The staff of the Institute is going to establish contacts with professional institutions and organizations abroad and in that way facilitate knowledge transfer to interested parties in the Republic of Croatia.

The institute, which should start work by the end of 2005, is going to have a promotional, educational, and advisory role. In Figure 9 it can be noticed that it is expected to have a positive effect on the elimination of a large number of identified causes of the current unsatisfactory condition.

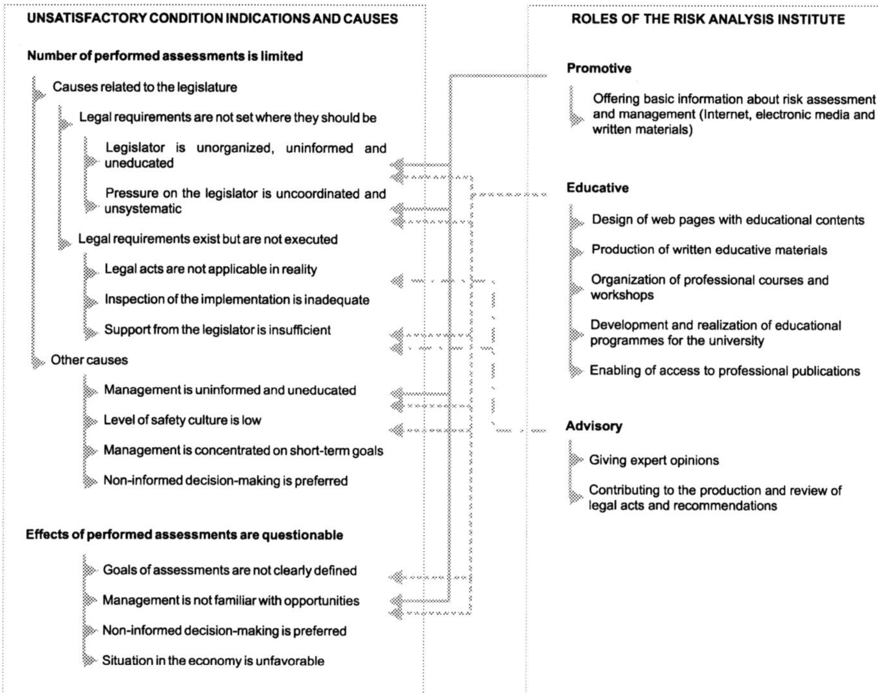

Figure 9: Unsatisfactory condition causes and links to the roles of the Institute.

4. Conclusions

Imposition of legal obligations is not sufficient for risk assessment to become recognized as a useful tool to support informed decision making. Aside from the clear motive for its implementation (which can be, but is not necessarily connected with the meeting of legal obligations), additional preconditions such as certain levels of information, education, economic, and social development have to be met. At the moment, in the Republic of Croatia, a large number of important preconditions are not fulfilled, which results in the infrequent application of risk assessment and the use of the results in an unsatisfactory way. The Risk Analysis Institute is being founded as one of the steps towards the improvement of the condition. It will be a nonprofit institution that is going to deal with technological and other risks and is going to have a promotional, educational, and advisory role.

5. References

1. Skanata, D., Sinka, D. (1998) Linking Risk Assessment to Strategic Environmental Assessment: Case Study Zagreb, Croatia, in N. Lee (editor), *Case Studies of Environmental Assessment in Countries in Transition*, CEU, Budapest, 33-42

2. Sinka, D., Mikulicic, V. (2000) Severe Accidents Risk Assessment as a Basis for Emergency Preparedness, *International Conference on Nuclear Option in Countries with Small and Medium Electricity Grids, Dubrovnik, June 19-22 2000 - Proceedings,* Croatian Nuclear Society, 647-653

3. Sinka, D., Jovic, J., Keco, M., Poljanac, M., Medakovic, S. (2004) Risk assessment of oil transport by JANAF pipeline from Gola to Omisalj, Enconet International, Zagreb

4. Sinka, D., Skanata, D., Veronek, B., Freo, D., Mesic, S. (2004) Gas Processing Plant Quantitative Risk Assessment, Abstract, Risk Analysis: The Profession and the Future, Society for Risk Analysis Annual Meeting, Palm Springs, December 5-8 2004 - Proceedings, 110-111

Part 5

Case Studies in Environmental Security Analysis

RISK ASSESSMENT OF THE EFFECT OF WASTEWATER TREATMENT AND DESALINATION PROCESSES ON THE COASTAL AQUIFER IN ISRAEL

Using Multi-Criteria Decision Analysis

N. HARUVY
Netanya Academic College
1 University Rd Netanya 42100, Israel

S. SHALHEVET
Shalhevet Consulting
126 Thorndike Street, Brookline, MA 02446 USA

Abstract

The coastal aquifer, one of Israel's major water resources, is becoming increasingly polluted. Several wells in major urban areas were closed down because they contained contaminants at levels considered harmful to human health. Groundwater contamination is exacerbated due to irrigation with treated effluents and underground water quality has become a major concern regarding future supply of water in the country. While secondary and tertiary wastewater treatment processes can decrease most pollutants including nitrates, salinity level is higher in effluents than in influents and should be treated. Desalination processes can be applied to saline aquifers, to National Carrier water, or to unconventional water sources including wastewater or seawater.

Our work examines different options of treating water supply sources. The assessment procedure was based on a multidisciplinary approach, including economic, hydrological, technological, agricultural, social, political, and environmental criteria. The economic criteria include the size of the initial investment and cost-benefit analysis. The hydrological criteria include the derived aquifer salinity through time. The technological criteria include the relevant treatment processes and their adaptation to the region's conditions. The agricultural criteria include the effect on crops. The social criteria include the effect on the employment rate. Finally, the environmental criteria include the effect on soil salinity and energy use that may causes greenhouse gas emissions. Other criteria include the effect on Israel's image in the world and on its ability to export agricultural products.

These criteria were combined using multi-criteria decision analysis to arrive at a risk assessment of the various treatment processes and to decide on the optimal

I. Linkov et al. (eds.), Environmental Security in Harbors and Coastal Areas, 369–378.

water treatment option that will ensure the health of the population as well as the quality and safety of the coastal water and the environment.

1. Background

The annual water resources in Israel are limited and amount to 2,000 MCM (Million Cubic Meter). Agricultural water use is approximately 1,300 MCM, which is 65% of the total water use. With the future increase of population, domestic demand for water will increase and reduce the freshwater supply available for agriculture. The Israeli Water Commission is planning to supply part of domestic water demand with desalinated seawater, and several seawater desalination plants are already in various stages of operation.

Agriculture will be based on marginal water sources, mainly on treated effluents. Reuse of treated effluents in agriculture provides an unlimited and cheap water source. Urban wastes should be treated and discarded anyway; hence, agricultural reuse of effluents serves also as environmental quality agent. A drawback to using wastewater for irrigation is that domestic and industrial effluents carry pollutants, including micro and macro organic and inorganic matter. These can be treated through appropriate secondary and tertiary treatment processes. Another drawback is that the salinity level is higher in effluents, causing potential soil and groundwater damage as well as reduced crop yield and quality from irrigation with saline water. Regular treatment processes do not get rid of salinity, unless the treatment process includes a relatively expensive desalination processes [4].

Pollution in Israel's coastal aquifer is increasing. Groundwater contamination is exacerbated due to irrigation with treated effluents. The groundwater quality has become a major concern regarding future supply of water in the country.

2. The Hydrological Model

We compared several scenarios regarding water supply alternatives for town and agriculture. The water supply alternatives included local groundwater, National Carrier water, and wastewater. The hydrological model predicts the flow of chlorides through the unsaturated zone of the subsoil, into the groundwater below.

The model is based on the following three assumptions. First, pumping capability is influenced by amount of leaching to groundwater. Second, desalination of groundwater and/or imported water is applied when reaching a predetermined threshold of chloride concentration in drinking and/or irrigation water. And third, part of the treated water source is then desalinated to a given salinity level and diluted with other water sources until reaching the permitted level.

2.1. APPLICATION TO A CASE STUDY

The hydrological model was applied to a specific case study in central Israel composed of eight hydraulic cells in the coastal aquifer. Region A, with an agricultural area of 8,000 ha, partly irrigated by wastewater, has an aquifer salinity higher than 230 mgl Cl. Region B has a big city, with 150,000 citizens and a wastewater treatment plant, and aquifer salinity of 75-164 mgl Cl.

Agriculture in Region A uses 24.5 MCM of irrigation water; of this, 61% is wastewater, 38% is from the local aquifer, and less than 1% comes from National Carrier water. Agriculture in Region B uses 34.9 MCM of irrigation water; of this, 31% is wastewater, 56% is from the local aquifer, and 34% comes from National Carrier water (Table 1). Region A needs very little water for urban use, which comes mostly from the local aquifer, and Region B's urban area uses water mostly from the National Carrier.

Table 1: Water supply sources (MCM).

	Region A	**Region B**	**Total**
Agriculture	24.46	34.88	59.34
National Carrier	0.22	11.96	12.18
Local aquifer	9.41	19.39	28.81
Wastewater	14.82	3.53	18.35
Town	2.57	24.71	27.29
National Carrier	0.02	21.76	21.78
Local aquifer	2.55	2.95	5.50
Initial salinity mg/l	239.82	179.64	192.17

3. Desalination Alternatives

Desalination processes can be applied to saline aquifers, to National Carrier water, or to unconventional water sources including wastewater or seawater. We examined the following desalination alternatives:

- Brackish groundwater
- National Carrier
- Treated wastewater
- Seawater

3.1. EVALUATION OF DESALINATION ALTERNATIVES

Decisions on major issues such as water supply are often made using Multi-Criteria Decision Analysis (MCDA). According to Linkov et al. [6], this relies on multidisciplinary information, incorporating natural science, physical science, economics and other social sciences, medicine, politics, and ethics. MCDA includes different methods to use information from different fields to reach a decision. Varis et al. [8] considered the trade-off of the environmental risks and the economic and social utility in evaluating river water quality. Wastewater treatment methods should be analyzed using a broad set of criteria, including social, economic, and environmental criteria. Bradley et al. [1] used social, economic, and environmental criteria to evaluate methods of wastewater treatment systems and to examine ways to reduce shortcomings through technological advancements and changes to management of the systems.

While in Haruvy [5] we focused mainly on economic considerations, in this research we combine scientific, technological, environmental, economic, social, and political considerations in choosing the optimal desalination method. For example, we consider the influence on unemployment as well as groundwater salinity in choosing a desalination method. This was done by quantitatively assessing the alternatives through a modified Delphi method of questionnaires eliciting responses from experts on desalination and water quality and supply.

Using questionnaires for experts to evaluate water resources according to multiple criteria was done by Stone et al. [7], Ghanbarpour et al. [2], and many others. The method in this paper is based on Wolf and Murakami [9]. They used a questionnaire given to water experts to evaluate desalination and other water supply methods for each measure (economic investment, economic cost-benefit, etc.), and rank them from 0 to 100; these measures were in turn given a relative weight to reach a general comparative ranking of water supply methods by multiplying the rank for each measure with its assigned weight. In our questionnaire, each expert ranks each scenario on each criterion from 0 to 100, and assigns a weight to each criterion. The weighted results were used to evaluate each scenario. The highest ranking is the most desirable, so that alternatives with higher cost have a lower ranking, and alternatives with higher benefit cost ratio have a higher ranking.

3.2. THE CRITERIA USED FOR EVALUATION

3.2.1. Economic criteria

The size of the initial investment and cost-benefit analysis. Lower costs receive higher ranking. We hypothesized that seawater desalination has a high initial investment, so it would receive a low ranking. Low-level wastewater desalination is the least expensive, in terms of initial investment, so it would get the highest ranking (Table 2). The possible range for the initial investment: from 1 (no investment) to 100 (extremely high investment), and for the cost-benefit analysis: from 1 (very low benefit/cost ratio) to 100 (extremely high benefit/cost ratio)

3.2.2. Hydrological criteria

The aquifer salinity through time. Higher grade represents lower groundwater salinity. Higher level desalination (desalination to a higher quality level) means that the water will be less saline, and therefore have a lower impact on groundwater salinity. Therefore, we hypothesized that higher level desalination will receive a higher rank as a more groundwater-friendly method (Table 2).

Table 2: Desalination cost and salinity.

Process	Investment Cent/CM	Total cost Cent/CM	Groundwater Salinity-year 100
Groundwater	13.0	36.0	716
National Carrier	14.6	29.4	453
Wastewater	3.3	41.6	357
Seawater	32.5	54.2	182

3.2.3. Technological criteria

The amount of water available (water quantity), quality, and reliability of supply. A ranking of 100 for water quantity would mean this method can provide a virtually unlimited amount of water; for water quality it would mean the method provides a very high water quality, and 100 for water source reliability would signify a completely reliable source of water supply. Seawater can potentially supply a high quantity, so we hypothesized it would get a high rank on quantity. Wastewater desalination is limited to the supply available, so it would get a low rank on quantity. The quality of desalinated seawater is high, while that of wastewater is low. The reliability of the supply source is high in seawater (always available), and lower in wastewater.

3.2.4. Environmental criteria

Energy use and greenhouse gas emissions; a high rank means very low energy use and very low greenhouse gas emissions. Seawater desalination is energy intensive and is associated with more greenhouse gas emissions, so we hypothesized it would receive a low rank. Lower quality desalination involves less energy use and emissions, so it would receive a higher rank as a more environmentally friendly method.

3.2.5. Agricultural criteria

The effect on the crop yield and quality. Higher grade represents better crop yield or quality. We hypothesized that as higher quality desalination is assumed to be better for crops, it will therefore receive a higher grade.

3.2.6. Social criteria

The effect on the employment rate. The employment in some peripheral areas of the country, where there are few alternatives available, is influenced by the cost of water supply to the agricultural sector. Higher grade represents lower negative impact on employment rate. This criterion has some correlation with economic and agricultural criteria—methods that have lower costs will have a lower impact on water price. A lower impact on water price means lower impact on agriculture, and less likelihood of people losing their jobs as farm workers due to increases in the price of water.

3.2.7. Other criteria

The effect on Israel's image in the world and on its ability to export agricultural products. Higher grade represents lower negative impact on marketing image. We hypothesized that irrigation with wastewater may have a negative image in export market and that therefore wastewater irrigation will receive a low rank on this measure.

4. Results and Discussion

Questionnaires were disseminated among experts dealing with water issues to examine the effectiveness of the questionnaire and the level of feedback. This questionnaire is still in the process of being disseminated, and further work is needed to increase the number of respondents. This paper presents the preliminary results of the questionnaire, which shed light on decision making regarding the allocation and treatment of water resources.

A major criteria for evaluation, the potential effect of water quality on human health, was omitted from the survey. Gurjar and Mohan [3] examined the short-term and chronic health effects due to water contamination. This subject should be added to future surveys regarding water treatment.

Table 3 shows the ranking according to the economic criteria.

Table 3a: Ranking of desalination alternatives according to economic criteria: treatment.

Process	Investment	Cost/benefit	Total
Weight	30%	70%	100%
Groundwater	66	73	71
National Carrier	61	82	76
Wastewater	80	51	60
Seawater	23	38	34

Table 3b: Ranking of desalination alternatives according to economic criteria: supply.

Process	Domestic threshold	Agricultural threshold	Total
Weight	40%	60%	100%
Groundwater	83	87	85
National Carrier	45	39	41
Wastewater	35	74	58
Seawater	43	27	33

This should be compared to the supply cost of desalinated level at the domestic water quality threshold and the agricultural water quality threshold (Table 4).

Table 4: Supply cost of desalinated water (million NIS).

Process	Domestic threshold	Agricultural threshold
Groundwater	380	404
National Carrier		
Wastewater		710
Seawater	385	828

The cost of seawater desalination is higher than the cost of groundwater desalination, and the ranking of seawater desalination is therefore lower on this criterion, although not in the same proportion as the real cost.

The ranking according to technological criteria is presented in Table 5.

Table 5: Ranking of desalination alternatives according to technological criteria.

Process	Quantity	Quality	Reliability	Total
Weight	23%	35%	42%	100%
Groundwater	19	60	68	54
National Carrier	32	85	75	69
Wastewater	56	27	41	40
Seawater	89	81	80	82

Seawater desalination is the highest ranked alternative, providing the greatest quantity and reliability of water, while wastewater is the lowest ranked alternative.

The ranking according to hydrological criteria is presented in Table 6.

Table 6: Ranking of desalination alternatives according to hydrological criteria.

Process	Salinity level	Additional salt	Total
Weight	34%	66%	100%
Groundwater	16	75	55
National Carrier	52	49	50
Wastewater	76	80	79
Seawater	86	31	50

Wastewater and groundwater desalination are ranked as the highest, providing the lowest amount of additional salt to the groundwater aquifers.

The ranking according to environmental criteria is presented in Table 7.

Table 7: Ranking of Desalination alternatives according to environmental criteria.

Process	Energy use	Greenhouse gas emissions	Total
Weight	57%	43%	100%
Groundwater	79	68	74
National Carrier	83	62	74
Wastewater	67	74	70
Seawater	48	25	38

National Carrier and groundwater desalination are ranked as the most environmentally friendly—the National Carrier water desalination as requiring the lowest amount of energy in the process, and wastewater desalination as causing the lowest amount of greenhouse gas emissions.

Table 8 presents the agricultural, social, and other rankings; the summary of the results of all criteria; and the final weighted rankings.

Table 8: Summary of all criteria and weighted average.

	Weight	Groundwater	National Carrier	Wastewater	Seawater
Economic treatment	13%	71	76	60	34
Economic supply	11%	85	41	58	33
Technological	16%	54	69	40	82
Hydrological	20%	55	50	79	50
Environmental	29%	74	74	70	38
Agricultural	4%	67	80	37	60
Social	6%	52	61	65	43
Other	1%	76	88	24	71
Total	100%	66	65	62	48

It can be seen that wastewater is ranked highest on the social criteria, but National Carrier water is the highest ranked alternative in combination of the last three criteria.

5. Conclusions and Recommendations

We presented a method in water allocation and treatment decision making. This should be based on assessing the various dimensions including economic, technological, hydrological, environmental, agricultural, and social. Other dimensions should also be taken into account including health and applicability.

Preliminary results in Table 8 show that the optimal alternative is brackish groundwater desalination. This alternative receives the highest rating, mainly due to the low cost of supplying the water, hydrological benefits from improving groundwater quality, and low negative environmental impacts. Desalination of National Carrier water desalination is second: its main comparative advantage is the low cost of treatment. Wastewater desalination is third: its environmental and economic advantages offset the hydrological and agricultural advantages of seawater desalination. Seawater desalination receives the lowest ranking and should be adopted only to supplement the quantity available from the other alternatives.

The results of the survey lead to the conclusion that desalination should start by exploiting brackish groundwater, which will provide a very limited quantity of water. Then desalination of National Carrier water and wastewater should be applied, and, finally, seawater desalination should be used only as a last resort. Other criteria can be taken into account, including the applicability and task of the relevant water source in the future planning of national water sources. We presented a comprehensive framework for decision making while increased survey is needed to get recommendations.

6. References

1. Bradley, B.R., Daigger, G.T., Rubin, R., Tchobanoglous, G. (2002) Evaluation of onsite wastewater treatment technologies using sustainable development criteria. *Clean Technologies and Environmental Policy*, 4(2), 87 - 99

2. Ghanbarpour, M.R., Hipel, K.W., Abbaspour, K.C. (2005), Prioritizing Long-term Watershed Management Strategies Using Group Decision Analysis. *International Journal of Water Resources Development*, 21(2), 297 – 309.

3. Gurjar, B. R. and Manju Mohan, M (2003). Potential Health Risks Due to Toxic Contamination in the Ambient Environment of Certain Indian States. *Environmental Monitoring and Assessment*, 82(2), 203 – 223.

4. Haruvy, N. (2004), Irrigation with treated wastewater in Israel- assessment of environmental aspects: 375-384 in : Linkov, I. and Ramadan, A. (eds.) "Comparative risk assessment and Environmental Decision Making" Kluwer.

5. Haruvy, N. (2005), Reuse of wastewater in agriculture- economic assessment of treatment and supply alternatives: 257-262 in: Morel, B. and Linkov, I. (eds.) "Environmental Security and Environmental Management as affecting aquifer pollution: The Role of Risk Assessment" Springer.

6. Linkov, I., Varghese, A., Jamil, S., Seager, T.P., Kiker, G., Bridges, T. (2004). Multi-Criteria Decision Analysis: A Framework for Structuring Remedial Decisions At Contaminated Sites. In Linkov, I. and Ramadan, A. (eds). *Comparative Risk Assessment and Environmental Decision Making*. Kluwer publishers, 2004, p. 15-54

7. Stone, R., Garber, W., Friedland, H. (1970). Water Quality: Costs Benefits of Irreducibles. *Journal of Sanitary Engineering Division, Proceedings of the American Society of Civil Engineers*, 96(Sa3), 691-697.

8. Varis, O' Kløve, B. and Juhani Kettunen (1993). Evaluation of a real-time monitoring system for river quality - a trade-off between risk attitudes, costs, and uncertainty. *Environmental Monitoring and Assessment*, 28(3), 201 – 213.

9. Wolf, AT & Murakami, M. (1995), Techno-political decision making for water resources development: the Jordan River watershed. *Water Resources Development*, 11(2), 147-162.

EVALUATION OF DRINKING GROUNDWATER FOR THE RURAL AREAS ADJACENT TO THE NEARBY DESERT OF GIZA GOVERNORATE OF GREATER CAIRO, EGYPT

M.M. EMARA, I. EL SABAGH, A. KOTB, A.S. TURKEY, D. HUSSEEN
Science Center for Detection and Remediation of Environmental Hazards
Al-Azhar University, Nasr City, Cairo, Egypt

Abstract

The choice of the area under investigation takes into consideration several important aspects: heavily populated villages; type of drinking water as groundwater; its situation on the west side of the Nile near the intersection point between the Nile and its Delta; its neighborhood to the desert of Giza governorate; and its closeness to highly industrial and agricultural activities. The present study is an extension of an early study on drinking groundwater in the southern part of the area under investigation, namely Kafr Hakim village. The current paper covers an area from Kafr Hakim in the south to Nekla in the north, surrounded by the Moheet drain in the east and the Mansourya canal in the west.

Twenty ground and surface water samples were taken from some wells of depths from 20 – 100 m and from some surface water drains in the same area. The exact positioning of each sample was precisely determined using GPS instruments. The samples were analyzed and environmentally characterized through various important cations such as Na, K, Ca, Mg, Cd, Zn, Pb, Fe and Mn, using atomic absorption technique. The anions determined using either spectrophotometric or ion-chromatographic methods; they included bicarbonate, sulphate, acetate, chlorides, nitrates, nitrites, and phosphates. All above-measured parameters were presented using the newly developed geographic information systems (GIS), which facilitates the presentation of final results. This manuscript sheds light on the evaluation of these water samples as potable according to international and the Egyptian regulations. Also a trial is made to explain any unusual and abnormal data with some recommendations for remediation.

1. Introduction

The Egyptian people have become very aware of the impact of the environment on drinking water. Needless to say, the relation between water quality and the is very

I. Linkov et al. (eds.), Environmental Security in Harbors and Coastal Areas, 379–394.

obvious. In an earlier study on the quality of the drinking water at Kafr Hakim village [1], the Giza governorate showed that the groundwater used for this purpose indicated the presence of some heavy element pollutants with concentrations above the allowable limits. When the problem was discussed with the decision makers in this area, they immediately abided with our recommendation at the time, namely increasing the depth of the source of the potable groundwater. This was carried out for three wells in this same village and the new depths were in the range of 60 – 90m. Also, they changed the general drainage line from an open type to a covered one.

Therefore, we decided as a research team to restudy the quality of the water after these new modifications. We also decided to extend our study to other geographical areas taking into consideration other parameters causing the source of possible pollution as well as the direction of such pollutants. Therefore we extended our monitoring to areas like Ghidan village, Abd EL Samad village, Mansouria village, Qumbera, Manchiat Radwan, Berkach, and Nekla.

The area under investigation is a part of Greater Cairo, west of the River Nile, Giza governorate (Figure 1).

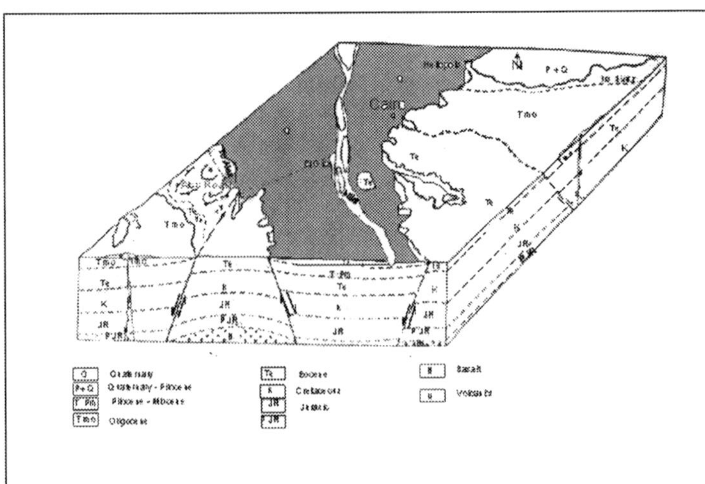

Figure 1: Study area of Greater Cairo, west of the River Nile, Giza governorate.

It covers an area of about 42.00 Km^2. The southern part of the study area is very close to the nearby desert, namely, the pyramids plateau and Abu Roash hills. The concerned rural areas are located in a floodplain that has a relative elevation of 17-21m. This floodplain occupies the banks of the River Nile. The surface of the floodplain consists of the fertile top clay-silt layer underlain by sand and gravel forming the alluvial aquifer.

2. Geologic Setting

According to the geology of Greater Cairo described by Shata [2], our study area is located on the southern tip of the Nile Delta. The Nile floodplain is formed by a tectonic depression, and probably bounded by faults. Cretaceous and Tertiary strata and Tertiary volcanics are exposed at the border of the floodplain. They are part of the uplifted Hing belt (Figure 2).

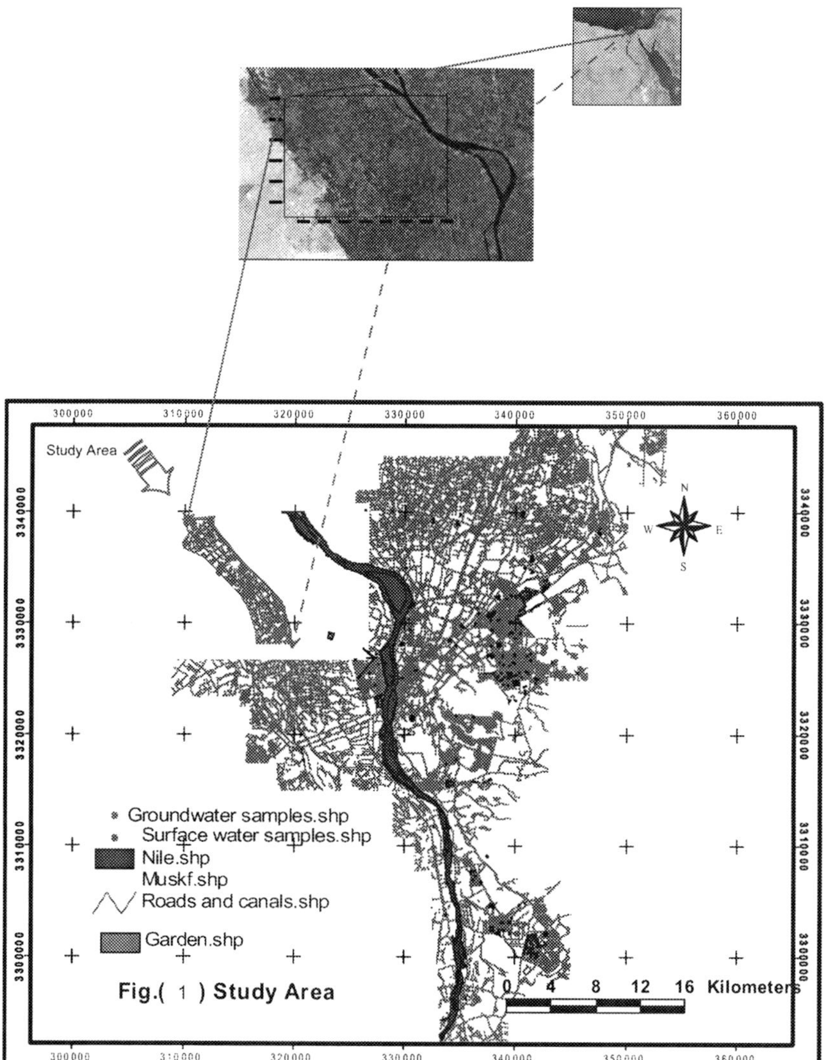

Figure 2: Modified block diagram showing the subsurface sequence in the study area (modified from [2]).

3. Groundwater Hydrology

In Greater Cairo, the main aquifer system consists of coarse massive sand and gravel intercalated by clay lenses, belonging to the late Pleistocene. The aquifer is covered by a layer of salty clay and/or fill deposits. This layer acts as a semipervious aquitard of thickness ranging from 5m to 20m and vanishing near the eastern edges of the floodplain. The aquifer thickness ranges from 20m to 140m. The real extent of the aquifer is bounded to the east and west by limestone escarpments of the Eocene and Cretaceous age [3, 4].

The general groundwater flow for the Quaternary aquifer in northward and westward directions is shown in Figure 3.

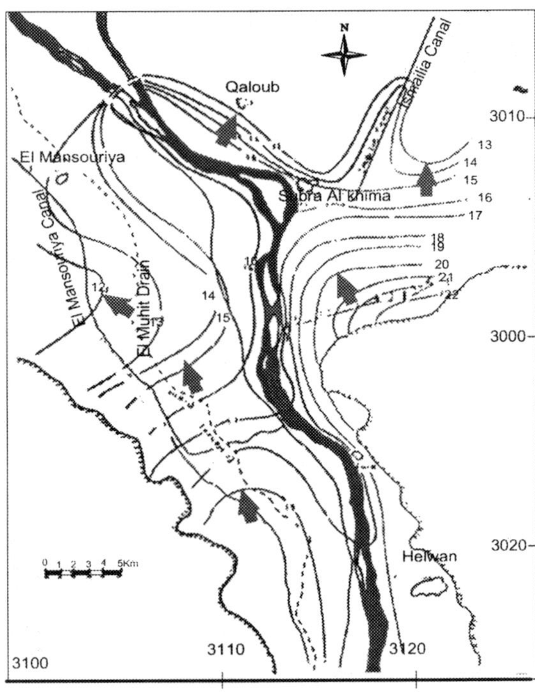

Figure 3: Modified groundwater flow map (from [3]).

The isopiesometric levels show a gradual decrease in the northward and westward directions. Two cones of depressions are locally developed, one to the north of Heliopolis and another one close to the Pyramids on the western side. The Pleistocene aquifer in the Cairo area is of infinite areal extent, as it is connected hydraulically with the same aquifer, both in the Nile valley in the southward direction and in the Delta in the northward direction [2]. The regional flow pattern in this aquifer is in the northern direction. A local anomalous flow pattern is, however, noticed within the Cairo area between Heliopolis and the Pyramids, and is either related to fault structure or to fold structure. In the study area, groundwater flows in a westward direction.

3.1. RECHARGE AND DISCHARGE

The aquifer underlying Greater Cairo is replenished through various means, depending mainly on land use. The most important sources consist of:

1. Seepage from the river and the main distribution canals.
2. Deep percolation from irrigation of agricultural lands.
3. Seepage from the drinking water supply network.
4. Infiltration/seepage from the sewage trenches/pits in unsewered regions.

On the other hand, discharge occurs as:

1. Groundwater return flow to the river.
2. Interception by the sewage system.
3. Groundwater withdrawals.

This regime affects the flow of groundwater, regionally as well as locally. Monitoring groundwater heads is thus an important means for the determination of the balance components, and direction of groundwater flow in relation to the other components of water.

3.2. GROUNDWATER USE

Groundwater, both shallow and deep, is intensively used in the Greater Cairo region. Public deep wells are generally owned by the Cairo potable water organization or local units, while private wells are used for both drinking and irrigation purposes. The private drinking wells are generally shallow and operated by hand pumps, while irrigation wells are deeper.

3.3. GROUNDWATER POLLUTION

Groundwater pollution is a result of surface as well as internal reasons. Internally, the pollution can increase due to the vulnerability of pollution and contamination by minerals. On the other hand, surface activities include various types of human activities.

3.4. INDUSTRIAL POLLUTION RISK

About 120 industrial activities and gas filling stations occupy about 1% of total Cairo land use [5]. The major industries existing in the region determine the expected pollutants in the effluent.

3.5. AGRICULTURAL POLLUTION RISK

The agricultural area covers about 15% (while the desert area and turtle back cover about 24%). The west bank of the River Nile is considered the main agricultural part of Greater Cairo together with the northern part. These agricultural zones are the main supply of vegetables for Greater Cairo. The main source of irrigation water is the Nile, while the groundwater irrigation amount is limited. As a result of agricultural activities, deep percolation from the water table to the groundwater is expected. Such water contains salts (result of evapotranspiration), fertilizer, and pesticide residues. From the water table to the groundwater, a change in the chemical content of the water occurs as a result of dilution and chemical reaction.

3.6. URBAN POLLUTION RISK

The urban area covers about 60% of the total area. The major urban areas are confined to the banks of the river and are generally served with an old sewerage system. The remaining urban areas are represented by scattered and unplanned settlements which are either served or unserved with a local network or by septic tanks.

4. Literature Review

A large number of papers have been written about Greater Cairo in general. In particular, the study area is discussed by many authors among them Hefny and Khalil [6]. They studied the hydrogeological and hydrochemical conditions of the Greater Cairo area and assumed that groundwater in the Quaternary aquifer is generally fresh to highly saline. The increase in salinity eastward is mostly due to the leaching of the older marine and fluriomarine strata located in the eastern cliffs. This is in addition to the high rate of seepage from the old sanitary network at the densely populated areas.

Awad [7] studied pollution of groundwater resources in the southern part of the Nile Delta, and its impacts on development plans. Pollution of groundwater with trace elements, especially iron, manganese in most of wells and lead in some wells was detected.

Diab [8, 9] studied groundwater occurrence in the southern sector of Alexandria-Cairo desert road. The same author studied the groundwater pollution of the Quaternary aquifer under the Nile Delta area. He found that this pollution is partly the result of agricultural fertilizers, manure, and biocides. Wastewater disposed in canals can diffuse in groundwater. The main pollutants include nitrates and phosphates, where their concentrations in the groundwater of the Quaternary aquifer are less than in surface water resources. Denitrification may be responsible for the flow concentration of nitrate in groundwater.

Lasheen et al. [10] studied the concentration of Cd, Zn, and Pb in the wastewater of the abu Zaabal fertilizer factory. Cd ranges from 2.3 to 600μg/ℓ. Zn between 4.6-646μg/ℓ, and Pb ranges from 16.2 to 600μg/ℓ were found. In the water of the Ismailia canal, the total content of Zn ranges from 1.36 to 10μg/ℓ with a mean value of 4.06μg/ℓ and Pb ranges from 0.5 to 6.9μg/ℓ with a mean value 0.76μg/ℓ. The Northwest Cairo area between Abou Rawash and Gebel Hamza area was studied by Ahmed [11] hydrogeochemically to detect the relation between surface water and groundwater in this area.

The chemical characters of surface water (represented by E-Mansouria Canal) revealed that the total dissolved solids (TDS) is 318mg/ℓ, which lies in the fresh water category. The dominant anions for this canal are in the order: $HCO_3^- > Cl^- > SO_4^{--}$, while the dominant cations are in the order of $Na^+ > Ca^{++} > Mg^{++}$. Therefore the dominant water chemical type is Na^{+} + HCO_3^{-}. For groundwater, the majority of the collected samples (east and northeast of the study area) reflect the fresh water category. The southwestern part of the study area reflects the brackish water category while the rest of the southern samples belong to the saline water category. This higher salinity may be attributed to leaching and solubility processes. Moreover, ionic exchange mixing and evaporation are also considered.

5. Aim of the Present Work

The present work is mainly devoted to evaluating the groundwater as the main source of drinking water. GIS technique is the main tool to achieve this goal.

5.1. FIELD WORK:

5.1.1. Sampling

Groundwater samples: Seventeen groundwater samples have been selected from Qumbera Kafr Hakim, Ezbet Ghidan, Ezbet Abd El Samd, Al Mansouriya, Menshat Radwan, Berqasha and Nekla in May 2004 (Figure 4). Five other groundwater samples were obtained for duplicate checks in January 2005.

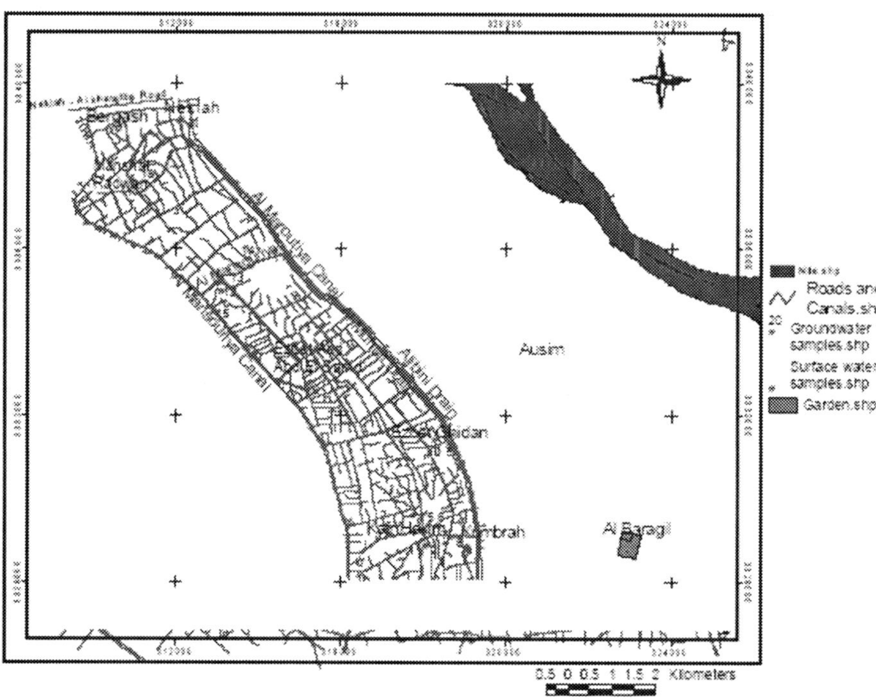

Figure 4: Well location map.

Surface water samples: Three surface water samples from the drains and canals were taken. The anions were analyzed using Ion Chromatography and the trace heavy elements were analyzed using Atomic Absorption Technique. All other determinations were carried out according to AOAC methods.

5.1.2. Instrumentation

pH and conductivity were analyzed using instruments (WTW Wissenshaftlich-Technische werkstatten G mb H Ion lab Multi Level 1, ba 12237 de) for measuring pH, conductivity, salinity, and dissolved oxygen. Other analytes were measured using with either an Atomic Absorption Spectrophotometer (Aurora, Canda) or an ion-chromatograph (D-120 Ion chromatograph [Dionex Ion pac]).

5.1.3. Procedures

The water samples were analyzed as received. The cations were obtained through wet method, spectrophotometry, or atomic absorption technique. All methods were recommended by USEPA, Egyptian EEAA, or AOAC methods.

5.1.4. Groundwater Samples

Multiple water samples were obtained from the study area. For each sample, 33 parameters were reported and each of these parameters is the average of two or three different measurements. The standard deviation within each reading is within the already available experimental error of the corresponding method. In general the errors in each case never exceeded 1 to 2%.

5.2. DISCUSSION OF RESULTS

All the above-measured parameters were presented using the new Geographic Information System (GIS). Arc GIS software was used to create the spatial distribution maps and analyze the different data obtained.

In order to handle remote geographic data with other geological and hydrogeological data, GIS was applied. GIS is a computer software package that allows the user to handle large amounts of spatially distributed data by combining a database management system with digital mapping capabilities. The ability to analyze and manipulate data in both spatial and tabular form makes GIS more powerful than a traditional tabular data system [12].

5.3. GIS AND DECISION MAPS:

5.3.1. GIS Map of Cadmium Concentration

We paid special attention to Cd as a trace element due to its environmental hazards and potential risk to human health. The previous study [1] on the southern part of investigated area (Kafer Hakim) revealed a high concentration above the allowable limits (0.005 ppm according to EPA and WHO). Cd concentration is ranged between 0.002 and 0.01 ppm. The first survey, in May 2004, indicated high concentrations at both Qumbera (in the east southern part of the study area) and Berkash (located in the west southern part). Kafer Hakim showed no increase in this very toxic element, (Figure 5), which leads to renal and liver failure.

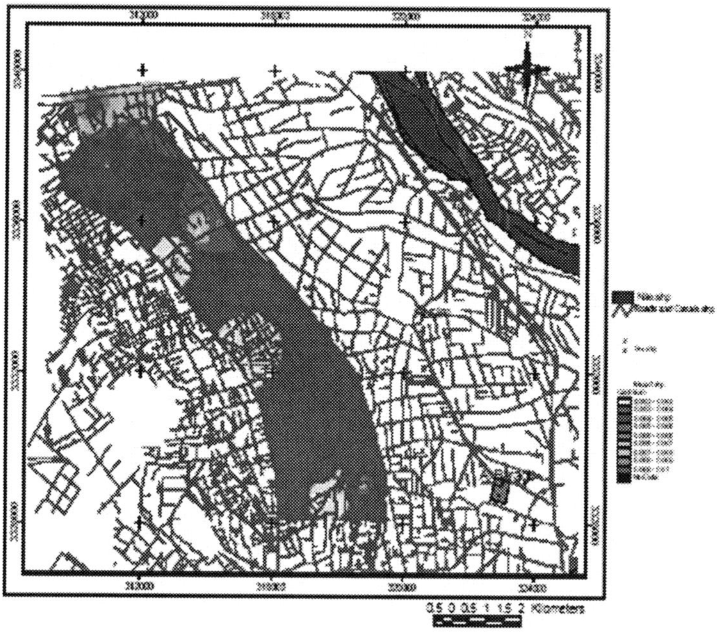

Figure 5: Distribution of cadmium (May 2004).

In Kafer Hakim, the governmental responsibles drilled three wells in the same area in response to community demand. Analysis of the groundwater for these three wells unexpectedly revealed worse results than for surface water (Figure 6).

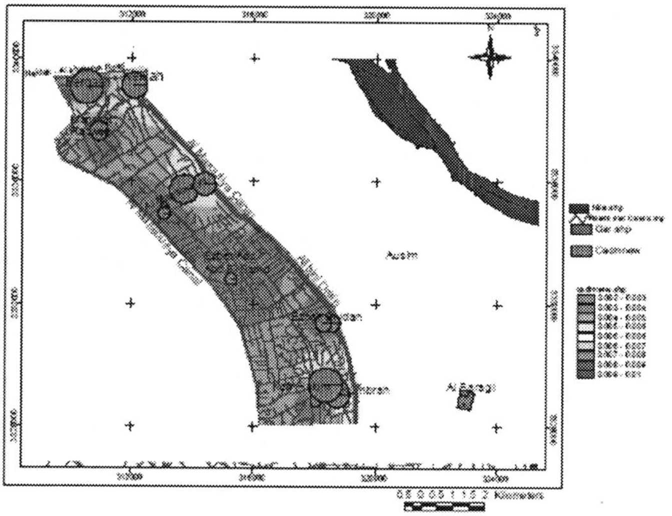

Figure 6: Groundwater cadmium results.

Concentrations of Cd [13, 14] as a trace element along the whole area show some differences. To the south, faulting and folding structures as well as Tertiary aged basalts exposed at Abou Rawash, which are very close to Quaternary aquifer, may be responsible for high Cd concentration in Qumbera. Although the sudden increase of Cd concentration in Berkash seems obscure, there are many reasons that may combine to explain this increase. Fertilizers and pesticides, open drainage and sewage, and leaching from the aquifer body strata itself, where Cd may be associated with alluvial deposits, all may give a convenient explanation especially this are generally represented as a meeting point of Nile River with its Delta. The idea of throwing down the alluvial load by Nile River when it loss its velocity and strength may enhance the deposition of heavy metals in alluvial deposits.

5.4. GIS MAP OF NITRATE CONCENTRATION

Nitrate concentration ranges between 0 and 32.9 ppm, with the highest value recorded at Ghidan village (Figure 7). This may be attributed to the presence of a canal linking El Muheet drain and El-Mansouria Canal. In addition, parallel drains (El-Libini and El- Muheet), which receive all both agricultural drainage and sewage drainage from Zenien plant, represent a dangerous source of possible nitrate contamination across percolation to the ground aquifer system. The allowable limit for nitrate is 10 to 50 ppm (according to WHO and EPA).

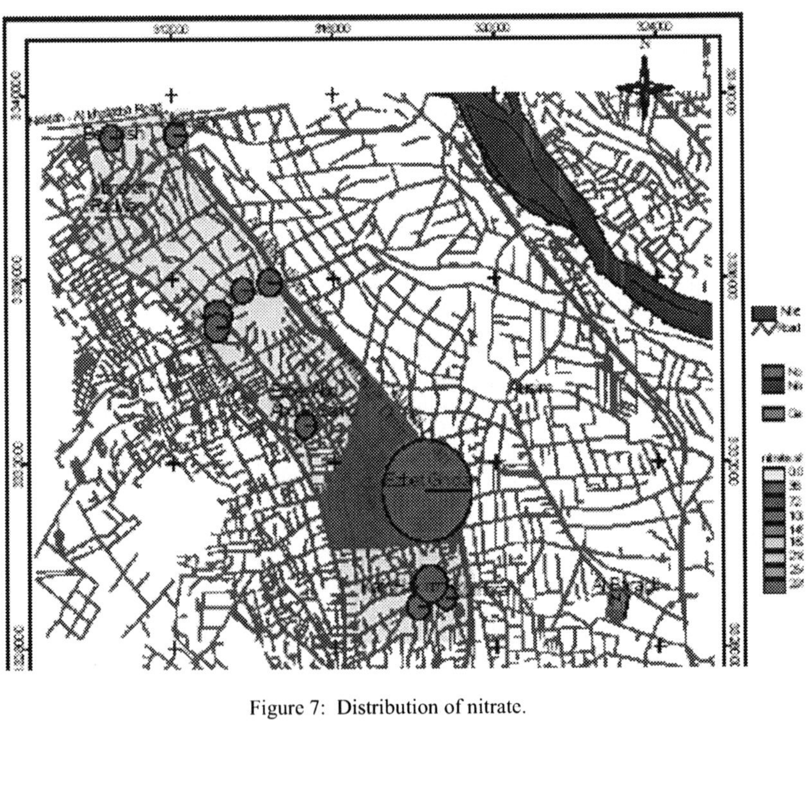

Figure 7: Distribution of nitrate.

5.5. GIS MAP OF MANGANESE CONCENTRATION

The southern part of the investigated area, which includes Qumbera, Kafer Hakim, and Ghidon villages, recorded the highest values for manganese (Figure 8).

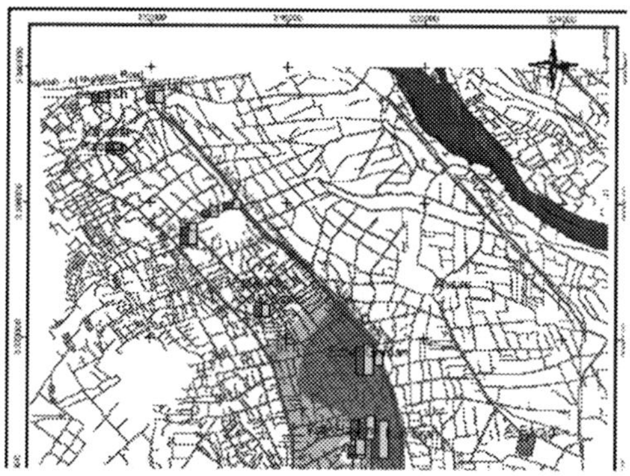

Figure 8: Distribution of manganese.

The range of manganese concentration is 0.058 to 2.78 ppm and the allowable limit is 0.05 ppm (WHO and EPA). An increase has been observed beside the complicated drain network where the three villages of Qumbera, Kafer Hakim, and Ghidan are located.

5.6. GIS MAP OF IRON CONCENTRATION

The iron concentration value lies within the range of 0.141 and 1.704 ppm (Figure 9) while the allowable limit is 0.3 ppm according to WHO and EPA. The highest values of iron are observed at the southern part of the study area (Qumbera, Kafr Hakim, and Ghidan) and the El Mansuria well.

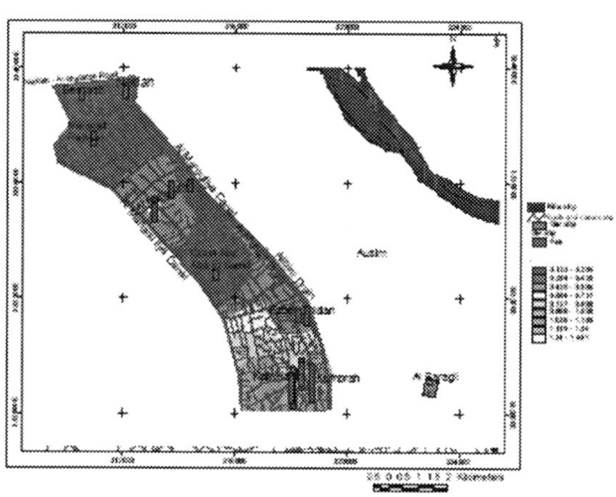

Figure 9: Distribution of iron.

This increase may be ascribed to the complicated drain network that feeds the aquifer with iron through a penetration process many long years ago. Also, we can add: the movement of contaminants during the up-down flow and south-north flow where industrial activities are conducted in the southern part of the Giza governorate.

5.7. GIS MAP OF LEAD CONCENTRATION

The recorded values of lead concentrations are found to be up to 0.15 ppm, while the allowable limit is 0.01 ppm. The highest value was recorded at Abd El-Samad village (Figure 10).

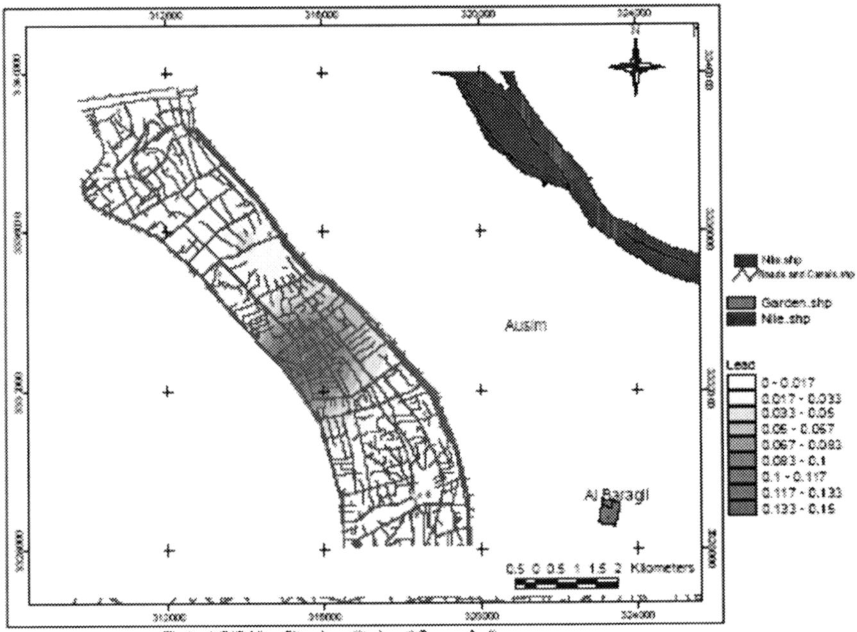

Figure 10: Distribution of lead.

5.8. GIS MAP OF TOTAL DISSOLVED SOLIDS

Distribution of TDS along the investigated area shows an observed high concentration along the southern part of the investigated area (Figure 11).

The range of TDS values lies between 363 and 1169 ppm while the allowable limit is 500 ppm. Structural geology and nearness from Eocene limestone may play a big role in increasing TDS values.

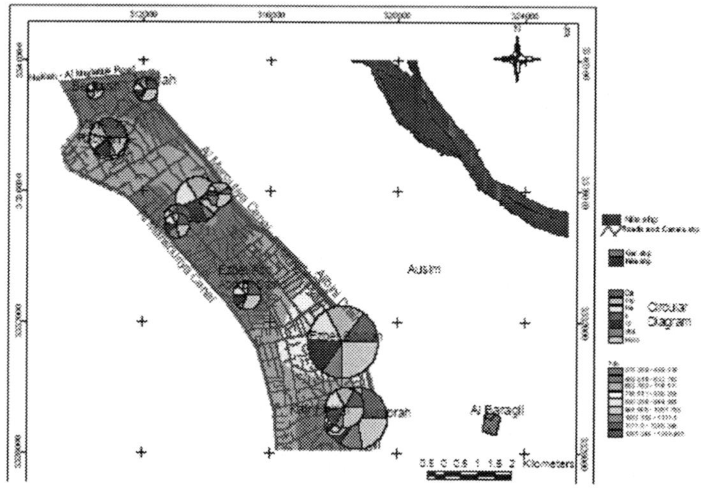

Figure 11: Distribution of total dissolved solids (TDS).

The circular diagram in Figure 12 shows an increase in Cl- and SO4 in the southern part of the investigated area, where the aquifer thickness decreases and the effect of marine deposits increase under the Quaternary aquifer. Figure 12 shows the proportional relationship between the TDS and the summation of trace elements concentrations (Cd, Mn, Pb, and Fe).

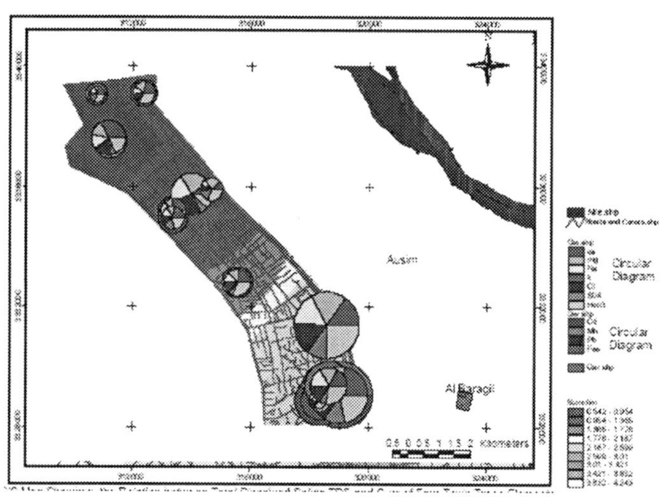

Figure 12: Relation between TDS and sum of selected toxic trace elements (decision map).

5.9. DECISION MAP

The integration of individual maps forms the decision map (Figure 13). The sum of Cd, Mn, Pb, and Fe is presented in the map to illustrate that the more risky area is the southern part. This may lead the scientific team to start water treatment in that area. Additionally, the governmental decision maker is urged to supply these rural areas with more healthy water till the treatment is carried out.

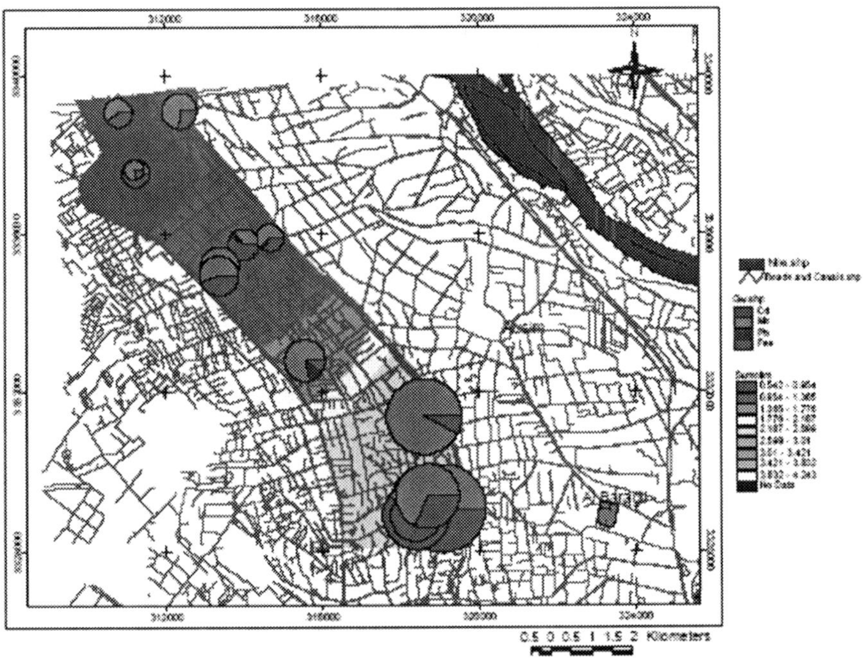

Figure 13: GIS decision map.

6. Conclusions

Evaluation of drinking groundwater quality as illustrated by GIS maps showed that cadmium, manganese, lead, iron, nitrate, and total dissolved solids concentrations in some wells are over the recommended limits.

The southern part of the study area (Kafr Hakim and Ezbet Ghidan) is the more risky area. This area is close to the structural plateau of Abu Roash, where basaltic flows could be having an effect, impacting the water-bearing formation in relation to aquifer thickness as well as water quality. Leaching of water with faulted limy rocks as well as extrusive rocks such basalt may also affect the water quality. High concentration of sulphate and chloride in these areas may be attributed to the limy formations in such areas. Trace elements such as cadmium, manganese, lead, and iron may be affected by the subsurface and structural situation for this area.

Industrial activities, wastewater drains, and fertilizers are the main sources of surface water and groundwater pollution. Hand pump and governmental drilled wells are situated in almost porous and permeable water bearing formations of Quaternary aquifer. This condition is accelerating the migration of pollutants within the study area.

7. Recommendations

This work prompted several recommendations to local authorities to improve environmental conditions and to implement an effective monitoring program. The recommendations included the following:

- Drilling sites in Kafr Hakim should be approved before drilling using the results of piesometer measurements other than those from more costly productive wells.
- The Zenien water treatment plan to drain its water to the open channel drain should be abandoned and not implemented.
- Under drainage system for the severe wastewater drains should be constructed.
- Treatment of drinking groundwater for the areas of Kafr Hakim and Ezbet Ghidan is strongly recommended.
- Complete periodic measurement of soil, plant, and water parameters as well as an environmental plan covering all medical and social surveys are recommended.
- Reconstruction of the sanitary system should be conducted using the tube system instead of drains to avoid water contamination.
- Villagers should be instructed to substitute groundwater with pure water until water purification methods are available.
- Screening tests should be used for early detection of diseases to avoid complications.

8. References

1. a. El Beialy,A.;Bersi,A.;Abdulmohesen,sh.;Kotb,A.;Emara,M.,(2003),Cadmium Toxicity and Human Health Impact:A possible Cause of Osteomalsia and Osteoporosis in Pollutet Areas, Sc.J., Az.,Med.,Fac.(Girls), Vol.24.No.2,May,2003:1069:1078.

b. Kotb, A.; Nasr, I. M. and El-Beialy, A. (2004): Groundwater pollution and its health impacts, El-Giza Governorate, Nile valley, Egypt. Egyt. J. Biophys. Biomed Engng.vol.5. No.1 pp.59-72.

2. Shata A.A. (1988): Bulletin of the Association of Engineering Geologists." Vol xxv, No2, pp. 149-183

3. RIGW (1989) "Hydrogeological Map of Egypt", National water Research Center, El-Kanater, Egypt.

4. El-Arabi N.E. (1999), "Problems of Groundwater Quality Related to the Urban Environment in Greater Cairo" Proceedings of International Symposium "Impacts of Urban Growth on Surface water and Groundwater Quality. (UGG)", Burmingham, UK, 18-30 July 1999. IAHS Publication no. 259, ISBN 1-901502-06-6, pp. 29-39.

5. Platenburg R.J., Hoencamp T.E. and El-Arabi N.E. (1997), "Soil quality management at industrial sites: Dutch experiences relevant for Egyptian context". In: Proc. Environment 97. First International Conference and Trade fair on Environmental Management and Technologies, (February 1997, Cairo. Egypt), vol 2.

6. Hefny K. and Khalil J.B. (1989), "General Hydrogeological conditions of Greater Cairo Area" J. of water Science, Water Research Center. 5 th Issue.

7. Awad S.R. (1993), Pollution of Groundwater Resources in the Southern Part of the Nile Delta, Egypt, and its Impacts on Development Plans". M.Sc. thesis. Fac. of Sc. Menoufia University.

8. Diab M.S., Abd El-Latif T. and Abd El-Baki A.A (1982), "Groundwater Occurrences in the southern sector of Cairo Alexandria desert Road" Ann. of Geol. Surv. Egypt, vol. x, p. 797-805.

9. Diab M.S. (1982) "Groundwater Pollution in the Quaternary Aquifer under the Nile Delta, Egypt". J. Geol, special volume, Part 2, pp. 135-144.

10. Lasheen M.R., Abdel Shafy H.I. and Ashmawy A.M. (1982), "Trace Metals in Ismailia canal in D. Stukey and A. Hamza (eds) Management of Industrail wastewater in Developing Nations." Proceedings of the International Symposium, Alex., March.

11. Ahmed, K. A. (2002), Hydrogeological Studies on the Groundwater aquifers in the northwest Cairo between Abou Rawash and Gebel Hamza", M.Sc. thesis, Al-Azhar University.

12. Lanfear, K. (1989), Guest editorial "Geological Information System and water resources applications". Water Resources. Bull. Vol. 25, pp. v-vi.

13. Emara, M. M., Kotb, A., El Beialy,A and Bersy, A., (2004) Evaluation of Bottled Drinking Water in Egypt, Biannual Conference on Chemistry (Chem03), Fac. of sc., Cairo, univ., Vol. 1, pp. 288-304.

14. Tolba , M. E., Kotb, A.M. and El-Beialy, A.(2004)Environmental Assessment of Groundwater in Qaret Um El-Saghir, North Western Desert, Egypt and its Health Effects, Egypt. Egyt. J. Biophys. Biomed Engng.Vol. 5, pp. 87-103.

IMPACT OF LITHUANIAN SEA HARBORS ON ENVIRONMENTAL STABILITY AND LOCAL ECOSYSTEMS

G. IGNATAVIČIUS
University of Vilniu
Center for Environmental Studies, Lithuania

V. VILUTIENĖ
Military Academy of Lithuania

Abstract

The Lithuanian continental seaside track is not long but there are plenty of wonderful places where people cut themselves from the daily life in city, to have a rest and spend their leisure. This region is strategically important for Lithuania from an economic point of view, as well as from a recreational and environmental perspective. In Lithuania there is only one state seaport and the Būtingė Oil Terminal. Most pollution incidents happen in these two places and, therefore, it is easier to make controls and find polluters. Environmental monitoring on Lithuania Baltic Sea coastal waters is carried out by two different types of monitoring programs: the National Environmental Monitoring Programme and the monitoring programs of Economic Units.

1. Introduction

The coast of Lithuania along the Baltic Sea extends 99 kilometers. There are well preserved sandy beaches that kept their primeval origin. 50 kilometers from the seacoast, the depth is 50 meters and the sea is shallow in comparison to the deepest places that reach up to 200 meters. Administratively, it is a dependency of Klaipėda region. The Lithuanian continental seaside track is not long but there are plenty of wonderful places where people cut themselves from the daily life in city, to have a rest and spend their leisure.

This region is strategically important for Lithuania from an economical point of view, as well as from a recreational and environmental perspective. The following factors could be identified as important for the environment and value of this region: water pollution, improvement of shipping and oil transportation, invasive species, and tourism. There are two especially environmentally important and

I. Linkov et al. (eds.), Environmental Security in Harbors and Coastal Areas, 395–407.

governmentally protected areas: Kuršių Nerija national park and Seaside regional Park (Figure 1).

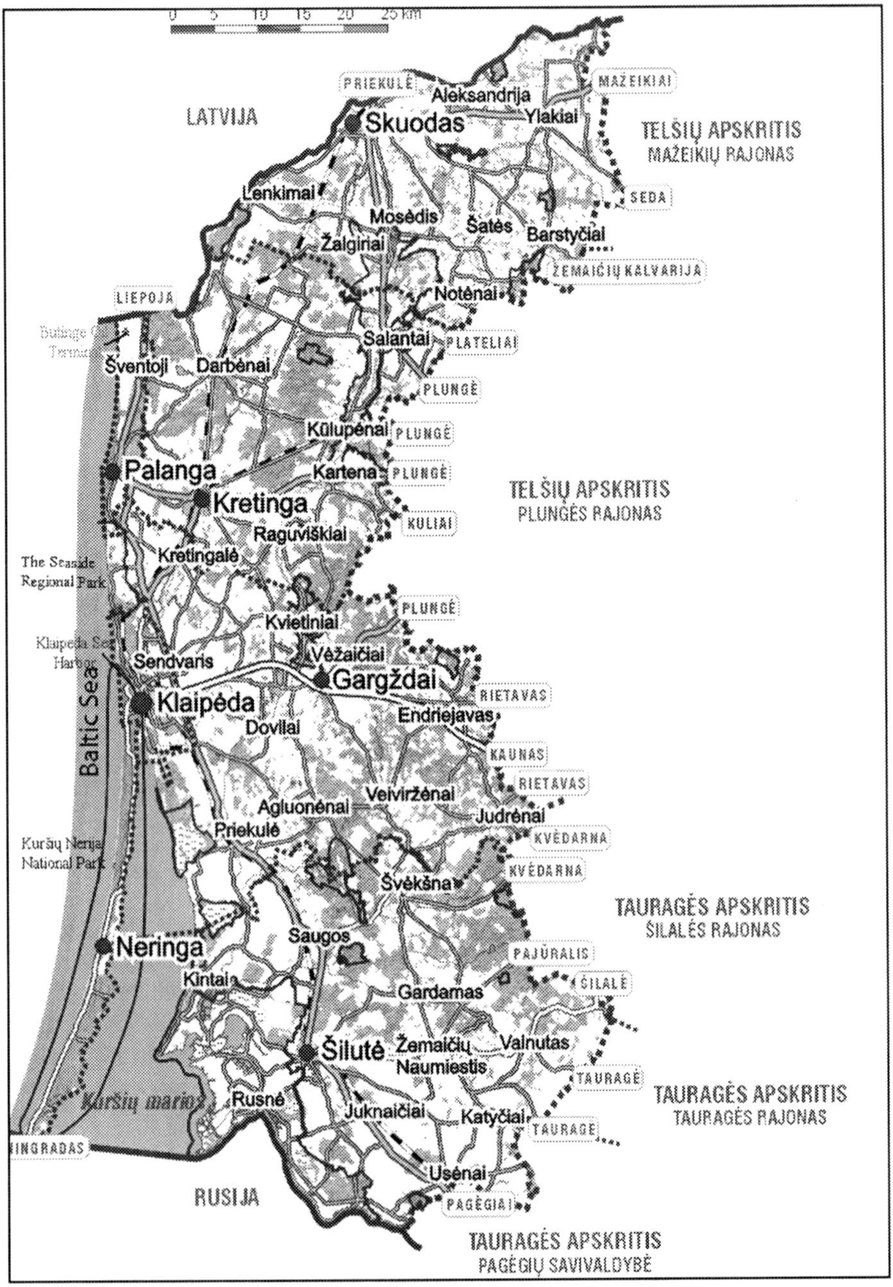

Figure 1: Lithuanian Baltic seacoast.

According to the Law of Protected Areas in the Republic of Lithuania, Kursiu Nerija National Park is protected by the state. In the classification of IUCN (The World Conservation Union), KNNP has been recognized as Category II. Kursiu Nerija National Park has been a member of EUROPARC federation since 1997. In 2000, the Kursiu Nerija National Park was included on the UNESCO World Heritage List as a valuable cultural landscape. There are 16 conservational areas, including two strict nature reserves and four nature conservation reserves. An estimated 900 plant species grow in the Kursiu Nerija National Park, with 31 of them on the Red List of Lithuania.

The Seaside Regional Park is governmentally protected territory of the Lithuanian seaside from Klaipėda city to Old Palanga. The park takes up 5,033 ha overland and 30 km in the water area of the sea. Seaside (Karkle) Regional Park was established in order to protect beautiful landscapes of the seaside, natural and cultural values, and biological Baltic Sea diversity; re-establish destroyed or broken natural and cultural values; and at the same time make conditions for cognitive tourism and lounge development.

Specific economic features of Klaipėda region are mostly conditioned by Klaipėda port, which is the only distant port in the North and which never freezes all year round in the East side of the Baltic sea.

Klaipėda port has big shipment capabilities in comparison to other ports nearby. Its capacity reaches 25-30 million tons per year. In the near future, the establishment of a modern logistics center next to Klaipėda port together with flexible multimodal transport operations, packing, and categorizing services, will turn the port into one of the best strategic economic engines in the northern part of the Baltic sea.

The Būtingė Terminal is the newest facility of Mazeikiu Nafta, situated in a year-round ice-free area of the Baltic Sea. Its history started in 1995, when the company Būtingės Nafta was established for the purpose of constructing and operating the terminal. In 1998, Būtingės Nafta was merged into Mazeikiu Nafta. The terminal can export up to 14 million tons of crude oil a year. As an import and export terminal, it is capable of not only exporting crude oil but also accepting import cargoes.

2. Pressure on Environment

Every year, there are increases in crude oil and oil production transportation. New oil terminals were constructed in the Baltic Sea; Būtingė, Primorsk, and Vysock extend oil terminals in the main East Baltic ports. Navigational risk assessments are very important for finding legal and organizational solutions to decrease navigational and environmental risk.

Keeping in mind that every year the number of incoming ships to the Būtingė Terminal is increasing and that the volume of oil loading is growing, the Būtingė Oil Terminal is a rather risky business in terms of pollution. Analysis of annual reports indicates problems that need to be solved: improvement of ship inspections and development of technical control possibilities.

The Baltic Sea at the Port of Klaipėda is in the zone of direct impact of continental runoff, where dissolved and particulate matter from the Curonian lagoon disperses. This water area is distinguished for active mixing of different chemical types of water, high input of organic matter, nutrients, oil products, heavy metals, and active biogeochemical processes.

The more dangerous risk for the environmental stability of Lithuanian Baltic Sea waters and beaches arises from water pollution by oil products.

Another very important impact factor on the marine environment is soil dumping in the Baltic Sea. Investigations revealed increasing deformations of the sea bottom in the dump sites. The concentrations of total organic carbon and total hydrocarbons are 1.2 and 2.2 times and heavy metals of nickel 2.3 times as high in the dumpsites as in the surrounding areas.

3. State of Environmental Monitoring on Lithuanian Baltic Sea Coast

The system of environmental monitoring on Lithuania Baltic Sea coastal waters are carried out along two different types of monitoring programs: The National Environmental Monitoring Programme and monitoring programs of Economics Units. Observations of the Baltic Sea along the National Environmental Monitoring Programme carried out for more than two decades enabled selection of the optimal network of monitoring stations, covering Lithuania's coastal waters and the open sea.

The Baltic Sea Monitoring Programme is closely connected with the Helsinki Commission (HELCOM) Programme carried out by the countries of the Baltic region and is maximally adjusted to national and international requirements, although general optimization of the marine monitoring system is still necessary. The goals of the National Environmental Monitoring Programme in this region are to:

- Collect and provide information on the natural processes in the coastal zone of the Baltic Sea and in the open sea.
- Collect and provide data on eutrophication levels and impact on the biota.
- Establish concentrations of the pollutants in the water and the sediments as well as their accumulation in the aquatic organisms.
- Establish biodiversity of animal species and population status.
- Prepare models for the assessment of long-term changes.
- The monitoring stations and the selected oceanological, hydrochemical, and hydrobiolological parameters ensure the observations of natural processes and the impact of human activities. The stations are distributed so that they provide comprehensive information on the status of the sea, the coastal zone, and the most polluted sites within the sea. Marine monitoring is based on 32 stations (Figure 2).

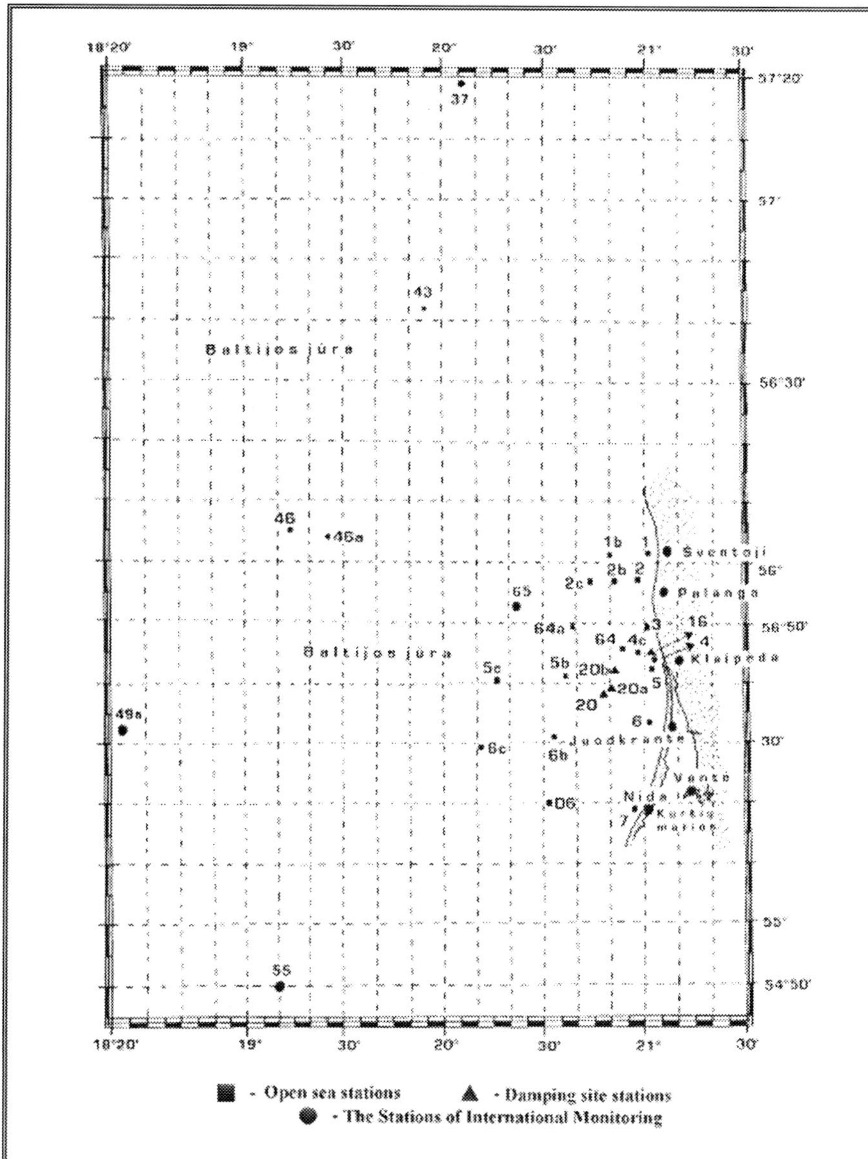

Figure 2: The network of National environmental monitoring stations in Lithuania's Baltic Sea zone.

Twenty-eight of them belong to the national monitoring network (1, 1b, 2, 2b, 2c, 3, 4, 4c, 5, 5c, 6, 6b, 6c, 6d, 7, 16, 20, 20a, 20b, 43, 46, 46a, 64, 64a, 65, 1K, 4K, 7K), and four to the international HELCOM Programme (J2-65, J1-37, K1-49a, L1-55). The national monitoring network includes:

- Ten open sea stations (46, 46a, 2c, 64a, 5b, 5c, 6b, 6c, D6, 43). Sampling rate: four times a year (once per season.)
- Fifteen coastal zone stations (1, 1b, 2, 2b, 3, 4, 4c, 16, 64, 5, 6, 7, 20, 20a, 20b).

This part of the network also includes:

- Four dumping stations (16, 20, 20a, 20b). Sampling rate: six times a year.

Three stations of the recreational zone or "hot spots" (1K, 4K, 7K). Sampling rate: 16 times a year.

The monitoring system of Baltic coastal zone dynamics consists of the network of measuring stations covering Lithuania's Baltic seacoast. The stations are set to represent fragments of the coastline that have the same morpholitological characteristics.

4. Pollution by Oil Products

Oil products are considered one of the most common polluting substances. In the natural environment they form thousands of organic compounds with different characteristics and different impacts on the marine environment: from nonharmful aliphatic to toxic and partly poisonous polycyclic aromatic and heterocyclic compounds.

In Lithuania there is only one state seaport and the Būtingė Oil Terminal. Most pollution incidents happen in these two places and, therefore, it is easier to make controls and find polluters. For almost all of the pollution incidents in Klaipėda State Seaport, the polluters are identified.

Till now the situation with oil pollutants in Lithuanian Baltic Sea zone is quite good. In 2003 pollution of the Baltic Sea region with oil hydrocarbons fluctuated from 0.00 mg/l to 0.16 mg/l. The highest concentrations exceeding the maximum allowable limits (0.05 mg/l) were observed in spring near Nida, Smiltyne, Melnrage, and Sventoji. Higher amounts of oil hydrocarbons were observed in the ground dumping site in the winter and summer. No difference was observed comparing the littoral zone with the open sea and the dumping region. Higher concentrations of pollutants were found only episodically with no major impact on the state of environment.

Basically, in the last 10 years, water pollution with oil hydrocarbons in the Baltic Sea remained the same. Increased concentrations of oil hydrocarbons were only observed in 1996-1997 (Figure 3).

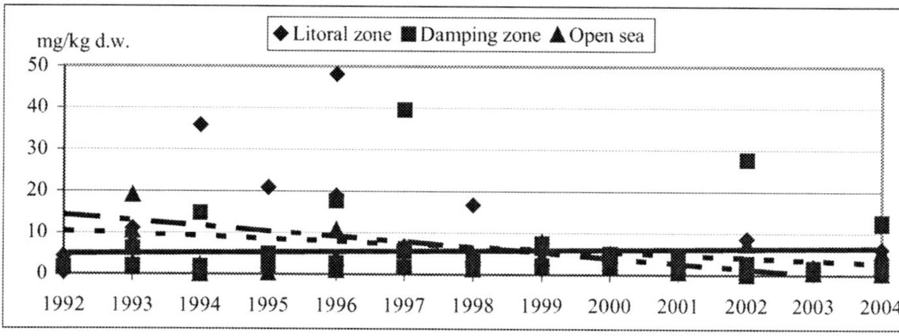

Figure 3: Concentrations of oil hydrocarbons in the Baltic Sea water in 1992-2004.

There was no pollution of the Baltic Sea bottom sediment with oil hydrocarbons in spring and autumn (<20 mg/kg of dry soil) (Figure 4). Based on the amounts of oil hydrocarbons observed in the bottom sediment of the dumping region in summer (20,61-40,48 mg/kg of dry soil), the soil in this region can be assigned to the second class of pollution.

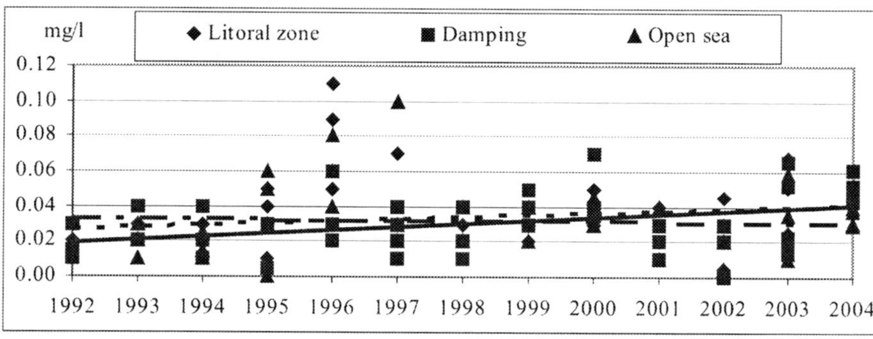

Figure 4: Amount of oil hydrocarbons in bottom sediments in 1992-2004.

In 1993-2002, bottom pollution with oil hydrocarbons remained the same in the dumping region and the littoral zone, and decreased in the open sea. Nevertheless, according to the results of chemical analysis, the amounts of oil hydrocarbons exceeding MAC and noncharacteristic to the eastern coastal zone of the Baltic Sea were observed in some areas of oil collection and drift of oil slick These amounts are caused by pollution due to anthropogenic impact and provides an assessment that operation of Būtingė terminal has direct connection with these pollutants reaching the marine environment. In addition, unlikely small concentrations of oil hydrocarbons in the surface water layer of the buoy area prove that oil dispergents were used. Assessment of impact of these materials to the environment is very contradictory. Pollution with oil is most often noticed in places where there is an intensive human activity, which is connected to shipping, shipbuilding, and repair and in places where there are many ships. There are three large dockyards and seven stevedoring companies in Klaipėda seaport territory. From 2000 to 2003, 187 reports of Baltic Sea pollution in the Lithuanian coast and Klaipėda seaport area were recorded. It is worth mentioning that the number of registered pollution cases has decreased: from 76 in 2000 to 12 in 2003.

4.1. INTRODUCED SPECIES

According to the data base of introduced species in the Baltic Sea, three zooplankton species acclimatized in the Baltic Sea: *Acartia tonsa, Ameira divagans* and *Cercopagispengoi.* These species were brought into the Baltic Sea with ballast water. According to zooplankton monitoring data of the Marine Research Center, only two introduced zooplankton species have been observed in the waters of our coastline: *A. tonsa* and *C. pengoi. A. tonsa* is a representative of euryhalinic waters mostly developing in estuaries. Firstly, this type of species was observed in zooplankton in of the Baltic Sea in 1925 and in 1952 *A. tonsa* was observed in Aistmares. Today *A. tonsa* is very abundant in the Baltic Sea, including bays and estuaries. It has perfectly acclimatized and has taken its ecological niche among zooplankters-filtrators. Due to certain difficulties in identification of this species, *A. tonsa* was described in the Lithuanian zone for the first time in 1999. Since 1999, during all seasons *A. tonsa* is constantly observed in small quantities (about 1% of total abundance) in zooplankton samples. *A. tonsa* constitutes from 0.09 to 5.1 thou. units/m3 in aquatory of the littoral zone and 0,7 to 2,3 thou. units/m3 in the open sea. There were no special changes observed in the Baltic Sea ecosystem in terms of the introduction of such species. The summer organism of zooplankton *C. pengoi* is an alien species from the Caspian Sea. It appears in phytoplankton when the water temperature reaches 17°C and disappears when the water temperature drops down to 13°C. This species was first observed in the Baltic Sea region in 1992 in Riga Bay, and in September 1999 in the littoral zone new Būtingė of the Lithuanian territorial waters. The density of *C. pengoi* organisms in this aquatory is so large that it obstructs the nets with individuals of this species and stops the fishery. Since 1999 C. *pengoi* is a constant element of the summer zooplankton association at the Lithuanian coast, and its abundance is close to 54.8 to 79.7 thou. units/m3 in zooplankton monitoring stations located in the regions of littoral zone, and 0.06 to 0.09 thou. units/m3 in the open sea. During the past two years *Cercopagis pengoi* has taken a stable position in the summer zooplankton — it constitutes 4 to 5% of the total abundance of mezozooplankton.

Seventeen introduced species of bottom fauna have been observed in the waters of the Lithuanian coastline: one species of hydroids and one of multi-bristle worms, 11 species of crustacean and four species of limpets. Most of the introduced benthos species at the Lithuanian coast originated in much earlier times: the 18th, 19th, or even 13th centuries. Some species originated at the beginning of the 20th century, 1 r — *Marenzelleria viridis*, from the Polychaeta class, originated in about 1985 in Lithuanian waters. Most of these species came to the Baltic region from the Ponto-Caspian basin attached to the hulls of ships or ballast waters.

5. Ship Control

Būtingė Oil Terminal in Lithuania (about 7 km from the coast line) does not have ship waste and ship cargo residue reception facilities. Therefore, prior to heading to Būtingė Oil Terminal tankers have to confirm that wastes and ship cargo residues were disposed of at the last port of call and/or that ship waste and ship cargo residues reservoirs are within 25 percent of total capacity.

A major reason for the practice of illegal dumping at sea, especially of fuel residues, is that considerable cost is incurred by disposal in port and that such cost is borne by ship operators. To create an incentive for disposing of the wastes in port reception facilities in spite of the costs incurred, the costs of disposal are not charged to any single ship but apportioned to all ships entering the port. Recommendation 19/8 of the Helsinki Convention therefore provides that the costs of waste disposal in ports are to be charged in accordance with a "no-special-fee-system". The aggregated cost of disposal is calculated by the port authority and is charged to each ship as a lump sum as part of the ship station fees regardless of whether the ship uses the disposal facilities. Therefore, the ships must always pay a waste charge regardless of whether ship-generated wastes are delivered or not.

At Klaipėda Seaport in Lithuania, ballast water, rinse water, chemically polluted water, residues of paint, and quarantine (infected) waste are managed under ship agreements with appropriate stevedoring companies with "no-special-fee-system" not applied. Part of this waste is classified as hazardous waste and, therefore, there is a risk that such waste can be left in the territory of the port illegally.

The Klaipėda seaport is constantly developing: some companies are introducing international quality and environmental standards. Nevertheless, with such intense activity in the Klaipėda seaport area, the Marine Environment Protection Agency has recorded pollution cases often. Selectively, the State Control of the Republic of Lithuania checked the information on the responses to pollution: in all cases necessary investigation was conducted, collecting of pollutants was arranged, but not in all the cases the monitoring institutions managed to identify polluters and to establish damage to nature (Figure 5).

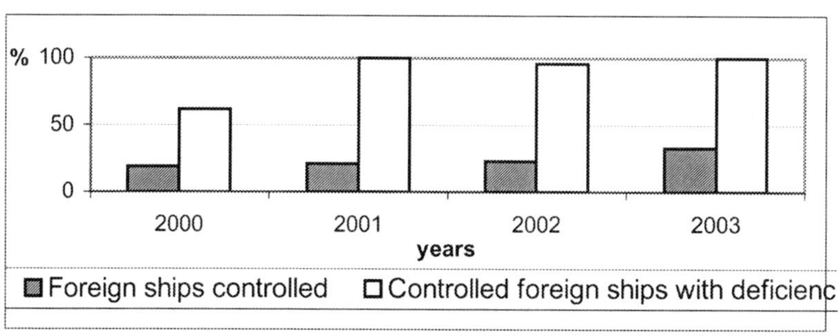

Figure 5: Oil spills in Baltic sea Lithuanian territorial waters in 2000 – 2003.

In Lithuania there is a large percent of controlled foreign ships with deficiencies (Figure 6). In Lithuania also ships with small deficiencies are held back in the port until the deficiencies are fixed. In Lithuania, the legislation does not involve fines for violation of the safety navigation rules, but the Government of the Republic of Lithuania is preparing the adjustments for the imposition of fines for violation of the safety navigation rules.

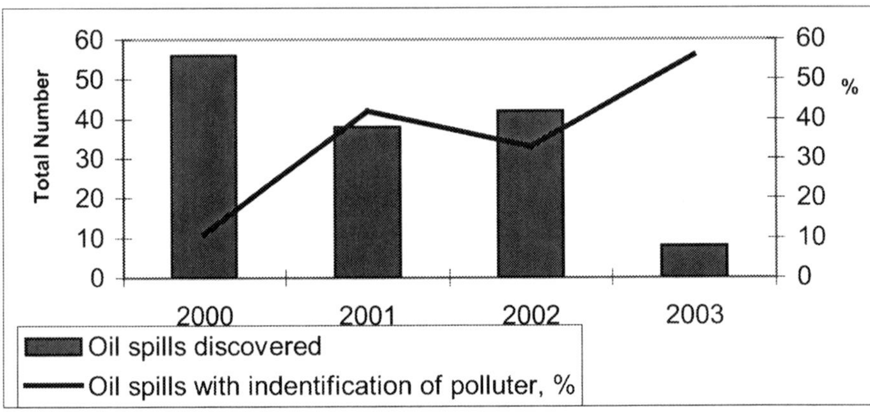

Figure 6: Results of ships control in Klaipėda Sea port and Būtingė oil terminal in 2000 – 2003.

In Lithuania the National Contingency Plan for Pollution Incidents at Sea is mainly designed for preparations for response to spillage of oil and petroleum. It does not cover pollution by other substances and no pollution prevention measures have been provided for in the plan.

6. Pollution Prevention

According to the National Oil Spillage Contingency Plan for Lithuania, pollution incidents in the Baltic Sea are divided into three levels depending on their extent and management scheme. The most complicated are third-level incidents of large extent. In the case of such incidents Lithuania has a right to ask for international aid from other states. In the same case forces of economic entities are mobilized, Emergency Response Committee to Oil Spills is immediately summoned and all the information is constantly provided to Emergency Management Centre.

In November 2001, a pollution incident in the Baltic Sea occurred while loading oil to a tanker in the Būtingė Oil Terminal. In extreme meteorological conditions and after a crack in an underwater feed pipe, 59 tons of oil was spilled into the sea. A few months earlier, a pollution incident occurred in the same terminal, but it was of a smaller extent: the oil spill was around three tons.

In 2001, 5 million tons of oil were loaded in the Būtingė Oil Terminal, 50 incoming ships were registered, and 79 percent of the incoming ships were inspected. Control of buoy and ships was carried out following the Lithuanian Law on Environmental Protection, the Law on Marine Environmental Protection, requirements of

MARPOL 73/78 and HELCOM Conventions. The above-mentioned cases showed that the incidents were caused by the malpractice of the Terminal itself.

When pollution incidents happened the following actions were taken:

- Extent of pollution was determined (specimens were taken).
- Possible spread of pollution was forecast.
- Neighboring states were informed.
- Polluters were identified
- Damage for nature was assessed.
- Collection of pollutants was arranged.
- Klaipėda District Prosecutor‘s Office brought a case on marine pollution in materials, causing damage for human health and marine fauna.

In both cases the Lithuanian National Contingency Plan for Pollution Incidents at Sea was followed, protocols of the breach of the environmental protection law were made, responsible staff of the Būtingė Terminal were punished, and working groups consisting of responsible employers from governmental institutions related to the Būtingė Oil Terminal were formed. The Government of the Republic of Lithuania passed a resolution which delegated provision of proposals on improving prevention of incidents in potentially dangerous objects (among them in Būtingė Oil Terminal) to various governmental institutions.

7. Conclusions

Systems of environmental monitoring on the Lithuanian Baltic Sea seaside, which involve national monitoring and environmental monitoring carried out by Economic Units allows to improve and optimize system for the observation of environmental situation in this region.

In Lithuania there is only one state seaport and the Būtingė Oil Terminal. Most pollution incidents happen in these two places and, therefore, it is easier to make controls and find polluters. For almost all of the pollution incidents in Klaipėda State Seaport, the polluters are identified.

Every year, there are increases in crude oil and oil production transportation. New oil terminals constructed in the Baltic Sea—Būtingė, Primorsk, and Vysock—extend oil terminals in the main East Baltic ports. Navigational risk assessments are very important for finding legal and organizational solutions to decrease navigational and environmental risk

At Klaipėda Seaport in Lithuania, ballast water, rinse water, chemically polluted water, residues of paint, and quarantine (infected) waste are managed under ship agreements with appropriate stevedoring companies with “no-special-fee-system” not applied. Part of this waste is classified as hazardous waste and, therefore, there is a risk that such waste can be left in the territory of the port illegally.

A major reason for the practice of illegally dumping especially of fuel residues at sea is that considerable cost is incurred by disposal in port and that such cost has to be borne by ship operators. Būtingė Oil Terminal does not have ship waste and ship cargo residue reception facilities. Therefore, prior to heading to Būtingė Oil Terminal tankers have to confirm that wastes and ship cargo residues were disposed of at the last port of call and/or that ship waste and ship cargo residues reservoirs are within 25 percent of total capacity.

A very important impact factor for the marine environment is soil dumping in the Baltic Sea. In this territory, the concentrations of total organic carbon and total hydrocarbons are 1.2 and 2.2 times and heavy metals of nickel 2.3 times as high in dumpsites as in the surrounding areas.

No significant changes of oil hydrocarbon concentrations were observed in the area of the Lithuania Baltic Sea territory. Direct negative impact on the environment from oil spills during the accidents was short-term due to high dilution of water, active hydrodynamics of the water mass, and biochemical decomposition processes. Long-term impact on marine organisms can be observed as they bioaccumulate mutagenic oil hydrocarbons through the food chain.

The dynamic analysis of chemical constituents of complex oil compounds and the examination of the composition of HC confirmed that oil products get into Lithuanian beaches from the sea. It was determined that the concentrations of chemical elements accumulating in the Lithuanian beach sediments by the dynamic coastline are very small. The variations of their concentrations in the beach sediments are related with the location of terrestrial sources of pollution and/or the composition of beach sediments: the concentrations tend to increase with an increasing amount of thin dispersed matter.

The number and population of introduced species in Lithuania seaside is growing. Although invasive species enrich the nutrition base of local ecosystems, they pose a threat to the established stability of local species populations. In addition, some of these species have toxic impact; thus, the appearance of invasive species and their prevalence in local waters should be considered a negative phenomenon. Unfortunately, this phenomenon is increasing.

8. References

1. Ministry of Environment of the Republic of Lithuania (MoE)., 2005. Programme of National Environmental Monithoring. Ministry of Environment, Vilnius. 42 p.

2. Ministry of Environment of the Republic of Lithuania (MoE)., 2002. National report on sustainable development. Ministry of Environment, Vilnius. 145 p.

3. Olenin S., Klovaitė K. 1998. Introduction to the marine and coastal environment of Lithuania. *on: RedList of marine and coastal Biotopes and biotope complexsesof the Baltic sea,Beltssea and Kattegat.* Baltic Marine Environment Protection Comision. Helsinki.

4. Zaromskis R. 1994. Hydrotechnical constructions influence on the coastal zone aquatic complex. *Geography*, T.30: 35-39. Vilnius.

5. Zilinskas G. 1995. Waves tide in the Swash Zone. *Annals of Geography*. T. 8: 235-255. Vilnius.

6. Ministry of Environment Lithuania. 2003. State of Environment 2002. 147 p. Vilnius.

7. Ministry of Environment Lithuania. 2004. State of Environment 2003. 164 p. Vilnius.

8. Smailys V., Tilickis B., 2001. Environmental problems of Lithuanian seaside. *Environmental Research, Engineering and Management*. No.3. (17). P 76-81. Vilnius. (On Lithuanian.).

9. Grigelis A. Implications of Accelerated Sea-Level Rise (ASLR) for Lithuania. *Proceeding of SURVAS Expert Workshop on European Vulnerability and Adaptation to impacts of Accelerated Sea-Level Rise (ASLR)* Hamburg, Germany, 19th-21st June 2000. P 99-104. Vilnius.

10. Dailidienė I., 2001. Risk factors of the oil products spillage in the Lithuanian economical zone. *Bulletin of the Maritime Institute in Gdansk*. Vol. XXVIII, No. 2. P. 101-108.

11. Dubra J., Dubra V., Styra D., 1998. Water Dynamics in the Klaipėda Strait. *Aplinkos inžinerija*. Technika, VI t., Nr. 3. Vilnius. P. 83-87.

12. Leppäkoski E., Gollasch, S. and Olenin, S. 2002. Alien Species in European Waters. *Invasive aquatic species of Europe - distribution, impact and management*. Dordrecht, Boston, London. Kluwer Academic Publishers, 1-6.

13. Leppäkoski, E., Olenin, S. 2001. The meltdown of biogeographical peculiarities of the Baltic Sea: the interaction of natural and manmade processes. Ambio, 30 (4-5): P. 202-209

14. Dubra J., Dubra V., 2001. Relation between the Marine Legislation and the State of the Coastal Environment According to the Lithuanian Example. Institute for Estuarine and Coastal Studies of the University of Hull. *The Marine Environment: Science and Law*. Scarborough. P. 167-180.

15. Dubra J., Olšauskas A. The Main Problems of the Protection of the Sand Coasts in Lithuania. LITTORAL 2002 6th International Symposium Posters Volume III EUROCAST. Porto. P. 45-48.

THE CASPIAN SEA REGION

Environmental Security and Risk Assessment Applications in Oil and Gas Development Projects

E. OSTROVSKAYA
Caspian Marine Scientific and Research Center
ul. Shiryaeva, 14, 414045 Astrakhan, Russia

A. KURAPOV
"LUKOIL-NighneVolghskneft" Ltd.
ul. Komsomolskaya, 16, 400066 Volgograd, Russia

L. ANISIMOV
"LUKOIL-VolgogradNIPImorneft" Ltd.
ul. Sovetskaya, 10, 400131 Volgograd, Russia

Abstract

The Caspian Sea is the largest inland water body in the world and of great importance for the socioeconomic development of bordering countries. The Caspian Sea's unique fish resources and oil and gas fields are projected to provide a significant source of food and economic prosperity to the Caspian region, as well as energy to many parts of the world. This paper illustrates the need for objective assessment of transboundary risks on a regional scale. On the whole, the main threats to environmental security in the Caspian region are well known, though not fully evaluated. Risk analysis can be an important tool for identifying different challenges, revealing new hazards, assessing risks and, consequently, providing decision makers with options for mitigating both short- and long-term negative impacts on the regional economy and deterioration of the environment.

1. Introduction

The Caspian Sea is the world's largest inland water body, which is of great importance for the socioeconomic development of the littoral states and the whole world. Its unique fish resources and oil and gas fields provide food and jobs to the Caspian state population and energy supply to many parts of the world. The hydrological regime of the northern part of the Caspian is determined mainly by the

I. Linkov et al. (eds.), Environmental Security in Harbors and Coastal Areas, 409–420.

river flow. The Volga River has a major hydrological impact on the Caspian Sea and other tributaries because of its contribution about 80% to total river flow discharge into the Sea. One of the most interesting features of the Caspian Sea is the multi-annual fluctuation of its level, constituting over 3 meters and produced by variations of the river flow and evaporation. Multiple changes of transgressive and regressive phases of level state have occurred throughout the Caspian Sea's existence. The biodiversity of the Sea and its coastal zone makes the region one of the most valuable ecosystems in the world. The level of aquatic species endemism reaches 80% formed due to the Sea's long-term isolation from other water bodies. The Caspian Sea harbors 54 endemic fish, 53 endemic mollusks, and one endemic mammal, the Caspian seal. In spite of the ecological hierarchy of the Sea's ecosystem, it is generally uniform because of very active water circulation and biota migrations covering all water areas. There are significant protected areas in all the Caspian countries—including the Astrakhan biosphere reserve—of international importance. The Caspian Sea is a highly productive water body, thanks to intensive solar radiation and wealthy biogenic flow. The total fish biomass constitutes about 2900 thousand tons. Marine and fluvial ichthyofauna consists of 123 species and subspecies including 25 commercial species. In addition to the exclusive fisheries of highly valuable fish like sturgeons, natural resources of the region include oil and gas.

2. Actual Threats to the Environmental Safety of the Caspian Sea

At present there are the following environmental problems in the region (Figure 1):

- Threats to unique biodiversity, decline in commercial fish stocks because of the introduction of exotic species, inadequate regulation of the river flow, overfishing and poaching, etc.
- Pollution of water objects and offshore petroleum development, increasing the risk of pollution by hydrocarbons.
- Deterioration of environmental quality.
- Negative impacts of sea-level fluctuations on the coastal landscapes, population, and coastal infrastructures; etc.

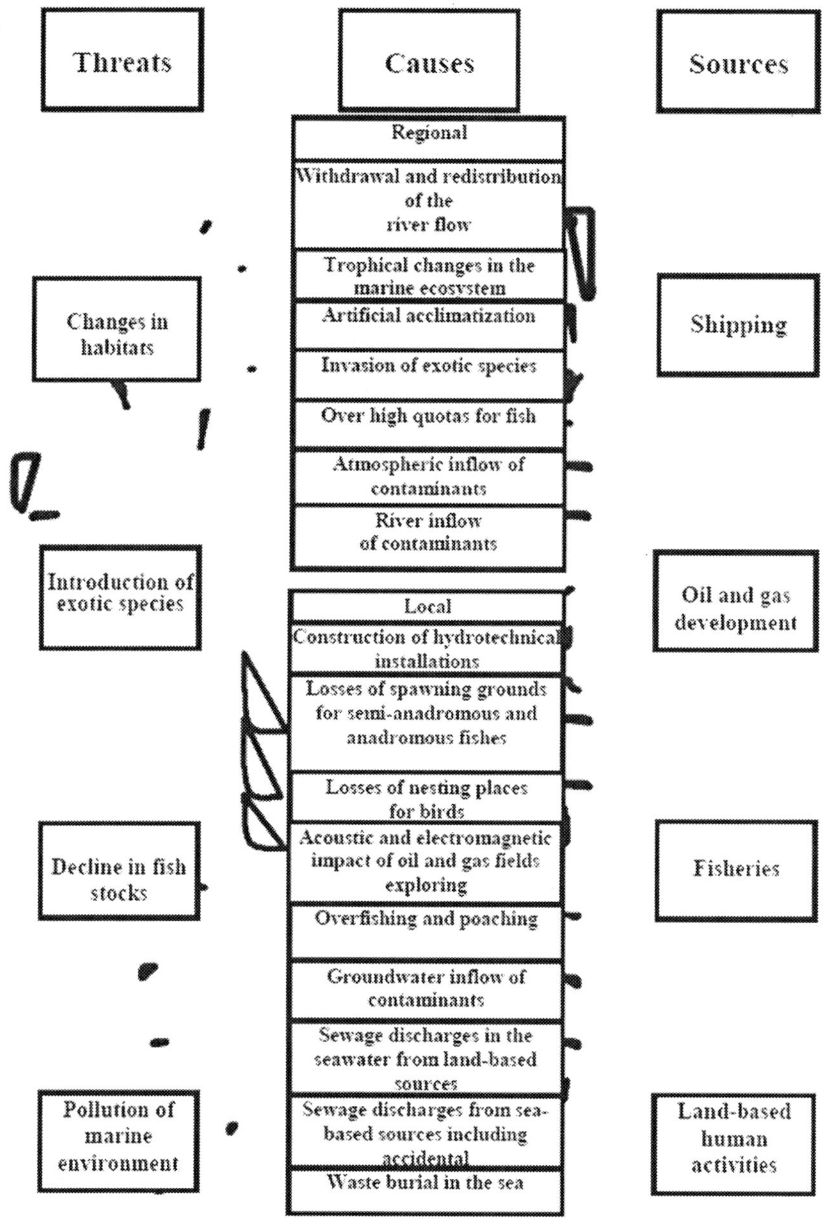

Figure 1: Direct threats to environmental safety of the Caspian Sea.

2.1. POLLUTION

Indices of water quality in the Caspian Sea in recent years show a rather stable hydrochemical regime. It may be noticed that the total pollution of the sea has decreased considerably in comparison with the 1980s. The major pollutants in this region are hydrocarbons, heavy metals, phenols, and synthetic surface-active substances. In addition to river flow, the basic sources are the shipping industry, agriculture, local industry, and urban and rural coastal territories. We lack reliable information on the atmospheric inflow of contaminants excluding may be sulfur oxides but it may have contributed to total pollution of the Caspian Sea considerably. Because of this lack, the pollution from and through atmosphere is not controlled practically in all Caspian countries. Studies of pollution by persistent organic chemicals such as polyaromatic hydrocarbons or harmful pesticides are limited and scarce. In addition, there is a pressing need to study pollution by heavy metals especially by mercury. These contaminants are carcinogenic and the most dangerous for biota and human health.

Transfer and accumulation of contaminants are closely connected with horizontal circulation of the Caspian Sea waters. Contaminants from land-based sources are associated with particle matter in the seawater. Results obtained from the original studies demonstrate that pollutants are accumulated in the detritus food chain including fishes feeding on benthic organisms when exchange of waters between the Northern and the Middle parts of the Caspian is decreased. That led to the increase of fish diseases through toxicoses. This phenomenon has been observed in the Caspian Sea over the last decade. Thus, pollution harmfulness to biological communities of the Sea depends not only on the concentrations of toxic compounds in the marine environment, but also environmental variability.

2.2. OVERFISHING AND POACHING

In the 1980s and 1990s, due to inadequate and too-intensive commercial fishery, including illegal fishing and the decline in both natural and artificial reproduction rates, the total stock of sturgeons in the Caspian basin was depleted. At present, the standing stock of sturgeons is still in a tense condition. On average, 60 to 65% of sturgeon population exists thanks to artificial reproduction. The Caspian states, including Russia, make significant efforts to prevent illegal fishing of sturgeon and other commercially valuable fishes. However, the problem is not completely solved yet.

2.3. INTRODUCTION OF EXOTIC SPECIES

The introduction of alien species, which have mostly affected the Caspian Sea ecosystem, should be mentioned first of all. In our opinion, they include the *Rhizosolenia*, fouling species and comb-jelly *Mnemiopsis*. These are "weed" species: useless both as food for fish and as competitors for endemic species. Until 1934, alga Pyrrophyta *Exuviaella cordata* prevailed in phytoplankton of the Caspian Sea. Penetration of alga Bacillariophyta *Rhizosolenia calcar-avis* from the Azov-Black Sea basin into the Caspian in the autumn of that year caused considerable changes in its phytoplankton dynamics. This species has spread through the entire Caspian over a short period of time, developed in large quantities changing radically the composition and distribution of phytoplankton. Areas of algae Cyanophyta, Chlorophyta, a partly *Peridiniaceae* were reduced remarkably and the result of interspecies competition was that dominating earlier *Exuviaella* had lost its previous importance. Among Bacillariophyta species, *Rhizosolenia* became dominant and important for the Caspian Sea production due to features like high ecological flexibility, ability to adapt relatively fast to changes in biochemical composition of water, biogenic flow of rivers, and pollution of coastal areas. *Rhizosolenia* is characterized by mass development the whole year round and depending on a season may constitute up to 95 to 97% of total phytoplankton. Introduction of fouling species from the Azov and Black seas—first *Mytilaster* and then *Balanus* and *Sagina*—led to intensive biofouling of vessels' bottoms, flumes, marine buoys, offshore drilling units, and port facilities. Over a short period of time the biomass of fouling increased by 15 times. Comb-jelly *Mnemiopsis leidyi* is also the Azov-Black Sea invader. It was brought in 1998 via the Volga-Don canal, most probably with ballast waters or on "dirty" bottoms of vessels. The consequence of the *Mnemiopsis* invasion is first of all declines in zooplankton biomass (according to [6] - by 4-6 times). *Mnemiopsis* reproduces rapidly and has no natural enemies at sea. That led to an uncontrolled outbreak of its abundance and distribution in practically the entire water area of the Caspian in 2000. Moreover, comb-jelly is a food competitor of many commercial fish species including sprat, one of the most mass species. Sprat catch used to constitute about 80 % of total fish catches in the basin. Decrease of sprat stock has led to decreases of total catches by two times in 2001 in comparison with 2000 (Figure 2). For example, incomes of the Astrakhan oblast budget have decreased by 24 million rubles because of the collapse of the fish economy [5].

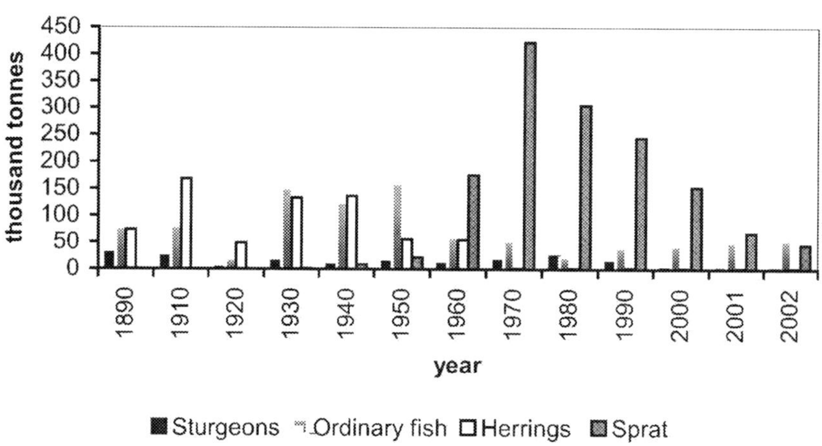

Figure 2: Fish catches in the Caspian region excluding Iran (using data published in [3]).

Food competition between sprat and comb-jelly is considered to be the main reason for the sprat decline. According to the specialists in fisheries food base of fish may be destroyed in five to ten years. Sprat in its turn constitutes a food basis for seals hence sprat is not yet the last victim of *Mnemiopsis*.

3. Risk Analysis Applications in Oil and Gas Development Projects

3.1. PRELIMINARY ANALYSIS AND NATURAL HAZARDS

By expert estimations, the Caspian's carbon reserves place third after the Near East and Siberia. The Caspian shelf reserves investigated make up 12 billion tons of conditional fuel and in case of splitting the sea into national zones Kazakhstan will get (in billion tons) 4.5, Azerbaijan – 4, Russia – 2, Turkmenistan – 1.5, and Iran – 0.9. Thus, the necessity for evaluation of the impact of oil and gas development activities on the marine environment is critical. Risk analysis is a part of a systematic approach to decision making and includes practical steps to solve the problems of prevention or reduction of industrial hazards and hence damage to human beings, property, and the environment. At the first stage of research we defined a set of risks, which are conditioned by:

- Natural features of the region
- Industrial activities
- Individual risks
- Environmental impact
- Transboundary hazards
- Social and political hazards

Prospecting and development of oil and gas deposits in the Caspian region is entailed with great natural hazards that bring about unfavorable developments in operational activities. Those are critical at laying oil wells and pipelines, construction of marine installations, and operational activities. The hazards are estimated by the following criteria: location, frequency, and amount of damage. For the Caspian water area, the specific set of hazards and their impacts is shown in Table 1.

Table 1: Natural hazards of the Caspian sea region and their impacts on offshore and coastal infrastructure.

Natural Hazards	Natural Hazards Impact
abnormally high geopressure (overpressure)	oil discharge and gas release
admixtures of hydron sulfide and mercaptans	air and water pollution
lythological peculiarities like hazardous geological structures, gas saturated areas, loose bottom sediments	surface subsidence, earthquakes, gas griffons, friable grounds
sea level fluctuations	flooding of marine and coastal installations and infrastructure
intensive strong winds	surge phenomena, currents, waves
hazardous glacial conditions	ice drifting

3.2. RISK ANALYSIS CONDUCTED BY OIL COMPANY "LUKOIL"

Since 1997 Oil Company «LUKOIL» has been oil prospecting in the Russian sector of the Caspian Sea. Several prospective deposits have been explored, the development of which will start in 2007 to 2008, with an estimated capacity of about 10 billion cubic meters of gas and 4 million tons of oil per year. Research in risk analysis has been done within the Environment Impact Assessment Program in compliance with the Federal Law «On Industrial Safety of Hazardous Industrial Facilities» (of July 21, 1997, No. 116-FZ). By law and according to normative documents of GOSTECHNADZOR of Russia, the results of comprehensive risk assessments should be submitted in the Industrial Safety Declaration (ISD). The analysis procedure is defined in the document to be guided by - RD 03-418-01 «Instructions on Conducting the Risks Analysis of Hazardous Industrial Facilities» (Figure 3).

Analysis of risks in Northern Caspian oil and gas developments is based mostly on the appropriate experience in similar physical, geographic, and geological conditions. It resulted in creating the first block of database to describe physical and geographical conditions of the Northern Caspian and natural hazards of the water area. The second block of the database generalizes the world experience of prospecting and development of continental shelf areas that have already been thoroughly studied and well developed like the Gulf of Mexico, the North Sea, and the South Caspian.

The third block of data features peculiarities of hazardous industrial facility development on the territories adjacent to the Northern Caspian water area—Astrakhan Gas & Condensate Development and oil development in Tengiz (Kazakhstan)—in the fluids of which high concentrations of hydrogen sulfide and mercaptans (transboundary risks) have been noted. Accidentally, the oil and gas deposits of the Russian geological sector of the Northern Caspian do not contain hydrogen sulfide or its concentration is so low, that the hazards conditioned by the gas presence may be neglected at the analysis of risks entailing the development. On the other hand, the availability of the huge oil developments in the Kazakhstan sector of the Northern Caspian in the close proximity of the Russian ones may seriously endanger the personnel and the environment and should be taken into account as part of risk estimation.

In accordance with its conception of environmental safety, to minimize the effects of unfavorable natural impact, Oil Company «LUKOIL» takes upon itself the obligation to perform drilling on the Caspian Shelf with the «zero» discharge of technological wastes into the sea. That is why the platforms are supplied with equipment for collection, storage, and transportation of waste products to the shore for treatment, with local treatment facilities and environmental control devices. The household effluents are supposed to undergo disinfections and then delivered to the

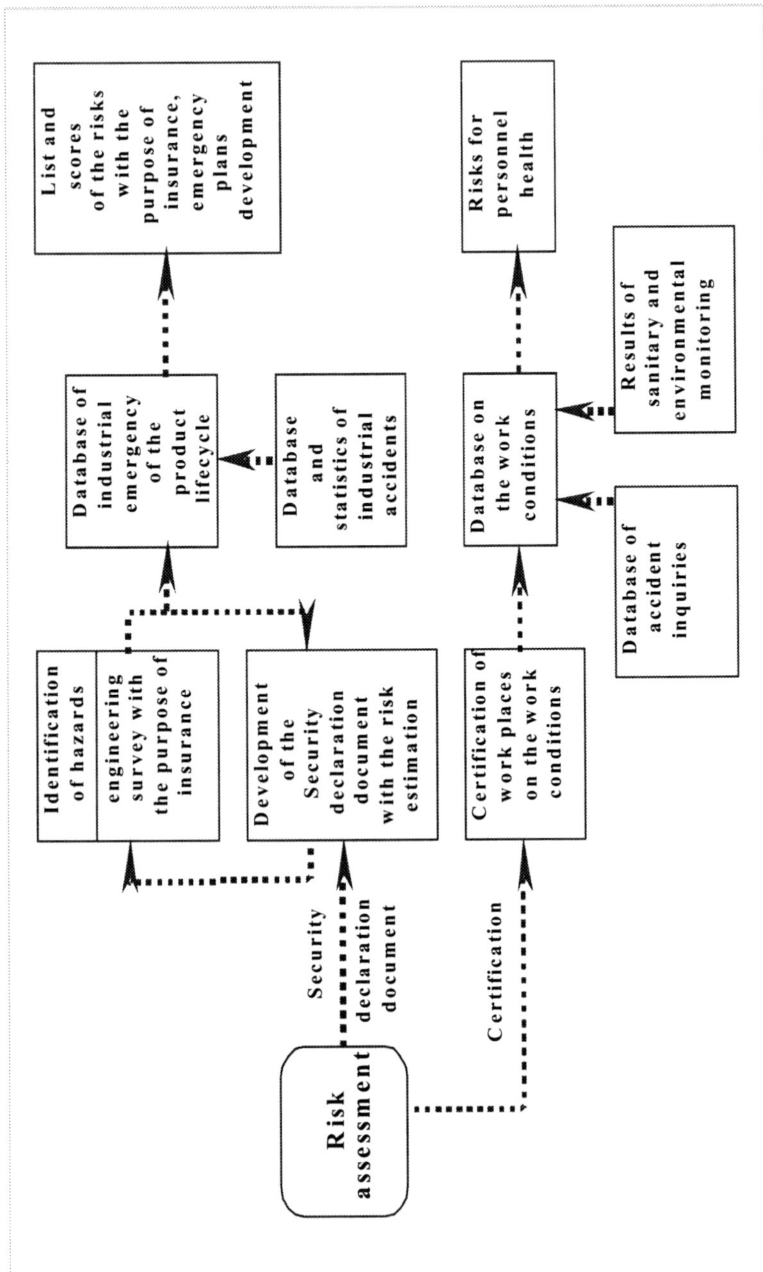

Figure 3 : Risk assessment procedure.

shore. The waste drilling mud is supposed to be treated and used again, the excess drilling mud is planned to be delivered to the shore treatment facilities for processing. Provided that all is done as planned, the most probable causes of environmental damage would issue from oil spills during its transportation, erection of drilling platforms, and laying of pipelines.

On the basis of qualitative risk assessment the following developments have been singled out that require more detailed consideration and more accurate evaluation:

1. Destruction of the platform resulting from ice drift.
2. Destruction of the platform resulting from «upper» gas release or griffon formation and from the destruction of the rigs.
3. Release of oil and gas at drilling.
4. Oil and gas release from the operating wells. Pollution of the sea waters and fire hazard on the platforms.
5. Oil and gas release at transportation by the tankers and pipelines. Sea pollution.

In the case study below, risk analysis enabled the company to opt for the safest means of oil transportation from the oil field to the shore. Two alternatives were considered: pumping oil over a pipeline or transportation by tankers. As a rule, oil transportation by surface or subsurface pipelines is less hazardous than by tankers. In the North Sea, two serious accidents have been registered on marine pipelines over more than 20 years of commercial use for oil and gas. Similar accidents took place in the Caspian Sea as well. The accident rate at Caspian offshore platforms is considerably higher than at similar facilities in the North Sea, the Gulf of Mexico, and the Persian Gulf. During five years of commercial use of the offshore development «Imeni 28 Aprelya» (Azerbaijan), six serious accidents took place on the main, 100-km-long, pipelines. Expert evaluation indicates that the average annual occurrence of marine pipeline destruction is 0.8 accidents per 1,000 km.

Pipeline construction and maintenance depends, to a great extent, upon natural conditions. The safety of oil/gas pipeline exploitation can be affected by ice abrasion and other hydrometeorological factors like winds, waves, icing, etc. That is why the expert estimation of natural hazardous impact may not prove accurate enough, since there has not been much experience of constructing pipelines in such a vast shallow area with complex ice conditions. Ice drift in ice areas can amount to 0.1 to 0.3 m/sec. The main factors influencing ice drift are the predominant wind directions during the ice period, sea currents, etc. [5]. A serious danger to subsurface pipelines comes from ice formations (stranded hummock) appearing in the Northern Caspian water area during winter. This is pack ice formed aground both as mighty hummocks and as barriers of up to 1 mile long, scores of meters wide and up to 10 meters high. They get formed at a depth of 5 m, but sometimes at depths up to 10 m. Stranded hummocks may form furrows on the sea bottom from scores of meters to 1 kilometer long and 50 to 100 meters wide. The furrow depth can reach 1.75 m, and by some observations, a maximum of 5 m [2]. It should be noted that the designed pipeline depth is 3 m.

The number of accidents in oil transportation by tankers is greater than that by pipelines. Worldwide statistics show that most lost oil (94.4 per cent) is spilled during shipping operations and loading and unloading in ports—53.6 per cent within the port area and 39 per cent while shipping. By other data, on the USA shelf in the Gulf of Mexico the average number of hazardous spills (of the volume over 1000 barrels) of oil produced or transported is: 0.79 from platform drilling; 1.82 from pipeline transportation; and 3.87 from shipping by tankers [4].

However, in the Northern Caspian case study, the risk-score comparison of different methods of oil transportation showed that the pipeline version poses greater risk. According to the analysis of statistical data used as the base for the subsurface pipeline accident rate modeling, pipeline maintenance life is a critical factor, constituting in a number of cases 40 to 50 years. To a significant degree, pipeline risk rates are critically affected by the specific for the given region potential impact of ice formations (stranded hummocks) at a considerable length of the main pipelines. Taking into consideration all those factors, the company has rejected the idea of constructing a subsurface pipeline and decided to ship the oil produced from the platform to the shore by tanker.

3.3. TRANSBOUNDARY RISKS

The increase of risk rates at oil and gas developments in the Russian sector of the Northern Caspian and the threat to its ecosystem as a whole is accounted for by the prospecting and development of subsalt deposits in North Kazakhstan. The expected volumes of oil production by 2010 are 100 million tons per year. The major volume is to be produced at Tengiz, Korolyovskoe and Kashagan oil fields, the latter situated in close proximity to the licensed area of the Oil Company «LUKOIL».

Tengiz oil field has been developed by an American company «Tengizchevroil» since 1993. During the period of 1993-2001 59 million tons of oil were produced, while the emission into atmosphere made up (in thousand tons): sulfur oxide – 136.37, carbon oxide –175.5, nitric oxide – 43.45, hydrogen sulfide – 0.35, carbon dioxide – 58.16, others – 1.4 [1]. For Tengiz, at the current production capacity of about 12 million tons per year and with its pollution rate, the total area under the real impact of its activities is estimated as 25,400 square km by the sum SO_2+NO_2. Taking into account the fact that the whole eastern part of the Northern Caspian is at sea level—27 m abs. is 51,700 square km—the operation of two (three at best) oil and gas complexes like Tengiz would be enough to subject the whole area of this part of the sea to systematic pollution. In addition, there aren't available reliable data on concentrations and distribution of some pollutants, mercaptans in particular, which are virulent toxic substances. Their Maximum Admissible Concentration is 0.000009 mg/m^3, and even one billionth of it in the air is harmful to human health. So far there has been no control of the spread of those highly toxic substances (mercaptans) in the air in Tengiz area, in the town of Atyrau, Karabotan, or East Kashagan areas. It is not clear what amount of gases, first of all hydrogen sulfide, remains in the mass of block elementary sulfur at its utilization (storage). There are no data on the processes that take place at the sulfur and air contact at the open storage of threateningly increasing bulk of elementary sulfur.

The most hazardous event for the offshore developments in the Russian sector is uncontrolled releases on Kashagan oil field. The daily oil output there can make from 1,000 to 10,000 tons with the release of great amounts of H_2S and mercaptans. To define the transboundary risk rate of the Kashagan well's open fountaneering for the licensed "LUKOIL" sites, a model of air pollution dispersion was considered with the use of associated gas formula like that of Tengiz development. By these estimations, the gas emissions associated with uncontrolled discharge of oil on Kashagan development would lead to air pollution of a vast territory of 400 km in radius that would cover the whole area of the Northern Caspian and south of the Astrakhan oblast including the Astrakhan Biosphere Reserve.

4. Conclusion

Risk analysis is a new kind of research for our country and it has an interdisciplinary character. However, large-scale work that has been done in our region to investigate hydrological conditions of the Northern Caspian, the structure and geological conditions of the upper zone of sedimentary bottom cover and to determine hydrogeological and geochemical parameters of open oil and gas developments enable us to give a full assessment of the natural hazards of the area.

As for the analysis and assessment of hazards and uncertainties, the things are best in oil and gas industry. A number of regulatory mechanisms exist to control the procedure of assessment of oil and gas development activities' environmental impact, which includes the analysis of risks and, first of all, the risk of emergencies. As the experience of the Russian oil company «LUKOIL» in this field shows, the risk analysis of some oil development projects in the Northern Caspian enables us to recommend measures for improving their environmental safety. In particular, that refers to the decision made by the company to reject the construction of subsurface pipeline because of the high probability of its destruction by drifting ice. In this connection, it should be noted that the present the capital projects of several trans-Caspian oil and gas pipelines from Kazakstan and Turkmenia are being planned. Will the hazards that may arise at their construction and operation be properly assessed? That part of the Caspian Sea is characterized by hazardous geological structures, weak ground, and high seismicity [7]. If those data are not taken into consideration while designing trans-Caspian pipelines, the ecosystem of the Caspian may suffer permanent damage in case of their destruction.

The objective assessment of transboundary risks is practically not available. However, for the near future it is possible to forecast a serious threat to both environmental safety of the Northern Caspian and Russian oil companies, since a gigantic Kashagan deposit is being prepared for development on the boundary with the licensed sites of the Russian sector of the Caspian Sea. Kashagan deposit is characterized by extremely high geopressure, high concentration of H2S, and mercaptans. This hazard requires special consideration and assessment.

The issue of risk assessment of new exotic species invasion is of no less importance. As shown above, the invaders have contributed significantly to the change in the Caspian Sea ecosystem. Introduction of *Rhizosolenia* and *Mnemiopsis* caused considerable damage to the pelagic food chain, which led to decline in commercial stocks of herrings and sprat by several times. Introduction of *Mytilaster* and *Balanus* caused essential damage to the detritus food chain, which gave rise to a decrease in food base for benthophags, including sturgeon fishes. Now, it may be concluded, the ecological niches of the Caspian ecosystem are filled with "weed" invaders virtually not contributing to fish production. Undoubtedly, the expected intensification of marine shipping to follow the oil boom will increase the probability of new introductions.

To our mind there is an urgent need to make environmental impact assessments of regional traditional activities like fisheries, navigation and recreation, which may affect the marine and coastal ecosystems considerably as well.

On the whole, the main threats to the Caspian Sea environment safety (natural and anthropogenic) are well known. However, they have not been thoroughly evaluated. The use of risk analysis procedure can become an important tool in assessing known risks and revealing new hazards, and, consequently, will enable us to mitigate negative impact on the regional environment and prevent its deterioration.

5. References

1. Diarov, M.D., Gilaghov, E.G. and Yergaliev, T.Zh. (2005) Prospects of utilization of hydrocarbon resources and expected negative consequences for the environment of the Kazakhstan sector of the Caspian Sea, Materials of the First International Scientific and Practical Conference "Problems of the Caspian Ecosystem Conservation under Conditions of Oil and Gas Fields Development" (February 16-18 2005, Astrakhan). Astrakhan, KaspNIRKh Publishers, 76-79.

2. Glukhovskoi, B.H., Lagutin, B.L. and Rzheplinsky, G.V., eds. (1986) Hydrometeorological conditions of the USSR sea shelf zones, Vol. II, The Caspian Sea, Leningrad, Hydrometeoizdat Publishers.

3. Ivanov, V.P. (2000) Biological Resources of the Caspian Sea, Astrakhan, KaspNIRKh Publishers.

4. Kaiser, M., Pulsipher, A. and Byrd, R. (2003) Study estimates Gulf of Mexico decommissioning costs, J. Oil & Gas, 6, 39-47.

5. Mazhnik, A.Yu. et al. (2003) Economic indicators of Russian fisheries development in the Caspian, Fisheries Researches in the Caspian: Scientific-research works for 2002, Astrakhan, KaspNIRKh Publishers, 531-544.

6. Sokolsky, A.F. et al. (2002) The state of the fish food base of the Caspian sea, in Fisheries Researches in the Caspian: Scientific-research works for 2001, Astrakhan, KaspNIRKh Publishers, 124-136.

7. Terziev, F.S., Kosarev, A.N. and Kerimov A.A., eds. (1992) The Caspian Sea. Vol 1. Hydrometeorological conditions Sankt-Petersburg, Hydrometeoizdat Publishers.

INVESTIGATION OF DISCHARGE OF FRESH WATER IN THE CANAKKALE STRAIT (DARDANELLES -TURKEY)

A. BABA, O. DENIZ
Canakkale Onsekiz Mart University
Faculty of Engineering and Architecture, Geological Eng. 17020 – Canakkale, Turkey

M. TURKOGLU
Canakkale Onsekiz Mart University
Faculty of Fisheries, 17020- Canakkale, Turkey

H. OZCAN
Canakkale Onsekiz Mart University
Faculty of Agriculture, Soil Dept. 17020 – Canakkale, Turkey

Abstract

The Canakkale Strait (Dardanelles) is a water passage connecting the Aegean Sea and the Marmara Sea. Its NE/SW trend is interrupted by a north-south bend between Eceabat and Canakkale. The width of the Strait varies from 1.35 to 7.73 km, the narrowest part located between Canakkale and Kilitbahir. The average depth of the Strait is approximately 60 m; the deepest part reaches more than 100 m. The aim of this study is to determine the locations of fresh water discharge into the Canakkale Strait. For this purpose, 52 sample points were selected in the study area. The temperature, pH and salinity were measured at different depths of water in the Canakkale Strait (1, 5, 10, 20 and 30 m). A Geographic Information System (GIS) was used to create a tabular and spatial database, with the aim of integrating the physical properties in the Canakkale Strait. The results of this study show that discharge of fresh water is seen in four locations on the southeastern coast between Dardanos and Canakkale, and that the Aegean Sea and Marmara Sea water are mixing between Canakkale and Kilitbahir. In the future, these sources of fresh water may be needed as population increases in the region. However, water pollution in the Canakkale Strait may limit the use of this fresh water.

1. Introduction

Seawater intrusion into coastal aquifers has been investigated by several researchers since the study of Ghyben and Herzberg in the 1890s (Bear and Dagan

I. Linkov et al. (eds.), Environmental Security in Harbors and Coastal Areas, 421–427.

1964; Frind 1982; Reilly and Goodman 1985; Gunay and Elhatip 1988; Voss and Anderson 1993; Back 1996; Elhatip 2003). Most recently, there have been several studies focusing on complicated aquifer conditions and the intrusion of saltwater (Croucher and O'Sullivan 1995; Huyakorn and others 1996; Sadeg and Karahanoglu 2001; Zhou and others 2003; Karahanoglu and Doyuran 2003). However, few studies have been done on the discharge of fresh water to the sea. Usually, water discharges from karstic rock to the sea have been studied. The discharge of the karstic water is high in the Aegean and Mediterranean seas (Gunay and Elhatip 1988; Elhatip 2003) Compared to other Neogene sedimentary sequences, the discharge of karstic water is easy to determine in sea water. Neogen sedimentary sequences are in an outcrop in the vicinity of the study area. For this study, the discharge of fresh water (coming from Neogen sequences), from urban areas of the Canakkale Strait is being investigated. A Geographical Information System (GIS) was used to ease monitoring of spatial and temporal changes (ESRI 1996; Kosmas and others 1997; Barazzuoli and others 1999; Thwaites and Slater 2000; Hudak and Sanmanee 2003). The distribution of physical properties of water is also monitored by GIS.

The Canakkale Strait (Dardanelles) is a very important water passage connecting the Aegean Sea and the Marmara Sea (Figure 1).

Figure 1: Location map of study area.

The Strait of Canakkale (Dardanelles) has two current systems. One of the currents is derived from the Aegean Sea, where the water density is high. The other comes from the Marmara Sea, characteristically of low density. Aegean water is typically flowing from the southwest to northeast under the Marmara Sea water.

The main goal of this study is to determine the locations of fresh water discharge into the Canakkale Strait.

2. Methods

Canakkale Onsekiz Mart University "R/V Bilim I" was used for this purpose. 52 points were selected within the study area for water samples (Figure 2). The temperature (T), pH, salinity, conductivity were measured at 1, 5, 10, 20 and 30 m depths. The physicochemical parameters were measured using a "6600 Model Multiple YSI Probe" connected to a computer, in situ. The Arc View 3.2 GIS (ESRI 1996) was employed to develop the maps. For the mapping procedure, the point themes were generated. The surface maps were then produced from the developed point themes by using Nearest Neighbor, an inverse distance interpolation technique, with 12 neighbors. The cell size for interpolation was 100 m. The study area border (Canakkale Strait border) was digitized from a topographical map scaled to 1/25,000 by using ArcInfo. Then data was exported to ArcView for mapping.

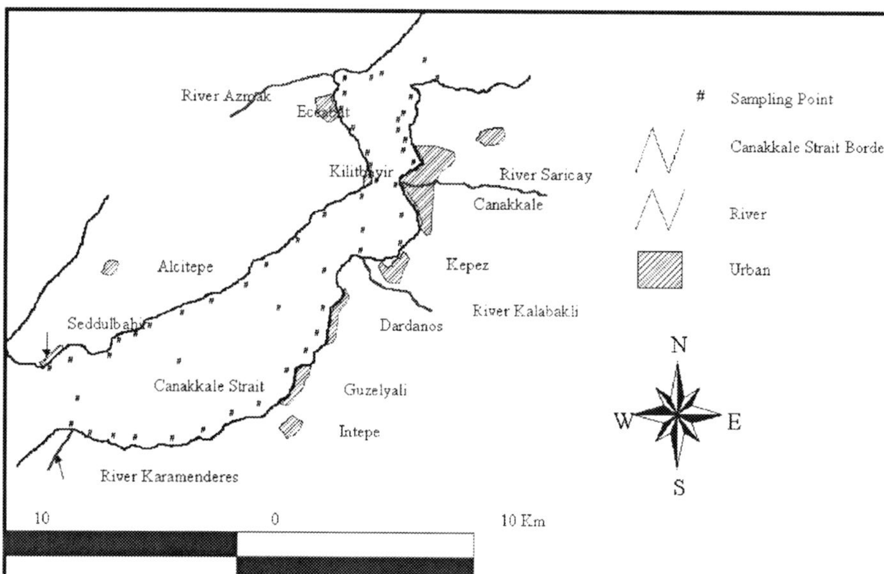

Figure 2: Study area and sampling points.

3. Results and Discussion

Two current directions are observed in the Canakkale Strait. The southwest flowing current from the Marmara Sea is of lower density than the northeast flowing current from the Aegean Sea.

3.1. TEMPERATURE AND SALINITY

Usually, temperature decreases with water depth, while salinity increases with water depth. However, in the Canakkale Strait, measurements show a different pattern. The temperature range is 6-20°C, and generally increases both with depth and with closer proximity to the mouth of the Marmara Sea (see Figure 3).

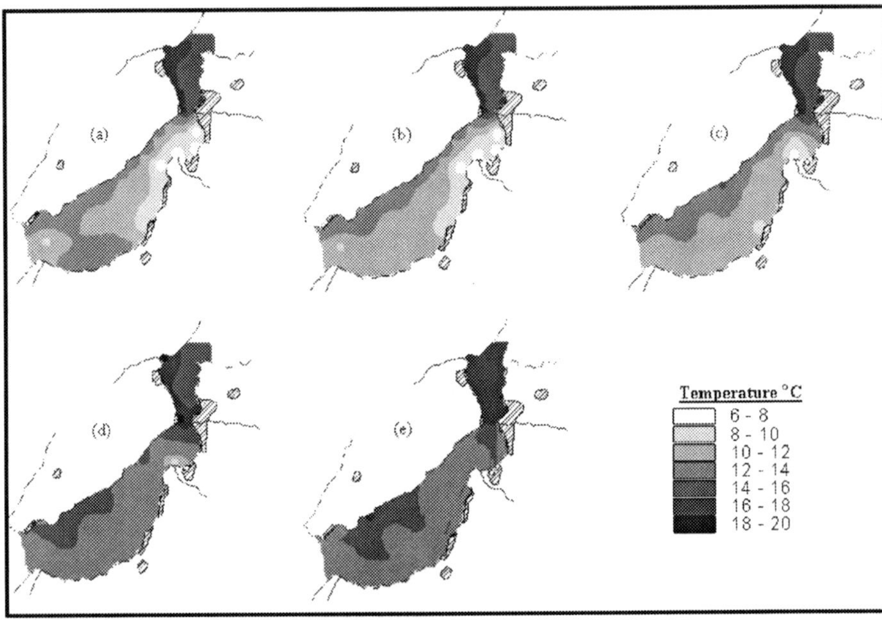

Figure 3: Distribution of temperature at different depths within the Canakkale Strait (a: 1m; b: 5m; c: 10m; d: 20m; e: 30m).

The salinity ranges between 20-40‰, and increases with water depth (see Figure 4).

The temperature increase with depth indicates turbidity in the Strait, and the mixing of the Marmara Sea and Aegean Sea waters. Temperature and pH measurements are highly variable at the narrowest part of the Canakkale Strait at Kilitbahir due to turbulence. The narrowing of the Strait also leads to different values to the northeast and southwest of Kilitbahir. The temperature and pH measurements to the northeast of Kilitbahir are higher than to the southwest (see Figures 3 and 5).

Temperature results show that discharge of the fresh water into the Canakkale Strait can be seen in the four sample locations between Dardanos and Canakkale at depths of 1 and 5 m; and at one location at 10 and 20 m depths. This point is affected by fluid discharge from the River Kalabakli (see Figure 3).

It is thought that the use of temperature techniques to measure fresh water discharges may give more accurate results depending on many internal and external factors, including the density equilibrium between fresh and sea waters, and other physical and structural properties of the coastal water-bearing formations.

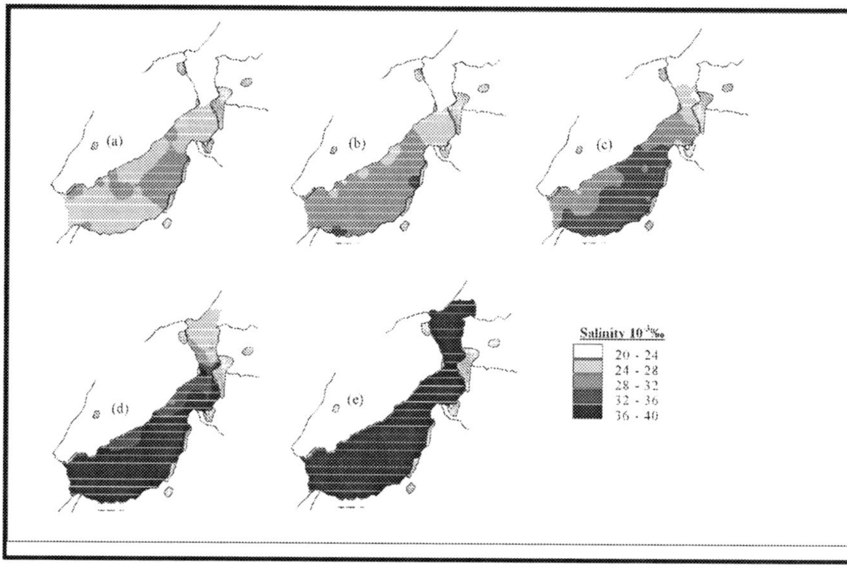

Figure 4: Distribution of salinity at different depths within the Canakkale Strait (a: 1m; b: 5 m; c: 10 m; d: 20 m; e: 30 m).

3.2. pH

The mean pH of Canakkale Strait is slightly basic at around 8.4. The average pH of Canakkale Strait is around 9 from 5 m to 20 m in the northeast of the Strait. During photosynthesis, hydrogen atoms are used by phytoplankton. The freshwater influx

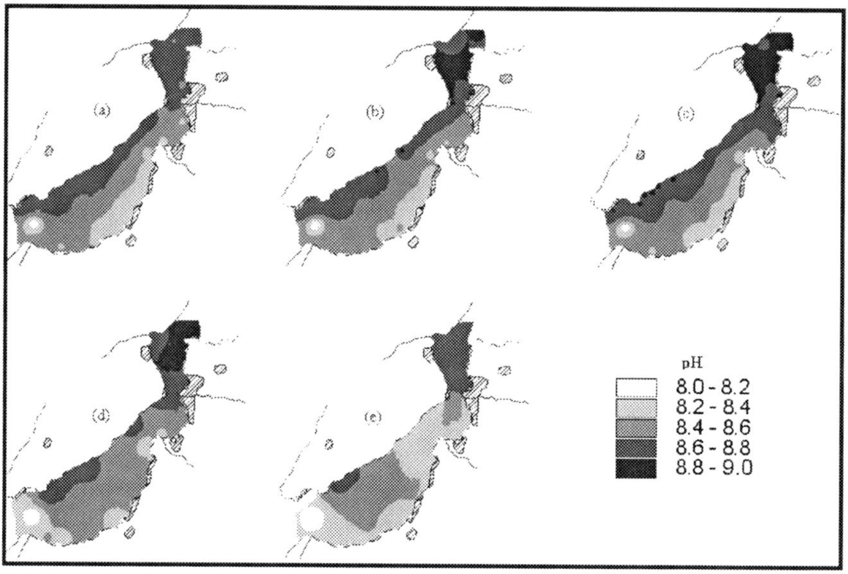

Figure 5: Distribution of pH at different depths within the Canakkale Strait (a: 1m; b: 5m; c: 10m; d: 20m; e: 30m).

is causing higher productivity levels, causing the water to become more basic in that area. Respiration and the breakdown of organic matter will lower the pH in the southwest of the Strait (Figure 5).

4. Conclusion

Population growth, urban expansion and economic development have persistently raised the demand of water supply, and consequently, greatly increased the exploitation of fresh water around the Canakkale Strait in the last several decades. This area is one of the important tourist sites in Turkey. In addition, the Canakkale Strait (Dardanelles) is a very important water passage connecting the Aegean Sea and the Marmara Sea. Therefore, this area will see expansion in the near future.

This study shows four points of fresh water discharge into the Canakkale Strait. This freshwater is coming from groundwater aquifers in the region. Much of the fresh water in the vicinity of the study area is discharged into the Canakkale Strait, Aegean and Marmara Sea. At present, the groundwater supplies appear to be sufficient enough for domestic and irrigation demands. But the population along the southeast coast of the Canakkale Strait is rising and the demand for freshwater increasing. This study has pinpointed the locations of freshwater discharge, so that if freshwater is needed in the future, it may be possible to extract it from these locations, prior to discharge into the strait or from the strait itself.

5. Acknowledgements

The authors would like to thank Maria-Victoria Gunnarsson for critical review of the manuscript.

6. References

1. Baba, A. Kavdır, Y., & Deniz, O., 2004. The impact of open waste disposal site on groundwater, International Journal of Environment and Pollution, Vol. 22, No:6 pp.676-687.

2. Back, W. 1996. Hydrochemical facies and groundwater flow pattern in northern part of Atlantic coastal plain. US Geol Survey, prof. paper 498-A, 1-42.

3. Barazzuoli, P., Bouzelboudjen, M., Cucini, S., Kiraly, L., Menicori, P. & Salleolini, M., 1999. Orogenic alluvial aquifer of the River Cornia coastal plain (southern Tuscany, Italy): database design for groundwater management. Environmental Geology 39 (2): 123-143.

4. Bear, J. & Dagan, G. 1964. Moving interface in coastal aquifers. J Hydraulic Div Am Soc Civil Eng 99 (HY4): p. 193–216, Congress, vol. xxi, Geology Publishing House, Beijing, 474-483.

5. Croucher, A.E. & O'Sullivan, M.J., 1995. The Henry problem for saltwater intrusion. Water Resources Research 31 (7): 1809–1814.

6. Elhatip, H. 2003. The use of hydrochemical techniques to estimate the discharge of Ovacik submarine springs on the Mediterranean coast of Turkey. Environmental Geology, 43: 714-719.

7. ESRI, 1996. Using Arcview GIS. Environmental System Research Institute, Inc., Redlands, CA. USA.

8. Günay, G. & Elhatip, H. 1988. Hydrogeological investigation of the karst submarine springs on the Mediterranean coast of Turkey by means of remote sensing techniques. Proceeding of the IAH 21 St., Congress, vol xxi. Geology Publishing House, Beijing, 474-483.

9. Frind, E.O. 1982. Simulation of long-term transient density-dependent transport in groundwater, Adv Water Resources 5: 73–97.

10. Hudak, P.F. & Sanmanee, S. 2003. Spatial patterns of nitrate, chloride, sulfate, and fluoride concentrations in the Woodbine Aquifer of North-Central Texas. Environmental Monitoring and Assessment, **82** (3): 311-320.

11. Huyakorn, P.S., Wu, Y.S. & Park, N.S. 1996. Multiple approaches to the numerical solution of a sharp interface saltwater intrusion problem, Water Resource Research 32 (1): 93–102.

12. Karahanoglu, N. & Doyuran, V. 2003. Finite element simulation of seawater intrusion into a quarry-site coastal aquifer, Kocaeli-Darica, Turkey. Environmental Geology, forthcoming paper (in press).

13. Kosmas, C.S., Danalatos, N.G. & Moustakas, N.K. 1997. The soils. Hydrobiologia 351 (1/3): p. 21-33.

14. Thwaites, R.N. & Slater, B.K. 2000. Soil-landscape resource assessment for plantations - a conceptual framework towards an explicit multi-scale approach. Forest Ecology and Management, 138: (1-3), 123-138.

15. Reilly, T.E. & Goodman, A. 1985. Quantitative analysis saltwater-fresh water relationships in groundwater systems, a historical perspective. J Hydrol 80: 125–160.

16. Sadeg, S.A. & Karahanoglu, N. 2001. Numerical assessment of seawater intrusion in the Tripoli region, Libya. Environ Geology 40: 1151–1168.

17. Voss, C. & Anderson, J. 1993. Regional flow in the Baltic Shield during Holocene coastal regression. Ground Water 31(6): 989–1006.

18. Zhou, X., Chen, M. & Liang, C. 2003. Optimal schemes of groundwater exploitation for prevention of seawater intrusion in the Leizhou Peninsula in southern China. Environmental Geology, 43: 978-985.

THE SOUTH MEDITERRANEAN COASTAL SUSTAINABLE DEVELOPMENT PROJECT

Environmental and Human Development Risk

B. RAISSOUNI
Al Akhawayn University
Badr Oulad Nassar Center of Environmental Issues and Regional Development, B.P 1884, Ifrane Morocco

M. EL BOUYOUSSFI
Agency of Promotion and Economic and Social Development of the North Provinces of Kingdom of Morocco

Abstract

The Tangier Mediterranean Project (TMP), a strategic priority for the sustainable economic and social development of the Northern Region of Morocco, requires the construction of a sea harbor handling domestic and business activities (passengers, containers, transshipments, cereals, other goods…), a 98-ha free zone for the stocking of goods and light processing, an industrial free zone targeting export-oriented production industries, and a 125-ha duty free/trade zone. The main objective is to create jobs by attracting export oriented enterprises, develop tourism, and generate sources of income for the harbor activities. The geographical region of the project lies in an area free of any industrial activity and was until now devoted to agriculture and fishery activities. After the general presentation of the TMP, this paper will discuss ecosystem changes (aquatic and terrestrial) that will occur following the development of the project and the risks of human and ecosystem changes.

1. Introduction

Morocco has known an increasing demographic and economic growth in the last decades that contributes to the sustainable development of the country. Within the new government approach, based on empowering the regional governmental representation with special attention to sustainable development of local communities and poverty alleviation, three main regional agencies responsible for assessing and managing development projects, have been set: first in the north part of Morocco, then in the south of Morocco, and more recently the Oriental Agency [1, 2]. Participation of the stakeholders in decision making in any development

I. Linkov et al. (eds.), Environmental Security in Harbors and Coastal Areas, 429–438.

project was among the innovative approach to be carried by these institutions in order to efficiently achieve the socioeconomic development goals.

Due to the geopolitical importance of the Moroccan Mediterranean coast and a growing interest to enhance development activities of this area, a special agency, TMSA (Tangier Mediterranean Special Agency) was created in 2002 [3] with the objective to design and manage an important development project aimed on constructing mainly a harbor and other economic/logistical sites located in the Mediterranean coast in the neighborhoods of Ksar Es Seghir – Ksar El Majaz in the new province Fahs Anjra, a little fishing port that constitutes the closest site of the coast to the North European Mediterranean Coast (9 miles), and halfway to the cities of Tangier, Fnideq, and Sebta.

This region is known for important historical events: the invasion of Spain in 711 by Tarik Ibn Ziyad and intensive economic and trade exchange between the two continents. The region also benefits from natural beauty and diversity, combining sea landscapes, wild beaches, and the rich natural reserves of its Djebel Moussa mountains. The population is mainly constituted of fishermen and farmers grouped into small villages and scattered farms.

The objective of this paper is to present an overview of the goals of this unique development project and to discuss the risks associated with environmental problems and human development issues.

2. The Region's Strategic Importance

2.1. OVERVIEW OF THE REGION

The concerned region can be classified in three levels:

- Region 1: Around the port; it is composed of 5 communes: 4 rural and 1 urban (Table 1).
- Region 2: The inland region; it is composed of 11 communes: 9 rural and 2 urban (Table 2).
- Region 3: The Tangier-Tetouan metropolis; it is composed of 2 cities (Tangier and Tetouan), 3 urban communes, and 25 rural communes (Fig 1).

Regions 1 and 2 are under the influence of two big metropolises: Tangier and Tetouan. In addition, they are affected by an "exchange economy", mainly fed by illegal trade commerce (where products are sold at a tax-free price), with the city of Sebta, which is still under the governance of the Kingdom of Spain. Consequently, this impedes the development of a local economy.

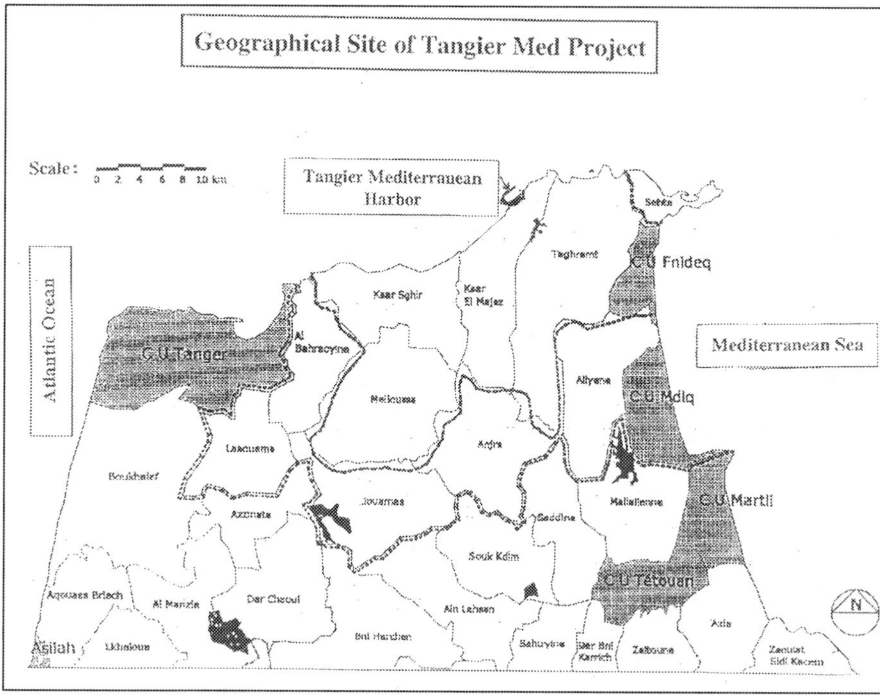

Figure 1: Geographical site of Tangier Mediterranean Project.

Some other main characteristics of this region are a continuous population movement from rural areas and from other regions of Morocco, all searching for a better livelihood.

The region of Tangier-Tetouan constitutes an axis of national growth and development, with many advantages (strategic location, beauty and diversity of natural resources, dynamic population) but also is confronted by some difficulties (low infrastructure, unemployment, parallel economy, and inability to control the urban growth).

2.1.1. *Population*

There are currently 104,986 inhabitants in Region 1 (Table 1, around the port), 209,646 inhabitants in Region 1 (The inland, Table 2), and 2.47 million inhabitants in region 3, according to the 2004 census. The results show that there has been a strong demographic growth in the last decade (from 1994 to 2004). Also, there is a constant rural exodus toward the cities of Tangier and Tetouan and urban communes such as Martil, Mdiq, and Fnideq, which explains the difference in growth rate between the urban communes from one side and rural communes from the other side.

Women and young people of Region 1 and 2 face many obstacles to earning a living. This is mainly due to the high illiteracy and unemployment rates.

Consequently, these social categories (women and young) started to look for alternatives, such as immigration to Europe.

Table 1: Area surrounding the port and its population distribution.

	Commune	Total Area (Km²)	Population in 1993	Population in 2004
4 rural communes	Taghramt	160	11,484	12,190
	Ksar El Majaz	85	6,609	7,420
	Ksar Es Seghir	83	8,818	9,705
	Melloussa	118	9,743	10,801
1 urban commune	Fnideq	24	34,486	64,870
Total	**ALL**	**470**	**71,140**	**104,986**

Table 2: Population growth in the communes.

Commune	Population in 1994	Population in 2004
Fnideq (U.C)*	34,486	64,870
Taghramt (R.C)**	11,484	12,190
Ksar El Majaz (R.C)**	6,609	7,420
Ksar Es Seghir (R.C)**	8,818	9,705
Melloussa (R.C)**	9,743	10,801
Allyene (R.C)**	5,654	5,044
Mdiq (U.C)*	21,093	46,366
Al Bahraouyine (R.C)**	7,258	10,501
Anjra (R.C)**	13,415	15,035
Jouamâa (R.C)**	6,765	7,173
Laaouama (R.C)**	10,286	20,541
Total	**135,611**	**209,646**

(U.C) Urban Commune (R.C)** Rural Commune*

2.1.2. Cultural and Natural Richness

The environmental diversity and richness in terms of natural resources, landscapes, historical monuments and archeological sites, such as Larache and Ksar Es Seghir, can make of this region a major tourist destination. In the past, there is Luxus, which contains the ruins of an old Roman city. Ksar Es Seghir contains an archeological site dating from the 12th century, when the King of Portugal took control of the city and started building forts. Even today, we can find ruins of an old Portuguese fort on the Ksar beach.

The region possesses some valuable natural sites such as the Site of Biological and Ecological Interest (SIBE) of Tahaddart in the south of Tangier, Jbel Moussa in Fahs Anjra, Talasmtane Park in Chefchaouen, and the Rmel forest in Larache. Additionally, the commune of Ksar Es Seghir has important groundwater resources of high quality, free from any kind of pollution.

The communes of Melloussa, Bahraouyine, Ksar Es Seghir, Ksar El Majaz and Taghramt constitute an important forest area, composed mainly of maritime pine and dwarf oak.

2.2. OVERVIEW OF THE TANGIER MED HARBOR

2.2.1. Geographical Location and Regional Development

The TMP is being implemented in the Gibraltar Strait, 35 km east of Tangier. This seaport will rely mainly on import/export of goods. TMP is a study whose nucleus is the construction of the port. However, the whole region surrounding the port (Region 1), which is 470 km² (Table 1), is directly or indirectly affected at the industrial, economic, commercial, and social levels.

As shown in Table 1, this region encompasses five main towns/zones. These include four rural communes: Melloussa (1 and 2), Ksar Es Seghir, Ksar El Majaz, Taghramt; and one urban commune: Fnideq. This latter is being seen as a future deposit zone for the merchandise whose destination is the port. Therefore, developing Ksar Es Seghir– Ksar El Majaz into a city and making it a tourist pole becomes a priority. Melloussa 1 and 2 are also given priority because of the presence of a commercial and industrial zone that will definitely play an important role once the port starts. With a budget of 18 billion Dirhams ($1.8 US) [3], TMSA's main function is to build the port. "Fond Hassan II pour le development" and Abu Dhabi contributed mainly to the financing of this project. It is expected that the durability of this port will be at least 100 years.

The Moroccan government takes care of basic infrastructures such as building roads, railways, and electricity supply networks (high tension lines). Actually, a highway from Terzel (15 km south of Tangier) to the port is being constructed. This highway (total of 54 km long) will pass through three strategic points [4]—Melloussa 1, Melloussa 2, and Ksar Es Seghir—with two main segments; two companies are working on each segment. In parallel with that, a railway is also being constructed, connecting Tangier with the Tangier Med harbor. Finally, another highway is planned to connect Fnideq to Tetouan. This way, the most important poles are being linked, in order to ease communication and transportations (of people and merchandise). A future plan would be to construct an international airport as a final touch in Cruche Blanco, midway between Tangier and Tetouan.

2.2.2. Social and Environmental Concern

As stated earlier, while implementing this project, new jobs are needed. In order to involve local population, some workers from the region are still being trained and recruited.

In order to minimize risks associated with environmental problems, mainly the fragility of the mountain ecosystem, preliminary studies on the inventory of concerned sites to be protected have been launched in Jbel Moussa and Punta Cerez in particular.

2.2.3. Implementation Planning

The implementation of the TMP started in 2002 and is expected to be partly functional at the end of the year 2006. The year 2005 is an important stage in the process of implementation, because many construction phases will be underway at the same time.

The main goals of TMSA are the creation of 150.000 new jobs, attraction of export-oriented enterprises in the free zones, and encouragement of the penetration of foreign markets and enterprises by endowing them with an efficient seaport and a contribution to tourism development. Furthermore, the project aims to balance regional development by providing the northern region with a large economic pole and qualitative infrastructure, hence reducing the pressure on the city of Tangier to turn its focus towards tourism and thus make of it a pole of cultural attraction.

2.2.4. Description of the Port Infrastructure

One of the features of the proposed harbor shown in Figure 2 is the possibility of transhipment of the big ships inside the port. In other words, merchandise is being spread over other ships that will take it to other destinations.

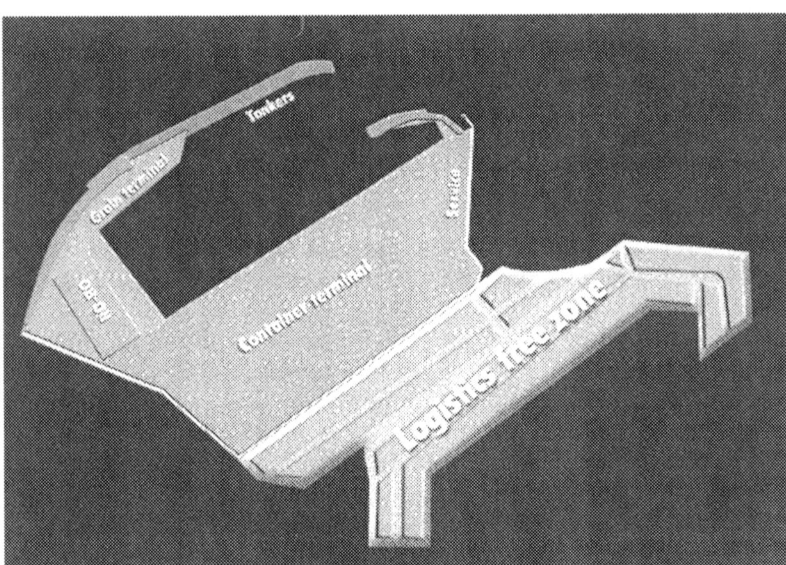

Figure 2: The four main components of the proposed Tangier Med Port Project.

Facilities will likely include:

- A container terminal: 1600 m long and 16 m deep, this terminal has a 90-ha surface area. This terminal has been conceived to be one of the biggest terminals in the world and has the capacity to receive the largest container ships in the world. Most traffic comes from "Transhipment", which is a fast-growing activity in the Mediterranean basin (more than 300 in 24 hours).

- A RO-RO terminal: With four roll-on/roll-off, this terminal offers an important capacity for passengers and merchandise TIR traffic (International Road Transport).
- Bulk and General Cargo terminal: 15-ha open area, where merchandise is expected to be delivered to the port of Tangier.
- A bunker terminal: This terminal will supply the hinterland (inland place that contains merchandise to be delivered to the economic/industrial zones) with refined oil products and offers the capacity for refueling activity for ships docking in the harbor and/or in transit in the Strait of Gibraltar.

At present (2005), the main container company that has invested in this port is "Maersk Sealand". The request for the International Ship and Port Facility Security (ISPS) of the port has been launched and construction adjustments and any security measures for accident prevention are being considered during the construction steps. Experts of ISPS are currently making sure that the proper norms and guidelines are being implemented.

3. Human Potential Assessment and Risks

As stated earlier, there is a high demographic growth. The most dominant activities for subsistence are the traditional ones: agriculture (mainly the cultivation of cereals) and fishing, completed by a commerce whose main income is from the illegal trade.

Region 1 and 2 are mainly of rural type, with unqualified human resources and poor equipment. Coastal zones have a strategic position (most traffic between Tangier and Sebta uses the coastal road) and are better equipped in terms of road quality, whereas the inland zones are wedged and the low-quality traffic roads amplify this isolation.

In this section, our main area of concern is Region 1 and 2 because the implementation of the TMP influences not only Region 1 but also Region 1. There are many reasons for that: first of all, Region 1 cannot fill the 150,000 new jobs needed by Tangier Med Project , because the number inhabitants is low and mostly unqualified, so a solution is to train some of them and bring other labor from Region 1. Furthermore, the commune of Melloussa, through its industrial zone, will directly or indirectly influence the neighboring communes that are part of Region 1. This latter is composed of nine rural communes (Taghramt, Ksar El Majaz, Ksar Es Seghir, Meloussa, and Al bahraouyine, Laaouamra, Jouamaa, Anjra, and Allyene) and two urban communes (Mdiq and Fnideq). The total surface area is 1000 Km^2. In 1994, there were 135,611 inhabitants, of which 40% were from Fnideq and Mdiq, and the current population density is 139 inhabitants/Km^2 (3.8 greater than the overall national density in 1994) [5]. In 2004, the total population of this region was 209,646 (Table 2).

The recent census in reported in Table 2 shows that a 1994 trend is recurring: the rural population decreasing whereas the urban one is increasing. This is due to the continued massive migration of rural inhabitants toward urban communes.

The illiteracy rate is high and affects mostly women (75.6%) in cities; in the rural zones it is much higher (90%). This is due to the low rate of schooling. In 1994's census, it was found that 1 girl out of 5 was going to school. Data from the 2004 census is not available.

From Table 3 we can observe that the unemployment rate differs according to gender, with 15.3 % for males and 19.9 % for females. It is expected that the unemployment rates reported in Table 4 will decrease due to the implementation of TMP, which expects to create 150,000 new jobs [6,7]. However, proper human capacity building programs should be planned and carried out (illiteracy programs, professional training…).

Table 3: Unemployment rate.

Gender	Unemployment rate
Male	15.30%
Female	19.9%

One challenge of the project is to contribute to local population development through capacity building and opportunities for improving livelihoods. This will make the local population accept the project and contribute to its success. If qualified labor is brought from urban cities (Tangier and Tetouan), strategies should be developed regarding any overpopulation settlement in the region. Special attention should also be given to environmental and natural resources conservation in addition to basic infrastructure, mainly waste water facilities.

Furthermore, transportation costs and means will be an issue to be dealt with. A highway and railways are being built in order to facilitate transit to/from this region, but this may contribute to environmental degradation, human insecurity, and health impacts due to traffic jams and accidents.

4. Environmental Assessment and Risks

The region has a rich but a fragile environment. It possesses important natural sites and landscapes. However these are vulnerable and will be exposed to higher degradation threats due to natural factors (erosion and climatic change), and mainly to human pressure (industrial and domestic pollution). The coastal side is a rocky zone, which limits the surface and the number of balneal stations that can be built in the northern coast region. Nevertheless, a couple of balneal stations are being built whose aim is to compete and become as dynamic as the one of Tetouan and Tangier. It is obvious that additional human pressure in new recreational areas will contribute to their degradation.

The region presents risks for earthquakes and floods. It is classified together with the city of Agadir as the most seismically active zone in Morocco. Tangier Med

Project managers have thought of an alternative against floods, which is the construction of a dam. This latter would also help in securing access to drinkable water.

Landslides damage roads and make the maintenance work costly and redundant. There is a hydraulic network made of rivers, Oueds such as Oued Aliane, Oued Ksar Es Seghir, Oued Rmel, and Oued Marsa, which is the main source of floods. Actually, Oued Rmel is being diverted from the port in order to avoid floods within the port area.

The forest area is dense and diversified (dwarf oak, pines…), especially in Ksar Es Seghir. The forest is particularly dense and presents risks for fires, which is the principal risk of forest ecosystem degradation. Clearings of land exist in the very inclined slopes.

The natural beaches are used by coastal birds during the migration and wintering periods. In spring and autumn, up to 200 species make their way across the strait. The cliffs are used by migratory birds as resting place, especially the cliffs located between Ksar Es Seghir and Oued Rmel. This development of the region may induce some perturbations to the ecosystem.

One of the most direct impacts of TMSA project is the destruction of one of the biggest beaches of the region at the profit of building the harbor. Also, because of some oil leaking, the water surrounding the port is being polluted.

The economic/industrial zones in Meloussa are built in the middle of a natural environment. The main concern here is to preserve the natural site. Some alternatives would be to follow international standards in order to control gas emissions and other forms of pollutions such as water pollution, where important water resources are under threat (underground water and superficial hydraulic network), such as Ksar Es Seghir‘s water resources (that are of very high quality and should be conserved). This pollution can increase with the development of the region.

The coastal sides have a multitude of beaches suitable for tourism. However, this latter should be protected and maintained properly to reduce solid waste pollution and littering. Local population and visitors should have sustainable environmental awareness and education campaigns. The forest zones are rich and contain rare species that need protection and preservation. The diverse landscapes of the region should be preserved because they are unique to the Mediterranean coast. Finally, the region includes four main archeological sites—Ras Leona grave, Ras Cirés headlight, the archeological site of Ksar Es Seghir, and Belyounech—that should be promoted and given more consideration. Local population using this coastal area for fishing and for tourism may suffer from the implementation of the project if no clear alternative or guidelines are offered to them.

5. Conclusions

The Tangier Med Project (TMP) looks very promising for the sustainable development of the Tangier Tetouan region. Although environmental concerns seem to be integrated into the project agenda, there is still no evidence of any key method and tools used for environmental risk assessment.

Data gaps and the lack of better scientific knowledge of the ecosystem changes during the implementation of the project may have an adverse effect on decision making for future development activities and planning. However, environmental risks have to be compensated by other means in order for local communities to accept changes and adapt to a new lifestyle.

Setting up effective environmental monitoring and implementation schemes is necessary in order to maintain the natural aspects of the region and the ecosystem balance. Proper planning for the new urban areas is needed, taking into consideration population flow and the social problems associated with population increase. Development and enhancement of economic activities and social programs, through close collaboration with European partners, may then contribute to stabilization of the local population. This may also contribute to a decline in illegal immigration from the South Mediterranean and Sahel countries to the North Mediterranean coast.

6. Acknowledgement

High appreciation to His Excellency Abdelaziz MEZIANE BELEFQUIH, his Majesty Mohamed VI counselor, Mr. Said Elhadi, General Director of TMSA for giving to the authors and the students the opportunity to visit the implementation site of Tangier Med and for providing the documentation. We are thankful to NATO for their valuable financial support for presenting this paper.

7. References

1. Benhima Driss (2005), *Communication*, Ifranc.

2. Chaoui Mohamed (2005), *le Souverain nomme Mbarki à la tête de l'agence de l'Orientale,* L'Economiste (1995).

3. Debbi F., M. Najib Mekouar (2004), *SDAU (Schéma Directeur d'Aménagement Urbain du Détroit), Intermediary Synthesis*, Ministère délégué chargé de l'habitat et l'urbanisme.

4. TMSA (2005), *communication.*

5. National Census (2004), http://www.recensement.hcp.ma

6. ELHADI S. (2204), Workshop on *Tangier Med A Regional Development Vector*, Tangier.

7. Zyat J. (2004), Workshop on Tangier Med A Regional Development Vector, Tangier.

LAKE SEVAN IN ARMENIA

Socioeconomic Analysis for Secure Development Policies

V. SARGSYAN
Yerevan State Institute of Economy
Nalbandyan 128, Yerevan, 375025 Armenia

Abstract

Agricultural growth, industrialization, population demands for drinking water, recreation, and other demands in coastal areas pose environmental risk and security difficulties. These pressures are well illustrated by the challenges facing Lake Sevan, the largest lake in the Transcaucasus Region of Armenia and one of the largest freshwater high-mountain lakes in Eurasia. The paper presents an investigation around the Lake, which integrates a set of criteria for performing socioeconomic valuation and cost-benefit analysis with the further aim of optimal control modeling and development of security policies.

1. Introduction

Our lives depend on both natural lakes and artificial reservoirs because they provide drinking water for millions, water for agricultural and industrial development and unique recreational opportunities. The traditional friendly relationship between lakes and humankind is now being jeopardized by unbalanced economic development. Agricultural growth together with the construction of irrigation and drainage systems, the building of factories and the use of fertilizers and pesticides in the coastal areas – all may cause problems concerning environmental risk and security. The environmental situation around Lake Sevan in Armenia is a typical example. Sevan is the greatest lake of Transcaucasus Region and one of the greatest freshwater high-mountain lakes of Eurasia. The Lake is situated 60 km to the North-East from Yerevan. In natural conditions, Lake Sevan has been located at an altitude of 1916.2 m a. s. with the surface 1416 km2 and volume 58.5 km3 (in 1934). It is the main water resource for Armenia and important source of freshwater for some neighbor countries. The paper presents an investigation around the lake, which integrates a set of criteria for performing socioeconomic valuation and cost-benefit analysis with the further aim of optimal control modeling and development of security policies.

I. Linkov et al. (eds.), Environmental Security in Harbors and Coastal Areas, 439–449.

2. Lake Sevan Problem: History and Background

The so-called Sevan Problem arose in the beginning of the twentieth century. In the range of environmental problems in Armenia, the problem of Lake Sevan is of special topicality, having originated in the 1930s and still remaining unresolved. Primarily, the problem was how to use the natural resources of the site. The high location of the lake compared to the fertile but arid Ararat Valley and lack of energy resources in the country attracted engineers to find methods to explore the water of the lake intensively. A hydroelectric system has been developed in the area on the River Hrazdan, which is the main outflow from the lake. Its water started to be used intensively for irrigation and electricity. Yet in the 1960s, the environmental consequences of extensive exploitation of the lake were evident. Lowering of the water level (currently about 18 meters) had an effect on the ecosystem of the lake from physical conditions to primary production and the fish community. In the 1966, a new project was presented (construction of 48.3 km long Arpa-Sevan water tunnel for restoration of water stock) to stabilize the level of the lake.

During the last reformation period, economic requirements triumphed over environmental considerations. The crash of the central planned economy and the problems of the transition period as well as the blockade of supplies and electricity from outside have worsened environmental conditions in Armenia. Under blockade conditions, the winters of 1992-93 and 1993-94 brought enormous hardship to a population lacking heat and electric power. Therefore, since 1993 the outflow from the lake has been increased. The Ministry of the Environment reported that the lake's water level had dropped by 50 centimeters in 1993. Experts said that this drop brought the level to within 27 centimeters of the critical point where flora and fauna would be endangered.

Currently resources of the lake are used for energy production and irrigation of Ararat Valley, the main agricultural area of Armenia, but the quality and stock of the water are important resources for fishery and recreation. From the other hand arises the problem of pollution of the lake and recreational area. The main source of pollution is agricultural production in the coastal areas of the lake. Here arises the problem of optimal distribution of environmental resources of Sevan for organizing agricultural production between the two main agricultural regions of Armenia: Ararat Valley, where water resources are directed for irrigation and the same time produce energy; and Gegharkounik District, where intensive agricultural production (application of toxic pesticides and fertilizers) decreases the environmental quality of the lake [4,5]. Economic activities negatively influenced coastal resources and the recreational value of the lake and surrounding area. In addition to economics, the problem also involves social and political issues.

The government has increased attention to the problem and is making a large investment towards saving the lake. It is important to mention the support of The World Bank "Lake Sevan Action Program" [2,3]. Multidisciplinary investigations around this problem and well-founded scientific approaches are considered necessary for development of secure management policies for sustainable development of the lake.

The problem of Lake Sevan is one of the central environmental problems of the region. It is at the intersection of several environmental related problems: management of freshwater resources, irrigation and optimal control of agricultural production and land resources, optimal control of usage of toxic substances in agriculture and industry, interactions of energy and environment, and management of coastal resources and recreation.

In the framework of environmental risk analysis and under discussion of a set of conditions, working out and implementing policies of sustainable and secure development of the lake is a necessary task. Investigation of the problem can include multidisciplinary approaches. For development of sustainable policies, economic and social analyses are important and could be realized by working out and realizing system of models.

3. General Framework and Methods

The bases of the presented investigation are the questions of *resource valuation, cost-benefit analysis, and optimal control of pollution.* Investigation of the system must take into account a set of criteria, which play an important role in the presented integrated problem. Environmental degradation and the inefficient use of natural resources require intervention to properly price environmental resources and internalize environmental costs. The economic restructuring offers an unparalleled opportunity to incorporate economic instruments into environmental policies. Integrating the true environmental costs and risks into economic activities will help make decision-makers aware of the implications of their policy decisions, although obviously the price of many environmental assets can never be exactly determined. Economic instruments should not be considered as replacing direct regulations and administrative interventions in environmental policy. They should be linked to and supplement clearly defined standards on emissions and environmental quality. Cost-benefit analysis will help answer the question of how much regulation is enough. From an efficiency standpoint, the answer to this question is simple: regulate until the incremental benefits from regulation are just offset by the incremental costs. In practice, however, the problem is much more difficult, mainly because of inherent problems in measuring marginal benefits and costs.

Valuation of the resources is central problem. In some cases environmental effects are immediate and well known; in others they are long term and subject to great uncertainty. Is important to consider not only *economic*, but also, *social and political aspects of the valuation problem.*

Evaluation of environmental resources of the lake (especially recreational) needs a social approach. Socioeconomic analysis starts with the basic value judgment that it is the preferences of individuals that count (willingness to pay). Moreover, individuals are assumed to regard environmental resources as commodities, which they would be willing to trade against material possessions. However, such value judgments could be introduced into a cost-benefit analysis in the form of constraints, or environmental standards, which must be met.

Lakes and reservoirs have not only economic and social significance, but also political and strategic importance. *Political valuations* must be considered also, either by obtaining a direct political decision, or by inputting values from past decision. As a direct substitute for "willingness to pay" -type data, this would seem to have little justification. On the other hand, it may be that there are some issues on which people believe that moral preferences are revealed in the market. Perhaps political leaders are the appropriate people to judge this issue.

4. Socioeconomic Valuation: Willingness to Pay (WTP) Survey

A Willingness to Pay (WTP) survey was taken as a tool for research and future cost-benefit analysis. This section presents a description and outputs of a survey[1] that was intended to generate information on recreational usage of the lake, attitudes of visitors and local residents toward possible management changes, and estimates of visitors' and residents' willingness to pay for water quality improvements at the lake.

4.1. SAMPLE SELECTION AND SURVEY STRUCTURE

Two groups of respondents were targeted to participate in the survey: recreational users (visitors) of the lake and local residents. Two different questionnaires were worked out for the residents and visitors. The Sevan, Gavar, Martuni, and Vardenis regional centers were selected for the residents' survey. The visitors' survey was organized in the recreational areas of the lake. A total of 574 people—including 256 residents and 318 visitors—were interviewed.

The surveys represented by interviewers to the two groups (visitors and residents) were very similar. The survey was designed to focus on how the respondent values different levels of water quality at Lake Sevan. In order to provide a baseline level of quality for the respondent, current conditions at Lake Sevan were presented in the questionnaires. First of all, it was necessary to define the measures of Lake Sevan Quality, which was done on the basis of similar studies [1,6] and materials and investigations around the environmental conditions at the lake. The quality of lake can be described in many ways. The main measure of quality for Lake Sevan is the level of water. The level of the lake, originally 1,925 meters above the sea, has fallen about 18 meters since the 1930s, which is the main reason for environmental problems at the site. The next measure of water quality is the clarity of the lake water. Water clarity is usually described in terms of how far down into the water an object is visible. Initially (1930) it was 15 to 20 meters and now it is 2.5 to 3 meters. Another measure of water quality is the amount of nutrients and

---

[1] Survey was organized by the support of MacArthur Foundation in May – August 2002.

other substances contained in the water. Water quality degradation can result from a number of sources including runoff from the surrounding community containing first of all agricultural chemicals. Currently the nutrients contribute to the occurrence to the algae blooms in the lake, usually 10-12 times per year. Under some circumstances blooms can be a health concern, causing skin rashes and allergic reactions. Concerns about bacteria can result in beach closings. The overall quality of the water can impact other conditions of the lake. Poor water quality results in an undesirable odor and color to the lake water. At this time the water has a mild odor that many describe as "fishy". Finally the quality of the water impacts the variety and quantity of fish in the lake. Thus, the condition of Lake Sevan can be summarized in terms of *water level, water clarity, algae blooms, water color, water odor, bacteriological situations, and fish diversity.*

Respondents were then presented with various plans, each describing a different overall condition of the lake, and were asked about their willingness to pay for the plan. In addition to the valuation questions, both versions of the survey also contained questions pertaining to the respondents' support for various projects for improving water quality, their opinion concerning various land use changes, and the water quality attributes most important to them. The visitors' and residents' versions of the survey differed in that the visitors' version collected information about recreation trips.

4.2. RESULTS

In this section summary statistics from the Lake Sevan survey are provided, focusing on (*a*) reported visitation and spending patterns, (*b*) attitudes toward various watershed and land use changes, and (*c*) implied valuations.

4.2.1. Visitation and Spending

On average, visitors reported high recreational usage. The average total number of trips taken was 3.6 during 2001. Of those trips, an average of 1.52 consisted of multiple-day visits (i.e., the respondents spent at least one night near Lake Sevan). The most popular recreation activities engaged in by visitors were swimming and beach use. Respondents reported spending an average of $34 on a typical visit. Respondents from Armenia reported spending an average of $28 per trip, while out-of-Armenia respondents reported spending an average of $64 per trip. Spending can also be categorized by the type of trip taken. Respondents who took only single-day visits reported spending an average of $18 per trip, while respondents who took only multi-day visits reported spending an average of $68 per trip. Multi-day visits formed 23% of total trips.

4.2.2. Opinions

Respondents were asked to allocate 100 importance points to the lake characteristics. Sanitary safety and purity of the water and surrounding areas is the most important characteristic for both visitors and local residents. In general, both visitors and local residents appear to either support or are indifferent to various water quality projects and land use changes presented in the questionnaires. As expected, nobody in either group opposes increasing of the level of the lake, as the decreasing of level is the main reason of environmental problems of the site.

Restrictions on residential development are supported by 81% of visitors surveyed, with fewer than 6% opposing restrictions, but only 23% of local residents supported this project as the region really needs residential development. The issue that generated the most opposition among respondents was the institution of no-motor boat days. Roughly 40 percent of visitors oppose no-motor boat days, with only 17 percent supporting them. In the case of local residents, roughly 33 percent oppose no-motor boat days, while about 37 percent support them as motor boats on the lake are used mainly by residents for local small businesses (fishing, recreation). Roughly 57 percent of local residents and only about 30 percent of visitors support limiting hotel building around the lake. There appears to be wide support for restoration of prairies and development of nature preservation areas in both groups. There is an understandable conflict between group opinions concerning development of land use under agriculture.

4.2.3. Valuation

One important goal of the survey was to estimate the value that both visitors and local residents place on the preservation and/or restoration of Lake Sevan. Armenia is not a rich country; conservation budgets are tight and there are more projects than there is money to fund them. Thus, society must decide where to focus the available resources, both private and public sources. To help with these decisions, a method was devised to measure the value people place on environmental goods as measured by their willingness to pay for the goods. Two of these techniques are employed in this study. The first method is based on observing the public use of a natural resource (visits to the lake) and inferring visitors' willingness to pay for the resource from their behavior. The second method is based on directly asking whether people are willing to pay various sums of money to support a particular project.

The first value estimated in this study is the willingness to pay for the existing level of visits to the investigated recreational site. This can be thought of as providing a baseline of the value visitors place on preserving the existing level of the resource in terms of how much enjoyment they get from Lake Sevan at its current level of water quality. Based only on the reported single-day trips data, the average recreational value per season of Lake Sevan is $27 per visitor. Next, the value of various water quality changes was estimated. Both the visitor's and resident's versions of the survey contained a scenario entitled Plan A. The description of the plan stated that if nothing is done to improve the water quality of the lake, it is likely to deteriorate over the next decade. Conditions at Lake Sevan after deterioration were presented in the survey. Responders were then asked the following question, "Would you vote yes on a referendum to maintain the current water quality of Lake Sevan and avoid the deteriorated water quality as described under Plan A? The proposed project would cost you $B (payable in five [$B/5] installments over a five-year period)." In this question, the value of "B" was varied so that different respondents were faced with different project costs. The value of "B" varied between $20 and $160 for visitors and between $30 and $240 (1.5 times higher) for local residents.

Figure 1 plots the relationship between the percentage of visitors indicating they would be willing to pay the stated amount along the horizontal axis. Roughly 89 percent would be willing to pay $20 toward this plan ($4 annually for five years), but only about 17 percent would be willing to pay $160. Based on these data the average willingness to pay is approximately $101 per visitor in support of Plan A.

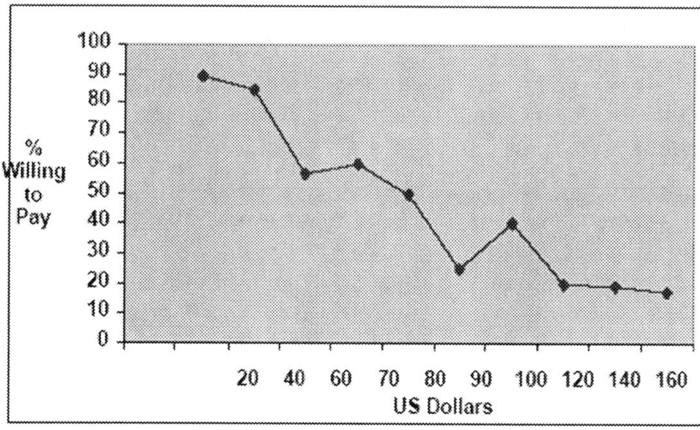

Figure 1: Willingness to pay for Plan A visitors.

Figure 2 plots the relationship between the percentage of local residents indicating they would be willing to pay the stated amount along the horizontal axis.

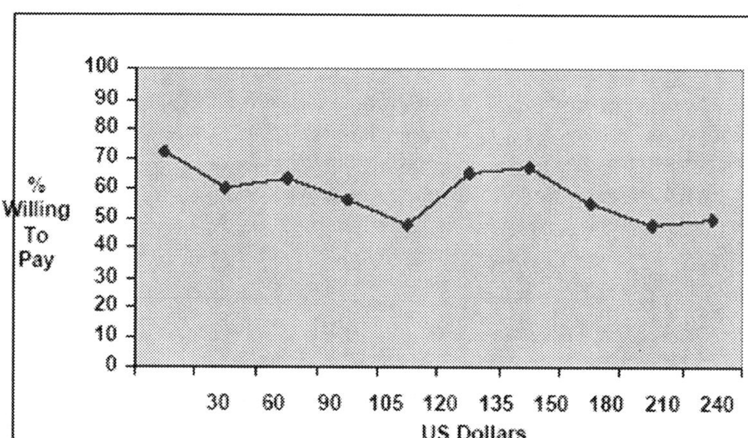

Figure 2: Willingness to pay for Plan A by local residents.

On average, local residents would be willing to pay approximately $168 in support of Plan A. This significantly higher value for residents is not surprising given their continuous exposure to the lake and its attributes. For this reason the magnitude of the value of for local residents was 1.5 times higher.

Plan B focused on willingness to pay for improvements in water quality. Two versions of Plan B were created: the first described a program that would result in a moderate improvement in water quality over the next five to ten years (Plan B1), while the second described a program that would result in a substantial improvement in water quality over the next ten to twenty years (Plan B2). Visitors would, on average, be willing to pay approximately $93 in support of the low quality improvement described in Plan B1. This value is actually less than the $101 visitors were willing to pay for Plan A, which simply maintained the current lake conditions. However, the two results are not statistically different, suggesting that visitors are willing to pay roughly $100 to maintain the lake, but little, if any, for modest improvements.

Local residents would, on average, be willing to pay approximately $159 in support of the low quality improvement described in Plan B1. Again, this value is slightly lower than the $168 local residents were willing to pay for Plan A, though the two are not statistically different. This indicates that local residents are willing to pay roughly $168 to maintain the lake, but little, in any, for modest improvements.

The same analysis was done from the visitor's survey for the high quality improvement. Based on these data, for visitors - $205 and for local residents the figure is $285. According to the fact of low income level and living standard in Armenia willingness to pay for high quality improvement for both visitors and residents is enough high.

In addition to the values described above, visitors indicated that the quality changes described in the survey would affect the number of trips they would expect to take. After each quality change plan was described, the respondent was asked to consider all the recreation trips they made to Lake Sevan in the past year, and report the number of trips they would have made if conditions were as described in the plan. The response to the decreased water quality described in Plan A is dramatic. With the decrease in water quality, average number of trips would decrease 2.4 times. Visitors also responded to the higher water quality scenarios by indicating that they would increase the number of trips they would take. With the low quality improvement, respondents would take an average of 5.2 trips, while with the high quality improvement respondents would take an average of 8.1 trips.

5. Monetary Value Analysis

Above presented survey created the basis for analysis of sustainable environmental management of the resources of Lake Sevan. Two models were worked out and realized: monetary value analysis of water resources and the integrated model of optimal environmental management of the lake[5].

Figure 3 represents the scheme of monetary analysis. For the water resources of Lake Sevan, such analysis was not done yet. The problem is to compare two directions of water usage—market use and recreational use—and find the optimal balance between them. For definition of recreational value (nonmarket value) of the lake, results of Willingness to Pay Survey were used. The nonmarket value of the lake suffers from nonoptimal market water use. From the other side it is impossible to stop the use of the water for irrigation purposes. The water of Lake Sevan has a market use and accordingly a market price, which is the sum of two components: price of produced electricity and price of irrigation water.

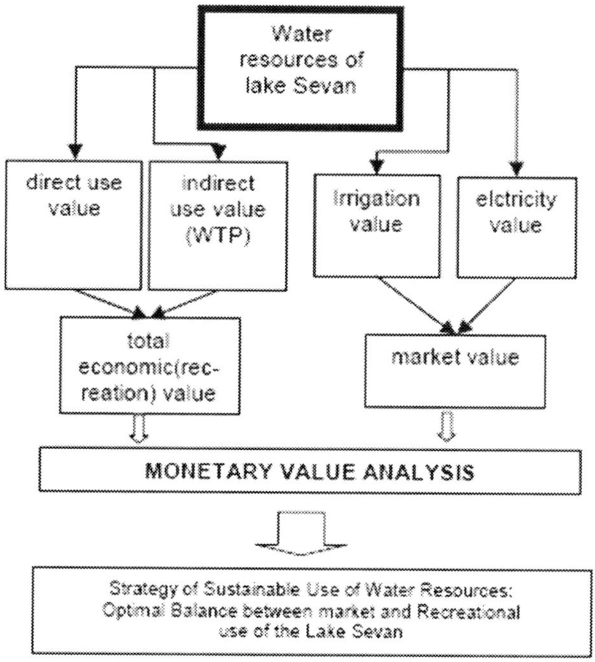

Figure 3: The scheme of monetary value analysis of Lake Sevan water resources.

During the investigation period, the average price of electricity on Sevan-Hrazdan cascade was 0.018 USD for one Kilowatt-Hour. One cubic meter of the water released from Sevan by River Hrazdan in Sevan-Hrazdan cascade in average produces 1.8 kWh electricity. There is an established price for irrigation water in Armenia of 0.011 USD for one cubic meter.

Even after approximate calculations and rough analysis, the dominance of the recreational value of the lake is evident above benefits from irrigation and energy production. The values were calculated on the basis of 2001 prices. As is plotted in Figure 4, total production of energy and irrigation water sales are decreasing. This is the result of government lake protection policy, which is being realized in recent years.

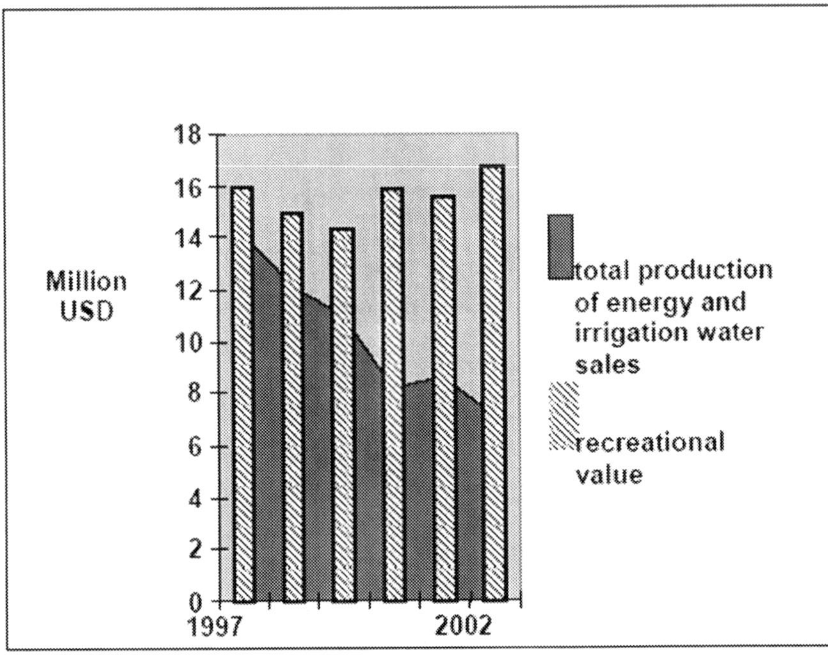

Figure 4: Values of Lake Sevan.

Presented monetary value analysis is integrated into the multi-staged system of models [5], which also considered optimal pollution control [4] and generates management policies.

6. Conclusions

At present, economic development is supposed to be accompanied by the increasing role of the environment. The environment, like an economic system, also requires optimal management based on multi-criteria analysis. The sustainable development of the environment is a real demand, especially for countries in transition, where market relations are at a foundational stage.

Water resources of Lake Sevan in Armenia are used for energy production and irrigation, which lowered the level of the lake. Lake Sevan is the main source of irrigation in Armenia. It provides about 25% of annual irrigation water for the region. Rainfall in the crucial period May-August is inadequate and irrigation is necessary to ensure satisfactory crop development. Meanwhile energy is produced on the Sevan-Hrazdan Cascade. This formed the market value of the water resources. From the other hand is important to discuss the recreational value of the lake. Sevan is one of the main recreational sites of Armenia. As expected, monetary value analysis of water resources revealed overbalance of recreational value against benefits from irrigation and energy production.

As presented above, currently it is impossible to stop water usage for irrigation, water outflow from the lake, but is on possible minimum level. Meanwhile, development of alternative irrigation sources is a necessary strategy. On the other hand, considering also benefits from tourism, the development of the lake as a recreational site will have an important role for the development of the economy of Armenia, where tourism is considered a dominant development direction.

7. References

1. Azevedo C., Herriges, J, and Kling, C. (2001) Valuing Preservation and Improvements of Water Quality in Clear Lake, Staff Report, Center for Agricultural and Rural Development, Iowa State University, Iowa, USA.

2. Armenia, Lake Sevan, Action Program, Main Report, (1999) The International Bank of Reconstruction and Development.

3. Armenia, National Program of Protection of Environment, Main Report, (1999), The World Bank.

4. Sargsyan, V.A. (2000) Risk Assessment for Agricultural Pollutants (Armania): Modelling and Optimal Control, Assessment and Managemet of Environmental Risks, Cost-efficient Methods and Applications, Edited by Igor Linkov and Jose Palma-Oliveira, Kluwer Academic Publishers, NATO Science Series, Netherlands, pp. 275-282

5. Sargsyan, V.A. (2002) Models of Optimal Environmental Management, The Model of Optimal Managementof the Resources of Lake Sevan (Armenia), book, "Nahapet" Publishers, Yerevan, Armenia

6. Yohe, G.W. (1977), "Comparisons of price and quantity controls: A survey", Journal of Comparative Economics 1, 213-33.

451-464

MULTI-CRITERIA DECISION ANALYSIS AND THE TAR CREEK SUPERFUND SITE

Initial Findings

C.J. BATCHELOR
U.S. Army Engineer Research and Development Center
3909 Halls Ferry Road, Vicksburg, MS 39180 USA

USA L72 Q51 Q25 Q57 C88 J15 Q58

Abstract

The Tri-State Mining District was formed to encompass areas of Oklahoma, Kansas, and Missouri where lead, zinc, and other metals were mined from the 1900s until the 1960s. Tar Creek in Ottowa County, Oklahoma was the recipient of much of the mining waste generated during this period. The Tar Creek watershed is an approximately 53.3-square-mile area, where 19,566 people reside. It is characterized by high heavy metal soil concentrations, contaminated surface and ground waters, air transport of contaminants, and exposed mining wastes. There are human health and ecological exposure hazards from these media. A need for evaluations of long-term solutions that could be constructed or implemented to improve the ecosystems is apparent. There has been a movement toward a more 'holistic' response to human health and wildlife risks at and adjacent to Tar Creek, including determining problems affecting residents and identifying appropriate remedial actions. In 1983, the area along Tar Creek was listed on the National Priority List (NPL) as a Superfund Site. The Environmental Protection Agency signed a Memorandum of Understanding with United States Army Corps of Engineers and the Department of Interior in 2003 to collaborate on assessment and remediation efforts with multiple stakeholders, which include tribal authorities, local interest parties, and other entities. Multi-Criteria Decision Analysis (MCDA) is a systematic and structured process beneficial to users during the pre- and post-phase of decision making. MCDA could prove an asset to the Tar Creek project, particularly when dealing with multiple stakeholders coupled with numerous remediation objectives and risk remedies, by applying decision processes such as Analytical Hierarchy Process (AHP) and Multi-Attribute Utility Theory (MAUT). Commercial software packages use decision processes as engines; for example, Expert Choice® utilizes AHP while Criterium DecisionPlus® exercises MAUT. MCDA, paired with decision-making tools, provides the results of modeling/-monitoring studies, risk analysis, cost, and stakeholder preferences so that risk managers are able to systematically evaluate and compare alternatives and actions supporting risk management and thus credibly prioritize resources. The

I. Linkov et al. (eds.), Environmental Security in Harbors and Coastal Areas, 451–464.

following sections will discuss the background and history of the Tar Creek Superfund Site, the MCDA framework/structure, commonly used MCDA tools in conjunction with theories, and a methodology for how MCDA can be effectively used at the site.

1. Introduction

The words 'Tar Creek' have become synonymous with words like mine shafts, chat piles, and orange-tainted waters. Tar Creek is in Ottawa County, Oklahoma, and for much of its length flows within the Tri-State Mining District. The Tar Creek Superfund Site encompasses approximately 53.3 square miles, where 19,556 people reside [20]. Principal cities within this area are Picher, Cardin, Commerce, Quapaw, and North Miami. Tar Creek flows southerly between Picher and Cardin, passes to the east of Commerce and Miami, and then flows on to its confluence with the Neosho River, one of the two major rivers in northeastern Oklahoma [19], and is also the primary drainage system for the Picher area. Mining in the area began in the 1900s and quickly began yielding massive tons of ore. The ore extracted from the mines within the principal cities was at depths 90 to 385 feet below the ground surface. The low-grade ore produced meant that about 95 percent of the crude ore was discarded on the surface in the form of mill tailings in enormous chat piles and large flotation ponds [13]. Mining ceased in the mid-1960s. Mining activities left a large area that is now characterized by contaminated surface and ground waters, high heavy metal soil concentrations, exposed mining wastes, and air transport of contaminants [20]. Because of their concentration, spatial extent, and/or toxicity, lead, zinc, and cadmium are the main contaminants of concern. Other related hazardous substances include copper and selenium. There are human health and ecological exposure hazards from these contaminants and thus the health and well being of the people in the Tar Creek area may be at risk (NRE No Date). For example, the blood lead levels of young children have tested above 10 μg/dL, which is a target safe level [3]. In August/September 1996, 38.3 percent (31 of 81) of the children tested in Picher had blood lead concentrations exceeding 10 μg/dL, 62.5 percent (10 of 16) of the children tested in Cardin had blood lead concentrations exceeding 10 μg/dL, and 13.4 percent (9 of 67) of the children tested in Quapaw had blood lead levels which exceed 10 μg/dL [19]. Natural resources potentially affected by contaminants at the site include, in part, federal and state threatened and endangered species; migratory birds; surface water; drinking water; plants: fish; biota; wildlife; and cultural, agricultural, and terrestrial resources [9]. The Natural Resources Damages Subcommittee also states legal basis for natural resource claims:

> The Comprehensive Environmental Response, Compensation, and Liability Act (CERCLA) provides that responsible parties may be held liable for damages for *injury, destruction, or loss of natural resources resulting from a release of hazardous substances, including the reasonable costs of assessing the damage.* (42 U.S.C. § 9607(a)(C)

There are a multitude of stakeholders and other interested parties that seek to provide input into the risk management process at the site. Tribal governments represented include not only the Quapaw tribe, which hosts most of the mine sites in the Tar Creek Superfund Site area, but also tribes such as the Miami, Wyandotte, Seneca-Cayuga, and Eastern Shawnee, who may have been impacted downstream. All tribes in the area and Indian people living in Ottawa and northern Delaware County, Oklahoma, are potentially impacted by site-related contaminants as they engage in their cultural practices of hunting; gathering; conducting ceremonial activities; and utilizing natural resources for crafts, medicines, and food. Living within any of the towns in Ottawa County could expose both adults and children to contaminants via the use of mine tailings as construction materials such as part of an asphalt mixture for pavement. Natural resources under CERCLA include "land, fish, wildlife, biota, air, water, ground water, drinking water supplies, and other such resources belonging to, managed by, held in trust by, appertaining to or otherwise controlled by the United States...any State or local government...or any Indian tribe..." (42 U.S.C. §9601(16). The main purported sources of ecosystem degradation at the site are two-fold: 1) the release of residual metal sulfides seeping from abandoned mine tailing and 2) mine tailings that were left uncovered and unstabilized [18]. These sources are the primary components of acid mine drainage for the Tar Creek site. Open mine shafts and boreholes remain mine hazards in populated areas. Water quality concerns include stream corridors subject to flooding with acid mine drainage potentially impacting both human health and wildlife. Flooding at Miami occurs frequently along Tar Creek and there is a need for evaluations of long-term solutions that could be constructed or implemented to reduce flooding and improve ecosystem health [18].

In 1983, the area along Tar Creek became a Superfund site when it was placed on the National Priority List [19]. A Record of Decision (ROD) followed in 1984. The ROD addressed two concerns: 1) The surface water degradation of Tar Creek by the discharge of acid mine water and 2) the threat of contamination of the regional water supply [19]. The United States Environmental Protection Agency (USEPA) signed a Memorandum of Understanding with United States Army Corps of Engineers (USACE) and the Department of Interior (DOI) in 2003 to collaborate on assessment and remediation efforts with multiple stakeholders including the Tribal Environmental Management Services, Quapaw tribe Environmental Office, the Oklahoma Department of Environmental Quality (ODEQ), and Congressional interests.

There are several ongoing remediation activities at the Tar Creek site and surface water and ground water are being monitored by ODEQ. Residential yard remediation, community education, and blood screenings have resulted in reductions from 31.2% in 1996 to 2.8% in 2003 for children between the ages of 1 and 5 living at the Tar Creek site [19]. Proper disposal of drums was part of an emergency removal action. Approximately 75 million tons of chat resides on the ground surface at the site. As a result, pilot studies of sub-aqueous disposal of chat piles and mill residue were conducted. The Bureau of Indian Affairs is focusing on the sale of marketable chat, possibly to pave streets, owned by Native Americans with other Federal and State agencies [18]. Despite the progress made, the lack of cooperation among federal, state, tribal and local entities has been an obstacle to effective and expedient risk management. An approach needs to be implemented that considers both the uncertainties and multiple dimensions of value and contributes to better decisions by helping managers to structure the problem, balance risks, and compare options based on outcomes and expressed stakeholder preferences.

Decision making in environmental projects is typically a complex and confusing exercise, characterized by trade-offs between sociopolitical, environmental, and economic impacts. Some environmental decisions involve many issues, which may be difficult to measure, compare, or compromise. Considerable research in the area of multi-criteria decision analysis (MCDA) has made available practical methods for applying scientific decision theoretical approaches to multiple criteria problems [5]. Potentially, MCDA could prove an asset the Tar Creek project. The Tar Creek project is multi-faceted and many different stakeholders are involved. The Oklahoma Plan for Tar Creek states: "Local cooperation is a key to success, and local stakeholders are and will continue to be involved" (ODEQ No Date). Lack of cooperation among federal, state, tribal and local entities has been an obstacle. A systematic and structured process would be beneficial during the planning phases of decision making because MCDA is designed to address uncertain outcomes and many potentially conflicting objectives from multiple stakeholders will be affected differently by the decisions made. Decision makers would benefit from obtaining four types of input that would aid management of risks at the site: the results of modeling/monitoring studies, risk analysis, cost or cost-benefit analysis, and stakeholder preferences. This chapter reviews the MCDA support framework, discusses MCDA tools along with theories and the advantages and disadvantages of each, and gives a path forward for users of MCDA as a reply to complex decision-making challenges and how MCDA could be used to more effectively manage stakeholder involvement and risks at the site.

2. MCDA Support Framework

Daily life is filled with decisions and some can be made relatively easily. But at times, decision making can be complicated and overwhelming. Decision making is a process described as being continuous, systematic and structured. Policy makers and managers must have diverse skills to address a wide variety of decision-making challenges [15]. Decision analysis is used in ecological restoration, water resources and ecological risk assessments where stakeholder interaction plays a major role in the decision-making process. Restoration project planning starts with the definition of existing problems, a clear statement of project objectives, and an understanding of uncertainty [12]. This decision process emphasizes: 1) the importance of defining objectives related to the appropriate ecosystem structure, function and spatial scale; 2) the role of ecological models, restoration hypotheses, and key ecological parameters; 3) explicit consideration of uncertainties in site processes and material performance in the restoration design; 4) guidelines for project design and feasibility analysis and the use of experimentation at this stage; and 5) monitoring and adaptive management of restoration projects after implementation [12]. There are also objectives within a structured risk assessment decision process. The five elements are process diagrams, long- and short-term goals linked to strategy, reliable information systems, risk assessment tools, and collaboration of environmental personnel across the organization [7]. Fostering improved risk management decisions: specifying '*learning*' for current and future decisions as one of several explicit objectives for the decision at hand, drawing on notions of applied decision analysis [8]. Including learning as an explicit objective of risk management policies enables participating stakeholders, institutions, and decision makers to recast difficult policy choices in a way that increases the opportunities for successful deliberation [8]. Stakeholder involvement is an important factor. Promoting robustness, decisions require synergy among different environmental programs different levels of managers, decision-makers, technical experts and stakeholders. A decision support system must be designed to meet the needs of stakeholders [4]. Stakeholders may request a path to guide their decision making process. Such paths offer a starting point and steps leading to an answer. It can also keep stakeholders on course, participative and in focus [4]. MCDA accomplishes all these things while effectively analyzing extraordinarily dissimilar information. For MCDA to be effective, the underlying assumption is that decision-makers are open and willing to discuss the decision-making components as they develop.

MCDA follows a framework similar to the traditional scientific method. The components are: problems, alternatives, criteria, evaluation, decision matrix, weights, synthesis, and decision. Problems are the initiators and some problems can be considered 'wicked' meaning they contain high-level conflict of objectives and values. Wicked problems also usually have a multitude of uncertainties[21]. The alternatives (often called options) are choices as they relate to the problem. Criteria are used in MCDA to judge the alternative solutions to the decision problem [21]. Criteria sets can and should be monitored throughout the decision process as well. Evaluation measurements are made for each criterion. Criteria are coherent and effective if they are concise, exhaustive, nonredundant, and clear [21]. Decision matrix summarizes the performance of each alternative for each criterion. Weights measure the relative importance of a criterion as judged by the decision maker. The synthesis takes all the decision framework efforts and prepares them for use in the final step of the decision support framework: the decision [21].

3. MCDA Tools

MCDA was developed in response to the need for effectively analyzing multiple sources of dissimilar information. Presently, there are several methods available to simplify the process of decision making by organizing large amounts of information into a manageable format. A significant advantage of multiple criteria methods is the allowance for evaluation of criteria measured either quantitatively (with numerical values) or qualitatively (rank ordering) [1]. Commercial software packages such as Expert Choice® (www.expertchoice.com) and Criterium DecisionPlus® (www.infoharvest.com) are two common examples. Such packages contain decision programs to highlight theories, which are used as engines. Analytical Hierarchy Process (AHP) and Multi-Attribute Utility Theory (MAUT) are approaches that work similarly. They both utilize optimization algorithms by applying numerical factors, as being representative of the importance of criteria, whether it is in respect to the entire goal or singly. In contrast, AHP uses a quantitative comparison method that is based on pairwise comparisons of decision criteria rather that utility and weighting functions thus the rationality assumption in AHP is more relaxed than in MAUT [6].

To begin the MCDA process, each goal should be researched and clearly defined and articulated. The criteria can be compared to the goal, the alternatives and/or each other, depending on project objectives. A systematic approach must be utilized at all times to avoid a breakdown in the decision-making process. The number of criteria developed is directly linked to the number of comparisons evaluated. For example, only six criteria result in 45 comparisons [21]. In essence, the more objectives you have, the more comparisons will need to be evaluated. This is why it is essential to be clear on the defining of necessary criteria within the decision-making process. The weighting process uses pair-wise comparison against the entire decision model. After analysis is completed, the output can be generated in a variety of graphical forms such as 2-D charts, sensitivity line graphs which display interactivity, and head-to-head criteria comparisons using bar graphs, and depends on the software package employed. Based on my experience, Expert Choice® and Criterium DecisionPlus® prove invaluable when addressing with multiple objectives because they both organize a wealth of information in a user-friendly manner.

Expert Choice® is a commercial decision-making tool that is based on AHP theory. It incorporates judgments and personal values in a logical way [14] and is suitable for decisions with both quantitative and qualitative criteria [2]. AHP is a methodology that capitalizes on the concept of building a hierarchy where one begins with an overall goal. It also works well in several settings. If raw data are lacking, Expert Choice® will function with minimal options, criteria, and decision-maker opinions [21]. Because of its capability to allow stakeholders to manually vote individually using keypads, Expert Choice® also can be effectively used at large stakeholder meetings. This capability allows stakeholders to view their decisions and how they fit into the overall decision-making process.

Criterium DecisionPlus® is another commercially available package, but unlike Expert Choice®, it is based on MAUT theory. MAUT is a quantitative comparison method used to combine dissimilar measures of cost, risks, and benefits, along with individual and stakeholder preferences, into high level, aggregated preferences [2]. Criterium DecisionPlus® strengths are its ability to address high levels of uncertainty and large quantities of raw data and thus manage more information than other commercially available programs. It can handle up to 200 alternatives [21]. There is a feature specifically for brainstorming, which allows users to make necessary modifications before or during the matrix compilation process. Criterium DecisionPlus® also produces rich graphical displays including a hierarchy model similar to that of Expert Choice®. In combination, these features enhance the robustness of a decision by effectively managing data and uncertainty.

4. Methodology

There are several ways to begin the decision-making process. A logical place to begin would be to gather together all stakeholders so that their input can be included early in the process. For the Tar Creek Superfund Site, appropriate stakeholders to assemble include: USEPA, USACE, DOI, ODEQ, the Tribal Environmental Management Services, and the Quapaw tribe Environmental Office. This allows the development of initial criteria, weights, and alternatives early in the decision-making process and with all stakeholders present, for the process to begin smoothly. A hierarchical template including criteria, sub-criteria (if necessary), and alternatives is generated and the resultant information is input into the decision program(s). Input modifications can be made at any time, an advantage inherent in both software packages. After revisiting, and possibly modifying the template, the decision program(s) are utilized. Time required to complete this task varies based on the complexity of the decision-making process. In a simple scenario such as the purchase of a new car, analyses can be explored within a matter of hours. On the other hand, the decision-making process could be prolonged (e.g. weeks) when assessing complex, multi-dimensional issues. Once the goals, criteria, subcriteria and alternatives are decided on, they are input into the MCDA program of choice. For comparison purposes, results of both Expert Choice® and Criterium DecisionPlus® programs are presented.

A scenario was developed to determine how to improve environmental quality at the Tar Creek Superfund Site, which is the overall project goal. Important criteria and sub-criteria are health effects, which include lead, cadmium, and arsenic blood levels, subsidence, chat use, water quality, natural resource damage, and mine shafts. Potential alternatives are no action with possible monitoring, beneficial reuse, phytoremediation, passive or pump water treatment, removal by excavation, containment and stabilization, and sale of marketable chat [18]. This information is input into the Expert Choice® and Criterium DecisionPlus® software packages to form working hierarchy models (Figures 1 and 2).

In Criterium DecisionPlus®, a graphic brainstorm allows users to enter a central goal, branch-like criteria and alternatives (Figure 2).

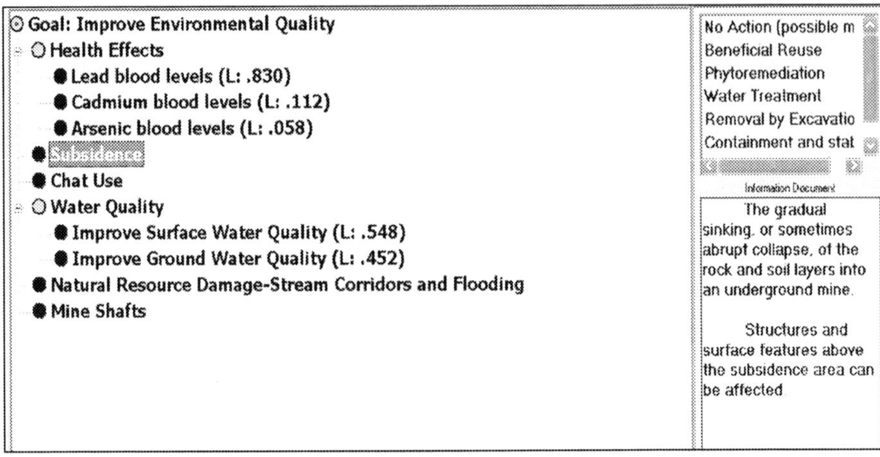

Figure 1: Expert Choice® hierarchy.

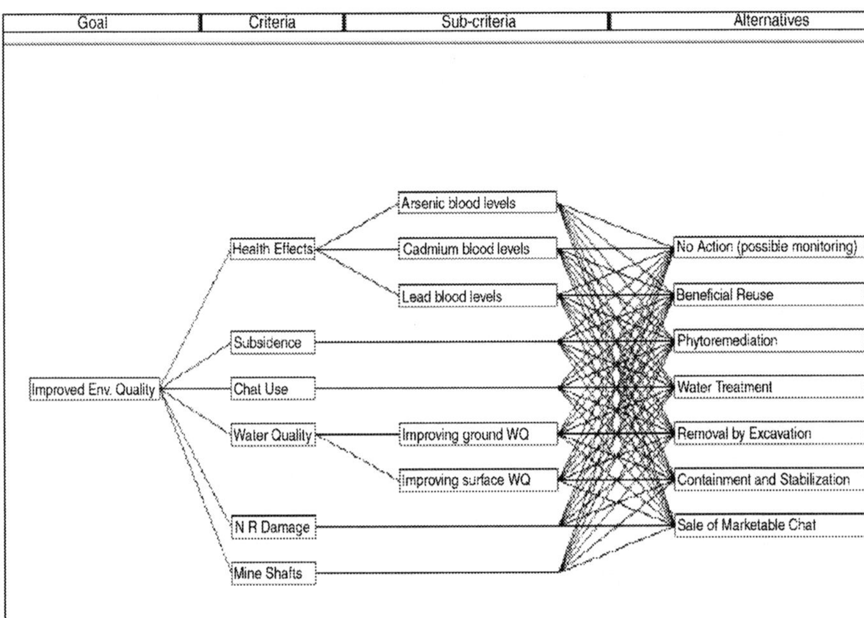

Figure 2: Criterium DecisionPlus® hierarchy model.

Setting ratings to cross-sectional entries incorporates weights. For example, "with respect to improving environmental quality, which is more important; health effects or natural resource damage?" In a hypothetical stakeholder meeting, pairwise comparisons and weightings were made in both Expert Choice® and Criterium DecisionPlus® (Figures 3 and 4).

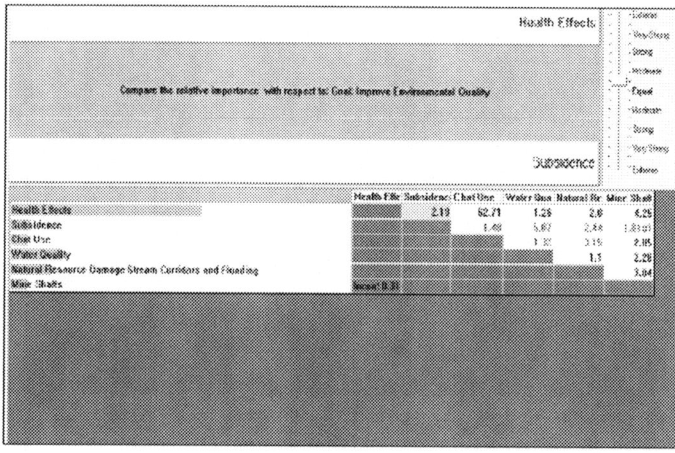

Figure 3: Expert Choice® pairwise comparison.

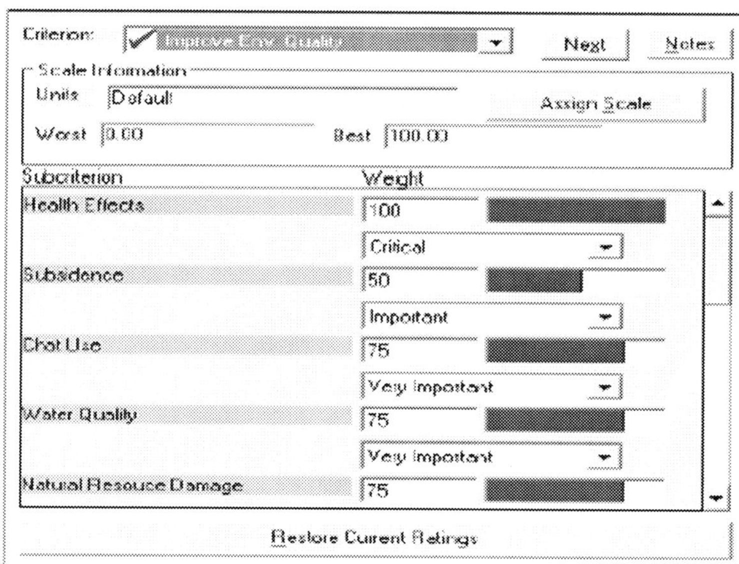

Figure 4: Criterium DecisionPlus® weighting.

Such comparisons allow stakeholders to discuss relative weightings, thus continuing the structured communication pattern throughout the decision-making process. On screen, users can click on the overall goal or criteria and select vehicles and displays where they may either choose by extremes, a manipulative pie chart, or numerical values to make other qualitative or quantitative comparisons. Weights will appear by each comparison. Weights were assigned where worst equals 0 and best equals 100 and were selected in regards to site-specific data. Composite scores were then calculated (Figures 3 and 8). With respect to the goal of improving environmental quality, water treatment (26.8%) had the highest ranking in both software packages (Figures 6 and 8) with beneficial reuse (18.7%) and phytoremediation (16.7%) having slightly lower rankings. The Expert Choice® synthesis (Figure 5) reflected human health effects (56.5%) to be of critical concern with water quality (15.3%) and natural resource damage (12.5%) posing threats as well. Sensitivity graphs for both programs (Figures 7 and 9) allow participants the ability to performs a simple sensitivity analysis, as the MCDA facilitator can relay to the stakeholder participants' predictions when evaluating shifting numbers showing different outcomes, especially when evaluating close ranking criteria. Again, monitoring by revisiting will permit the group to discuss eliminating continuous, low-rating criteria.

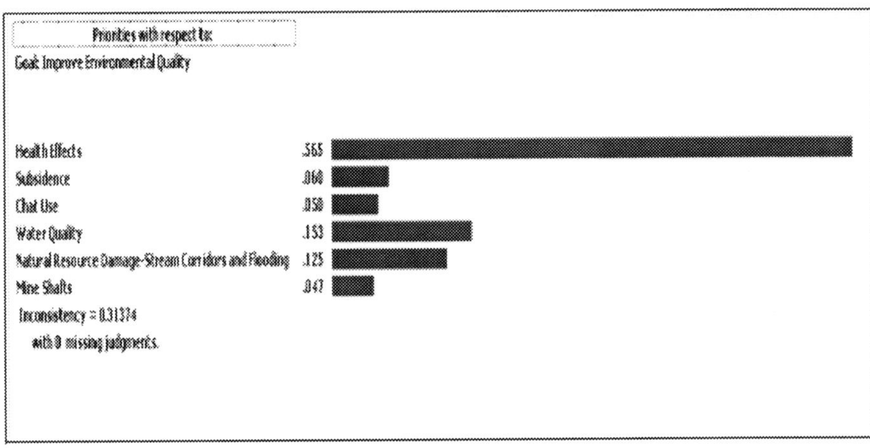

Figure 5: Expert Choice® results.

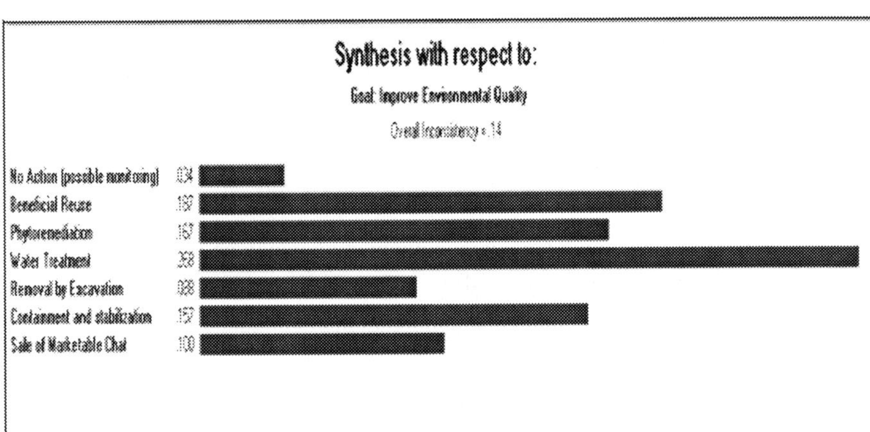

Figure 6: Expert Choice® synthesis.

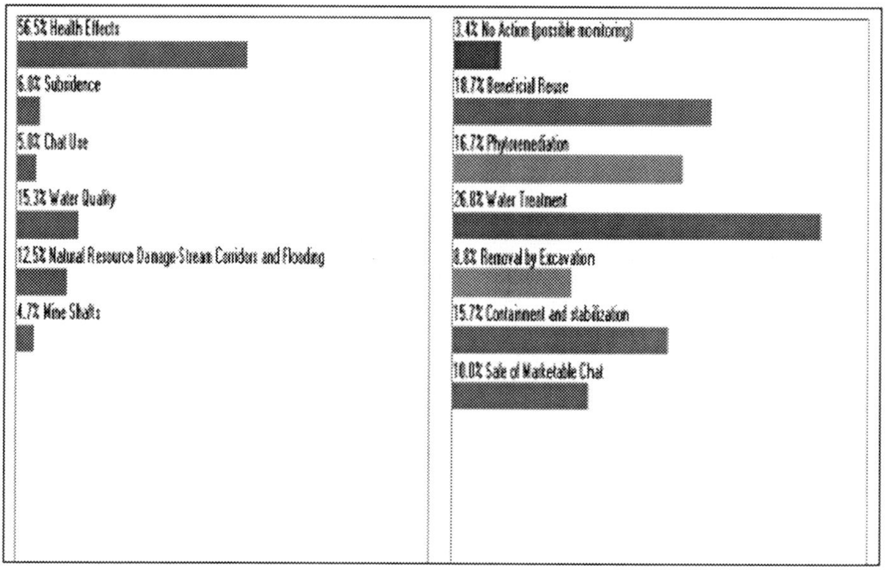

Figure 7: Expert Choice® sensitivity graph.

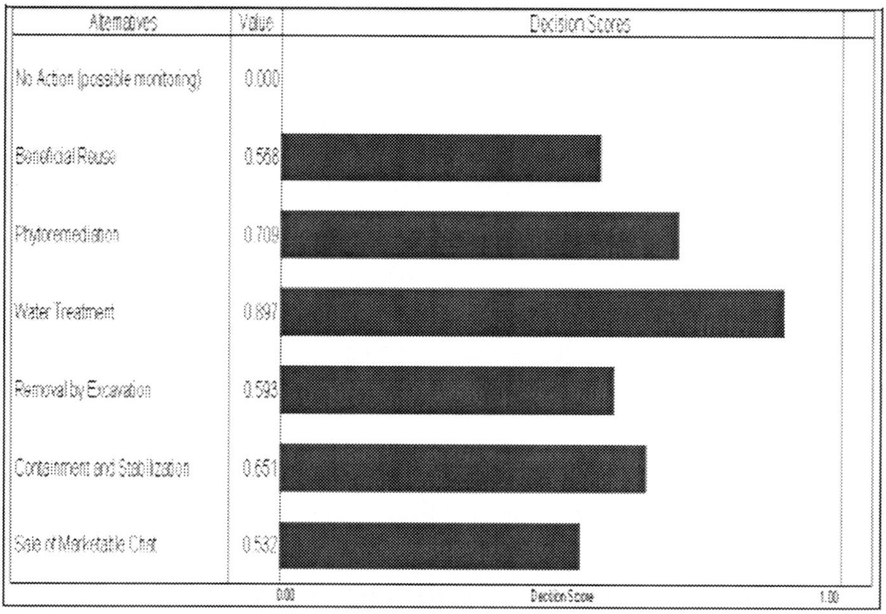

Figure 8: Criterium DecisionPlus® decision scores.

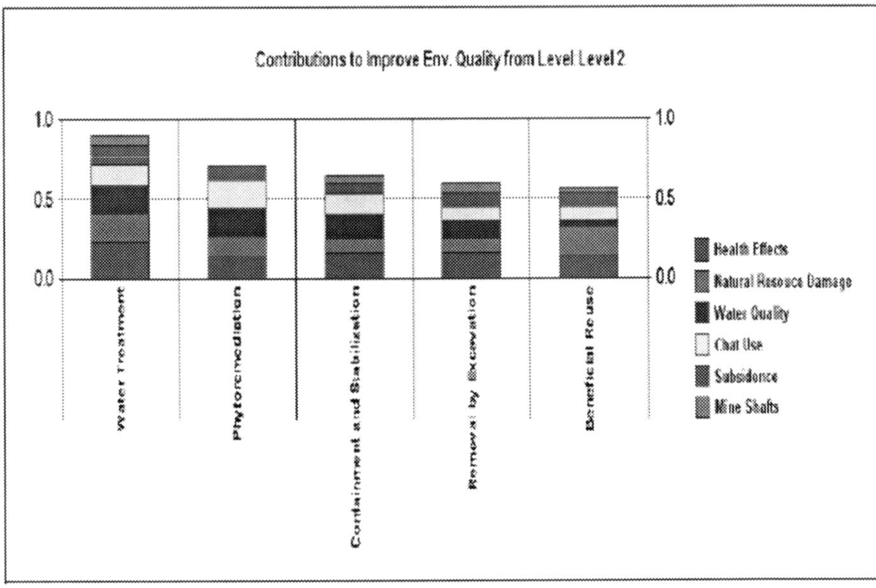

Figure 9: Criterium DecisionPlus® contributions by criteria.

There are a number of options to pursue when deciding on how to implement the application of decision making; i.e., contract negotiations and/or on-the-job training of in-house personnel. There are also companies that specialize in assisting with such cases, which can be hired to operate the software package. Technical and nontechnical experts can provide mental models or conduct stakeholder meetings as facilitators. Some companies simply enter data and work out complex uncertainty issues. The most cost-effective means to gain expertise in MCDA might be to train interested employees in-house so that they can serve in the capacity again in the future.

5. Conclusions

In the Tar Creek Superfund Site project, there are a multitude of issues plaguing risk management; i.e., human and ecological health, Native American concerns, water quality, and natural resource damage. While some progress has been made to remediate the site, the lack of cooperation among federal, state, tribal, and local entities has been an obstacle to effective and expedient risk management. However, there are approaches that can be implemented to address these complex decision-making situations. Problems often present themselves as being massive and unattainable, but approaches such as MCDA can provide a systematic structure to add organization, particularly with multiple objectives and stakeholders by using decision-making tools. Expert Choice® and Criterium DecisionPlus® are decision software programs that work well in such situations. Each package has unique strengths depending on the amount of available data and uncertainty. Large stakeholder groups, where decision-makers lack raw data, plays to the strengths of

Expert Choice®; whereas data-rich scenarios with high levels of uncertainty are situations that can be served well by using Criterium DecisionPlus®. These methodologies are all under the premise that a decision needs to be made following a complex and critical problem and decision-makers are open and willing to discuss the decision-making components as they develop. Although remediation activities are ongoing, risk management at the Tar Creek Site could greatly benefit by utilizing a planning and decision-making process such as MCDA. Both software packages indicated water treatment as the highest ranking risk management option for improving environmental quality, with phytoremediation and beneficial reuse having slightly lower rankings. Synthesis in both commercial packages also displayed human health effects as the primary concern with the highest ranking, with water quality and natural resource damage having lower rankings. Such rankings reflecting stakeholder inputs can lead to more effective, efficient and credible decision making at the Tar Creek Site. Tar Creek has been an environmental as well as economic and social dilemma for decades and using MCDA as a process, assisted by decision-making tools that are commercially available, provides a framework for managing such complex environmental challenges.

6. Acknowledgements

The author would like to express gratitude to Mr. Douglas Lilly, USACE-Tulsa District, for his permission to publish this document. Special thanks to Drs. Elizabeth Ferguson and Burton Suedel for their instructive comments.

7. Disclaimer

The work presented by the authors was performed in their private capacities and does not reflect the policies or views of their parent institutions.

8. References

1. Annandale D and R Lantzke. 2000. Making Good Decisions: A Guide to Using Decision-Aiding Techniques in Waste Facility Siting. Murdoch University. Perth, Australia.

2. Baker D, D Bridges, R Hunter, G Johnson. 2001. GuideboOklahoma to Decision-Making Methods. Department of Energy (WSCR-IM-2002-00002).

3. Center for Disease Control-Morbility and Mortality Weekly Report. 2003. Surveillance for Elevated Blood Lead Levels Among Children---United States, 1997-2001. 52(SS10): 1-21.

4. Chen CW, J Herr, L Weintraub. 2004. Decision Support System for Stakeholder Involvement. Journal of Environmental Engineering. 130(6): 714-721.

5. Linkov I, A Varghese, S Jamil, TP Seager, GA Kiker, TS Bridges. 2005. Multi-criteria decision analysis: a framework for structuring remedial decisions at contaminated sites. In: Linkov I and Bakr

Ramadan A (Eds.) Comparative risk and environmental decision-making. pp. 15-54. Kluwer Academic Press, The Netherlands. (in press).

6. Kiker, G, TS Bridges, A Varghese, TP Seager, I Linkov. 2005. Application of Multicriteria Decision Analysis in Environmental Decision Making. Integrated Environmental Assessment and Management. 1(2): 95-108.

7. Matthews DH, GC Christini, CT Hendrickson. 2004. Five Elements for Organizational Decision-Making with an Environmental Management System. Environmental Science Technology. 38: 1927-1932.

8. McDaniels TL and Gregory R. 2004. Learning as an Objective within a Structured Risk Management Decision Process. Environmental Science and Technology. 38(7): 1921-1926.

9. Natural Resources Damages Subcommittee. 2000. Alternatives for Assessing Injuries to Natural Resources at the Tar Creek Superfund Site.

10. Natural Resources and Environment (NRE) 492 (Group 5). ND. The Results of Mining at Tar Creek. *http://www.umich.edu/~snre492/cases_03-04/TarCreek/TarCreek_case_study.htm*

11. Oklahoma Department of Environmental Quality. ND. Oklahoma Plan for Tar Creek. *http://inhofe.senate.gov/superfun1.pdf*

12. PastorOklahoma RA et.al. 1997. An ecological decision framework for environmental restoration projects. Ecological Engineering. 9: 89-107.

13. Polk, J. 2005. Tulsa District leads multi-agency team in Tar Creek Cleanup. *http://hq.environmental.usace.army.mil/Corps_Environment/story3.htm*

14. Saaty, T. 1995. Decision Making for Leaders. RWS Publications.

15. Seely MK, J Zeidler, JR Henschel, P Barnard. 2003. Creative problems solving in support of biodiversity conservation. Journal of Arid Environments. 54:155-164.

16. Stahl, RG et.al., 2001. Risk Management: Ecological Risk-Based Decision-Making. Pensacola, FL. SETAC Press.

17. Tar Creek Native American Issues. ND. Final Report.

18. United States Army Corps of Engineers. 2004a. Tar Creek and Lower Spring River Watershed Management Plan. Draft. August.

19. United States Environmental Protection Agency Region 6. 2000. Five Year Review Tar Creek Superfund Site Ottawa County, Oklahoma. April. http://www.epa.gov/earth1r6/6sf/pdffiles/5yeartar-creek.pdf

20. United States Environmental Protection Agency Region 6. 2005. Tar Creek (Ottawa County). May 4, 2005. http://www.epa.gov/earth1r6/6sf/pdffiles/0601269.pdf

21. Yoe C. 2002. Trade-off Analysis Planning and Procedures GuideboOklahoma. U.S. Army Corps of Engineers, Institute of Water Resources Report (IWR 02-R-2).

UNMANNED SEA VEHICLES FOR HARBOR PATROLLING

Low-Cost Solutions to Enhance Coastal Security

S. CHIESA, N. VIOLA
Politecnico di Torino
Department of Aeronautic and Space Engineering, C.so Einaudi 40, 10129 Torino, Italy

A. MANNINI
Dyria Sistemi S.r.l.
via Veglia 57/5/d, Torino Italy

S. OLIVERO, F. STESINA
SiTI
Via Lagrange 35, 10123 Torino, Italy

Abstract

Harbors can be considered one of the most appealing targets for terrorist threats. Considering that small ships or swimming commandos cannot be easily tracked by on-ground surveillance and that on-site surveillance, by means of manned ships, is quite expensive and difficult to conceal from enemy eyes, the interest in Unmanned Sea Vehicles (USVs) is quite clear. USVs may be one or two meters long, equipped with proper surveillance sensors. If state-of-the-art technology is taken into account, USVs can also be thought of as low-cost vehicles. The paper presents a system low-cost solution and it highlights the usefulness of tools like Comparative Risk Analysis (CRA), Multi-Criteria Analysis (MCA), and simulation techniques.

1. Introduction

In times of serious terrorist threats, the nation's security is a priority. Harbors can surely be considered one of the most appealing targets to terrorists. There are many reasons including:

- Harbors have high relevance from both social and territorial points of view.
- Harbors represent a concentration of great value in a very narrow area. Serious damage can be easily caused within a small area (see Figure 1).
- Harbors are characterized by a significant density of infrastructure and facilities, generally higher than that of airports or other major transportation.

I. Linkov et al. (eds.), Environmental Security in Harbors and Coastal Areas, 465–483.

- Harbors generally neighbor civil areas. This implies that any damage that occurs in harbors can reach city zones (see Figure 2).
- Like airports, railway stations, and industrial areas, harbors can be damaged by ground and aerial attack, but, unlike the those others, harbors can be attacked also by sea forces (see Figure 3).

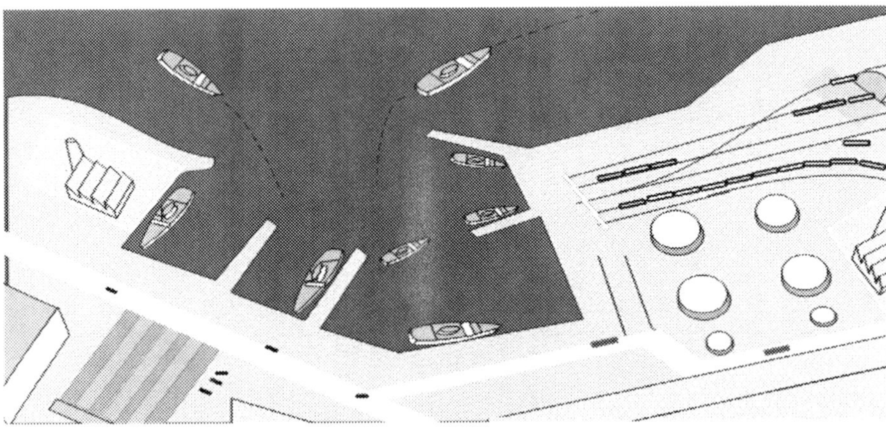

Figure 1: Harbors concentrate high-value targets in a small area. There is remarkable density of infrastructures and facilities.

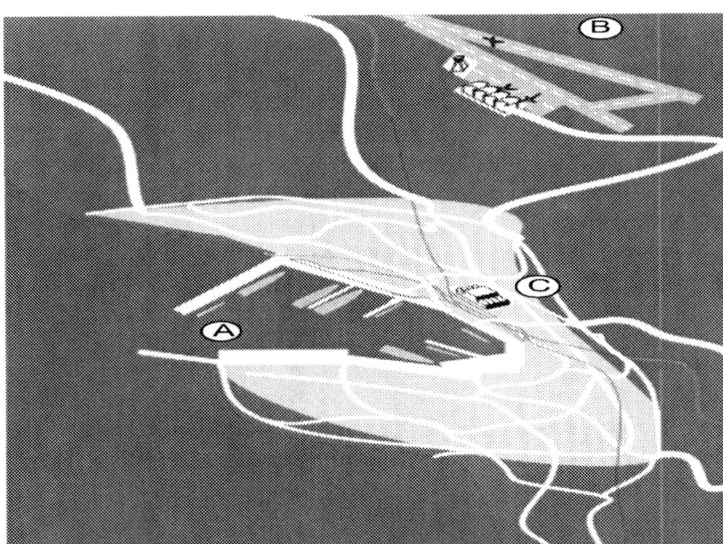

Figure 2: Harbor (A), airport (B), railway station (C) and city. In many areas, these facilities are located close to one another.

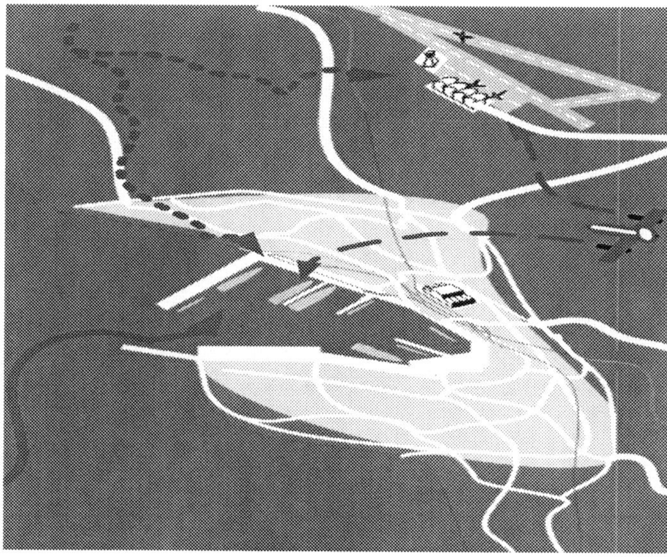

Figure 3: Harbor and airport potential injuries.

During the 20th century, military bases in harbor areas have been targets of high value because of ships moored there, air bases, logistic, petroleum, and command and control facilities. All these have been shown to be highly vulnerable to attack from the sea (Tables 1a and 1b).

Table 1a: Historical examples: harbors attacked by sea.

Harbor/date	Attack forces	Damages
Trieste/12-10-1917	1 torpedo-boat (MAS)	1 battleship sunk
Premuda/O6-10-1918	2 torpedo-boat (MAS)	1 battleship sunk
Pola/11-01-1918	2 men swimming with a contact mine	1 battleship sunk
Alexandria (Egypt)/12-18-1941	6 men on 3 "slow speed torpedoes"	2 battleship and 1 oil-tanker damaged

Table 1b: Historical examples: harbors attacked by air.

Harbor/date	Attack forces	Damages
Taranto/11-13-1940	21 torpedo planes	3 battleships damaged
Pearl Harbor/12-07-1941	About 300 drop-bombers and torpedo planes	2 battleship sunk and 4 battleships damaged

2. The Problem

Small ships or swimming commandos cannot be easily tracked by on-ground surveillance. In order to deal with this problem, nations have developed efficient detection systems in relevant coastal areas and harbors.

In Table 2, systems for surveillance and security, which are currently implemented or under study, are listed.

Table 2: Sensor types.

Type	Usefulness	Challenges	System Problems
Optical sensors (TV/IR Camera) installed in coastal facilities or in floats	Detection of surface boats and swimmers	▪ Fog ▪ Small targets	High number of sensors required
Radar sensors installed in coastal facilities or in floats	Detection of surface boats and swimmers	▪ Small targets ▪ Difficult identification	▪ High number of sensors required ▪ False alarm
Passive acoustic sensors	Detection of underwater targets	Difficult detection of surface targets and "sub-skimmers" air/water interface	▪ High number of sensors required ▪ False alarm
Active acoustic sensors	Detection of underwater targets	Small targets	Self revealing
Combination of optronic and underwater sensors	Detection of surface and underwater targets	▪ Different sensors' data fusion ▪ System integration	▪ Complexity and costs ▪ Difficult re-deployment
Piloted boats	▪ Inspection and identification ▪ Possibility of countervailing	Interference with normal activity	Cost

From the table above it is clear that:

- The concurrent employment of different types of sensors is unavoidable.
- The use of patrol boats is necessary for inspection, identification, and counteracting possible actions by enemy forces.
- Patrol boats have three main advantages.
  - They can carry different types of sensors, thus reducing the number of sensors that have to be placed into a fixed position.
  - They can bring sensors, installed on board, closer to the targets, thus enhancing the identification capabilities.

- On the other hand, on-site surveillance, by means of manned ships, may have the following drawbacks:
  - Patrol boats are easily detected, so they cannot be concealed from enemies.
  - Patrol boats are expensive.
  - Patrol boats interrupt normal harbors activity.

Taking all these considerations into account, as shown in Figure 4, it appears quite reasonable to conclude that Small Unmanned Boats (SUBs) represent a good alternative to manned patrol boats.

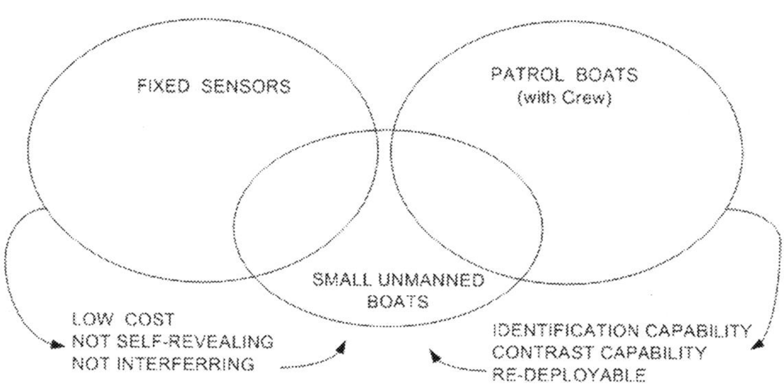

Figure 4 : Advantages of small unmanned boats.

The above thoughts lead to the interest in Unmanned Sea Vehicles (USVs): only a few meters long, equipped with proper surveillance sensors, propelled by silent electrical motors, and hardly avoidable by potential terrorists. If state-of-the-art technology is taken into account, USVs can also be thought of as low-cost vehicles. Low-cost USVs guarantee the availability of the fleet. In fact, the battery limitation of these relatively inexpensive vehicles can be overcome by simply increasing their number. In this way, when one vehicle goes out of use because of a low battery, others are ready to replace it.

A USV can also be equipped with acoustic alarm (and/or bright rocket), in order to:

- Raise other defenses in case of emergency, danger, or situations.
- Confuse enemies.
- Make security forces aware of potential attacks by men or machines.

It is important to remember that USVs can be equipped with GPS and proximity sensors, thanks to which they would be able to move along very narrow paths (e.g., between anchored ships) and along complex trajectories, characterized by sudden and unpredictable drifts and interruptions.

3. A New Idea of SUBs

According to what have been said in the previous paragraphs, examples of studies and manufacturing of USVs or, restricting our talk, of SUBs, are not lacking.

It has to be noted that in order to turn the potential advantage of difficult detectability into a real feature of USV/SUB (see Figure 4), in many cases they are characterized by half-submerged configurations.

A new idea of SUB has been conceived by the Working Group, of which the authors of the present paper are part. As summed up in Figure 5, in order to enhance cost reduction, the proposed SUB is characterized by a quite simple configuration and it has very small dimensions, thus implying the possibility of employing silent electric motors.

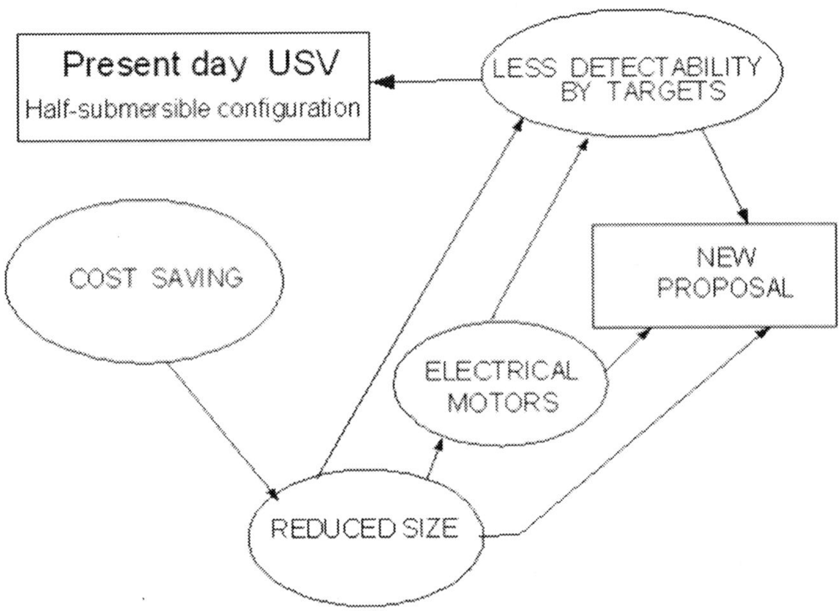

Figure 5: New SUB configuration.

The research strategy chosen by our Working Group is a step-by-step approach, characterized by the development of a prototype prior to the new SUB's manufacture. Notwithstanding its very low cost, the prototype has a very important role in the development of the new SUB as it is able to demonstrate the future SUB's design choices.

As far as the prototype's development is concerned, it has been decided to use a commercial amateur radio-controlled boat and to adapt it to accomplish its new role.

After a detailed analysis of what is available on the open market, we have chosen a boat with very small dimensions and high performance characteristics. It is driven by electric motors and it can house additional components, if needed. No problems of extra weight and volume arise. Table 3 shows the boat chosen and lists its main technical data.

Table 3: Prototype and technical data.

Main Technical Data	
Length:	935 mm
Width:	260 mm
Height:	190 mm
Weight	2600 g
Engines:	two 550 size fan cooled with water jackets
Propeller:	diameter 42 mm x pitch 1.2, shaft size: 4 mm
Battery:	2 x 7.2 V 1400-3300 mAh
Radio:	2 channel

Figure 6 presents a detailed view of the prototype's onboard systems; in particular the electrical motors and the actuator of the rudder.

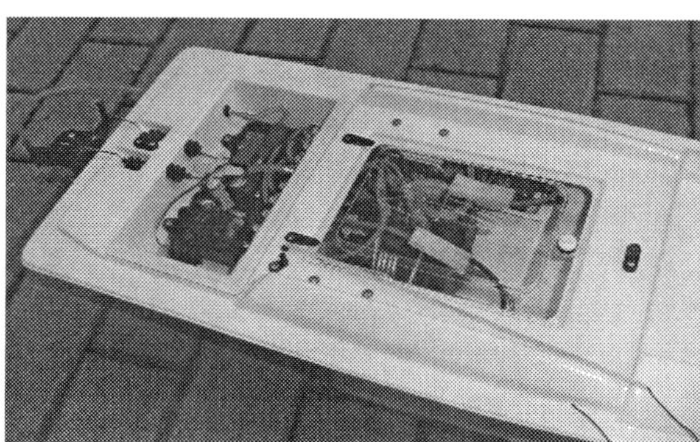

Figure 6 : Propulsion and steering systems.

One of the first steps to transform the boat from a common commercial model to a survey SUB technological prototype was to substitute the original radio-control for a higher performance model with more channels (six instead of two), which are necessary not only for guiding the boat (power of the motors and position of the rudder) but may also be used for managing sensors and possible alert devices (i.e., flares). Table 4 shows the new radio-control and its main technical data.

Table 4: Radio control and technical data.

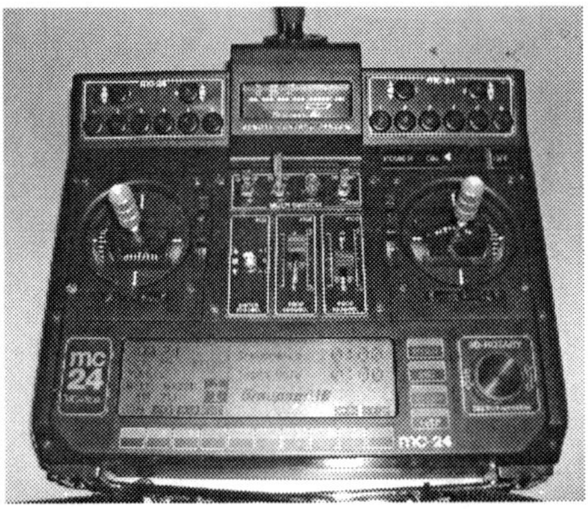

Main Technical Data	
Channel	6 (FM/PCM signal)
Model memory	8
Trim	Digital
Band	40 MHz
Flight Mode	3 (Manual, semiautomatic, automatic)
Size	190 mm x 175 mm x 40 mm

The next steps were:

- Installing onboard mission systems, consisting of one TV camera (equipped with audio sensor) and dedicated telemetry, as well as one GPS receiver, which, by means of the general telemetry package, transmits position data to the on-ground remote control station. By means of the radio-control the operator is able to guide the prototype by acting on the motors' power and rudder's position. Thanks to both the transmission of the GPS data and the TV camera images, the operator can pilot the prototype even when it is hidden from his/her view.
- Increasing the number of battery packages, useful for both feeding the Mission System and enhancing the operational duration.
- Adding a sensor able to reveal the battery's low-charge state. The transmission of this sensor's signal by means of general telemetry to the operator can significantly help manage the boat's resources.

As Figure 7 illustrates, both the basic and the mission systems are integrated by means of the Power Supply electronic card and the general telemetry, which, apart from transmitting GPS and battery state data, can be useful to transmit information from other sensors that may be integrated in the future.

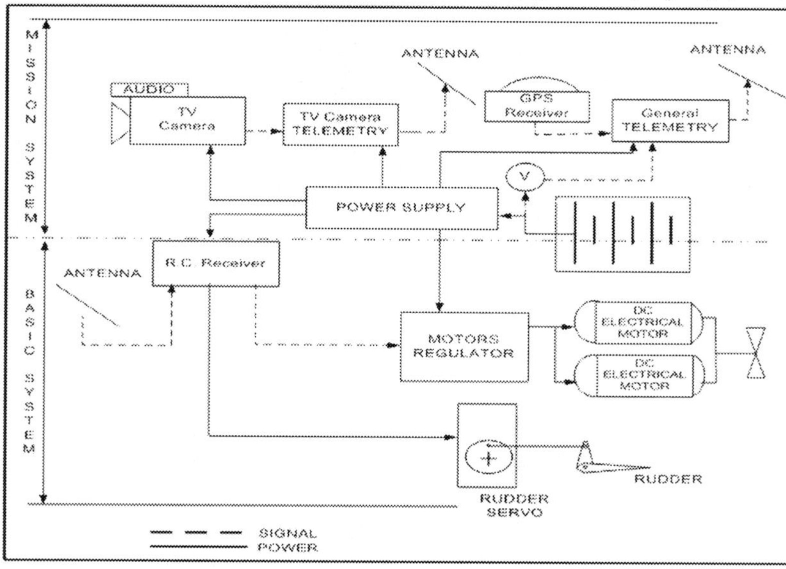

Figure 7: System layout.

The survey SUB technological prototype has also a Remote Control Station, manufactured to be easily transportable and illustrated in Table 5.

Table 5: Remote control station.

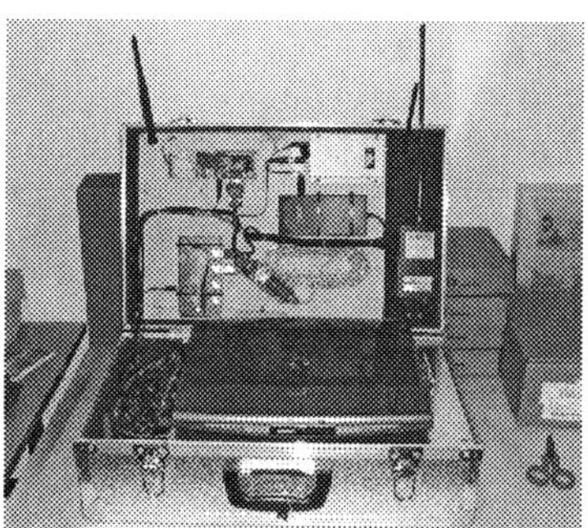

Radio-Control Transmitter

General Telemetry Receiver

TV Camera Receiver

PC

Battery (rechargeable)

Battery Charter

Future Improvements

Both Figure 8 and Figure 9 are pictures taken during the first test: in particular, Figure 8 shows an image taken from the TV camera on board the SUB vehicle and Figure 9 shows the prototype sailing rapidly.

From Figure 9, it appears quite clearly that another field to investigate concerns the prototype's camouflage, necessary to further reduce its chances of being detected by possible targets.

Figure 8: TV camera image.

Figure 9: SUB with TV camera installed. Camouflage efficacy.

By comparing Figure 9 with Table 3, it is possible to view the TV camera's system, installed on board the prototype. It is important to verify the TV camera's capability, considering its simplicity and low cost. A new solution, which is likely to become the standard solution for SUB, is already under development: it is characterized by a TV camera, which is able to move along training and elevation arcs and gyro-stabilized about roll and pitch axes. Figure 10 illustrates the study's progress for a gyro-stabilized system, which can be critical to its usually high cost.

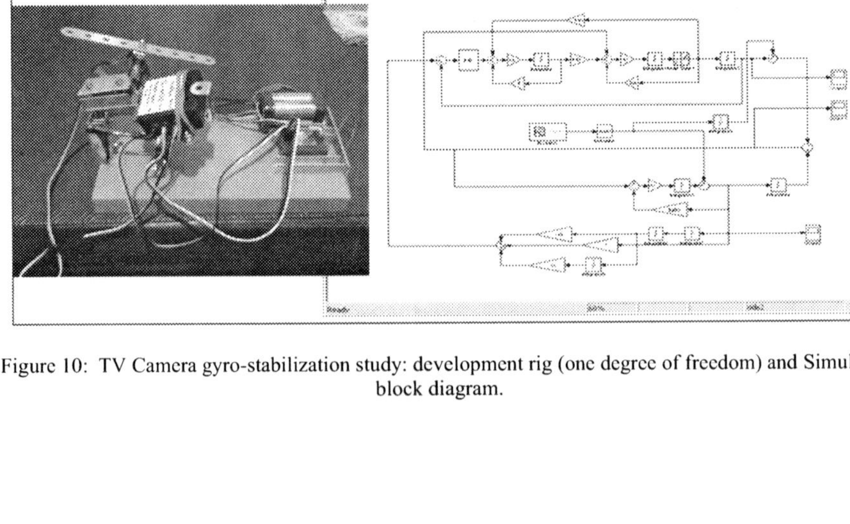

Figure 10: TV Camera gyro-stabilization study: development rig (one degree of freedom) and Simulink block diagram.

4. Other Applications of SUB Technologies

In addition to the security applications against terrorist attacks, SUB technology can be exploited in other fields, including environmental protection.

Quite often manufacturing and chemical industries are located in the neighborhood of harbors for the obvious advantage of easy transport of goods and direct availability of water for machine cooling. Unfortunately such industries may be a potential source of pollution.

Of course, other sources of potential pollution to the sea or the rivers are oil leakages and other toxic substances that could be discharged from the boats following an accident or because of crew behavior.

In order to monitor the environment, the SUB can be equipped with appropriate instrumentation, according to the specific mission needs.

Several kinds of missions can be foreseen:

- Patrolling seashore or river banks with environmentally critical industry installations. The SUB will be equipped with appropriate devices to take water samples and, if needed and possible, to accomplish real-time chemical analyses. Analysis results can be both radio-transmitted to a station (on the ground or onboard a mother-ship), recorded onboard, and then carried by the SUB itself to the station. Of course, if real-time analysis is not needed, the samples can be taken to the station and then analyzed.
- Water analysis or environment observation in areas (where it is hard to get manned machines) which can be reached only trough sea or rivers.
- Search and location of shoals for fishing or scientific purposes. For such missions, the SUB will be equipped with sonar and/or an underwater video camera. Also, in this case, the operation can be performed by several SUBs at the same time in different areas, coordinated by a mother ship.
- Patrolling of sea areas against illegal fishing and clandestine immigration. For this purpose, the SUB can be specifically designed and manufactured with special materials in order to avoid radar detection.

An important feature of all SUB applications is the SUB's high flexibility and ability to perform requested missions in virtually any situation and environment, considering contained cost and full safety conditions. The use of SUBs avoids the need of a fixed network of sensors with their related supporting facilities and allows for performance of the requested checks immediately when and where they are needed.

Figure 11 provides examples of applications of SUBs for environmental monitoring.

Figure 11a: Search and location of shoals for fishing or scientific purposes.

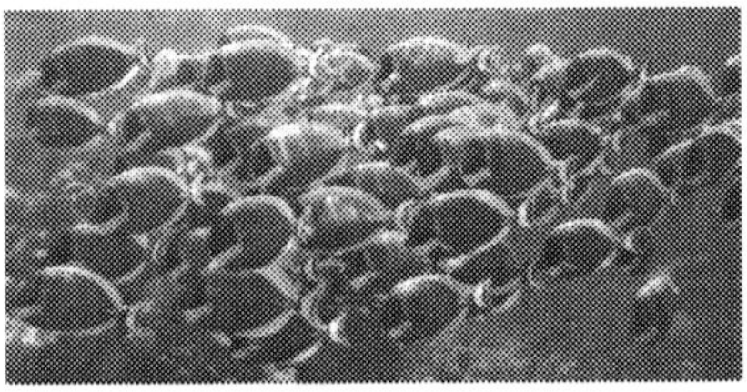

Figure 11b: Monitoring of polluted areas in consequence of accidents.

Figure 11c: Monitoring of polluted areas in consequence of accidents.

5. Further Developments

The development of the technological prototype, described in the previous sections of the paper, aims at both gaining basic information about SUB performance and expressing hypotheses about how the various design characteristics may affect the surveillance's efficacy. As an example, Figure 12 and Figure 13 show that both different performance (i.e., different ranges for onboard sensors or different values for SUB speed) and patrol strategies can influence mission efficacy. As far as patrol strategies are concerned, one of the most important factors is the number of SUBs simultaneously available and employed. This is highly dependent on Availability [1], which is on its turn dependent on Reliability and Maintainability as well as on the adopted logistic strategy (Maintenance Policy, spares supplies, number and skill of Maintenance Technicians). Please note that the logistic strategy can be considered as an extension of the operative strategy [2].

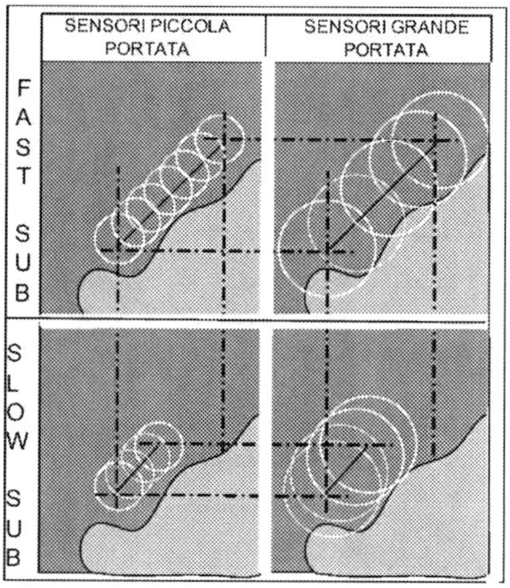

Figure 12: Example of influence of sensors and boat performance.

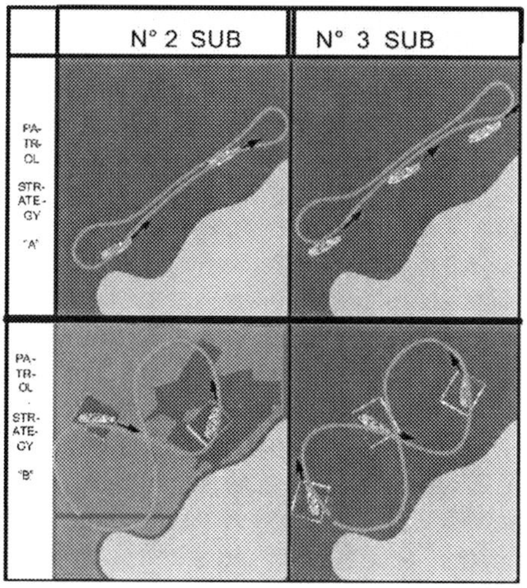

Figure 13: Example of influence of number of SUBs and patrol strategy.

It is important to observe that the parameters defining the patrol efficacy are not independent of each other but, in general, have mutual negative influence. Moreover each parameter so far considered is also a fundamental characteristic of the system's cost, either as Purchasing Cost or, in a wider view, as Life Cycle Cost (LCC). This implies the possibility of defining the System Effectiveness as a ratio between System Efficacy (i.e., for instance, the probability of intercepting an intruder in a certain environment) and System Cost. Figure 14 illustrates these concepts.

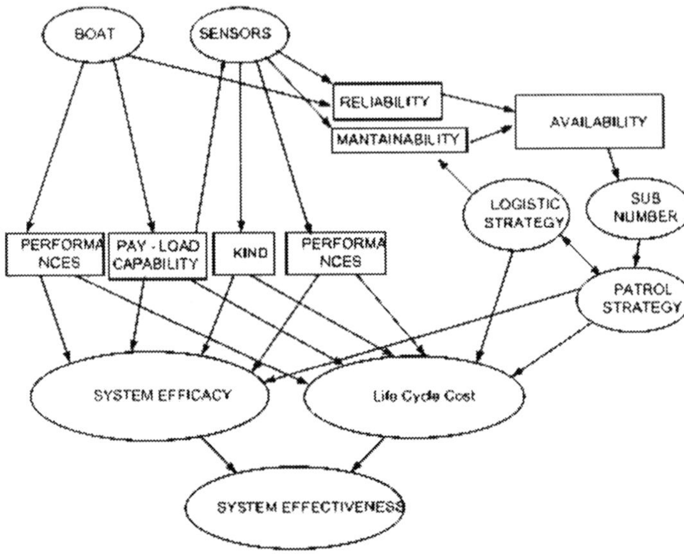

Figure 14: System effectiveness and influencing parameters.

Finding the best solution for complex problems, characterized by a great number of variables whose value may generally change (please note that, apart from the parameters in Figure 14, other variables defining the threat and the environment have to be considered), is not an easy task. Therefore standard algorithms do not suit these problems.

When dealing with such a problem, it is necessary to establish a rigorous procedure which helps the analyst to tackle the subject matter through a standard and verifiable approach. The main target is to make the analyst able to choose from time to time the most appropriate tools, according to the kind of problem, its level of definition, the type of available information, etc. For example, the development of a fleet of SUBs and its relative operative and logistic strategies may represent a different problem, which needs to be dealt with using different tools, depending on whether or not the SUB vehicles' characteristics have already been defined. In the former case, when an already available system is deployed in another operative theatre, the problem presents a smaller number of variables but, quite obviously, also more constraints in order to optimize the final result.

Basically, a more defined problem is suitable for being solved through an algorithmic approach, whereas a qualitative method is better for a less defined problem.

Figure 15 illustrates the methodology usually adopted by our Research Group to deal with a complex system [3].

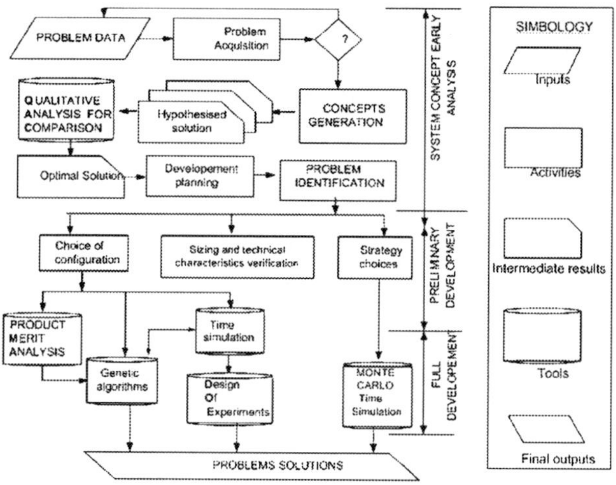

Figure 15: Methodology for system development.

In the more general situation, this methodology first of all provides for the detailed definition of the initial problem's data and constraints and then it provides for a qualitative approach, being the definition level of the problem still low and new. Afterwards the methodology proceeds to elaborate a few hypothesized systems concepts, among which the best one can then be chosen by means of the Qualitative Analysis of Comparison (QAC). As an example, Figure 16 shows how the QAC can be applied to accomplish a surveillance system for harbors. As can be seen below, all the different employable vehicles are compared from the point of view of efficacy, risk to human life, system availability, and cost. If the pictured table, which sums up all comparisons, has been edited according to previously defined criteria, in order to guarantee the objectivity of the comparison itself, it constitutes evidence and valid support for decision making (choosing which kind of vehicle to consider first).

Patrol and surveillance platform	Human Risks	Efficacy	Environmental Impact	Availability in near future	Cost	N° of STARS
Artificial satellite	NO *	Medium	Nothing *	YES *	Very High	3
Manned aircraft	YES	High *	Medium	YES *	Very High	2
HALE aircraft	NO *	Medium	Nothing *	NO	Very High	2
Micro UAV	NO *	Low	Heavy	Perhaps	Medium	1
Manned boats	YES	High *	Heavy	YES *	High	2
USV/SUB	NO *	High *	Nothing *	Perhaps	Very Low *	4
Airship unmanned	NO *	Medium	Medium	Perhaps	Medium	1
Fixed balloon	NO *	Low	Medium	YES *	Low	2

Figure 16: Qualitative analysis of comparison (simplified example).

After the first qualitative approach has been completed (maybe in repeated steps), one or more system concepts are put through the successive activities of development planning and problem identification. Please note that the latter implies also the choice of the tool best suited to deal with the problem itself.

Identified problems and their relative solutions may be quite different one from another. In some cases qualitative approaches may apply; in others numerical methods may be better suited.

As an example of qualitative approach, one of the most used qualitative analysis is the Product Merit Analysis, which is typical of the Value Analysis [4]. Table 6 shows a simple example of the Product Merit Analysis applied to the choice of the SUB propulsion system.

Unlike the Qualitative Analysis of Comparison, shown in Figure 16, we notice that the Product Merit Analysis, illustrated in Table 6, compares the two power systems from the point of view of several aspects, i.e., several technical characteristics.

When the level of problem definition is higher than in the previous case, the Product Merit Analysis is more detailed and assigns two different evaluations to each considered characteristic:

- Merit G, a numerical evaluation which determines the rate at which the requirement has been met.
- Importance Degree P, a numerical evaluation which weights the requirement.

Table 6: Product Merit Analysis (PMA) example.

Comparison between Electrical (DC motor + batteries) and Endothermic Propulsion System for a Small SUB

Technical Characteristics	Importance Degree "P"	Electrical Propulsion Merit "G"	Electrical Propulsion P X G	Endothermic Propulsion Merit "G"	Endothermic Propulsion P X G	REMARKS
Environmental compatibility	2	3	6	2	4	The degree of importance has been decided with a small, touristic harbor in mind.
Low noise	3	2	6	1	3	
Range	1.5	1	1.5	3	4.5	The Merit Criteria have been established on three levels: poor, medium, good (i.e., 1, 2, 3), directly expressed by analyst judgement thanks to the simplicity of the case.
Complexity	1.5	2	3	1	1.5	
Easy to start and re-supply	1.5	3	4.5	1	1.5	
Power/Weight	0.5	1	0.5	3	1.5	The Cost (lower for Electrical propulsion), has not been considered here in order to allow consideration of both Efficacy and Effectiveness.
	10.0	/	21.5	/	16	

In this way it is possible to express the Merit of the Product as weighted mean of the degree of fulfillment of the various requirements.

As in the case of the Qualitative Analysis of Comparison, both the evaluation's criteria and the importance of all the considered requirements have necessarily to be previously defined, in order to guarantee the objectivity of the results.

So far we have discussed qualitative approaches, which generally apply to problems whose level of definition is weak. When dealing with more detailed and defined problems, the mathematical tools used to solve them can be classified into three groups:

1. Simulation Models [5] and Optimization Algorithm. These methods are useful for both the preliminary identification of the system's quantitative characteristics and the completion of layout choices. It is necessary to develop, as soon as possible, mathematical models describing what we are willing to design and then to test them by means of extensive simulations in the time domain. Once the model has been tested, it is possible to run simulations characterized by different system parameters in order to search for the best system solution: quite obviously, in this way, the simulation becomes an important design tool.

   As Figure 15 shows, according to the situation, it is possible either to simulate different parameter sets (previously hypothesized) and then, on the basis of the Carpet of Solutions, to choose the best solution or, starting from a trial parameter set (initially hypothesized), it is possible to run more and more simulations, changing parameter values each time according to the results previously obtained, until the best configuration is achieved.

   In the latter case, Genetic Algorithms [6] represent one the most used optimization methods. They are highly efficient, especially when different parameters, potentially characterized by both a great (typically the characteristics which can be defined by a single number) and a very limited (architectural choices) number of values, may vary. In the former case, the adoption of the methodology of Design Of Experiments (DOE) [7], [8] is recommended in order to preliminarily elaborate the parameters' set and thus elicit the most information out of the fewest simulations. This kind of approach is appropriate not only for simulation models but also for physical models, when it is easy to match the mathematical model to a good and affordable physical model (Rapid Prototyping);

2. The classical engineering methods of analysis, like C.F.D., F.E.M., Dynamics Analysis, etc., which are essential for the verification and final sizing of configurations (they are already sufficiently defined).

3. Simulations characterized by stochastic variables (Monte Carlo Method), which are particularly interesting and useful when dealing with complex system models, representing the system as realistically stressed in its environment. In the present work, which addresses the problem of harbor security by means of SUB vehicles, the simulation model should include the

environment, the threats, the SUB fleet, and both operative and logistic strategies. It is important to remember that the Monte Carlo Method requires iteration of the simulation of each examined configuration until the stabilization of the global result (i.e., the percentage of success against a certain hostile attack) is attained. The process may obviously be repeated for a new configuration to search for the best one. For example, Figure 17 shows the Monte Carlo Method applied to the model of a fleet of anti-fire vehicles with its relative environment, threat, and operative-logistic strategy.

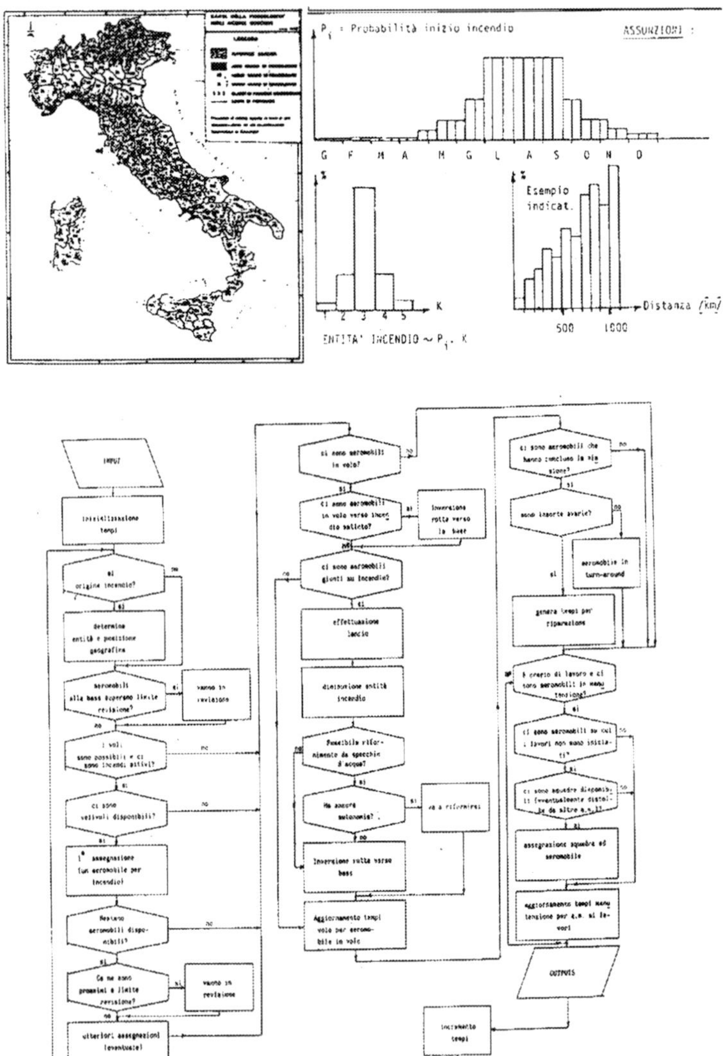

Figure 17: Monte Carlo Simulation example (anti-fire a.c. fleet [9]).

6. Conclusions

In the former chapter we have outlined a system development methodology which, starting from basic lessons learned from the SUB development prototype, is useful to design systems defense and surveillance for harbors, coastlines, and anchored ships. Considered threats are both terrorist and environmental and the systems of defense can be modular as well as integrated, as subsystems, in larger defense systems based on a plurality of devices.

Even if the development of affordable SUB, characterized by a good efficacy to cost ratio, is only in its initial phase, we believe that the prospects are very interesting and promising.

7. References

1. Chiesa, S., Gianotti, P., (1996), Operational Availability Of Airborne Surveillance Systems, N.A.T.O. 37th DRG SEMINAR on NON-LETHAL WEAPONS, c/o C.I.S.A.M., San Piero a Grado (Pisa), Italy.

2. Chiesa, S., Gianotti, P., Maggiore, P., (1996), Simplified Reliability Centered Maintenance Methodology for Industrial Context, SOLE 96, Washington, USA.

3. Camatti, D., Chiesa, S., Luda Di Cortemiglia, G., (1998), Design to System Effectiveness Methodology", SOLE '98 Int. Logistics Symposium, Seattle, USA.

4. Miles, D.L., (1972), Techniques of Value Analysis and Engineering, Mc. Graw Hill, New York.

5. Chiesa, S., Kostandinidis, K., (1987), Simulation for aircraft fleet efficacy evaluation, SOLE 3rd International Logistics Congress, Firenze, Italy.

6. Davis, L., (1991), Handbook of Genetic Algorithms, Van Nostrand Reinhold.

7. Taguchi, Genichi, (1987), System of Experimental Design, Don Clausing New York– UNIPUB/Krass Internationals Publications, volume 1 & 2.

8. Chiesa, S., Maggiore, P., Olivero, L., (1991), Ottimizzazione dei parametri di controllo di un servocomando: utilizzo di modello simulativo con applicazione di tecniche di Design Of Experiments, NT Tecnica & Tecnologia, n° 1, gennaio/febbraio.

9. Chiesa, S., Olivero, L., (1985), Simulazione operativa di una flotta di aeromobili anti-incendio, XXXIII Convegno Internazionale delle Comunicazioni, Genova.

10. Chiesa, S., (1988), Qualità di progetto e progettazione sistemistica, Notiziario Tecnico AMMA-Tecnica & Tecnologia, Torino, marzo-aprile.

LIST OF PARTICIPANTS

Angelides, Demos	Aristotle University of Thessaloniki Department of Civil Engineering Thessaloniki 54124 Greece	T: +30-2310-995702 F: +30-2310-995740 dangelid@civil.auth.gr
Apitz, Sabine E.	SEA Environmental Decisions Ltd. 1 South Cottages The Ford Little Hadham Hertfordshire SG 11 2AT United Kingdom	T: +44-0-1279-771890 F: +44-0-1279-771403 drsea@cvrl.org
Baba, Alper	Canakkale Onsekiz Mart Univeristy Department of Geological Engineering Terzioglu Campus Canakale, Turkey 17100	T: +9 0286218 0018/1212 F: +9 0286213 0155 alperbaba@comu.edu.tr
Belluck, David	U.S. Department of Transportation, FHWA 400 7th Street SW Washington, DC 20590 USA	T: +1 202-366-6549 F: +1 202-366-3409 david.belluck@fhwa.dot.gov
Biggs, Bryce G.	Prof. Training & Facilitation Services P.O. Box 238 Richmond KZN, South Africa 3780	T/F: +2733 330 7329 biggs@associates.co.z
Bridges, Todd S.	U.S. Army Engineer Research and Development Center Waterways Experiment Station EP-R, 3909 Halls Ferry Rd. Vicksburg, MS 39180 USA	T: +1 601-634-3626 F: +1 601-634-3120 fax todd.s.bridges@erdc.army.mil
Bronfman, Roman	The Democratic Choice Faction The Knesset, Israel	T: +9 7226753832 dan.goldenblatt@gmail.com
Chiesa, Sergio	Politecnico di Torino Dipartimento di Ingegneria Aeronautica e Spaziale Corso Duca Degli Abruzzi 24 10120 Torino Italy	T: + 39 0115646816 prof.s.chiesa@email.it

Citron, Marta	Venice Port Authority Dorsoduro 1401 Venice, Italy 30123	marta.citron@port.veni ce.it
Cooke, Roger M.	Delft Institute for Applied Mathematics Delft University of Technology Resources for the Future PO Box 5031 2600 GA Delft The Netherlands	T: +31-15278-2548 F: +31-15278-7255 cooke@rff.org
Demin, Vladimir	RRC Kurchatov Institute NSI Kurchatov sq. 1 Moscow, Russia	T: +7-095-196-9503 demin@nsi.kiae.ru
Didenko, Vladimir	Obninsk State Technical University of Nuclear Power Engineering (IATE) Kaluga Region Obninsk Studgorodoc-1 IATE 249039 Russia	T: +7(08439) 78560 vladimir@iate.obninsk. ru
Emara, Mostafa	Science Center for Detection & Remediation of Environmental Hazards (SCDREH) Al-Azhar University Nasr City, Cairo, Egypt	
Figueira, Jose	University of Coimbra and INESC-Coimbra Faculty of Economics Av. Dias da Silva 165 Coimbra Portugal 3004-512	T: +351-239-790-590 F: +351-239-790-514 Figueira@Fe.vc.pt
Filip, Zdenek	Institute of Chemical Technology Department of Biochemistry and Microbiology Technicka 5 16628 Prague 6 Dejvice Czech Republic CZ 166 28	T: +420-220-445-141 F: +420-220-445-167 zdenek.filip@vscht.cz
Ganoulis, Jacques	Aristotle Univeristy of Thessaloniki Department of Civil Engineering Thessaloniki, Greece 54124	T: +30-2310-99-5682 F: +30-2310-99-5681 iganouli@civil.auth.gr

Giove, Silvio	Department of Applied Mathematics University of Venice Dorsoduro, 3825/E Venice, Italy	T: +39-41-2346912 F: +39-41-5221756 sgiove@unive.it
Goodwin, R. Andrew	U.S. Army Engineering Research & Development Center Fisheries Engineering Team Portland District (CENWP-EC-HD) 333 SW 1st Ave. Portland, Oregon 97204 USA	T: +1 503-808-4872 F: +1 503-808-4875 Andrew.R.Goodwin@usace.army.mil
Grebenkov, Alexandre J.	Joint Institute for Power & Nuclear Research 99 Akademik Krasin St. Sosny, Minsk 220109 Belarus	T: +375 17 299-4542 F: +375 17 299-4355 greb@sosny.bas-net.by
Haruvy, Nava	Netanya Academic College One University Street Netanya, Israel 42100	T: +972-8-946-3189 F: +972-8-936-5345 navaharu@netvision.net.il
Ignatavieius, Gytautas	University of Vilnius Ozo 7-8 Vilnius, Lithuania 05129	T: +370 5265 0814 F: +370 5265 0855 gytisi@takas.lt
Lawrence Kapustka	ecological planning and toxicology, inc. 5010 SW Hunt Street Corvallis, OR 97333-9540 USA	T: +1 541-752-3707 F: +1 541-753-9010 Kapustka@ep-and-t.com
Kiker, Gregory A.	University of Florida Dept. Agricultural & Biological Engineering P.O. Box 110570 Gainesville, FL 32611-0570 USA	T: +1 352-392-1864 gkiker@ufl.edu
Kiker, Mindy K.	Zama Environmental 224 SW 40th Terrace Gainesville, FL 32607 USA	mindykiker@earthlink.net

Lahdelma, Risto	University of Turku Department of Information Technology Lemminkaisenkatu 14A Turku, Finland 20520	T: +358-40-503-1030 F: +358-2-333-8600 risto.lahdelma@cs.utu.fi
Levner, Eugene	Holon Academic Institute of Technology 52 Golomb Street Holon, Israel 58102	T: +972-3-502-6741 F: +972-3-502-6733 fax levner@hait.ac.il
Linkov, Igor	INTERTOX, Inc. 83 Winchester Street, Suite 1, Brookline, MA 02446 USA	T: +1 617-225-0812 ilinkov@intertox.com
Lotov, Alexander	State University - Higher School of Economics Myasnitskaya str., 20 Moscow, Russia 101978	T: +7-095-928-2094 F: +7-095-921-9730 Lotov1@ccas.ru
MacDonell, Margaret	Argonne National Laboratory 9700 S. Cass Avenue Argonne, IL 60439 USA	T: +1 630-252-3243 F: +1 630-252-4336 macdonell@anl.gov
Mannini, Achille	Dyria Sistemi S.r.l. via Veglia 57/5/d Torino Italy	T: +39 333760 1779 F: +39 011 354537 achille.mannini@fastwebnet.it
Neves, Ramiro	Instituto Superior Tecnico Av. Rovisco Pais, 1 Lisboa, Portugal 1049-001	T: +351-917224732 / 218417397 F: +351-218419423 ramiro.neves@ist.utl.pt
Nicholson, Scott	US Army Corps of Engineers NY District 26 Federal Plaza, Room 2021 New York, NY 10278-0090 USA	T: +1 917-790-8216 Scott.R.Nicholson@usace.army.mil
Ostrovskaya, Elena	Caspian Marine Scientific & Research Center Russian Federal Service for Hydrometeorology and Environmental Monitoring ulitsa Shiryaev, 14 RF, Astrakhan, 414045 Russia	T: +7 (8512) 30-34-40 F: +7 178512 301163 kaspmniz@astranet.ru eostrovskaya@mail.ru

Payne, Barry	U.S. Army Engineer Research & Development Center Environmental Laboratory, CEERD-EE-A 3909 Halls Ferry Road Vicksburg, MS 39180-6199 USA	T: 601-634-3837 Barry.S.Payne@erdc.usace.army.mil
Quartano, Katie	Aristotle University of Thessaloniki Department of Civil Engineering Thessaloniki, Greece 54124	T: +30-2310-99-5682 F: +30-2310-99-5681 iganouli@civil.auth.gr
Raissouni, Bachir	Center for Environmetnal Issues & Regional Development AUI CEIRD B.P. 1884 Ifrane Ifrane 53000 Morocco	T: 212-55-86-2330 B.Raissouni@aui.ma
Ramadan, Abou Bakr	Atomic Energy Authority Nasr City P.O. Box 7551 Cairo, Egypt	T: 202 27 48 787 F: 202 27 40 238 ramadan58@yahoo.com
Sargsyan, Vardan	Yerevan State Institute of Economy Department of Information Technologies & Systems 128 Nalbandyan Str. Yerevan 375025 Armenia	T: (374-9) 32-92-58 vsarg@freenet.am
Satterstrom, F. Kyle	Cambridge Environmental Inc. 58 Charles Street Cambridge, MA 02141 USA	T: +1 617-225-0810 F: +1 617-225-0813 Satterstrom@Cambridgeenvironmental.com
Sayed, Mohamed Farghaly	Arab Academy for Science and Technology 551 El-Horria Avenue, Gleem Alexandria, Egypt	T: 203 548 7785 F: 203 549 7882 mofarghali@yahoo.com mofarghali@aast.edu
Sinka, Davor	Enconet International Risk Assessment Department Miramarska 20 10.000 Zagreb Croatia	T: +385 1 6312 370 F: +385 1 6312 380 davor.sinka@enconet.hr

Skei, Jens	Norwegian Institute for Water Research (NIVA) Gaustadalleèn 21 0344 Oslo, Norway	T: 47 2218 5100 F: 47 2218 5200 jens.skei@niva.no
Wenning, Richard J.	ENVIRON 6001 Shellmound Street, Ste. 700 Emeryville, CA 94608 USA	T: 510-420-2556 rjwenning@environcorp.com
Xenidis, Yiannis	Aristotle University of Thessaloniki Department of Civil Engineering Thessaloniki, Greece 54124	T: +30-2310-99-5856 F: +30-2310-99-5740 ioxen@civil.auth.gr
Yatsalo, Boris	Obninsk State Technical University of Nuclear Power Engineering (IATE) Kaluga Region Obninsk, Studgorodok-1 IATE, 249030 Russia	T: +7(08439) 78560 F: +7(08439) 70822 yatsalo@prana.obninsk.org
Yoe, Charles	College of Notre Dame of Maryland 4701 N. Charles Street Baltimore, MD 21210-4010 USA	T: 410-744-0940 cyoe1@comcast.net
Zaidi, Mohammed K.	Radiological Environmental Sciences Laboratory 1955 Fremont Avenue MS-4149 Idaho Falls, ID 83401 USA	T: 208-526-2132 F: 208-526-2548 zaidimk@id.doe.gov

AUTHOR INDEX

A

Agostini, P. 267
Alcorn, J. 3
Almeida Dias, J. 217
Angelides, D. 251
Anisimov, L. 409
Apitz, S.E. 19, 85

B

Baba, A. 421
Batchelor, C.J. 451
Belluck, D.A. 3, 19
Benayahu, Y. 161
Benjamin, S.L. 3
Bridges, T.S. 37, 195

C

Chiesa, S. 19, 465
Cooke, R.M. 149
Critto, A. 267

D

Della Sala, S. 329
Demin, V. 317
Demnerova, K. 343
Deniz, O. 421

E

El Bouyoussfi, M. 429
El Sabagh, I. 379
Emara, M.M. 379

F

Figueira, J. 19, 217
Filip, Z. 19, 343

G

Ganoulis, J. 161
Gardner, K. 195
Giove, S. 267
Goodwin, R.A. 293
Grebenkov, A. 111

H

Haruvy, N. 369
Hull, R.N. 3
Husseen, D. 379

I

Ignatavičius, G. 19, 395

K

Kay, R. 175
Kiker, G.A. 37, 195, 293
Kiker, M.K. 19
Kim, J.B. 195, 293
Kotb, A. 379
Kurapov, A. 409

L

Lahdelma, R. 19, 217, 233
Levner, E. 161
Linkov, I. 3, 37, 161, 175, 195
Lotov, A.V. 275

M

MacDonell, M. 19
Magar, V. 329
Mannini, A. 19, 465
Marcomini, A. 267
Miller, A.C. 127

O

Oen, A. 85
Olivero, S. 465
Ostrovskaya, E. 409
Ozcan, H. 421

P

Pandey, V. 293
Payne, B.S. 127

R

Raissouni, B. 19, 429

S

Salminen, P. 217, 233
Sargsyan, V. 439
Satterstrom, F.K. 195
Seager, T.P. 175, 195
Semenzin, E. 267
Shalhevet, S. 369
Sinka, D. 19, 355
Skanata, D. 355
Skei, J. 121
Stesina, F. 465
Suedel, B. 127

T

Tervonen, T. 217
Tkachuk, A. 195
Tuler, S.P. 175
Turkey, A.S. 379
Turkoglu, M. 421

V

Vilutienė, V. 395
Viola, N. 465

W

Wenning, R.J. 19, 329
White, S. 85

X

Xenidis, Y. 19, 251

Y

Yakushau, A. 111
Yatsalo, B. 195
Yoe, C. 97

Z

Zaidi, M.K. 309

Printed in the United States
73895LV00001B/20

9 781402 058004